Nanotechnology for Sustainable Agriculture, Food and Environment

Nanotechnology has the potential to drastically transform the agri-food sector with its significant applications to improve agricultural productivity and the efficiency of agrochemicals. The food sector has benefitted from the inclusion of nanoparticles in food matrixes and the nanoencapsulation of nutraceuticals. Smart packaging materials designed with the help of nanotechnology have been used for increasing the shelf life of stored food products. Nanomaterials have been extensively used for the delivery of important agrochemicals to enhance their bioefficacy, prevent their degradation, and control their release. Various nanomaterials have been explored for remediation of arising environmental issues. Nanotechnology has also made a useful contribution to the utilization of huge agricultural and food wastes for production of valuable products. The existing and emerging applications of nanotechnology will contribute to environmental sustainability.

Nanotechnology for Sustainable Agriculture, Food and Environment has been structured to provide a widespread coverage and up-to-date progress of nanotechnology and its applications in the agri-food sector and environmental remediation. Synthesis of value-added nanomaterials from agri-food wastes and their potential applications in environmental remediation have been explored. In addition, toxicity issues with nanomaterials have also been discussed.

Features:

1. Elaborated information on the use of nanotechnology for sustainable agriculture
2. In-depth study about valorization of agri-food waste
3. An overview of applications of nanotechnology in environmental remediation
4. Toxicity analysis of nanotechnology-based products

We aim to satisfy the need for a reference book for scientists, researchers, academicians and students in nanotechnology, agricultural, food, nutraceuticals, environmental and material sectors.

Nanotechnology for Sustainable Agriculture, Food and Environment

Edited by Avnesh Kumari, Rajni Garg and
Rishav Garg

CRC Press
Taylor & Francis Group
Boca Raton London New York

CRC Press is an imprint of the
Taylor & Francis Group, an **informa** business

First edition published 2024
by CRC Press
6000 Broken Sound Parkway NW, Suite 300, Boca Raton, FL 33487-2742

and by CRC Press
4 Park Square, Milton Park, Abingdon, Oxon, OX14 4RN

CRC Press is an imprint of Taylor & Francis Group, LLC

Library of Congress Cataloging-in-Publication Data
Names: Kumari, Avnesh, editor. | Garg, Rajni, editor. | Garg, Rishav, editor.
Title: Nanotechnology for sustainable agriculture, food and environment /
edited by Avnesh Kumari, Rajni Garg and Rishav Garg.
Description: First edition. | Boca Raton, FL : CRC Press, 2024. |
Includes bibliographical references and index.
Identifiers: LCCN 2023026433 (print) | LCCN 2023026434 (ebook) |
ISBN 9781032503011 (hardback) | ISBN 9781032503066 (paperback) |
ISBN 9781003397861 (ebook)
Subjects: LCSH: Nanobiotechnology. | Sustainable agriculture. |
Agricultural innovations. | Agricultural wastes. | Nanostructured
materials—Toxicity testing.
Classification: LCC TP248.25.N35 N3693 2024 (print) | LCC TP248.25.N35 (ebook) |
DDC 660.6—dc23/eng/20230614
LC record available at https://lccn.loc.gov/2023026433
LC ebook record available at https://lccn.loc.gov/2023026434

ISBN: 9781032503011 (hbk)
ISBN: 9781032503066 (pbk)
ISBN: 9781003397861 (ebk)

DOI: 10.1201/9781003397861

Typeset in Times
by codeMantra

Contents

Preface

Nanotechnology deals with particles in the size range of 1–1,000 nm. It has an influence on every discipline of science and technology by providing various methods, protocols and nanoparticles (NPs) for improving the performance of existing methods and protocols. The global scientific community has realized that nanotechnology has great potential in the agri-food sector. They are heavily funding nanotechnology projects related to the agri-food sector. The United States Department of Agriculture (USDA) has set out motivated plans to achieve targeted goals in this sector. The Food Standards Agency of the United Kingdom has carried out widespread studies to measure the potential of nanotechnology in food packaging. The major focus is on the development of functional foods, nutrient delivery systems and methods for optimizing food appearance, such as colour, flavour, and consistency. The Government of India has also initiated work in this direction by allocating special funds. The Department of Science and Technology is the nodal agency for implementing the nanomission. Other government agencies, such as Department of Biotechnology and Indian Council of Agricultural Research, are also funding research proposals in this area.

Nanotechnology has the potential to drastically transform the agri-food sector with its significant applications. It has the potential to improve agricultural productivity and the efficiency of agrochemicals. Crops have been genetically improved through the use of NPs. Gene expression analysis in plants and animals under stress conditions has also been improved with the help of nanoarrays. Nanotechnology has also played a role in reducing postharvest losses of grains, fruits and vegetables. The food sector has benefitted from the inclusion of NPs in food matrixes and the nanoencapsulation of nutraceuticals. Smart packaging materials designed with the help of nanotechnology have been used to increase the shelf life of stored food products. Nanomaterials (NMs) have been extensively used for the delivery of important agrochemicals to enhance their bioefficacy, prevent their degradation and control their release. NMs can also be used as bioactive compounds in functional foods. Worldwide, functional foods and food supplements containing NPs are becoming available. A key development area seems to be the progress of "nanometals" and nutraceutical supplements.

Agrochemicals have been encapsulated in NMs, resulting in much better activity than their native counterparts. Essential oils loaded in NMs have shown improved physical stability and reduced the rate of oil evaporation. NMs have also been used to deliver DNA, genes and other beneficial chemicals to target plant parts. NMs possess the tendency to penetrate the plants through the roots and get translocated to other plant parts through the vascular system to further affect the various physiological processes in plants. Nanotechnology can also be used for accelerating the competence of microarrays. There are reports that NMs have been used for improving the specificity and sensitivity of sensors. NM-based biosensors have been successfully developed for the detection of pesticide residues in water or soil samples. NM biosensors have massive merits over other techniques, including sensitivity of detection in the nanomolar (NM) range for many analytes. NM-based kits can also be used for the detection of viruses and other plant pathogens, and have also been used for plant disease management in crops. Biosensors based on nanotechnology have been used in the detection of foodborne pathogens and toxins.

Various nanomaterials have been explored for remediation of arising environmental issues. The proper disposal, recycling and reuse of waste discarded by the agri-food-based industries as well as local vendors is a critical matter of concern. Nanotechnology has also made a useful contribution to the utilization of huge agricultural and food wastes for production of valuable chemicals such as biosurfactants, biofuels and nanomaterials. These biosynthesized nanomaterials have been extensively utilized for water and soil remediation in addition to applications in catalysis, energy production, pharmaceutics and diagnostics because they are non-toxic, cost-effective and eco-friendly. The existing and emerging applications of nanotechnology will contribute to environmental sustainability.

Despite the incredible prospects of nanotechnology, it also has certain limitations, such as toxicity issues, which may pose danger to the environment and human health. Toxicity studies of NPs used in the agri-food sector need to be evaluated and carried out at an extensive level before recommending their use. Despite the fast progress in nanotechnology, we are aware of the shortage of good-quality books in the area of nanotechnology and its applications in agriculture and food sectors. Although there are outstanding books, reviews and book chapters dealing with one or several topics in the area of applications of nanotechnology in agriculture and food sectors, a book containing a widespread coverage and up-to-date progress of nanotechnology and its application in the agri-food sector and environmental remediation is not available.

This book aims to talk about the applications of nanotechnology in agriculture and environmental remediation. This book mainly focuses on applications of nanomaterials to plant physiology and function, enhancement of crop yield and productivity, preventing postharvest losses, packaging, nanoencapsulation and nanodiagnostics. Synthesis of value-added materials from agri-food wastes and their potential applications in environmental radiation will be explored. In addition, discussions on the toxicity issues of nanotechnology-based products in the agriculture and food sectors and commercialization of nanotechnology-based products will also be discussed. Important references will be included in each chapter for the assistance of readers who wish to continue further reading on any of these topics in a greater depth.

We aim to satisfy the need for a reference book for scientists, researchers and academicians in the nanotechnology, agricultural, food, nutraceutical, environmental and material sectors. It will also serve as a brilliant resource book for undergraduate and postgraduate students in agriculture, nutraceutical sciences, food sciences and nanotechnology.

Editor Biography

Dr. Avnesh Kumari obtained her MSc (Hons.) Chemistry in 2001 and PhD in 2007 from the Department of Chemistry, Panjab University, Chandigarh. Subsequently, she worked as postdoctoral fellow at CSIR-IHBT, Palampur (2008–2010). She joined CSIR-Institute of Himalayan Bioresource Technology, Palampur in 2011 and is presently working there as Assistant Professor and Senior Technical Officer (2). She has been working in the areas of nanobiology and electron microscopy. In nanobiology, her main work is on the improvement of solubility, bioavailability and efficacy of important phytomolecules possessing activities like antioxidant, antimicrobial and anticancer. Secondly, she has been working on the isolation of nanocellulose from different plant species and using it in different applications. She has published 49 research articles in peer-reviewed journals, 5 international patents (1 granted and 4 filed) and 12 book chapters so far. Early in her career, she was a recipient of JRF and SRF fellowships from CSIR, India. She is also a recipient of CSIR Quick Hire Scientist Fellow scheme. She later received a DST-Women Scientist Scheme – a postdoctoral fellowship. She has served as a reviewer for several journals and is a life member of several scientific societies.

Dr. Rajni Garg is working as an Associate Professor in the Department of Applied Sciences, Galgotias College of Engineering and Technology, Greater Noida (UP), India. She received her B.Sc. (Medical) at the D.A.V. College, Abohar (Punjab), India; M.Sc. (H.S.) in Chemistry, Panjab University, India; M.Phil. in Chemistry at CDLU, India; and Ph.D. (Chemistry) at the GKU, India. Her areas of interest include nanochemistry, green chemistry and supramolecular chemistry. She has authored and co-authored many scientific papers printed in international and national journals and in the proceedings of international and national scientific conferences. She has also published five patents on the potential applications of nanomaterials in cementitious materials. Currently, she is working on green synthesis and potential applications of nanoparticles. She serves as a reviewer for many journals and has authored one textbook with Mc-Graw Hill Publications. She is a part of a large social media network on LinkedIn and Facebook with reputed scientists, academicians and researchers. She has recently successfully edited a handbook of research on *Green Synthesis and Applications of Nanomaterials* published by IGI Global in 2022 with contribution of authors across the world.

Dr. Rishav Garg is working as Head, Department of Civil Engineering at Galgotias College of Engineering and Tech, Greater Noida (UP), India. He received his B.Tech. in Civil Engineering at the RIC, Kurukshetra, India; M.Tech. in Structural Engineering at PTU, Kapurthala, India; and Ph.D. in Civil Engineering at PTU, Kapurthala, India. His research interests include applications of nanotechnology and composite materials in sustainable construction. He has authored or co-authored several technical papers in international and national journals. He is actively engaged in developing techniques for the strength and durability enhancement of cementitious materials. He has also presented his work at many international and national scientific conferences. He has published patents on sustainable construction practices. He is actively engaged in strategic development as a member of various academic and professional bodies with his rich industrial, research, and academic experience. In addition to his academic and administrative contributions, he is actively providing his services in consultation for innovative projects. He has edited a handbook of research on *Green Synthesis and Applications of Nanomaterials* published by IGI Global in 2022 with contribution of authors across the world.

List of Contributors

Oluwatosin Adekunle Adeluwoye
Lead City University
Ibadan, Nigeria
and
Omics Logic Africa, Pine Biotech Inc.

Oluyomi Stephen Adeyemi
Landmark University
Omu-aran, Nigeria
and
Tohoku University
Sendai, Japan

Jana Wissam Al-Ashkar
Beirut Arab University
Debbieh, Lebanon

Wardishan Kassem Alrifai
Department of Analytical Chemistry, Faculty of Pharmacy
Erciyes University
Kayseri, Turkey

Paul Ocheje Ameh
Department of Chemistry and Forensic Science
Nigerian Police Academy
Wudi, Nigeria

Mathias Maduakolam Aniobi
Landmark University
Omu-aran, Nigeria

Ansar Anjum
Department of Applied Sciences
Galgotias College of Engineering & Technology
Greater Noida, India

Rotdelmwa Maimako Asalaye
SDG 03 Group – Good Health & Well-being
Abuja, Nigeria

Piotr Barbaś
Department Agronomy of Potato
Plant Breeding and Acclimatization Institute – National Research Institute
Branch Jadwisin, Poland

Venus Borgohain
Department of Botany, School for Basic and Applied Sciences
Central University of Punjab
Bathinda, India

Kamel Fahmi Bou-Hamdan
Beirut Arab University
Debbieh, Lebanon

Emeka ChimaOgoko
Department of Chemistry, Faculty of Sciences
National Open University of Nigeria
Lagos, Nigeria

Nnabuk Okon Eddy
Department of Pure and Industrial Chemistry
University of Nigeria
Nsukka, Nigeria

Tobiloba Christiana Elebiyo
Landmark University
Omu-aran, Nigeria

Oluwayemi Elizebeth
Landmark University
Omu-aran, Nigeria

Gamze Ersan
Arizona State University
Tempe, Arizona

Mahmut Selim Ersan
Arizona State University
Tempe, Arizona

Rajni Garg
Department of Applied Sciences
Galgotias College of Engineering & Technology
Greater Noida, India

Rishav Garg
Department of Civil Engineering
Galgotias College of Engineering & Technology
Greater Noida, India

Abu Barkat Md Gulzar
Plant & Microbial Molecular Biology Laboratory, Department of Biotechnology
Assam University
Silchar, India

Ali Samir Hamdan
Norwegian University of Science and Technology
Trondheim, Norway

Prerna Higgins
Department of Chemistry
Sam Higginbottom University of Agriculture Technology and Sciences
Prayagraj, India

Israt Jahan
Department of Health Care Services, Vocational School of Health Services
Mardin Artuklu University
Mardin, Turkiye

Atanu Jana
Division of Physics and Semiconductor Science
Dongguk University
Seoul, South Korea

Henrietta Ijeoma Kelle
Department of Chemistry, Faculty of Sciences
National Open University of Nigeria
Lagos, Nigeria

Saima H. Khan
Department of Chemistry
Sam Higginbottom University of Agriculture Technology and Sciences
Prayagraj, India

Birsen Korkmaz
Department of Analytical Chemistry, Faculty of Pharmacy
Erciyes University
Kayseri, Turkey

Barbara Krochmal-Marczak
Department of Plant Production and Food Safety
Carpathian State College
Krosno, Poland

Vinay Kumar
School for Basic and Applied Sciences
Central University of Punjab
Bathinda, India

A.S.M. Islamul Hoque Laskar
Plant & Microbial Molecular Biology Laboratory, Department of Biotechnology
Assam University
Silchar, India

Monika Mahajan
Punjab Agricultural University
Ludhiana, India

Pranab Behari Mazumder
Plant & Microbial Molecular Biology Laboratory, Department of Biotechnology
Assam University
Silchar, India

Mona Mittal
Department of Applied Sciences
Galgotias College of Engineering and Technology
Greater Noida, India

Ismail Ocsoy
Department of Analytical Chemistry, Faculty of Pharmacy
Erciyes University
Kayseri, Turkey

Anduang Odiongenyi
Department of Chemistry
Akwa Ibom State University
Ikot Abasi, Nigeria

Raphael Ogbodo
Department of Chemistry
University of Iowa
Iowa City, Iowa

Oludipe Olorunleke
Landmark University
Omu-aran, Nigeria

Piotr Pszczółkowski
Experimental Department of Cultivar Assessment
Research Centre for Cultivar Testing
Uhnin, Poland

Rameswar Singh Rattan
Mahindra Summit Agriscience Ltd.
Mumbai, India

Reetu
School for Basic and Applied Sciences
Central University of Punjab
Bathinda, India

Damilare Rotimi
Landmark University
Omu-aran, Nigeria

Eze Ifeanyi Samson
Department of Pure and Industrial Chemistry
University of Nigeria
Nsukka, Nigeria

Barbara Sawicka
Department of Plant Production and Commodities Science
University of Life Sciences
Lublin, Poland

Rajat Sharma
Department of Chemistry, USS
Rayat Bahra University
Mohali, India

Shaziya H. Siddiqui
Department of Chemistry
Sam Higginbottom University of Agriculture Technology and Sciences
Prayagraj, India

Dominika Skiba
Department of Plant Production and Commodities Science
University of Life Sciences in Lublin
Lublin, Poland

Swati
School for Basic and Applied Sciences
Central University of Punjab
Bathinda, India

Richard Alexis Ukpe
Department of Chemistry
Federal University
Otuoke, Nigeria

Rafael Vargas-Bernal
Universidad de Guanajuato
Irapuato, Guanajuato, México

Ebubekir Yılmaz
Department of Restorative Dentistry, Faculty of Dentistry
Nuh Naci Yazgan University
Kayseri, Turkey

1 Introduction to Nanotechnology and Nanomaterials

Rajni Garg and Rishav Garg
Galgotias College of Engineering & Technology

Rajat Sharma
Rayat Bahra University

Ansar Anjum
Galgotias College of Engineering & Technology

Nnabuk Okon Eddy
University of Nigeria

CONTENTS

1.1 INTRODUCTION

Nanotechnology is a nano-scaled technology that involves fundamental particles in the size range of 1–100 nm. The name "Nano" has its roots in the word, "Nanos," a Greek word for "dwarf," which refers to an order of magnitude in the range from 1 to 100 nm (1 nm = 10^{-9}m) (Chums-ard et al. 2019). Nanotechnology is viewed as a crucial technology that has the ability to bring about fundamental improvements in a wide variety of technological fields and has ultimately generated improved and

DOI: 10.1201/9781003397861-1

diverse system innovation. Considering the present rate of engagement of nanomaterials in our society, it is predicted to have substantial economic, ecological, and societal repercussions shortly (Zheng et al. 2022). As a result, not only are fine-sized items controlled in nanotechnology, but they are also distinguished by the fact that distinct characteristics and material qualities exist in nanoscale. Nanotechnology is a prevailing multidisciplinary science that has opened many substantial new horizons. Integrating the research strengths of several fields and solving key scientific difficulties through this method is an urgent necessity, as is the development of nanotechnology from its rudimentary stages up to its eventual industrial applications (Cardoso et al. 2022).

Innovations in the cross- and peripheral domains have historically characterized the progress of modern science and technology. In this approach, nanotechnology is ripe with potential for truly novel research and development. When the theories and concepts developed via research in this area are widely applied to our production and living, they will substantially enhance our mental world and bring conceptual changes to human civilization (Zheng et al. 2022). A new industrial revolution will be sparked by nanotechnology, which will revolutionize how humans perceive the external world. Materials built on the nanoscale, for instance, have vastly superior and state-of-the-art characteristics with several potential applications (Ethaib et al. 2022). The information economy and allied sectors will undergo a radical transformation as a result of this. Similarly, nanotechnology is having an impact on life science technologies. Many therefore consider nanotechnology to be the future backbone upon which both chemical and biotechnology might function simultaneously (Chugh, Siddique, and Solaiman 2021).

The unique properties offered at nanoscale include optical, mechanical, and electrical properties that are drastically different from the bulk material and allow for new foreseen functions, including revolutions in batteries, construction materials, vaccines, electronics, agriculture, etc. (Rajni Garg et al. 2021). It has also proven to be an important component in combating the degradation of the environment and in nanomedicine. As a form of nanomedicine, the development of therapies and drug carriers based on nanotechnology is allowing for targeted drug delivery without the body's immune system or the blood-brain barrier being compromised. The use of bendable electronic sensors to track nerve activity, high-speed nanopore (Lata, Bhardwaj, and Garg 2022), DNA sequencing technology, micro-nanosensor-based cancer diagnostics with liquid biopsy technology, etc. are all relatively new advancements in the field of nanodiagnostics. Nanomedicine's therapeutic effect can be tracked with the help of contrast imaging, which can be used in conjunction with diagnosis and medical function (Vazquez-Munoz and Lopez-Ribot 2020). The field of biomedical technology is making great strides towards the creation of new drug designs that will eventually replace standard medicinal techniques with better responses (Luiz et al. 2022).

There has been a rush of cutting-edge nanotechnology research carried out in scientific and technical areas, and many of the findings have shown significant potential for industrialization. Carbon nanotubes, nanopolishing solutions, and nanosensors, among other things, have attained large-scale manufacture (Bhardwaj, Lata, and Garg 2022a). Incorporating innovative nanomaterials with the ability to transform chemical catalysis and manufacturing procedures, nanotechnology has the potential to inspire innovation in energy technology. With these considerations, this chapter not only presents an introduction to the innovative aspects of nanotechnology but also explores the history, synthesis, characterization, and applications of nanomaterials in various sectors.

1.2 HISTORY OF NANOMATERIALS

Materials in the nanodimension have gained considerable scientific and industrial attention, spurred by the fact that materials' macro characteristics can be treated as components of multiple microsystems. These nanoparticles, due to their small size, possess a high aspect surface-to-volume ratio as compared to bulk material, which leads to reduced lattice constants and depressed melting points (Eddy and Garg 2022). The nanosize particles consist of fewer atoms comprising the solid, which leads to a smaller number of orbitals that form the band and hence a change in band structure. Nanomaterials were used by the Ancient Egyptians in their dyes and paints over 4,000 years ago, according to archaeological discoveries (Rajni Garg et al. 2021). Many types of metallic nanoparticles were used in ceramic and glass structures throughout the Bronze Age, marking the beginning of the era of manufactured metallic nanoparticles. Michael Faraday's synthesis of colloidal gold suspension in 1857 is often cited as the beginning of the modern age, although the formal presentation of the notion of nanotechnology can be attributed to Richard Feynman in 1959 (Sharma, Garg, and Kumari 2020). But the formal adoption of the word "nanotechnology" by Norio Taniguchi in 1974 and the introduction of the scanning tunnelling microscope (STM) by Gerd Binning and Heinrich Rohrer in 1981 advanced the field (Mehnath et al. 2021). Later, in 1985, fullerenes were discovered by Harold Kroto, Robert Curl, and Richard Smalley; in 1991, carbon nanotubes were synthesized by SumioLijima and his colleagues; and in 2004, graphene was discovered by Andre Geim and Kostya Novoselov (Withers et al. 2015).

1.3 CLASSIFICATION OF NANOMATERIALS

Nanomaterials can be elaborated into three distinct categories based on where they were created: (i) naturally occurring nanomaterials found in living creatures; (ii) incidental nanomaterials created by chance during a process; and (iii) artificial nanomaterials created in a laboratory (manufactured by human beings) (Rajni Garg et al. 2021). Nanoparticles, nanowires, nanotubes, nanosheets, nanostructured materials, complicated extensions, flutter nanostructures, and even nano-sea-urchins are all examples of recently developed nanomaterials with widely varied origins, morphologies, and chemical compositions. Biomolecules, including glucose, cellulose, chitin, DNA, RNA, proteins, and lipids, are all examples of natural nanomaterials (Dolez 2015). Nanomaterials are created accidentally through both natural processes (such as forest fires, dust storms, volcanic eruptions, and cosmic reactions) and human activity (burning, engine exhausts, industrial and agricultural operations). Alternatively, engineered nanomaterials are created by humans with a tailored morphology in mind (Rajni Garg et al. 2022).

Nanomaterials are categorized as either 0D (dimensionless space), 1D (one axis), 2D (two axes), or 3D (all three axes) based on the number of spatial dimensions they exhibit at the nanoscale. Quantum dots, metal and metal oxide nanoparticles, and other 0D nanomaterials have been the subject of intense research due to their potentially vast range of applications (Sharma, Garg, and Kumari 2020). Nanotubes, nanorods, nanowires, and nanofibres are all examples of ID nanomaterials that have found extensive use in nanoelectronics and the energy industry. Nanosheets, nanoplates, and nanoprisms are all examples of 2D nanomaterials that have found new uses as sensors, reactors, and catalysts (Baig et al. 2021). There has been a recent surge in interest in the synthesis of 3D nanomaterials with controlled morphological features, such as nanoflowers, nanoballs, and nanocoils. Nanomaterials are typically divided into four different types based on their chemical composition: carbon-based, inorganic, organic, and composite (Nikalje 2015). Examples of carbon-based nanomaterials include fullerenes, graphene, carbon black, carbon nanotubes, and carbon nanofibres, while examples of inorganic-based nanomaterials include metal and metal oxide nanoparticles (Singh et al. 2018). Nanomaterials that are not made of carbon or inorganic compounds are classified as organic. Blends of the first three types with various architectures make up the composite-based nanomaterials (Thangadurai, Ahuja, and Sangeetha 2021).

1.4 SYNTHESIS OF NANOMATERIALS

Due to its potential to address problems that have slowed progress, nanotechnology has grown rapidly as an academic topic in recent decades. There has been a worldwide shift towards a more eco-friendly attitude, which has led to increased support for practises that preserve the planet's natural capital for future generations (Vasyukov et al. 2013). Because of this, the future of nanotechnology rests on its ability to "go green." A new era of "green synthesis," a branch of nanotechnology that employs trustworthy, sustainable, low-cost, nonhazardous, and environmentally friendly technologies, is currently emerging (Mehnath et al. 2021). Standard methods of synthesis for nanoparticles are often utilized in a wide variety of applications, but they come with a number of drawbacks that can be mitigated with the help of green synthesis (Bhardwaj, Lata, and Garg 2022b). Using sustainable materials, "green" nanotechnology strives to develop improved, nontoxic, and biocompatible nanomaterials with long-term benefits. Many biocompatible methods have been reported in the literature as having been used in the production of nanoparticles and/or nanomaterials (Rajni Garg et al. 2022).

Environmental and energy applications of biosynthesized nanomaterials in solar cells, photocatalysis, and biohydrogen generation have been reported. Consequently, researchers may make the connection between biosynthesized nanomaterials and their wide range of potential uses in ecological remediation (Saikumari, Dev, and Dev 2021). Green synthesis of metallic nanoparticles employing natural entities including microbes, algae, plants, and plant-derived materials has been shown to be a cost-effective and environmentally benign technology in recent years (Rajni Garg et al. 2022). Green nanotechnology is a promising approach to creating safe nanomaterials for the environment, but it has several drawbacks, especially when it

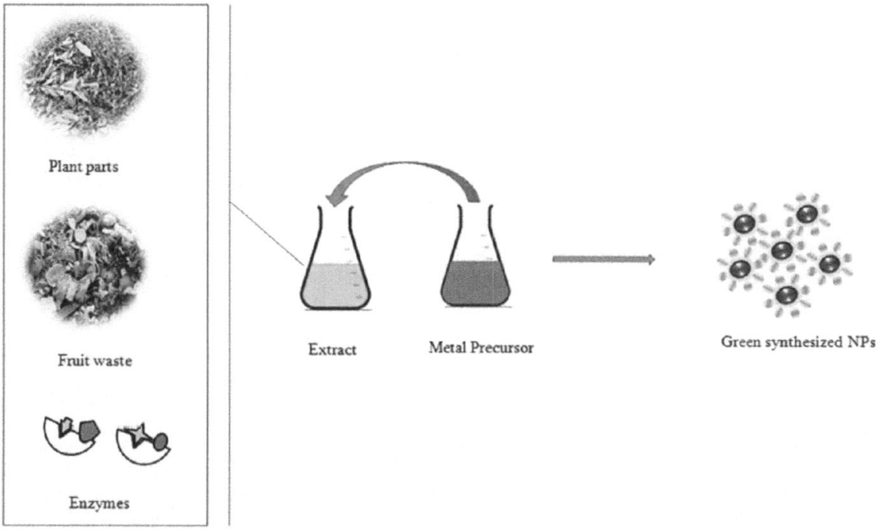

FIGURE 1.1 Green synthesis of nanoparticles.

comes to microbial treatment and process optimization (Rishav Garg, Garg, Khan, et al. 2022). However, phytocompounds isolated from plant extracts have attracted a lot of attention since they permit a one-step synthesis technique to produce high yield of nanoparticles of a certain size and form. Plants store a wide range of biochemicals with a high hyperaccumulation capacity, which is essential for the conversion of metal ions into nanoparticles (Lata, Bhardwaj, and Garg 2022). Biosynthesis of metal and metal oxide nanoparticles as well as nanocomposites (elements like platinum, gold, silver, titanium oxide, iron oxide, copper oxide, zinc oxide, and palladium) using food and agro-waste (Figure 1.1) is currently being explored across the globe using various environmentally friendly approaches.

1.5 CHARACTERIZATION OF NANOMATERIALS

The science of nanotechnology has led to an upsurge in the field of research lately in different domains, although the study of the surface-function relationship of nanoparticles is challenging due to the absence of suitable reference materials for calibration of analytical tools, interdisciplinary research, and challenges in sample preparation methodologies (Terban and Billinge 2022). However, to investigate the physicochemical properties of nanoparticles, different characterization techniques are used that are often exclusively applied to study the specific properties of nanoscale particles (Table 1.1). Due to their growing popularity, improved quantitative and qualitative characterization methods are essential (Ramirez-Perez et al. 2022). The complexity of these instruments increases as they detail different properties of nanomaterials, such as their elemental makeup, shape, crystal structure, magnetic properties, and strength (Santhosh et al. 2018).

TABLE 1.1
Some of the Techniques Used for Characterization of Nanoparticles

Name of Technique	Basic Principle	Information Achieved	Reference
X-ray diffraction	Constructive interference of monochromatic X-rays and crystalline sample	Crystal composition and structure, crystal size	Rishav Garg et al. (2023)
X-ray absorption spectroscopy	Excitation of core electrons in lower shell/s by X-rays absorption	Chemical state of species, interatomic distances for noncrystalline nanoparticles as well	Terban and Billinge (2022)
Small-angle X-ray scattering	Collision between X-ray and surface particle results in scattering of the waves in all directions	Particle size, size distribution, growth kinetics	Ramirez-Perez et al. (2022)
X-ray photoelectron spectroscopy	Electrons in the sample after absorption of energy eject from the sample with certain kinetic energy	Elemental composition of nanomaterial and oxidation states	Blažeka et al. (2022)
Fourier transform-nfra red spectroscopy	Dipole moment changes by IR interaction that correspond to a specific vibrational energy	Functional groups present on surface	Rishav Garg, Garg, Okon Eddy, et al. (2022)
Brunauer–Emmett–Teller	For a homogeneous surface the uppermost layer is in equilibrium with vapour phase. All surface sites have the same adsorption energy for adsorbate and adsorption on the adsorbent occurs in infinite layers	Specific surface area	Yousefi-Limaee et al. (2023)
Thermogravimetric analysis	The thermal property of sample is monitored against time or temperature, as temperature of sample, in specified atmosphere, is programmed	Thermal stability	Onyszko et al. (2022)
UV-Vis spectrosocpy	Electronic transition on absorption of UV-Vis radiation	Optical properties, concentration and agglomeration	Bhardwaj, Lata, and Garg (2022b)
Dynamic light scattering	Brownian motion of dispersed particles	Hydrodynamic size distribution profile	Saikumari, Dev, and Dev (2021)

(Continued)

TABLE 1.1 (*Continued*)
Some of the Techniques Used for Characterization of Nanoparticles

Name of Technique	Basic Principle	Information Achieved	Reference
Differential centrifugal sedimentation and nanoparticle tracking analysis	Particles of different densities or sizes in a suspension settle at different rates	Particle size and size distribution	Minelli et al. (2018)
ICP-MS	Fine aerosol of sample is subjected to hot argon plasma forms analyte atoms that are ionized. Ions are then sorted and quantified based on their mass-to-charge (m/z) ratio in the mass analyser	Elemental composition, size, size distribution, NP concentration	Santhosh et al. (2018)
Superconducting quantum interference device (SQUID-nanoSQUID)	Josephson effect and the flux quantization in a superconducting ring	Magnetization saturation, magnetization remanence, blocking temperature	Vasyukov et al. (2013)

The fundamental characterization methods used for nanomaterials can be categorized into two categories: microscopy (used for imaging) and spectroscopy (used for analysis). The techniques including X-Ray Diffraction (Rishav Garg et al. 2023), Scanning Electron Microscopy (Rishav Garg, Garg, Khan, et al. 2022), Transmission Electron Microscopy, High-Resolution TEM (Jassal et al. 2020), and Atomic Force Microscopy (Cavaliere et al. 2015) provide information about the size, morphology, and crystal structure of nanomaterials, while the specific surface area is determined by Brunauer–Emmett–Teller (Yousefi-Limaee et al. 2023). The SQUID (*superconducting quantum interference device*) (Vasyukov et al. 2013) measures extremely subtle magnetic fields; ferromagnetic resonance (FMR) is used to examine the magnetization of ferromagnetic materials; and X-Ray Magnetic Circular Dichroism (XMCD) determines the magnetic moment and is specially used for characterization of certain materials (Hernández-Hernández et al. 2020). Spectroscopic techniques used for characterization of nanoparticles include FT-IR (Rishav Garg, Garg, Okon Eddy, et al. 2022), Raman Spectroscopy (Song et al. 2017), XPS (Blažeka et al. 2022), ICP-MS (Santhosh et al. 2018), and UV-Vis Spectroscopy (Bhardwaj, Lata, and Garg 2022b). The information about the functional groups present on the surface of nanomaterials can be investigated using FT-IR spectroscopy, and the required modification of the sample can then be suggested. On the other hand, the intra- and interbond vibrations can be investigated using Raman spectroscopy (Song et al. 2017). For determination of the particle size of nanocolloids and the size distribution of nanoparticle dispersion, dynamic light scattering (Saikumari, Dev, and Dev 2021) is used in addition to differential centrifugal sedimentation and nanoparticle tracking analysis (Minelli et al. 2018). The in vivo studies of ocular retention, irritancy, and drug release patterns show promising results in favour of nanomaterials prepared by the methodology adopted (Luiz et al. 2022). Thus, it is observed that spectroscopic techniques play an important role in the characterization of nanoparticles. These techniques can be used in conjunction with microscopy techniques to extract complete information about nanomaterials (Yousefi-Limaee et al. 2023).

1.6 APPLICATIONS OF NANOMATERIALS

The high stability, affinity, and selectivity of nanomaterials have attracted a lot of attention from scientists. Nanomaterials' potential applications in society can be increased by learning more about their unique properties. With their new functional and physicochemical properties, nanomaterials have captivated the scientific community's interest (Eddy and Garg 2022). They have high-end optical-electronic properties and can have their surfaces functionalized in a variety of ways depending on the programmer's wishes. Due to these merits, they have found widespread use in industries as diverse as the pharmaceutical, cosmetic, and electrical product industries (Onyszko et al. 2022). Extensive research has been conducted over the past few decades on a wide range of metal and metal oxide nanomaterials to determine their important properties, such as morphology, surface mass aspects, and reactivity, for the development and use of nanomaterials and nanoscale devices in biomedical fields (Karthik et al. 2022). Antimicrobial, anticancer, biosensor and imaging, and tissue engineering, in addition to gene and drug delivery, are just a few of the many biofunctional applications that hold great promise. Nanoparticles' usage in medical diagnostics, tissue engineering, targeted drug delivery, regenerative medicine, and biomedical packaging, among others, has grown exponentially and attracted a lot of attention in recent years (Kabir, Islam, and Asaduzzaman 2022). Biological molecule functionalized nanoparticles in particular display novel behaviours and seem to be extremely useful in clinical diagnosis. Moreover, nanoparticles have a wide range of therapeutic applications, including nanodrug delivery system development, the formulation of novel pharmaceuticals, and the manufacture of therapeutic materials (Nikalje 2015). Especially the applications of nanomaterials in the agri-food and environmental sectors are worth mentioning (Figure 1.2).

1.6.1 Applications in Environmental Remediation

Public and scientific interest in pollution remediation was ignited by the constant discharge of contaminants into the environment and the resulting damage to the living world. Harmful pollutant discharges are disruptive to aquatic ecosystems as well as soil ones (Rishav Garg, Garg, Khan, et al. 2022). Because of the disruption in soil and water ecosystems, our ability to produce food is threatened. Therefore, it is critical to reduce or remove environmental hazards to provide a healthy and productive work environment. It is currently common practise to use a variety of methods, including the widespread application of nanoparticles, to purge the environment of harmful substances (Rishav Garg, Garg, Okon Eddy, et al. 2022). It is highly beneficial to clean up polluted soil and water with the use of nanoparticles and

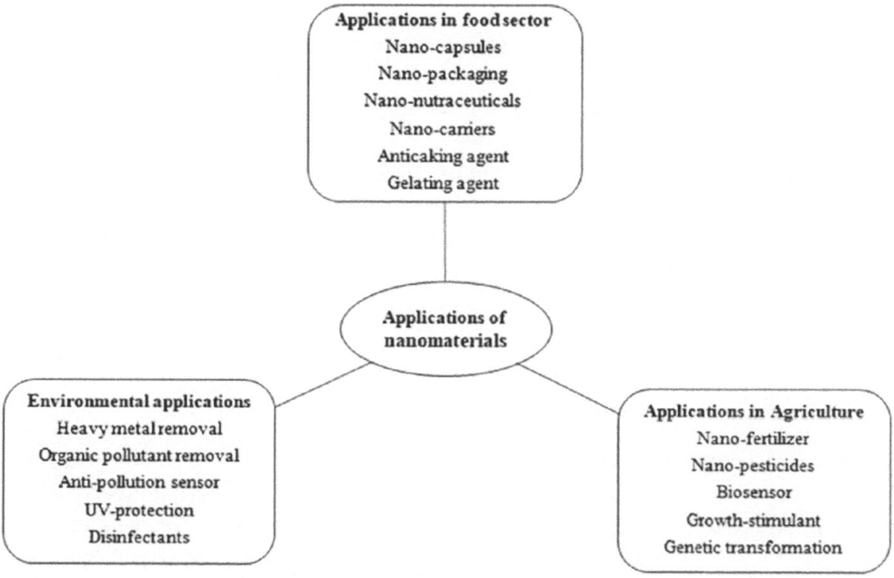

FIGURE 1.2 Applications of nanomaterials.

other environmental remediation techniques. The use of nanoparticles for pollution control has been the subject of extensive study over the past few decades. Surface charge, area, pore size, and surface functional groups of nanoparticles, as well as the parameters of aqueous water, determine the treatment methods (e.g., organic matter types, pH, ionic strength, surfactants, and temperature) (Eddy et al. 2022). Literature reports the use of various nanoparticles in environmental research, along with the mechanisms that are at play in the control of pollution mitigation utilizing these technologies. Adsorption is usually recognized as one of the most effective strategies for eliminating contaminants from water (Singh et al. 2018). Besides worsening the water shortage problem, water contamination can also lead to other environmental problems and health risks. Considerable research has been undertaken on the application of various nanoparticles to remove contaminants from water such as dyes, heavy metal ions, and other organic/inorganic compounds (Eddy and Garg 2022).

1.6.2 Applications in Agri-Food Sector

As a result of their exceptional properties, including as resistance to heat degradation, chemical impermeability, and superior mechanical capabilities, nanomaterials find applications in many different fields, including sensing and packaging (Lata, Bhardwaj, and Garg 2022). Nowadays, barrier coatings are made from a wide variety of nanomaterials and serve several functions, including extending the service life of the enclosed product (Onyszko et al. 2022). The unique morphological characteristics of nanomaterials give rise to their remarkable properties, such as high thermal and electrical conductivity, high specific surface area, and enhanced tensile strength (Bhardwaj, Lata, and Garg 2022b). Biopolymers are environmentally safe because they are biodegradable, biocompatible, and nontoxic. Their nanoformulations are mechanically stable, chemically inert, and can be tailored to have desired characteristics (Rishav Garg et al. 2023). Nanocoating of these materials prevents molecules of gases and liquids from passing through its surface, making it a promising material for use in packaging, barrier protection, and corrosion inhibition coatings in sectors such as the agri-food sector, medicine, industry, and environment (Bhardwaj, Lata, and Garg 2022a). Nanomaterials have been reported as a potential candidate for making biopolymer-based active, intelligent, and biodegradable packaging materials that provide sensing applications in addition to prevention of microbe invasions. Studies are in progress where these materials can be used to prepare edible coatings as well (Šuput et al. 2021).

1.6.3 Antimicrobial Potential

Antibiotic-resistant infections are a major cause of death in modern society. When antibiotics are overused or misused, they lose their efficacy, and new, more expensive, and more effective medications and therapies must be developed (Zheng et al. 2022). Consequently, the patient experiences a number of immediate and future complications. However, nanoparticles have shown antibacterial efficacy against a range of diseases; this is likely due to their diminutive size and their capacity to destroy cells in a number of ways (Onyszko et al. 2022). Nanomaterials present a novel approach to preventing the spread of bacteria that can cause human infections, in contrast to the conventional practise of using antibiotics to treat such conditions. In response, scientists have developed a wide range of antibacterial materials, such as nanoparticles, that may actively combat microbes (Karthik et al. 2022). Figure 1.2 illustrates the mechanisms by which nanoparticles serve as an efficient alternative antibacterial agent.

In response to issues with sustainable development, nanotechnology can be employed to lessen risks to human health and the natural environment. The application of nanomaterials to the therapy of viral infections and the development of methods to combat COVID-19 are two related instances (Patel, Mondal, and Ghosh 2020). On January 30, 2020, the novel coronavirus pandemic was declared a worldwide human health danger by the World Health Organization. Since then, the SARS-CoV-2 virus (novel COVID-19) has grown into an infectious sickness, causing a significant global public health problem that necessitates the collaboration of multiple research projects (Medhi et al. 2020). There is evidence that COVID-19 can cause severe cases of acute respiratory distress syndrome and coagulation abnormalities. Traditional treatments for viral infections are rapidly losing their effectiveness as a result of viral mutations and the appearance of new viral strains. In order to detect, treat, and prevent the spread of SARS-CoV-2, nanomaterials have been implemented (Vazquez-Munoz and Lopez-Ribot 2020). The numerous uses and advancements being made by scientists around the world are evidence of its potential. Nanotechnology-based diagnostics, medicines, and treatments have been shown to be effective against viruses comparable to SARS-CoV-2. It is anticipated that nanotechnology-based approaches will aid in the development of preventative, diagnostic, and therapeutic strategies for SARS-CoV-2 and its variants (Karthik et al. 2022).

1.6.4 Applications in Biomedical Sector

Biomedical sectors stand to greatly benefit from the opportunities afforded by nanotechnology for studying and altering a wide range of biomedical methods at nanoscales. Another well-liked field where nanotechnology is being put to use is nanomedicine (Nikalje 2015). Nanomedicine is a rapidly evolving field since it draws from many different areas of study. Through drug development, biomedicine, diagnostics, and medical imaging, nanomedicine has made its way into the biomedical business (Onyszko et al. 2022). Tissue engineering, nanosensor-based quick diagnosis of infectious disease, nanodrugs targeting tumours, and nanodrug delivery systems for regulated medication administration are all examples of common nanomedicine uses (Patra et al. 2018). Rapid progress in nanomedicine development thanks to advances in nanotechnology has boosted cancer treatment. With nanomedicine, cutting-edge, multipurpose, and precisely focused medical procedures are now possible. These days, nanoparticles may be found being used in a vast array of scientific disciplines (Khan et al. 2021). Nanoparticles have been widely documented to play an important role in the control of modern medicine, enabling a range of therapeutic procedures like the delivery of medication, the transfer of genetic information to malignant tumours, and the use of radiography as a contrast media agent (Chandraker et al. 2022). Novel cancer therapies have made use of a wide variety of nanomaterials, including those produced from organic and inorganic materials, lipids and glycans, and synthetic polymers (Luiz et al. 2022).

Nanocarriers for targeted drug delivery systems, biosensing, and imaging for clinical diagnostics are just some of the unique uses made possible by this cutting-edge technology (Luiz et al. 2022). Drug delivery and cancer therapy have already made use of a wide range of nanomaterials with important biomedical applications, such as magnetic nanoparticles, polymeric nanoparticles, carbon nanotubes, nanowires, nanocapsules, nanofluids, nanogel-based compounds, and liposomes (Mehnath et al. 2021). With their suitability and biocompatibility for broad and safer use, scientists have explored the potential applications of different nanomaterials for therapeutic reasons such as drug delivery, cancer detection, and treatment (Kabir, Islam, and Asaduzzaman 2022).

1.7 CONCLUSION

The increasing attention towards nanotechnology goes beyond its nanoscale but is influenced by the role of nanotechnology in pushing major changes in human society, which has been explored in this study. Nanomaterials that are safe for humans and the environment can be synthesized using green nanotechnology, which entails reducing the amount of resources used in the manufacture of the nanomaterials and limiting their negative effects on the environment. The enormous aspirations for nanotechnology are based on the controlled production of nanostructures with novel sorts of frameworks and processes by manipulation of the features of macroscopic entities to develop new applications in practically all industries of science and technology. It is anticipated that integrated scientific studies into nanotechnology therapeutic studies may be used in the prevention and/or treatment of viral infections, thereby overcoming these limitations and enhancing antiviral medicines. When it comes to nanotechnology, the world is not only concerned with its many applications, notably in the medical field, but also apprehensive about the eco-toxicity of nanomaterials. The toxicological evaluation of nanomaterials is highly desirable in order to provide innovative tools to reduce negative impacts on human health and the environment, leading to sustainable development.

REFERENCES

Baig, Nadeem, Irshad Kammakakam, Wail Falath, and Irshad Kammakakam. 2021. "Nanomaterials: A Review of Synthesis Methods, Properties, Recent Progress, and Challenges." *Materials Advances* 2 (6): 1821–71.

Bhardwaj, Sweta, Suman Lata, and Rajni Garg. 2022a. "Application of Nanotechnology for Preventing Postharvest Losses of Agriproducts." *The Journal of Horticultural Science and Biotechnology* 98: 1–14.

Bhardwaj, Sweta, Suman Lata, and Rajni Garg. 2022b. "Phyto-Mediated Green Synthesis of Silver Nanoparticles Using Acmella Oleracea Leaf Extract: Antioxidant and Catalytic Activity." *Pharmacognosy Magazine* 18 (77): 22.

Blažeka, Damjan, Rafaela Radičić, Dejan Maletić, Sanja Živković, Miloš Momčilović, and Nikša Krstulović. 2022. "Enhancement of Methylene Blue Photodegradation Rate Using Laser Synthesized Ag-Doped ZnO Nanoparticles." *Nanomaterials* 12 (15): 2677.

Cardoso, Rafael M., Tamires S. Pereira, Murilo H.M. Facure, Danilo M. dos Santos, Luiza A. Mercante, Luiz H.C. Mattoso, and Daniel S. Correa. 2022. "Current Progress in Plant Pathogen Detection Enabled by Nanomaterials-Based (Bio)Sensors." *Sensors and Actuators Reports* 4: 100068.

Cavaliere, Emanuele, Sebastiano De Cesari, Giulia Landini, Eleonora Riccobono, Lucia Pallecchi, Gian Maria Rossolini, and Luca Gavioli. 2015. "Highly Bactericidal Ag Nanoparticle Films Obtained by Cluster Beam Deposition." *Nanomedicine: Nanotechnology, Biology, and Medicine* 11 (6): 1417–23.

Chandraker, Sandip Kumar, Mishri Lal, Farheen Khanam, Preeti Dhruve, Rana P. Singh, and Ravindra Shukla. 2022. "Therapeutic Potential of Biogenic and Optimized Silver Nanoparticles Using Rubia Cordifolia L. Leaf Extract." *Scientific Reports* 12 (1): 1–15.

Chugh, Gaurav, Kadambot H.M. Siddique, and Zakaria M. Solaiman. 2021. "Nanobiotechnology for Agriculture: Smart Technology for Combating Nutrient Deficiencies with Nanotoxicity Challenges." *Sustainability* 13 (4): 1–20.

Chums-ard, Wisut, Derek Fawcett, Chun Che Fung, and Gerrard Eddy Jai Poinern. 2019. "Biogenic Synthesis of Gold Nanoparticles from Waste Watermelon and Their Antibacterial Activity Against Escherichia Coli and Staphylococcus Epidermidis." *International Journal of Research in Medical Sciences* 7 (7): 2499.

Dolez, Patricia I. 2015. "Nanomaterials Definitions, Classifications, and Applications." In *Nanoengineering: Global Approaches to Health and Safety Issues*, edited by Patricia I. Dolez, 3–40. Elsevier.

Eddy, Nnabuk Okon, and Rajni Garg. 2022. "CaO Nanoparticles: Synthesis and Applications in Water Remediation." In *Handbook of Research on Green Synthesis and Applications of Nanomaterials*, 247–68. IGI Global.

Eddy, Nnabuk Okon, Richard Alexis Ukpe, Paul Ameh, Rapheal Ogbodo, Rajni Garg, and Rishav Garg. 2022. "Theoretical and Experimental Studies on Photocatalytic Removal of Methylene Blue (MetB) from Aqueous Solution Using Oyster Shell Synthesized CaO Nanoparticles (CaONP-O)." *Environmental Science and Pollution Research*. doi. 10.1007/s11356-022-22747-w.

Ethaib, Saleem, Sarah Al-Qutaifia, Nadhir Al-Ansari, and Salah L. Zubaidi. 2022. "Function of Nanomaterials in Removing Heavy Metals for Water and Wastewater Remediation: A Review." *Environments - MDPI* 9 (10):123.

Garg, Rajni, Priya Rani, Rishav Garg, Mohammad Amir, Nadeem Ahmad, Afzal Husain, and Juliana Heloisa. 2022. "Biomedical and Catalytic Applications of Agri-Based Biosynthesized Silver Nanoparticles." *Environmental Pollution* 310 (June): 119830.

Garg, Rajni, Priya Rani, Rishav Garg, and Nnabuk Okon Eddy. 2021. "Study on Potential Applications and Toxicity Analysis of Green Synthesized Nanoparticles." *Turkish Journal of Chemistry* 45 (6): 1690–06.

Garg, Rishav, Rajni Garg, Nnabuk Okon Eddy, Abdulaziz Ibrahim Almohana, Sattam Fahad Almojil, Mohammad Amir Khan, and Seung Ho Hong. 2022. "Biosynthesized Silica-Based Zinc Oxide Nanocomposites for the Sequestration of Heavy Metal Ions from Aqueous Solutions." *Journal of King Saud University: Science* 34 (4): 101996.

Garg, Rishav, Rajni Garg, Nnabuk Okon Eddy, Mohd Amir Khan, Afzal Husain Khan, Thamer Alomayri, and Parveen Berwal. 2023. "Mechanical Strength and Durability Analysis of Mortars Prepared with Fly Ash and Nano-Metakaolin." *Case Studies in Construction Materials* 18 (December 2022): e01796.

Garg, Rishav, Rajni Garg, Md. Amir Khan, Manjeet Bansal, and Vinod Garg. 2022. "Utilization of Biosynthesized Silica-Supported Iron Oxide Nanocomposites for the Adsorptive Removal of Heavy Metal Ions from Aqueous Solutions." *Environmental Science and Pollution Research*. 1–10.

Hernández-Hernández, Aldahir A., Gabriel Aguirre-Álvarez, Raquel Cariño-Cortés, Luis H. Mendoza-Huizar, and Rubén Jiménez-Alvarado. 2020. "Iron Oxide Nanoparticles: Synthesis, Functionalization, and Applications in Diagnosis and Treatment of Cancer." *Chemical Papers* 74 (11): 3809–24.

Jassal, Prabhjot S., Robinka Khajuria, Ronit Sharma, Pritam Debnath, Sonal Verma, Abey Johnson, and Sumit Kumar. 2020. "Photocatalytic Degradation of Ampicillin Using Silver Nanoparticles Biosynthesised by Pleurotus Ostreatus." *Biotechnologia* 101 (1): 5–14.

Kabir, Syed Rashel, Farhadul Islam, and A. K. M. Asaduzzaman. 2022. "Biogenic Silver/Silver Chloride Nanoparticles Inhibit Human Cancer Cells Proliferation in Vitro and Ehrlich Ascites Carcinoma Cells Growth in Vivo." *Scientific Reports* 12 (1): 1–14.

Karthik, C., K. A. Punnaivalavan, S. Pandi Prabha, and D. G. Caroline. 2022. "Multifarious Global Flora Fabricated Phytosynthesis of Silver Nanoparticles: A Green Nanoweapon for Antiviral Approach Including SARS-CoV-2". *International Nano Letters*. 12: 313–44.

Khan, Suliman, Mohammad Mahdi Nejadi Babadaei, Anwarul Hasan, Zehra Edis, Farnoosh Attar, Rabeea Siddique, Qian Bai, Majid Sharifi, and Mojtaba Falahati. 2021. "Enzyme-Polymeric/Inorganic Metal Oxide/Hybrid Nanoparticle Bio-Conjugates in the Development of Therapeutic and Biosensing Platforms." *Journal of Advanced Research* 33: 227–39.

Lata, Suman, Sweta Bhardwaj, and Rajni Garg. 2022. "Nanomaterials for Sensing and Biosensing: Applications in Agri-Food Diagnostics." *International Journal of Environmental Analytical Chemistry*. 1–18.

Luiz, Marcela Tavares, Jessyca Aparecida Paes Dutra, Larissa Bueno Tofani, Jennifer Thayanne Cavalcante de Araújo, Leonardo Delello Di Filippo, Juliana Maldonado Marchetti, and Marlus Chorilli. 2022. "Targeted Liposomes: A Nonviral Gene Delivery System for Cancer Therapy." *Pharmaceutics* 14 (4): 821.

Medhi, Riddhiman, Pannaree Srinoi, Nhat Ngo, Hung Vu Tran, and T. Randall Lee. 2020. "Nanoparticle-Based Strategies to Combat COVID-19." *ACS Applied Nano Materials* 3 (9): 8557–80.

Mehnath, Sivaraj, Ashok Kumar Das, Sandeep Kumar Verma, and Murugaraj Jeyaraj. 2021. "Chapter Nine - Biosynthesized/Green-Synthesized Nanomaterials as Potential Vehicles for Delivery of Antibiotics/Drugs." In *Biosynthesized Nanomaterials*, edited by Sandeep Kumar Verma and Ashok Kumar Das, 94:363–432. Comprehensive Analytical Chemistry. Elsevier.

Minelli, Caterina, Aneta Sikora, Raul Garcia-Diez, Katia Sparnacci, Christian Gollwitzer, Michael Krumrey, and Alex G. Shard. 2018. "Measuring the Size and Density of Nanoparticles by Centrifugal Sedimentation and Flotation." *Analytical Methods* 10 (15): 1725–32.

Nikalje, Anna Pratima. 2015. "Nanotechnology and Its Applications in Medicine." *Medicinal Chemistry* 5 (2): 81–89.

Onyszko, M., A. Markowska-Szczupak, R. Rakoczy, O. Paszkiewicz, J. Janusz, A. Gorgon-Kuza, K. Wenelska, and E. Mijowska. 2022. "The Cellulose Fibers Functionalized with Star-like Zinc Oxide Nanoparticles with Boosted Antibacterial Performance for Hygienic Products." *Scientific Reports* 12 (1): 1321.

Patel, Priyank Pravin, Sayoni Mondal, and Krishna Gopal Ghosh. 2020. "Some respite for India's dirtiest river? Examining the Yamuna's water quality at Delhi during the COVID-19 lockdown period". *Science of the Total Environment* 744: 140851.

Patra, Jayanta Kumar, Gitishree Das, Leonardo Fernandes Fraceto, Estefania Vangelie Ramos Campos, Maria Del Pilar Rodriguez-Torres, Laura Susana Acosta-Torres, Luis Armando Diaz-Torres, et al. 2018. "Nano Based Drug Delivery Systems: Recent Developments and Future Prospects." *Journal of Nanobiotechnology* 16 (1): 71.

Ramirez-Perez, Javier Christian, Tatiana A. Reis, Cristiano L.P. Olivera, and Marcia A. Rizzutto. 2022. "Impact of Silver Nanoparticles Size on SERS for Detection and Identification of Filamentous Fungi." *Spectrochimica Acta Part A: Molecular and Biomolecular Spectroscopy* 272 (May): 120980.

Saikumari, N., S. Monish Dev, and S. Avinaash Dev. 2021. "Effect of Calcination Temperature on the Properties and Applications of Bio Extract Mediated Titania Nano Particles." *Scientific Reports* 11 (1): 1–17.

Santhosh, Chella, Arumugam Malathi, Ehsan Daneshvar, Pratap Kollu, and Amit Bhatnagar. 2018. "Photocatalytic Degradation of Toxic Aquatic Pollutants by Novel Magnetic 3D-TiO2@HPGA Nanocomposite." *Scientific Reports* 8 (1): 1–15.

Sharma, Rajat, Rajni Garg, and Avnesh Kumari. 2020. "A Review on Biogenic Synthesis, Applications and Toxicity Aspects of Zinc Oxide Nanoparticles." *EXCLI Journal* 19:1325–40.

Singh, N. B., Garima Nagpal, Sonal Agrawal, and Rachna. 2018. "Water Purification by Using Adsorbents: A Review." *Environmental Technology & Innovation* 11: 187–240.

Song, Chunyuan, Boyue Yang, Yu Zhu, Yanjun Yang, and Lianhui Wang. 2017. "Ultrasensitive Sliver Nanorods Array SERS Sensor for Mercury Ions." *Biosensors and Bioelectronics* 87 (January): 59–65.

Šuput, Danijela, Senka Popović, Nevena Hromiš, and Jovana Ugarković. 2021. "Degradable Packaging Materials: Sources, Application and Decomposition Routes." *Journal on Processing and Energy in Agriculture* 25 (2): 37–42.

Terban, Maxwell W., and Simon J.L. Billinge. 2022. "Structural Analysis of Molecular Materials Using the Pair Distribution Function." *Chemical Reviews* 122 (1): 1208–72.

Thangadurai, Devarajan, Vishal Ahuja, and Jeyabalan Sangeetha. 2021. "Nanomaterials and Nanoprocesses for the Removal and Reuse of Heavy Metals." In *Handbook of Nanomaterials and Nanocomposites for Energy and Environmental Applications*, edited by Oxana Vasilievna Kharissova, Leticia Myriam Torres-Martínez, and Boris Ildusovich Kharisov, 2649–60. Cham: Springer International Publishing.

Vasyukov, Denis, Yonathan Anahory, Lior Embon, Dorri Halbertal, Jo Cuppens, Lior Neeman, Amit Finkler, et al. 2013. "A Scanning Superconducting Quantum Interference Device with Single Electron Spin Sensitivity." *Nature Nanotechnology* 8 (9): 639–44.

Vazquez-Munoz, Roberto, and Jose L. Lopez-Ribot. 2020. "Nanotechnology as an Alternative to Reduce the Spread of COVID-19." *Challenges* 11 (2): 15.

Withers, Freddie, Osvaldo Del Pozo-Zamudio, Artem Mishchenko, Aiden P. Rooney, Ali Gholinia, K. Watanabe, Takashi Taniguchi, et al. 2015. "Light-Emitting Diodes by Band-Structure Engineering in van Der Waals Heterostructures." *Nature Materials* 14 (3): 301–6.

Yousefi-Limaee, Nargess, Mehdi Ghahari, Kumars Seifpanahi-Shabani, Alireza Naeimi, and Samaneh Ghaedi. 2023. "Evaluation of Adsorptive Efficiency of Calcium Oxide Nanoparticles for the Elimination of Cationic Dyes: Combustion Synthesis, Adsorption Study and Numerical Modeling." *Progress in Color, Colorants and Coatings* 16 (1): 1–20.

Zheng, Yuanxun, Yahui Zhang, Jingbo Zhuo, Peng Zhang, and Weixing Kong. 2022. "Mechanical Properties and Microstructure of Nano-Strengthened Recycled Aggregate Concrete." *Nanotechnology Reviews* 11 (1): 1499–1510.

2 Nanotechnology in Agri-Food Production

Current Trends and Future Prospects

Abu Barkat Md Gulzar, A.S.M. Islamul Hoque Laskar, and Pranab Behari Mazumder
Assam University

CONTENTS

2.1 INTRODUCTION

The question of how to ensure food security for a rapidly expanding global population is the greatest challenge facing humanity as a whole. Projections for 2050 show that the global population will have increased to 9 billion, at which point the need for food will increase from 59% today to 98%. Agriculture is the one and only source of food for the entire human population, and thus the development of new strategies to improve agricultural productivity has become an urgent task (Neme et al. 2021). To boost agricultural output, farmers all over the world widely utilize fertilizers, pesticides, herbicides, etc. However, excessive utilization of chemically produced fertilizers has been linked to major environmental problems such as declination of soil fertility, soil hardening, and air, soil, and water pollution that may kill ecofriendly microorganisms and also cause ecosystem instability. To overcome the challenges, newer approaches are needed that can boost agricultural productivity by minimizing the adverse effects on the environment. In this

DOI: 10.1201/9781003397861-2

regard, nanotechnology has shown promising results in agriculture by minimizing the harmful impacts on the environment and living organisms (Hotessa Halake and Muda Haro 2022). Nanotechnology is the science that studies nanometer-sized particles and their applications for the betterment of human life. Nanofertilizers, nanopesticides, and nanobiosensors are all examples of how nanotechnology is being put to use in agriculture. Increased yields and production rates, better resource use, and less waste generation are all possible outcomes of using nanotechnology in agriculture. Use of nanomaterials for soil remediation is another application of nanotechnology to make polluted land suitable for agricultural practices. Besides these, the food industry is one of the many fields where nanotechnologies have found useful applications (Vijayakumar et al. 2022). This chapter discusses the potential benefits of nanotechnology in boosting plant growth, protection, and yield under both stressed and non-stressed conditions. Besides these, this chapter also highlights the negative impacts of widespread use of nanotechnology on the agroecosystem and human health. Likewise, the discussion of positive and negative impacts of nanotechnology will help us understand the present state and future potential of nanotechnology in agriculture.

2.2 NANOTECHNOLOGY IN AGRICULTURE

In this section, we discussed some of the various ways in which nanotechnology might be used to improve soil quality and boost crop production in the agricultural sector (Figure 2.1). Here, we highlight some of the most recent developments regarding the application of nanotechnology in farming (Table 2.1), particularly in the areas of nanofertilizers and nanopesticides, nanobiosensor production, and the cleanup of polluted soils.

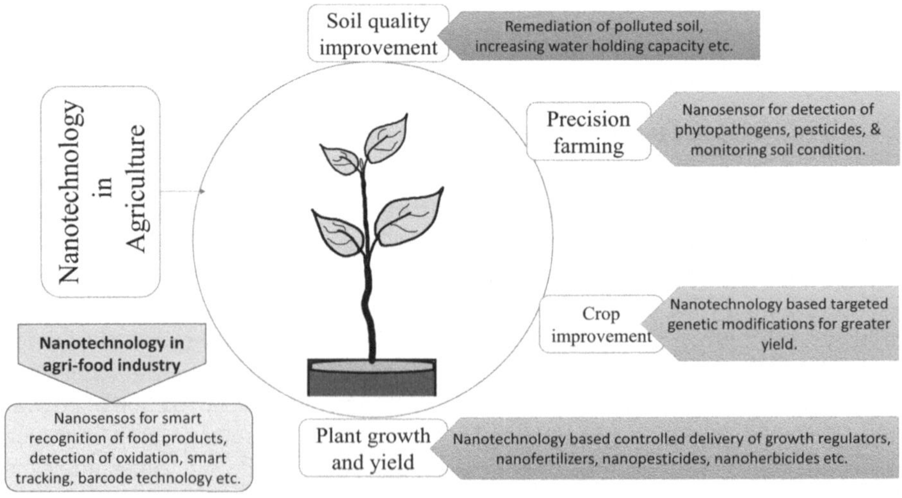

FIGURE 2.1 Nanotechnology applications in agriculture and agri-food industries.

TABLE 2.1
Impact of Certain Nanoparticles on Plant Development and Plant Defense

Nanoparticles	Crop Plant	Effects on Plants	Reference
Calcium hydroxyapatite	*Oryza sativa*	Increased cellulase and xylanase activity which further increased the production of reducing sugar	Dutta et al. (2014)
ZnO, FeO, and ZnFeCu-oxide	*Vigna radiata*	Foliar application improved plant growth and biomass as compared to the control plants. However, traces of nanoparticles inside the plant leaves indicates its accumulation in food chain	Dhoke et al. (2013)
Chitosan-silicon nanofertilizer	*Zea mays*	Showed control release of Si into the soil and increased seedling vigor index, growth, antioxidant enzyme activity and photosynthetic parameters in plants	Kumaraswamy et al. (2021)
nZnO, nFe_2O_3, $nTiO_2$	Hordeum vulgare	Application of nanofertilizer improved plant growth, spike length, number of grains, grain mass, and chlorophyll content as compared to the non-treated control	Janmohammadi et al. (2016)
Zinc oxide NP	*Arachis hypogaea*	Increased germination percentage, vigor index, root and stem length, chlorophyll content, and also increased productivity	Prasad et al. (2012)

(*Continued*)

TABLE 2.1 (*Continued*)
Impact of Certain Nanoparticles on Plant Development and Plant Defense

Nanoparticles	Crop Plant	Effects on Plants	Reference
Fe nano-oxide	*Triticum aestivum*	Foliar application improved plant growth, photosynthetic pigments, and respiration	Feng et al. (2022)
Nanoparticles	**Phytopathogen**	**Effect on Phytopathogen**	**References**
AgNPs	*Colletotrichum gloeosporioides*	Under laboratory condition, 90% growth inhibition of phytopathogen was observed after exposure of Ag NPs (5–24 nm) with 56 μg Ag/mL dose concentration in potato dextrose agar media	Aguilar-Méndez et al. (2011)
AgNPs	*Rhizoctonia solani, Sclerotinia sclerotiorum, S. minor*	Ag NPs (4–8 nm) severely inhibits hyphal growth of phytopathogenic fungi under laboratory condition	Min et al. (2009)
TiO_2 NPs	*Xanthomonas perfonans*	TiO_2 NPs (<8 nm) doped with Ag or Zn showed significant reduction of CFUs	Paret et al. (2013)
ZnONPs	*Botrytis cinerea, Penicillium expansum*	Exposure of ZnONPs (70 ± 15 nm) alter cellular functions and also damages fungal hyphae and restricts growth of fungus	He et al. (2011)
CuNPs	*Xanthomonas axonopodis* pv. *punicae* (*Xap*)	CuNPs (0.2 ppm) suppressed the growth of *Xanthomonas axonopodispv*., the bacterium responsible for the blight of pomegranate (*Punica granatum*).	Mondal and Mani (2012)

2.2.1 Nanofertilizers

A substantial growth in agricultural productivity, particularly in grains, during the last few decades has contributed immensely to meeting the nutritional demands of people all over the globe. One key factor contributing to increased agricultural productivity is the increased usage of chemical fertilizers. However, volatilization and leaching reduce the efficiency with which chemical fertilizers may be employed, leading to pollution and higher production costs. For instance, around half to two-thirds of nitrogen fertilizers used are emitted into the atmosphere (Subramanian et al. 2015). In this regard, nanotechnology is used to lessen nutrient leaching, facilitate controlled release fertilizers, and increase the availability of scarce nutrients. In this context, "nanofertilizers" refer to nanomaterials that are either nutrients (micro- or macro-) and also serve as carriers or additives for the nutrients. Encapsulating nutrients in nanomaterials creates nanofertilizers (Vijayakumar et al. 2022).

Nanosized particles, including hydroxyapatite, zeolite, clay minerals, polyacrylic acid, and chitosan, are used to create nanofertilizers. They may be sprayed on the leaves of plants or dug into the soil as fertilizer (Iqbal, Umar, and Mahmooduzzafar 2019). Nanofertilizers with modified urea nanoparticles (NPs) of hydroxyapatite release nitrogen to plants for 60 days, unlike traditional fertilizers. This is because hydroxyapatite has the highest surface area, and its strong bonds with urea allow the latter's nitrogen to be slowly released (Tarafder et al. 2020; Vijayakumar et al. 2022). In a report, it was shown that coating potassium fertilizer with polyacrylamide polymer resulted in a gradual release of the potassium. Leaching may be reduced by using a slow-release potassium fertilizer on sandy soils (Sim et al. 2021). Nanotechnology may improve compost output and organic waste breakdown in agriculture. While the technology has not made any significant strides yet, it will become the standard method in the near future (Vijayakumar et al. 2022).

The ability of NPs to protect crops and food production remains largely untapped. Agricultural fields and their products are subject to a variety of abiotic stressors, and it is well acknowledged that NPs may exert significant influence in mitigating these effects and ensuring agricultural production. Treatment with tin dioxide nanoparticles (TiO_2 NP) on maize, as described by Gohari et al. (2020), significantly influenced growth and mitigated salt stress effects on Moldavian balsam (*Dracocephalum moldavica* L.). Agronomic yield of Moldavian balm was shown to increase when 100 mg/L TiO_2 NP was applied to plants when they were growing in a salt stress environment (100 mM NaCl). In their study, they found that treatment with TiO_2 NP improves light absorption and energy transmission in salinity-stressed plants (Gohari et al. 2020). Another study found that when silver dioxide (SiO_2) and TiO_2 NPs were applied to soybeans, the plant's capacity to absorb water and fertilizer was enhanced, and nitrate reductase enzyme activity was boosted (Khodashenas and Ghorbani 2016). Rao (2019), also observed that combining nanosized TiO_2 and SiO_2 NP might improve germination, growth, and water and fertilizer use efficiency in *Spinacia oleracea*, *Triticum aestivum*, *Solanum lycopersicum*, and *Zea mays*. Thus, these NPs can be utilized to improve plant growth under water-limited conditions (Rao 2019).

Many secondary metabolites are now being researched as possible new sources for nutraceuticals, medicines, and agrochemicals (Abobatta 2018). But plants only make trace amounts of such secondary metabolites. As revealed by War et al. (2020), NPs' innovative and distinctive features contribute significantly to the production of secondary metabolites. He found that aloe vera cell suspension cultures produced more secondary metabolites, including flavonoids and phenols, after being exposed to TiO_2 NPs.

The data show that nanofertilizers are very effective, require less fertilizer, and have other positive environmental effects. Similarly, from an economic point of view, nanofertilizers are crucial for achieving profitable and sustainable agriculture.

2.2.2 Nanopesticides

Crop yield-reducing factors, including plant diseases, weeds, and damaging insects, are the focus of agrochemical technology. Pesticide use is the fastest and most cost-effective method of preventing the spread of disease and pests in modern agriculture. But unchecked pesticide usage has had serious consequences for people's health, livestock, pollinators, and ecosystems overall, leading to an uptick in interest in cutting-edge solutions (Vijayakumar et al. 2022). It's possible that nanopesticide formulations will provide advantages over chemical pesticides. Plant disease preventatives made from NPs are referred to as "nanopesticides." The nanopesticide formulations may preserve agrochemicals from environmental degradation while also increasing their water solubility and bioavailability, therefore revolutionizing the management of diseases, weeds, and insects in crops. On the other hand, nanopesticides have a high capacity for diffusion and bioaccumulation in soil, water, and food, which might have negative effects on water quality and human health if used carelessly (Mukhopadhyay 2014). In that case, a nanoencapsulation strategy for crop protection might be the answer to this problem. Nanoencapsulation is a process wherein biologically active substances are enclosed in capsule walls and then released into the environment at a regulated rate in response to an external stimulus. Nanocapsules, nanospheres, and nanovesicles are all examples of nanostructured matrix systems that might be created and used for pesticide delivery (Hotessa Halake and Muda Haro 2022). Deka et al. (2021) recently described the use of NPs as nanocarriers in the administration of pesticides, with the goal of dispersing active components to a targeted area over a specified time frame while minimizing harm to non-target species. When it comes to controlling *Aphis nerii*, Shaimaa and Mostafa (2015) looked at the efficacy of both pure silver and Ag-ZnNPs at varying concentrations as insecticides. The results demonstrated that silver NPs are an effective method for controlling *A. nerii* populations. In addition, the research discovered that the most effective insect death impact was achieved with NPs at 700 mg/L and imidacloprid at 1 μL/mL (Shaimaa and Mostafa 2015). At 1,000 ppm, a nano-sulfur (50–90 nm) was proven to be more efficient than a commercial sulfur product in controlling the growth of *Erysiphe cichoracearum* (Mukhopadhyay 2014). It has been established that CuNP and ZnONP exhibit potential antibacterial and antifungal activities against phytopathogens at low concentrations, as reported by War et al. (2020). CuNPs (0.02 ppm) were reported to limit the development of *Xanthomonas axonopodis* pv., the bacterium responsible for the blight of pomegranate (*Punica granatum*) (War et al. 2020). In another study, it has been reported that treatment with ethanol-based nanosilver colloid (<20 ppm) was effective in eliminating case-bearing cloth moth larvae, *Tinea pellionella* (Win, Khan, and Fu 2020).

2.2.3 Nanotechnology for Soil Quality Improvement

Degradation of soil quality is another major reason for reduced plant growth and crop yield. In this regard, nanotechnology has shown promising potential to improve soil quality through suitable agricultural practices. There have been reports that hydrogels, nanozeolites, and nanoclays may improve soil water retention ability. This enables the soil to function as a reservoir for water, slowly releasing moisture as needed, which may help farmers avoid dry spells in the middle of the growing season (Fraceto et al. 2016; Sekhon 2014). In contrast, interest in using NMs for the remediation of polluted soils, primarily by sorption, reduction, or chemical oxidation, has increased because of their reduced particle size, greater surface area, and well-established reactivity (Guerra et al. 2018). The activity and severity of both inorganic and organic soil pollutants are primarily determined by their sorption and desorption actions with soil components. Consequently, the toxicity and mobility of soil pollution may be controlled by supplementing the polluted soils with appropriate amendments. Adsorption onto mineral surfaces, ion exchange, organic ligand complexation, and surface precipitation are only some of the sorption mechanisms involved. There are two types of soil amendments: (i) by increasing pollutants' solubility and bioavailability, mobilization agents make them more amenable to washing and plant absorption. (ii) Immobilizing chemicals bind to soil contaminants, reducing their solubility and bioavailability and blocking them from entering the food chain. For instance, the addition of nZVI increased the soil washing efficiency of soil contaminated with metal(loids) due to the fact that the metal(loids) was first adsorbed before being magnetically isolated. Nevertheless, nano-hydroxyapatite particles have effectively immobilized metal concentrations in contaminated sediments and soils by reducing their exchangeable fraction, hence lowering the concentration of metals in the pore water.

2.2.4 Nanosensors in Agriculture

Soil qualities, nutrient monitoring, ambient conditions, phytopathogen and weed identification, and other aspects of plant development may all be tracked in real-time with the use of nanosensors, a smart sensing approach incorporated with nanotechnology. With the aid of these signals and data, farmers will be able to determine the optimal planting and harvesting times, as well as the water requirement, plant deficiencies, agricultural fertilizers, nutritional status, pesticides, herbicides, and other remedies required for specific plant morpho-physiological, patho-physiological, and environmental conditions, and thus reduce yield loss. When it comes to identifying fungal, bacterial, and viral pathogens in farming, nanosensors are becoming more useful. With the goal of detecting the pathogenic fungus *Sclerotinia sclerotiorum* in oil seeds, scientists used nanosensor techniques to measure salicylic acid byproducts (Wang et al. 2016). Similarly, Etefagh et al. produced CuO and nanolayer NPs and used for developing nanosensors for detection of *Aspergillus niger* (Etefagh, Azhir, and Shahtahmasebi 2013). Due to plant-pathogen type alterations, on-time, onsite, and sensor-based phytopathogen monitoring is growing rapidly.

2.3 AGRI-FOOD INDUSTRIAL USE OF NANOTECHNOLOGY

The agri-food industry's use of nanotechnology may be broken down into two primary categories. The *Nano inside* approach indicates the use of NPs or the structures derived through nanotechnology for nutrients, food supplements or food additives; the "nano outside" strategy refers to the implementation of nanotechnology in the realms of food production, packaging, and sensing technologies (Hossain et al. 2020). Oxidation of food items is one of the primary causes of food waste as it causes deterioration of food quality, including discoloration, off-odors, and off-flavors. In this regard, nanoplastics developed using nanotechnology can restrict the entry of oxygen and thus prevent the oxidation of food items. Improved gas barrier characteristics, thermal stability, and mechanical strength in polymer-silicate nanocomposites have also been reported. Nanocoatings produced using nanotechnology are used to cover fruits to prevent shrinkage and weight loss (Shaimaa and Mostafa 2015). Several "nanosensor"-equipped packaging materials for detecting the oxidation of food have been developed and put into use in recent years. The basic principle of operation is that NP-based sensors show a color change when oxidation happens in the food packaging, providing insight into the contents of the container (Prasad, Bhattacharyya, and Nguyen 2017). Nanosensor techniques are nowadays widely used for packaging food items such as meat, fruits, and milk (Sekhon 2014; Bumbudsanpharoke and Ko 2015). The NPs have been shown to exhibit antibacterial activity. Thus, food-contact surfaces that have been nanocoated may serve as a barrier or have antibacterial qualities. Embedding silver NPs into plastic for use in food storage containers has proven beneficial, and this works like disinfection of containers, reducing the likelihood of hazardous bacterial development. Therefore, nanotechnology is a future-oriented method; it provides biosecurity for agriculture and its products (Bumbudsanpharoke and Ko 2015).

Nanosensors aid (reactive nanolayers and nanosensor-based smart ink) in food labeling may give smart identification of the appropriate food product. Highlights such as temperature, time, freshness, pathogens, and humidity may be printed on the labels of packaged foods (Li et al. 2016; Bumbudsanpharoke and Ko 2015).

2.4 HARMFUL EFFECTS OF NANOTECHNOLOGY ON AGRO ECOSYSTEM AND HUMAN HEALTH

Although nanotechnology offers many potential uses in agriculture, it also poses risks to humans, animals, and the environment. Understanding agro-ecological toxicity requires studying the behavior of NPs at varying sizes and concentrations in soil, plants, and water. Nanofertilizers are more likely to be absorbed by the soil and damage soil ecosystems and soil structure. Excess concentrations of nanofertilizers in soil adversely affect beneficial soil microfauna and microbes, and also damage the gastrointestinal system of earthworms. NPs in soil showed reductions in growth, survival and fertility rates and increased mortality rates for earthworms and other soil microbes. In that case, the size of NPs played a role in ecotoxicity. Roh et al. (2010) performed an experiment to assess the relationship between the size of the NP and its toxicity. The researchers applied CeO_2 (15 and 45 nm) and TiO_2 (7 and 20 nm) NPs to *Caenorhabditis elegans* and found that the NPs with smaller sizes, i.e., CeO_2 (15 nm) and TiO_2 (7 nm), were more toxic compared to the larger NPs, i.e., CeO_2 (45 nm) and TiO_2 (20 nm) (Roh et al. 2010). According to the research of Hu et al. (2010), increased dose concentration and number of doses also showed ecotoxicological effects. In their study, when they increased the dose concentration of ZnO NPs from 1 to 5 g/kg soil, it bioaccumulated in the earthworm's body and caused DNA damage (Hu et al. 2010).

After being absorbed from the soil, NPs may be toxic to plants directly, or they may induce toxicity via the production of toxic ions during decomposition. The toxicity of NPs to plants, however, depends on a variety of factors, including the kind of plant, the NP's size, and its concentration during particular growth stages. When present in high enough concentrations, NPs may have a deleterious influence on plant development, including germination, biomass, root development, and leaf expansion, as documented by Lin and Xing (2007). NPs may cause a variety of harmful biological responses, such as enhanced ROS generation owing to additional electrons or biomolecule interactions, triggering oxidative stress, and preventing plants from properly absorbing and transporting water, thereby disrupting metabolic processes and photosynthesis (Kumar et al. 2019; Wang et al. 2015). Hong et al. (2016) documented that foilar application of CuO_2 and CeO_2 NPs with concentrations of 200 mg/L reduced photosynthesis in *Cucumus sativus L.* (Hong et al. 2016). However, changes in transcriptome, DNA damage, and genotoxic effects like micronuclei formation, chromosomal aberration, and DNA fragmentation were found in plants after exposure to NPs (Karami Mehrian and De Lima 2016). Du et al. (2011) reported that exposure to ZnO and TiO_2 NPs showed a negative effect on soil enzymes and reduced wheat plant mass (Du et al. 2011).

The NPs are readily dispersed into the soil, air, or water and make entry into living organisms, posing a threat to all forms of life. NPs' entry into the human body mainly occurs through breathing (inhalation), ingestion, and absorption by the skin during the production, processing and utilization of these particles. Inhalation of airborne NPs is the most prevalent route of exposure. The olfactory epithelium is another route via which NPs may enter the brain. Exposure to ultrafine airborne NPs may increase mortality and morbidity from respiratory and cardiovascular causes. From the food chain, NPs also make their way into the human gastrointestinal tract. NPs with a smaller size (i.e., $\leq$50 nm) are easily absorbed by the gastrointestinal tract and transported to the liver, blood, spleen, and bone marrow. TiO_2 in food additives or other forms may persuade inflammation through genotoxicity and oxidative stress, leading to chromosomal instability (Fu et al. 2014). NPs less than 10 nm in size are particularly harmful when absorbed via the skin. There is a risk of erythema, edema, and eschar development from particles of this size (Dubey and Mailapalli 2016).

A comprehensive set of rules, regulations, and policies is crucial for the secure use of NPs in agriculture and in food production sectors. In the US, the FDA regulates nanotech-based foods and nanotech-based packaging, whereas the European Union (EU) regulates nanotech-based food additives. On the other hand, the majority of nations that produce nanomaterials do not have adequate legislation pertaining to nanotechnology. Hence, for legitimate nanotechnological uses, comprehensive government rules and regulations, along with rigorous toxicity monitoring procedures, are required (Neme et al. 2021).

2.5 CONCLUSION AND FUTURE PROSPECTIVE

In conclusion, nanotechnology is one of the fastest-developing fields that has the capability to be used as a game-changing technology to alleviate global food insecurity in the near future. In the future, many scientists believe that nanotechnology will be used to create high-tech farms, complete with smart nanotools that will enable farmers to maximize their yields while minimizing their resources. It has been believed that nanotechnology will help to develop sustainable smart agriculture through the development of innovative and effective agrochemicals, such as nano-based fertilizers and pesticides, for feeding and protecting plants. Therefore, future research should concentrate on the constant interaction and environmental impact of plants and nano-agrochemicals. Future investigations must concentrate on finding solutions to the challenges posed by NPs. The broad use of nanotechnology cannot be ensured by studies of NP production and the approval of a few specialized applications appropriate only for usage in a controlled lab environment. As a result, scientists need to collaborate to enhance future research with a more grounded approach.

REFERENCES

Abobatta, Waleed Fouad. 2018. "Nanotechnology Application in Agriculture." *Acta Scientific Agriculture* 2 (6): 99–102.

Aguilar-Méndez, M. A., San Martín-Martínez, E., Ortega-Arroyo, L., Cobián-Portillo, G., & Sánchez-Espíndola, E. (2011). "Synthesis and Characterization of Silver Nanoparticles: Effect on Phytopathogen Colletotrichum Gloesporioides." *Journal of Nanoparticle Research* 13 (6): 2525–2532.

Bumbudsanpharoke, Nattinee, and Seonghyuk Ko. 2015. "Nano-Food Packaging: An Overview of Market, Migration Research, and Safety Regulations." *Journal of Food Science* 80 (5): R910–23.

Deka, Bhabesh, Azariah Babu, Chittaranjan Baruah, and Manash Barthakur. 2021. "Nanopesticides: A Systematic Review of Their Prospects with Special Reference to Tea Pest Management." *Frontiers in Nutrition* 8 (August):686131.

Dhoke, S. K., P. Mahajan, R. Kamble, and A. Khanna. 2013. "Effect of nanoparticles suspension on the growth of mung (Vigna radiata) seedlings by foliar spray method." *Nanotechnology Development* 3 (1): 1.

Du, Wenchao, Yuanyuan Sun, Rong Ji, Jianguo Zhu, Jichun Wu, and Hongyan Guo. 2011. "TiO_2 and ZnO Nanoparticles Negatively Affect Wheat Growth and Soil Enzyme Activities in Agricultural Soil." *Journal of Environmental Monitoring* 13 (4): 822.

Dubey, Alpna, and Damodhara R. Mailapalli. 2016. "Nanofertilisers, Nanopesticides, Nanosensors of Pest and Nanotoxicity in Agriculture." In *Sustainable Agriculture Reviews. Sustainable Agriculture Reviews*, edited by E. Lichtfouse, vol 19, 307–30. Cham: Springer.

Dutta, N., Mukhopadhyay, A., Dasgupta, A. K., & Chakrabarti, K. 2014. "Improved Production of Reducing Sugars from Rice Husk and Rice Straw Using Bacterial Cellulase and Xylanase Activated with Hydroxyapatite Nanoparticles." *Bioresource Technology* 153: 269–277.

Etefagh, Raihane, E. Azhir, and Nasser Shahtahmasebi. 2013. "Synthesis of CuO Nanoparticles and Fabrication of Nanostructural Layer Biosensors for Detecting Aspergillus Niger Fungi." *Scientia Iranica* 20 (3): 1055–58.

Fraceto, Leonardo F., Renato Grillo, Gerson A. De Medeiros, and Viviana Scognamiglio. 2016. "Nanotechnology in Agriculture : Which Innovation Potential Does It Have ?" *Frontiers in Environmental Science* 4 (March): 1–5.

Feng, Y., V. D. Kreslavski, A. N. Shmarev, A. A. Ivanov, S. K. Zharmukhamedov, A. Kosobryukhov, M. Yu, S. I. Allakhverdiev, and S. Shabala. 2022. "Effects of Iron Oxide Nanoparticles (Fe_3O_4) on Growth, Photosynthesis, Antioxidant Activity and Distribution of Mineral Elements in Wheat (Triticum aestivum) Plants." *Plants* 11 (14): 1894.

Fu, Peter P., Qingsu Xia, Huey-Min Hwang, Paresh C. Ray, and Hongtao Yu. 2014. "Mechanisms of Nanotoxicity: Generation of Reactive Oxygen Species." *Journal of Food and Drug Analysis* 22 (1): 64–75.

Gohari, Gholamreza, Asghar Mohammadi, Ali Akbari, Sima Panahirad, Mohammad Reza Dadpour, Vasileios Fotopoulos, and Seisuke Kimura. 2020. "Titanium Dioxide Nanoparticles (TiO_2 NPs) Promote Growth and Ameliorate Salinity Stress Effects on Essential Oil Profile and Biochemical Attributes of Dracocephalum Moldavica." *Scientific Reports* 10 (1): 912.

Guerra, Fernanda, Mohamed Attia, Daniel Whitehead, and Frank Alexis. 2018. "Nanotechnology for Environmental Remediation: Materials and Applications." *Molecules* 23 (7): 1760.

He, L., Liu, Y., Mustapha, A., & Lin, M. 2011. "Antifungal Activity of Zinc Oxide Nanoparticles against Botrytis cinerea and Penicillium expansum." *Microbiological Research* 166 (3): 207–215.

Hong, Jie, Lina Wang, Youping Sun, Lijuan Zhao, Genhua Niu, Wenjuan Tan, Cyren M. Rico, Jose R. Peralta-Videa, and Jorge L. Gardea-Torresdey. 2016. "Foliar Applied Nanoscale and Microscale CeO_2 and CuO Alter Cucumber (Cucumis Sativus) Fruit Quality." *Science of the Total Environment* 563–564 (September): 904–11.

Hossain, Kaizar, Syed Zaghum Abbas, Akil Ahmad, Mohd Rafatullah, Norli Ismail, Gaurav Pant, and Maruthi Avasn. 2020. "Nanotechnology: A Boost for the Urgently Needed Second Green Revolution in Indian Agriculture." In *Nanobiotechnology in Agriculture*, edited by K. Hakeem and T. Pirzadah, 15–33 Cham: Springer.

Hotessa Halake, Niguse, and Jara Muda Haro. 2022. "Role of Nanobiotechnology towards Agri-Food System." edited by B. Kumar. *Journal of Nanotechnology* 2022 (May): 1–7.

Hu, Changwei, M. Li, Yibin Cui, Dingsheng Li, J. Chen, and L.Y. Yang. 2010. "Toxicological Effects of TiO_2 and ZnO Nanoparticles in Soil on Earthworm Eisenia Fetida." *Soil Biology and Biochemistry* 42 (4): 586–91.

Iqbal, Muhammad, Shahid Umar, and Mahmooduzzafar. 2019. "Nano-Fertilization to Enhance Nutrient Use Efficiency and Productivity of Crop Plants." In *Nanomaterials and Plant Potential*, edited by A. Husen and M. Iqbal, 473–505. Cham: Springer International Publishing.

Janmohammadi, M., Amanzadeh, T., Sabaghnia, N., &Dashti, S. 2016. "Impact of Foliar Application of Nano Micronutrient Fertilizers and Titanium Dioxide Nanoparticles on the Growth and Yield Components of Barley under Supplemental Irrigation." *Acta AgriculturaeSlovenica* 107 (2): 265–276.

Karami Mehrian, Saeed, and Renata De Lima. 2016. "Nanoparticles Cyto and Genotoxicity in Plants: Mechanisms and Abnormalities." *Environmental Nanotechnology, Monitoring & Management* 6 (December): 184–93.

Khodashenas, Bahareh, and Hamid Reza Ghorbani. 2016. "Optimisation of Nitrate Reductase Enzyme Activity to Synthesise Silver Nanoparticles." *IET Nanobiotechnology* 10 (3): 158–61.

Kumar, Amit, Kiran Gupta, Sonal Dixit, K. Mishra, and Sudhakar Srivastava. 2019. "A Review on Positive and Negative Impacts of Nanotechnology in Agriculture." *International Journal of Environmental Science and Technology* 16 (4): 2175–84.

Kumaraswamy, R. V., V. Saharan, S. Kumari, R. Chandra Choudhary, A. Pal, S. S. Sharma, S. Rakshit, R. Raliya, and P. Biswas. 2021. "Chitosan-silicon Nanofertilizer to Enhance Plant Growth and Yield in Maize (Zea mays L.)." *Plant Physiology and Biochemistry* 159: 53–66.

Li, Hongchao, Chao Shan, Yanyang Zhang, Jianguo Cai, Weiming Zhang, and Bingcai Pan. 2016. "Arsenate Adsorption by Hydrous Ferric Oxide Nanoparticles Embedded in Cross-Linked Anion Exchanger: Effect of the Host Pore Structure." *ACS Applied Materials & Interfaces* 8 (5): 3012–20.

Lin, Daohui, and Baoshan Xing. 2007. "Phytotoxicity of Nanoparticles: Inhibition of Seed Germination and Root Growth." *Environmental Pollution* 150 (2): 243–50.

Min, J.-S., K.-S. Kim, S.-W. Kim, J.-H. Jung, K. Lamsal, S.-B. Kim, M.-Y. Jung, and Y.-S. Lee. 2009. "Effects of Colloidal Silver Nanoparticles on Sclerotium-Forming Phytopathogenic Fungi." *The Plant Pathology Journal* 25 (4): 376–380.

Mondal, K. K., and Mani, C. 2012. "Investigation of the Antibacterial Properties of Nanocopper Against Xanthomonas axonopodispv. punicae, the Incitant of Pomegranate Bacterial Blight." *Annals of Microbiology* 62 (2): 889–893.

Mukhopadhyay, Siddhartha S. 2014. "Nanotechnology in Agriculture: Prospects and Constraints." *Nanotechnology, Science and Applications* 7: 63–71.

Neme, Kumera, Ayman Nafady, Siraj Uddin, and Yetenayet B. Tola. 2021. "Application of Nanotechnology in Agriculture, Postharvest Loss Reduction and Food Processing: Food Security Implication and Challenges." *Heliyon* 7 (12): e08539.

Paret, M. L., G. E. Vallad, D. R. Averett, J. B. Jones, and S. M. Olson. 2013. "Photocatalysis: Effect of Light-Activated Nanoscale Formulations of TiO_2 on Xanthomonas perforans and Control of Bacterial Spot of Tomato." *Phytopathology* 103 (3): 228–236.

Prasad, Ram, Atanu Bhattacharyya, and Quang D. Nguyen. 2017. "Nanotechnology in Sustainable Agriculture: Recent Developments, Challenges, and Perspectives." *Frontiers in Microbiology* 8 (June). doi.10.3389/fmicb.2017.01014.

Rao, Pasupuleti Visweswar. 2019. "Nanobiotechnology Advances in Medicine, Agriculture and Other Important Areas: Applications and Future Perspectives." *Biotechnology* 18 (2): 89–97.

Roh, Ji-Yeon, Young-Kwon Park, Kwangsik Park, and Jinhee Choi. 2010. "Ecotoxicological Investigation of CeO_2 and TiO_2 Nanoparticles on the Soil Nematode Caenorhabditis Elegans Using Gene Expression, Growth, Fertility, and Survival as Endpoints." *Environmental Toxicology and Pharmacology* 29 (2): 167–72.

Sekhon, Bhupinder S. 2014. "Nanotechnology in Agri-Food Production: An Overview." *Nanotechnology, Science and Applications*, 7 (2): 31–53.

Shaimaa, H. Abd Elrahman, and M.A.M. Mostafa. 2015. "Applications of Nanotechnology in: An Overview." *Egyptian Journal of Soil Science* 55 (2): 197–214.

Sim, Daniel Hong Heng, Ivy Ai Wei. Tan, Leonard Lik Pueh. Lim, and Bassim H. Hameed. 2021. "Encapsulated Biochar-Based Sustained Release Fertilizer for Precision Agriculture: A Review." *Journal of Cleaner Production* 303 (June): 127018.

Subramanian, Kizhaeral S., Angamuthu Manikandan, Muthiah Thirunavukkarasu, and Christopher Sharmila Rahale. 2015. "Nano-Fertilizers for Balanced Crop Nutrition." In *Nanotechnologies in Food and Agriculture*, edited by M. Rai, C. Ribeiro, L. Mattoso, N. Duran, 69–80. Cham: Springer International Publishing.

Tarafder, Chaitaly, Mahbuba Daizy, Md. Morshed Alam, Md. Ripon Ali, Md. Jahidul Islam, Rakibul Islam, Md. Sohel Ahommed, Mohamed Aly Saad Aly, and Md. Zaved Hossain Khan. 2020. "Formulation of a Hybrid Nanofertilizer for Slow and Sustainable Release of Micronutrients." *ACS Omega* 5 (37): 23960–66.

Vijayakumar, M. D., G. J. Surendhar, L. Natrayan, Pravin P. Patil, P. M. Bupathi Ram, and Prabhu Paramasivam. 2022. "Evolution and Recent Scenario of Nanotechnology in Agriculture and Food Industries." edited by R. Lakshmipathy. *Journal of Nanomaterials* 2022 (July): 1–17.

Wang, Lu, Yanzhi Liu, Jinliang Liu, Yanhua Zhang, Xianghui Zhang, and Hongyu Pan. 2016. "The Sclerotinia Sclerotiorum FoxE2 Gene Is Required for Apothecial Development." *Phytopathology* 106 (5): 484–90.

Wang, Xiaoping, Xiyu Yang, Siyu Chen, Qianqian Li, Wei Wang, Chunjiang Hou, Xiao Gao, Li Wang, and Shucai Wang. 2015. "Zinc Oxide Nanoparticles Affect Biomass Accumulation and Photosynthesis in Arabidopsis." *Frontiers in Plant Science* 6: 1243.

War, John Mohd, Mohammad Afaan Fazili, Waseem Mushtaq, Abdul Hamid Wani, and Mohd Yaqub Bhat. 2020. "Role of Nanotechnology in Crop Improvement." In *Nanobiotechnology in Agriculture*, edited by K. Hakeem and T. Pirzadah, 63–97. Cham: Springer, 63–97.

Win, Theint Theint, Sikandar Khan, and PengCheng Fu. 2020. "Fungus- (Alternaria Sp.) Mediated Silver Nanoparticles Synthesis, Characterization, and Screening of Antifungal Activity against Some Phytopathogens." edited by Marco Rossi. *Journal of Nanotechnology* 2020 (November): 1–9.

3 Role of Nanotechnology in Preventing Postharvest Losses of Agriproducts

Tobiloba Christiana Elebiyo
Landmark University

Rotdelmwa Maimako Asalaye
SDG 03 Group – Good Health & Well-being

Damilare Rotimi, Oludipe Olorunleke, Oluwayemi Elizebeth, and Mathias Maduakolam Aniobi
Landmark University

Oluwatosin Adekunle Adeluwoye
Lead City University

Oluyomi Stephen Adeyemi
Landmark University

CONTENTS

3.1 INTRODUCTION

Agriculture has always been the steadiest and most important sector due to the fact that it is involved in raw material provision and production for the food and feed manufacturing industries for a better life globally. The agricultural field has been

DOI: 10.1201/9781003397861-3

facing a wide range of challenges, such as soil contamination with several detrimental environmental pollutants such as pesticides and fertilizers, unpredictable climatic changes, inefficient use of agrochemicals, and the unavailability of water, resulting in global food insecurity (Mittal et al. 2020). About 815 million individuals are currently malnourished due to incidences of food inaccessibility, unaffordability, and underutilization. Based on the projected surge in the global population to about 9 billion people by 2050, the number of malnourished individuals is expected to rise to 2 billion by 2050 if food production is not increased by 50%, particularly in developing countries (Verma et al. 2022).

Unintended reductions in the quantity and quality of food produced for human consumption at varying stages along the agricultural supply chain are known as postharvest losses (Bendinelli et al. 2020; Sawicka 2019). According to Sawicka (2019), annual food losses total over 1.3 billion tonnes, with root crops and fruit and vegetables accounting for 50% of those losses, followed by seafood (35%), grains (30%), and nuts, meats, and dairy products (20%) (see Figure 3.1). Poor storage facilities, inappropriate handling, ineffective processing methods, and microbial attacks are the primary causes of postharvest losses (Kumar and Kalita 2017).

Recent studies have shown the potential of nanotechnology to advance the agricultural industry by improving the efficiency of agricultural inputs, therefore bringing sustainable solutions to the issues of postharvest loss and global food insecurity (Usman et al. 2020). Nanotechnology is a multidisciplinary science that deals with the synthesis, modification, characterization, and generation of nanomaterials in the size range of one to a hundred nanometers (Elebiyo et al. 2022). It is an evolving and rapidly growing area with novel and exclusive food research and agricultural applications that have indicated a promising potential to maintain and promote sustainable

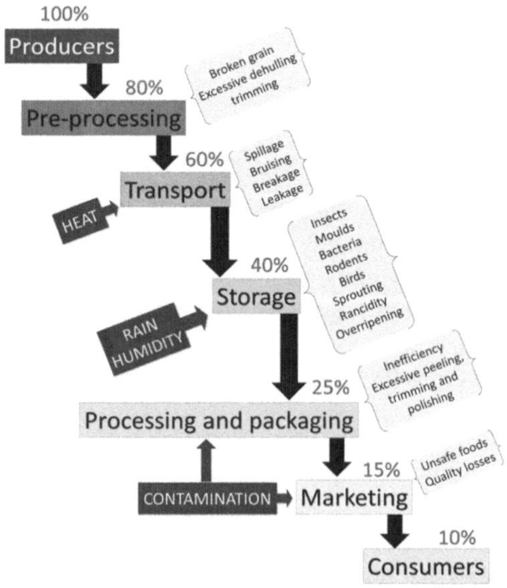

FIGURE 3.1 Postharvest losses at different points of the agricultural supply chain.

agriculture, which is vital to achieving Goal 2 of the sustainable development goals, which is aimed at attaining a hunger free world by 2030 (Usman et al. 2020).

Nanotechnology offers efficient techniques for regulating pathogenic attacks, which decrease food deterioration and promote food safety through the invention of improved sensors to monitor chemical, physical, or biological aspects of food throughout processing, packing, transport, and storage (Kah et al. 2018). Additionally, nanotechnology has also been applied in the fabrication of devices that enhance the soil surface for adequate water distribution and resource recovery. In order to increase crop productivity and prevent insect infestation, nanomaterials are currently being used for the targeted delivery of agrochemicals (Rodrigues et al. 2017). The nanocarrier, nanoemulsion and nanoencapsulation technologies are now employed to maintain the quality and quantity of agri-products as they transit down the food supply chain (Neme et al. 2021).

3.2 ROLE OF NANOTECHNOLOGY IN SUSTAINED FOOD QUALITY AND SAFETY

Sustaining global reductions in postharvest losses, particularly for perishable foods such as vegetables, fruits, seafood, and meat, has become a primary focus for both public and corporate institutions (Pushparaj et al. 2022). Seeing as traditional approaches to reducing postharvest loss and addressing food safety concerns, such as complex storage structures and analytical laboratory testing, are insufficient, novel food coatings and fast analytical techniques with low detection of pathogens, chemicals and temperature change have been developed (Fadiji et al. 2022).

Coating has been used for a long time in the food industry to improve the organoleptic and sensory qualities of food, reduce the risk of mechanical injury, prevent food quality loss due to dehydration, prevent ripening, and chemical contamination, and extend the shelf life of food by preventing the growth of spoilage microbes (Yadav et al. 2022). The antimicrobial properties of naturally occurring substances such as chitosan and carnauba wax have been used as food coats to extend the shelf life and improve the organoleptic properties of perishable foods such as apples, cabbage, cucumber, strawberries, and lettuce (Duan et al. 2019; Devi et al. 2022). Plant extracts and essential oils have also been explored as edible coatings to fortify and maintain the nutritional value of food (Mohd Israfi et al. 2022).

Nanoencapsulation technologies have been used to incorporate bioactive compounds into edible food coatings to prolong shelf life and improve nutritional food quality, which significantly reduces postharvest losses (Zhang et al. 2020; Sampathkumar, Tan, and Loo 2020). In a study conducted by Liu et al. (2021), coating with chitosan films encapsulating litchi peel extract and titanium dioxide nanoparticles (TiO_2) inhibited significant browning and weight loss in water-colored apples after 180 days of storage. Nanoemulsions containing carnauba wax and ginger essential oil were found to improve papaya quality over a 10-day period by delaying ripening, color changes, and natural disease severity (Miranda et al. 2022). It has been reported that coating pork with chitosan-gelatin nanoencapsulated with tarragon essential oil-loaded nanoparticles extends the meat's shelf life by 12 days by preventing lipid oxidation and microbial activity (Zhang et al. 2020). Nanoemulsions with gelatin-chitosan, cinnamon essential oil, and rosemary extract have also been used to extend the shelf life of chicken patties for days (Qiu et al. 2022). Similarly, nanoliposomes of Xanthan-gum and essential oil from *Litsea cubeba* were reported to deter oxidation and microbial contamination in salmon (Cui et al. 2022). Essential oil nanoemulsions of *Trachyspermum ammi*, *Carum copticum*, and *Ferulago angulata* could increase the shelf life of turkey fillets, lamb loins and rainbow trout fillets, respectively (Fallah et al. 2022; Shokri et al. 2020).

3.3 APPLICATIONS OF NANOTECHNOLOGY IN DETECTION OF FOOD CONTAMINANTS

Nanotechnology has been incorporated into the design of transducers that aid in the real-time monitoring of the physical, chemical, and biological properties or quality of food products, especially during transport and storage (Shawon, Hoque, and Chowdhury 2020). By detecting changes in these parameters such as humidity, temperature, pH, alterations in chemical concentrations, and microbial growth, nanosensors can allow for timely intervention to prevent spoilage and reduce postharvest loss (Neme et al. 2021). Nanosensors are currently being used to detect the presence of microorganisms such as *E. coli* 0157:H7, *Salmonella* spp. and *Listeria monocytogenes* in food products (Shawon, Hoque, and Chowdhury 2020). Figure 3.2 shows the applications of nanosensors in prevention of postharvest loss.

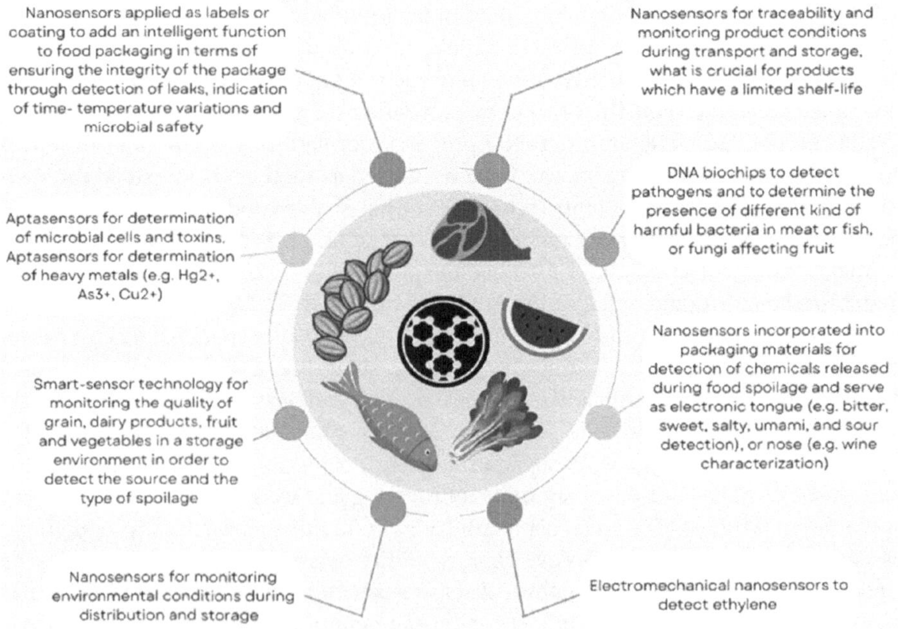

FIGURE 3.2 Application of nanosensors in prevention of postharvest loss.

A major accomplishment in the attempt to reduce postharvest losses is the possibility of integrating nanosensors into food packages, which makes the role of nanotechnology indispensable in the agri-food industry. Active NPs with antioxidant, antimicrobial, and/or preservative capabilities are capable of exerting or releasing their active qualities when incorporated into food packages. According to Motelica et al. (2020), to improve the quality and shelf life of stored food, active nanocomposites are made by combining active substances like antimicrobials and preservatives with standard nondegradable or biodegradable packaging polymers. Metals [Gold (Au), Copper (Cu) and Silver (Ag)] and metal oxides [Zinc oxide (ZnO), Silicon oxide (SiO_2) and Titanium dioxide (TiO_2)] are the most often employed NPs in active nanopackaging (Anvar, Ahari, and Ataee 2021). These active nanopackages also help to control and absorb moisture, oxygen, water vapor, carbon dioxide, and ethylene, and function as thermal insulators in addition to boosting microbiological safety (Drago et al. 2020).

The most frequent NPs used in food packaging are nanosilver due to its great stability and severe toxicity to a broad spectrum of pathogens. Although the exact mechanism by which AgNPs exert their antimicrobial effects is still unknown, many investigations have pointed to the following as likely candidates. Ag^+ release from silver NPs causes the membrane to become more permeable in the cytoplasm. As free silver ions are taken up by bacterial cells, they inhibit respiratory enzymes and produce reactive oxygen species (ROS), but they also halt the production of adenosine triphosphate (ATP). DNA damage and bacterial mortality may result from ROS production (Yin et al. 2020). Antimicrobial AgNPs could be used to extend the life of a variety of food goods. According to research by Mahdi et al. (2017), silver nanoparticle (NP)-coated trays may increase the shelf life of minced beef from 3 to 7 days when stored in the refrigerator. Stable silver NPs generated may be transported by several kinds of cellulose. On chicken exudates, hybrid materials made by combining UV light and heat were efficient against harmful microorganisms (Fernández et al. 2009).

Antimicrobial activity was shown by their films against *Staphylococcus aureus* and *Escherichia coli* (Becaro et al. 2016). Freshly cut carrots wrapped in low-density polyethene (LDPE) with low amounts of AgNPs showed antibacterial effectiveness against mesophilic aerobic and coliform bacteria, while the level of ascorbic acids was also preserved (Becaro et al. 2016). LDPE/AgNP films were shown to be more effective against fungus and gram-negative bacteria (Brito et al. 2020). Furthermore, carrageenan/AgNPs/Laponite nanocomposites were shown to have antibacterial activities against *E. coli* and *S. aureus* (Vishnuvarthanan and Rajeswari 2019). Mathew, Mathew, and Radhakrishnan (2019) found that films made of polyvinyl alcohol and AgNPs were more effective against *S. typhimurium*. This distinction is due to the fact that the cellular structure of gram-negative bacteria is different from that of gram-positive bacteria. In fact, the cell walls of gram-positive bacteria contain layers of peptidoglycan, which makes it difficult for silver NPs to penetrate (Shankar and Rhim 2017).

Copper nanoparticles (NPs) have shown promise as a safe and effective antibacterial agent with potential use in both medical and food applications. Although copper has poor biocidal capability, copper NPs are chosen over silver NPs in active nanopackaging due to their cheaper price, greater compatibility with polymers, and greater variety of physical properties. Copper has been shown to have an effective antibacterial effect against several bacteria and other pathogens. The effects of CuNPs on *E. coli* have been studied using three distinct techniques, including the minimum inhibitory concentration (MIC), the disc diffusion test, and the minimum bactericidal concentration (MBC) developed by Ruparelia et al. (2008).

TiO_2 is safe for consumption and has been authorized for use in pharmaceuticals, cosmetics, and food contact products by the FDA in the United States. The self-disinfecting characteristic of TiO_2 is now drawing a lot of attention as a means of achieving hygienic design goals on packaging surfaces and in food processing. TiO_2 has been shown to inhibit several different types of bacteria and fungi, including *Escherichia coli*, *Diaporthe actinidiae*, *Pseudomonas aeruginosa*, *Listeria monocytogenes*, *Staphylococcus aureus*, *Vibrio parahaemolyticus*, *Salmonella choleraesuis*, and *Penicillium expansum*.

The Food and Drug Administration (FDA) includes zinc oxide on its list of GRAS (generally recognized as safe) ingredients (Espitia et al. 2012). An active food packaging made of gallic-acid films containing chitosan and zinc oxide NPs has been suggested recently for use with foods including black fruits and vegetables (Yadav, Mehrotra, and Dutta 2021). Kalia et al. (2021) created a nanocomposite film out of chitosan and nettle leaf extract-synthesized CuO and ZnO NPs to increase the storage life of guava fruits. CuONPs > ZnONPs > nettle extract was shown to have the highest antioxidant and antibacterial properties among biologically produced NPs (Kalia et al. 2021). White brined cheese is a perfect medium for the growth of foodborne pathogens (Al-Nabulsi et al. 2020). It has also been observed that *C. jejuni* in raw chicken flesh may be controlled by using functionalized absorbent pads containing ZnONPs, which reduce the *C. jejuni* count to an undetectable level due to the immobilization of the bacteria (Anvar, Ahari, and Ataee 2021).

3.4 ROLE OF NANOTECHNOLOGY IN THE DETECTION CHANGES IN STORAGE CONDITIONS

Temperature and humidity levels in storage are crucial for maintaining food quality and shelf life. According to Kumar, Guleria, and Mehta (2017), the use of temperature-sensitive nanosensors embedded in food packaging was effective in preventing food spoilage by monitoring the temperature of perishable products such as meat and dairy during storage and transportation. With the aid of nanotechnology, time-temperature indicators (TTIs) have been designed to help track and translate temperature-related information during storage and transportation. This is especially crucial when nonideal storage conditions are used to store food.

Humidity indicators have also been designed to help monitor and maintain the storage conditions of agricultural products. Zhang et al. (2013) devised an iridescent humidity indicator to address this problem; he did this by casting nanocrystalline cellulose film into a thick iridescent film and tuning the film's color to cause it to interact with an electromagnetic field (HI). The dry film has a blue-green hue, but when exposed to high humidity; it changes to a reddish orange. According to Zhang et al. (2013), the resilience of photonic crystals to photobleaching and deterioration makes them an ideal encoding ingredient. A photonic crystal's reflection wavelength may be altered by binding a target analyte, which causes a change in the crystal's structural period.

Quantification and identification of microorganisms have been made easy by creating different gas sensors based on their gas emissions. As a result of their excellent sensitivity and stability, metal oxide gas nanosensors have become one of the most widely used kinds of sensors (Wang et al. 2010). Currently, nanosensors based on conducting polymers are being employed for microbial gas emission quantification and/or identification (Riu, Maroto, and Rius 2006). Insulating polymer matrices surround conducting particles in these compounds. Microorganism gas will cause a change in resistance, which is how the sensors will react (Jiang et al. 2018). To identify microbes, one common method uses gold nanoparticles that have been modified to include enzymes (Shedbalkar et al. 2014). Nanofibrils of perylene-based fluorophores can detect gaseous amines, which are indicative of fish and meat deterioration. The nanocomposites of zinc oxide and titanium dioxide are useful for detecting VOCs (Shedbalkar et al. 2014). Nanocrystalline SnO_2 as a photosensitizer is another O_2 sensor (Mills and Hazafy 2009). When exposed to even a trace quantity of oxygen, the color of one of these detectors begins to shift subtly.

Jung, Puligundla, and Ko (2012) developed a chitosan-based CO_2 detector to detect deterioration in the freshness or quality of packaged meals over time. The indicator signal was amplified by adding 2-amino-2-methyl-1-propanol (AMP) to the chitosan solution. Easy sachet packaging allows for the indicator solutions of chitosan and AMP to be included in gaseous CO_2-permeable food products. A detailed analysis of the sensor material's sensitivity to CO_2 and cross-sensitivity to O_2 allowed for its optimization for use in food packaging. Optical CO_2 gas sensors, in particular dry optical sensors with pH-sensitive dye indicators, were shown to be promising by Puligundla, Jung, and Ko (2012) for use in food packaging.

Food safety testing is an ongoing area of study because of the prevalence of foodborne diseases caused by viruses, fungus, bacteria, and toxins. About 48 million instances of foodborne disease are recorded annually in the United States, as reported by the CDC in 2011. About half a million people might benefit from a 1% drop in foodborne illness rates (Scallan et al. 2011). There is still a public health concern over food poisoning caused by bacteria, fungus, or viruses. Quick, accurate, and cheap detection methods are urgently needed (Duncan 2011). Immunological tests based on the selective interaction between antigen and antibodies are the mainstay of modern detection techniques. Similar methods are used by microbial sensors that investigate nanoparticles, which have their own set of advantages thanks to their size, shape, and electrical and optical properties.

Rather than relying on traditional microbiological analytical techniques, disease detection efforts are increasingly turning to nanomaterials themselves to make use of their unique electrical, magnetic, luminescent, and catalytic capabilities (Liang et al. 2014). To aid in this fight, more rapid, sensitive, and cost-effective diagnostic tests are being created. Nanotechnologists have focused their efforts on creating systems based on nanotechnology that are not only sensitive and fast but also cheap, durable, and repeatable. Recently, Ayala-Zavala et al. (2014) published a review of nanosensors for measuring microbial proliferation.

Sensor arrays make up the "Electronic Tongue" technology that reports the food's freshness and temperature. The gadget is an array of nanosensors that are particularly sensitive to gases generated by rotting microbes, resulting in a color shift that shows whether the food has degraded. These nanosensors, acting as an "electronic nose" or "electronic tongue," might be integrated directly into the packing material and used to detect volatile organic compounds (VOCs) produced by microorganisms during food spoilage. In electronic nose nanosystems, conducting polymers have been employed as detectors. Conducting organic polymer sensors alters their resistance as they absorb gas, allowing for detection and output (Bai and Shi 2007). The Kraft Foods Company has created a little electrical gadget shaped like a tongue for use in food packaging. With an array of nanosensors sensitive to gases emitted by decaying microbes, this innovative gadget may change color to signal if the food has degraded (Alam et al. 2021).

3.5 SAFETY ASSESSMENT/TOXICITIES OF NANOMATERIALS APPLICATION IN THE FOOD INDUSTRY

Over the years, nanotechnology's application has broadened across various fields, including agriculture, as it is employed in crop production, food processing, nutrition and product manufacturing (Biswas et al. 2022). The application of nanotechnology in the food system has given rise to a great diversity of products with improved stability, texture, functionality, shelf life, sensory attributes, and taste. The recent attention given to the application of nanotechnology in the food sector and the fast-growing pace and acceptability of this technology pose a question of its safety, especially in food consumables. By applying nanosensors to detect adulteration or infections in food during manufacture, processing, storage, packing, and transport, nanoscience is employed in the food industry to improve food security. However, using modified NPs in food products undoubtedly increases human oral nanoparticle exposure (Biswas et al. 2022). According to certain research, ingesting NPs has been associated with many biological effects, including oxidative stress response stimulation, DNA damage, and protein denaturation (Hong et al. 2017). Furthermore, numerous studies have connected oral exposure to nanomaterials, especially solid NPs, to damage to secondary organs and the gastrointestinal tract (Biswas et al. 2022).

Certain factors influence the risk associated with a nanomaterial, including its chemical nature, chemical and physical properties, interaction with tissues, and level of exposure. However, despite the perceived risk of nanoscience application in the food sector, only a few investigations have been conducted on its health impact (Duncan 2011). Certain factors must be considered in assessing the likely safety of a nanomaterial in the food sector. First, a material's physicochemical properties make it an NP. These nanoscale characteristics confound a nanomaterial and its distinct properties. Once a nanoparticle thus loses its nanoeffects either via disintegration or deterioration, its behavior is anticipated to be similar to that of its parent material, if such exists. Therefore, toxicity issues with ingestion of nanomaterials are primarily focused on those materials that might escape the gastrointestinal tract (GIT), thus letting NPs translocate to tissues and organs or producing local harmful effects in the GIT concentration (Biswas et al. 2022).

The World Health Organization (WHO), the European Union (EU), and the United Nations Food and Agriculture Organization (FAO) have emphasized the need to ensure NPs food safety even though there is still very little data on the carcinogenicity, teratogenicity, or genotoxicity of NPs in food (Biswas et al. 2022). The Scientific Committee Guidance on Nano-RA (SC Guidance on Nano-RA) has investigated key scientific investigations that provide perceptions into the hazard characterization, exposure evaluation, and physicochemical properties of nanomaterials and their applicable regions. However, several doubts regarding nano-engineered materials, NP parts, and materials with nanoscale properties could endanger consumers (Biswas et al. 2022).

Although NPs are reported to theoretically migrate to food products from food packages, migration tests, and risk evaluations are unavailable. The migration into the food matrix can be considered a process of mass transfer wherein elements of lesser molecular mass are stuck in packaging and then released into the food item (Souza and Fernando 2016). Studies that examine the migration probability of any undesired or dangerous compounds moving into foods are one of the most important stages in developing novel food packaging materials. The study by Souza and Fernando (2016) examined challenges that may arise during NP dispersion from polymer-based food packages to consumers. The study observed a lack of good analytical techniques from previous studies for identifying NPs with a smaller diameter and lower content and monitoring their possible movement to foods from packaging materials. Therefore, they highly advised that investigations that found no evidence of NP migration provide data on their assessment method detection limit, which consists of particle size or mass and number concentration.

Bronchoalveolar and hepatocellular carcinoma were both observed from the leaching of metallic NPs from food packaging into the food. The migration of these NPs into food was attributed to the physicochemical properties of the food, such as its pH, temperature, and solubility (Britto et al. 2023). In contrast, cationic NPs were also reported to pass through the cell by making the cell wall permeable, making the cell unstable, and causing disruption. van der Meulen et al. (2014) recommended a case-by-case approach to accepting and approving food-engineered NPs and not a generalized classification as safe or unsafe. As such, a thorough safety assessment of each nano-engineered food to determine its safety and approval individually is necessary.

There is advancement in the study of food nanotechnology and its societal concern regarding its safety for use and human consumption. Before nanofoods are introduced commercially, there is a need to study their possible toxicity to humans thoroughly. There are still no guidelines for the safety assessment of nanofoods. Also, the chemical synthesis of NPs by several chemical techniques affects the environment negatively through the production of pollutants. Thus, there is a dire need to account for regulatory regimens, biosafety, and public concerns during the manufacture, packaging, and consumption of nanofoods (Biswas et al. 2022). Due to these factors, the public's acceptance of food products that contain or use NPs will primarily depend on their confidence in industries and the government's ability to safeguard them against unknown risks (Duncan 2011).

Exposure to AgNPs has been indicated as a possible threat to male reproductive health, which indicates toxicity (Braydich-Stolle et al. 2005). Several studies have shown that AgNPs are cytotoxic to various mammalian cell types in vitro. However, the results of in vivo research into the long-term effects of oral exposure to AgNP are unclear. For instance, even though rats fed a steady diet of NPs contained AgNPs in almost every organ, few harmful consequences were seen except at the highest administered concentrations (Ferdous and Nemmar 2020). Another study reported the accumulation of AgNPs in the liver of weaning pigs orally administered with NPs, although no acute toxic effect was observed. In another study, liver lymphocyte inflammation and infiltration were noted in mice fed nano- and micro-sized AgNPs. The NP was observed to show a more adverse effect when compared to the microparticles of silver (Cha et al. 2008).

3.6 CONCLUSION AND FUTURE PROSPECTIVE

Although there may be advantages, the future of "nanofoods" is still up in the air. Nanotechnology is no exception to the general rule that people are far less willing to receive novel technologies in their food than other products. Although the public currently views generic applications of nanoscience in an unbiased or even positive light, some studies indicate that people are still skeptical about the use of nanotechnology in foods. Consumers are worried about ethical or moral issues such as the genuineness of the food supply, the equitability of socioeconomic distribution, privacy, engineering living matter, improving human capabilities, labeling, and the right to choose, in addition to the health and environmental effects of consuming nanofoods. Due to these factors, the public's acceptance of food products that contain or use NPs will primarily depend on their confidence in businesses and the government's ability to safeguard them against unidentified risks. How regulatory organizations handle this cutting-edge technology will also affect the future of nanofoods. The great potential benefits of nanotechnology must be evaluated against the hazards associated with improper usage and handling of nanomaterials, many of which are still being assessed.

REFERENCES

Al-Nabulsi, Anas, Tareq Osaili, Amani Sawalha, Amin N Olaimat, Borhan Aldeen Albiss, Ghadeer Mehyar, Mutamed Ayyash, and Richard Holley. 2020. "Antimicrobial activity of chitosan coating containing ZnO nanoparticles against *E. coli* O157: H7 on the surface of white brined cheese." *International Journal of Food Microbiology* 334:108838.

Alam, Arif U., Pranali Rathi, Heba Beshai, Gursimran K. Sarabha, and M. Jamal Deen. 2021. "Fruit quality monitoring with smart packaging." *Sensors* 21 (4):1509.

Anvar, Amir Ali, Hamed Ahari, and Maryam Ataee. 2021. "Antimicrobial properties of food nanopackaging: A new focus on foodborne pathogens." *Frontiers in Microbiology* 12.

Ayala-Zavala, Jesús Fernando, Gustavo Adolfo González-Aguilar, María Roberta Ansorena, Emilio Alvarez-Párrilla, and Laura de la Rosa. 2014. "Nanotechnology tools to achieve food safety." *Sensors* 341–353.

Bai, Hua, and Gaoquan Shi. 2007. "Gas sensors based on conducting polymers." *Sensors* 7 (3):267–307.

Becaro, Aline A., Fernanda C. Puti, Alan R. Panosso, Juliana C. Gern, Humberto M. Brandão, Daniel S. Correa, and Marcos D. Ferreira. 2016. "Postharvest quality of fresh-cut carrots packaged in plastic films containing silver nanoparticles." *Food and Bioprocess Technology* 9 (4):637–649.

Bendinelli, William Eduardo, Connie Tenin Su, Thiago Guilherme Péra, and José Vicente Caixeta Filho. 2020. "What are the main factors that determine post-harvest losses of grains?" *Sustainable Production and Consumption* 21:228–238.

Biswas, Rahul, Mahabub Alam, Animesh Sarkar, Md Ismail Haque, Md Moinul. Hasan, and Mominul Hoque. 2022. "Application of nanotechnology in food: Processing, preservation, packaging and safety assessment." *Heliyon* 8 (11):e11795.

Braydich-Stolle, Laura, Saber Hussain, John J. Schlager, and Marie-Claude Hofmann. 2005. "In vitro cytotoxicity of nanoparticles in mammalian germline stem cells." *Toxicological Sciences* 88 (2):412–9.

Brito, Sabrina da Costa, Joana D. Bresolin, Kátia Sivieri, and Marcos D. Ferreira. 2020. "Low-density polyethylene films incorporated with silver nanoparticles to promote antimicrobial efficiency in food packaging." *Food Science and Technology International* 26 (4):353–366.

Britto, Claudia Girthie John, Amal Dhivahar Sahaya Antony John, Archanna Jayakumar, Pavidharshini Selvasekaran, Ramalingam Chidambaram, and Haotian Zheng. 2023. "Nanotechnology applications for food safety: Benefits and risks." In *Nanotechnology Applications for Food Safety and Quality Monitoring*, edited by Arun Sharma, P.S. Vijayakumar, Er. Pramod K. Prabhakar, Ritesh Kumar, 3–30. Amsterdam: Elsevier.

Cha, Kyungeun, Hye-Won Hong, Yeon-Gil. Choi, Min Joo Lee, Jong Hoon Park, Hee-Kwon Chae, Gyuha Ryu, and Heejoon Myung. 2008. "Comparison of acute responses of mice livers to short-term exposure to nano-sized or micro-sized silver particles." *Biotechnology Letters* 30 (11):1893–9.

Cui, Haiying, Mei Yang, Ce Shi, Changzhu Li, and Lin Lin. 2022. "Application of xanthan-gum-based edible coating incorporated with litsea cubeba essential oil nanoliposomes in salmon preservation." *Foods* 11 (11):1535.

Devi, L Susmita, Sweety Kalita, Avik Mukherjee, and Santosh Kumar. 2022. "Carnauba wax-based composite films and coatings: Recent advancement in prolonging postharvest shelf-life of fruits and vegetables." *Trends in Food Science & Technology* 129:296–305.

Drago, Emanuela, Roberta Campardelli, Margherita Pettinato, and Patrizia Perego. 2020. "Innovations in smart packaging concepts for food: An extensive review." *Foods* 9 (11):1628.

Duan, Chao, Xin Meng, Jingru Meng, Md Iqbal Hassan Khan, Lei Dai, Avik Khan, Xingye An, Junhua Zhang, Tanzina Huq, and Yonghao Ni. 2019. "Chitosan as a preservative for fruits and vegetables: A review on chemistry and antimicrobial properties." *Journal of Bioresources and Bioproducts* 4 (1):11–21.

Duncan, Timothy V. 2011. "Applications of nanotechnology in food packaging and food safety: Barrier materials, antimicrobials and sensors." *Journal of Colloid and Interface Science* 363 (1):1–24.

Elebiyo, Tobiloba Christiana, Damilare Rotimi, Rotdelmwa Maimako Asaleye, Boluwatife Boluwatife Afolabi, Bukola Taiwo Atunwa, and Oluyomi Stephen Adeyemi. 2022. "Innovative and biocompatible approaches for nanomaterial production and application: Eco-friendly nanoscience." In *Handbook of Research on Green Synthesis and Applications of Nanomaterials*, edited by Rajni Garg, Rishav Garg, Nnabuk Okon Eddy, 1–26. Hershey, PA: IGI Global.

Espitia, Paula Judith Perez, Nilda de Fátima Ferreira Soares, Jane Sélia dos Reis Coimbra, Nélio José de Andrade, Renato Souza Cruz, and Eber Antonio Alves Medeiros. 2012. "Zinc oxide nanoparticles: Synthesis, antimicrobial activity and food packaging applications." *Food and Bioprocess Technology* 5 (5):1447–1464.

Fadiji, Ayomide Emmanuel, Doctor Mziwenkosi Nhlanhla Mthiyane, Damian C Onwudiwe, and Olubukola Oluranti Babalola. 2022. "Harnessing the known and unknown impact of nanotechnology on enhancing food security and reducing postharvest losses: Constraints and future prospects." *Agronomy* 12 (7):1657.

Fallah, Aziz A., Elham Sarmast, Saied Habibian Dehkordi, Abbas Isvand, Hossein Dini, Tina Jafari, Mohammad Soleimani, and Amin Mousavi Khaneghah. 2022. "Low-dose gamma irradiation and pectin biodegradable nanocomposite coating containing curcumin nanoparticles and ajowan (Carum copticum) essential oil nanoemulsion for storage of chilled lamb loins." *Meat Science* 184:108700.

Ferdous, Zannatul, and Abderrahim Nemmar. 2020. "Health impact of silver nanoparticles: A review of the biodistribution and toxicity following various routes of exposure." *International Journal of Molecular Science* 21 (7):2375.

Fernández, Avelina, Eva Soriano, Gracia López-Carballo, Pierre Picouet, Elsa Lloret, Rafael Gavara, and Pilar Hernández-Muñoz. 2009. "Preservation of aseptic conditions in absorbent pads by using silver nanotechnology." *Food Research International* 42 (8):1105–1112.

Hong, Fashui, Nan Wu, Yingjun. Zhou, Li Ji, Ting Chen, and Ling Wang. 2017. "Gastric toxicity involving alterations of gastritis-related protein expression in mice following long-term exposure to nano TiO (2)." *Food Research International* 95:38–45.

Jiang, Yong, Xufei Yang, Peng Liang, Panpan Liu, and Xia Huang. 2018. "Microbial fuel cell sensors for water quality early warning systems: Fundamentals, signal resolution, optimization and future challenges." *Renewable and Sustainable Energy Reviews* 81:292–305.

Jung, Junho, Pradeep Puligundla, and Sanghoon Ko. 2012. "Proof-of-concept study of chitosan-based carbon dioxide indicator for food packaging applications." *Food Chemistry* 135 (4):2170–2174.

Kah, Melanie, Rai Singh Kookana, Alexander Gogos, and Thomas Daniel Bucheli. 2018. "A critical evaluation of nanopesticides and nanofertilizers against their conventional analogues." *Nature Nanotechnology* 13 (8):677–684.

Kalia, Anu, Manpreet Kaur, Ashwag Shami, Sukhjit Kaur Jawandha, Mousa A Alghuthaymi, Anirudh Thakur, and Kamel A. Abd-Elsalam. 2021. "Nettle-leaf extract derived ZnO/CuO nanoparticle-biopolymer-based antioxidant and antimicrobial nanocomposite packaging films and their impact on extending the post-harvest shelf life of guava fruit." *Biomolecules* 11 (2):224.

Kumar, Deepak, and Prasanta Kalita. 2017. "Reducing postharvest losses during storage of grain crops to strengthen food security in developing countries." *Foods* 6 (1):8.

Kumar, Vineet, Praveen Guleria, and Surinder Kumar Mehta. 2017. "Nanosensors for food quality and safety assessment." *Environmental Chemistry Letters* 15 (2):165–177.

Liang, Hao, Xiao-Bing Zhang, Yifan Lv, Liang Gong, Ruowen Wang, Xiaoyan Zhu, Ronghua Yang, and Weihong Tan. 2014. "Functional DNA-containing nanomaterials: Cellular applications in biosensing, imaging, and targeted therapy." *Accounts of Chemical Research* 47 (6):1891–1901.

Liu, Zitian, Meijun Du, Huiping Liu, Kai Zhang, Xiangxin Xu, Kai Liu, Jianqiu Tu, and Qingrun Liu. 2021. "Chitosan films incorporating litchi peel extract and titanium dioxide nanoparticles and their application as coatings on watercored apples." *Progress in Organic Coatings* 151:106103

Mahdi, Karrar N.M., Ruud J.B. Peters, Erwin Klumpp, Steffi Bohme, Martine Van der Ploeg, Coen Ritsema, and Violette Geissen. 2017. "Silver nanoparticles in soil: Aqueous extraction combined with single-particle ICP-MS for detection and characterization." *Environmental Nanotechnology, Monitoring & Management* 7:24–33.

Mathew, Shiji, Jyothis Mathew, and EK Radhakrishnan. 2019. "Polyvinyl alcohol/silver nanocomposite films fabricated under the influence of solar radiation as effective antimicrobial food packaging material." *Journal of Polymer Research* 26 (9):1–10.

Mills, Andrew, and David Hazafy. 2009. "Nanocrystalline SnO2-based, UVB-activated, colourimetric oxygen indicator." *Sensors and Actuators B: Chemical* 136 (2):344–349.

Miranda, Marcela, Xiuxiu Sun, Anna Marin, Luana Cristina dos Santos, Anne Plotto, Jinhe Bai, Odílio Benedito Garrido Assis, Marcos David Ferreira, and Elizabeth Baldwin. 2022. "Nano- and micro-sized carnauba wax emulsions-based coatings incorporated with ginger essential oil and hydroxypropyl methylcellulose on papaya: Preservation of quality and delay of post-harvest fruit decay." *Food Chemistry* X 13:100249.

Mittal, Deepti, Gurjeet Kaur, Parul Singh, Karmveer Yadav, and Syed Azmal Ali. 2020. "Nanoparticle-based sustainable agriculture and food science: Recent advances and future outlook." *Frontiers in Nanotechnology* 2:579954.

Mohd Israfi, Nur Aisyah, Muhamad Israq Amir Mohd Ali, Sivakumar Manickam, Xun Sun, Bey Hing Goh, Siah Ying Tang, Norsharina Ismail, Ahmad Faizal Abdull Razis, Soo Ee Ch'ng, and Kim Wei Chan. 2022. "Essential oils and plant extracts for tropical fruits protection: From farm to table." *Frontiers in Plant Science* 13:999270.

Motelica, Ludmila, Denisa Ficai, Anton Ficai, Ovidiu Cristian Oprea, Durmuş Alpaslan Kaya, and Ecaterina Andronescu. 2020. "Biodegradable antimicrobial food packaging: Trends and perspectives." *Foods* 9 (10):1438.

Neme, Kumera, Ayman Nafady, Siraj Uddin, and Yetenayet B. Tola. 2021. "Application of nanotechnology in agriculture, postharvest loss reduction and food processing: Food security implication and challenges." *Heliyon* 7 (12):e08539.

Puligundla, Pradeep, Junho Jung, and Sanghoon Ko. 2012. "Carbon dioxide sensors for intelligent food packaging applications." *Food Control* 25 (1):328–333.

Pushparaj, Karthika, Wen-Chao Liu, Arun Meyyazhagan, Antonio Orlacchio, Manikantan Pappusamy, Chithravel Vadivalagan, Asirvatham Alwin Robert, Vijaya Anand Arumugam, Hesam Kamyab, and Jiří Jaromír Klemeš. 2022. "Nano-from nature to nurture: A comprehensive review on facets, trends, perspectives and sustainability of nanotechnology in the food sector." *Energy* 240:122732.

Qiu, Liqing, Min Zhang, Bimal Chitrakar, Benu Adhikari, and Chaohui Yang. 2022. "Effects of nanoemulsion-based chicken bone gelatin-chitosan coatings with cinnamon essential oil and rosemary extract on the storage quality of ready-to-eat chicken patties." *Food Packaging and Shelf Life* 34.

Riu, Jordi, Alicia Maroto, and F Xavier Rius. 2006. "Nanosensors in environmental analysis." *Talanta* 69 (2):288–301.

Rodrigues, Sónia M., Philip Demokritou, Nick Dokoozlian, Christine Ogilvie Hendren, Barbara Karn, Meagan S. Mauter, Omowunmi A. Sadik, Maximilian Safarpour, Jason M. Unrine, and Josh Viers. 2017. "Nanotechnology for sustainable food production: Promising opportunities and scientific challenges." *Environmental Science: Nano* 4 (4):767–781.

Ruparelia, Jayesh P., Arup Kumar Chatterjee, Siddhartha P. Duttagupta, and Suparna Mukherji. 2008. "Strain specificity in antimicrobial activity of silver and copper nanoparticles." *Acta Biomaterialia* 4 (3):707–716.

Sampathkumar, Kaarunya, Kei Xian Tan, and Say Chye Joachim Loo. 2020. "Developing nano-delivery systems for agriculture and food applications with nature-derived polymers." *iScience* 23 (5):101055.

Sawicka, Barbara. 2019. "Post-harvest losses of agricultural produce." In *Zero Hunger*, edited by Walter Leal Filho, Anabela Marisa Azul, Luciana Brandli, Pinar Gökcin Özuyar and Tony Wall, 1–16. Cham: Springer International Publishing.

Scallan, Elaine, Robert M. Hoekstra, Frederick J. Angulo, Robert V. Tauxe, Marc-Alain Widdowson, Sharon L. Roy, Jeffery L. Jones, and Patricia M. Griffin. 2011. "Foodborne illness acquired in the United States: Major pathogens." *Emerging Infectious Diseases* 17 (1):7.

Shankar, Shiv, and Jong-Whan Rhim. 2017. "Preparation and characterization of agar/lignin/silver nanoparticles composite films with ultraviolet light barrier and antibacterial properties." *Food Hydrocolloids* 71:76–84.

Shawon, Zayed Bin Zakir, Md Enamul Hoque, and Shiplu Roy Chowdhury. 2020. "Nanosensors and nanobiosensors: Agricultural and food technology aspects." In *Nanofabrication for Smart Nanosensor Applications*, edited by Kaushik Pal, Fernando Gomes, 135–161. Amsterdam: Elsevier.

Shedbalkar, Utkarsha, Richa Singh, Sweety Wadhwani, Sharvari Gaidhani, and Balu Ananda Chopade. 2014. "Microbial synthesis of gold nanoparticles: Current status and future prospects." *Advances in Colloid and Interface Science* 209:40–48.

Shokri, Sajad, Karim Parastouei, Maryam Taghdir, and Sepideh Abbaszadeh. 2020. "Application an edible active coating based on chitosan- Ferulago angulata essential oil nanoemulsion to shelf life extension of Rainbow trout fillets stored at 4 degrees C." *International Journal of Biological Macromolecules* 153:846–854.

Souza, Victor Gomes Lauriano, and Ana Luisa Fernando. 2016. "Nanoparticles in food packaging: Biodegradability and potential migration to food—A review." *Food Packaging and Shelf Life* 8:63–70.

Usman, Muhammad, Muhammad Farooq, Abdul Wakeel, Ahmad Nawaz, Sardar Alam Cheema, Hafeez ur Rehman, Imran Ashraf, and Muhammad Sanaullah. 2020. "Nanotechnology in agriculture: Current status, challenges and future opportunities." *Science of the Total Environment* 721:137778.

van der Meulen, Bernd, Harry Bremmers, Kai Purnhagen, Nidhi Gupta, Hans Bouwmeester, and L. Leon Geyer. 2014. *Governing Nano Foods: Principles-Based Responsive Regulation: EFFoST Critical Reviews# 3*, Vol. 3. Cambridge: Academic Press.

Verma, Krishan K., Xiu-Peng Song, Abhishek Joshi, Dan-Dan Tian, Vishnu D Rajput, Munna Singh, Jaya Arora, Tatiana Minkina, and Yang-Rui Li. 2022. "Recent trends in nano-fertilizers for sustainable agriculture under climate change for global food security." *Nanomaterials* 12 (1):173.

Vishnuvarthanan, Mayakrishnan, and Natesan Rajeswari. 2019. "Preparation and characterization of carrageenan/silver nanoparticles/Laponite nanocomposite coating on oxygen plasma surface modified polypropylene for food packaging." *Journal of Food Science and Technology* 56 (5):2545–2552.

Wang, Chengxiang, Longwei Yin, Luyuan Zhang, Dong Xiang, and Rui Gao. 2010. "Metal oxide gas sensors: Sensitivity and influencing factors." *Sensors* 10 (3):2088–2106.

Yadav, Ajay, Nishant Kumar, Ashutosh Upadhyay, Shruti Sethi, and Anurag Singh. 2022. "Edible coating as postharvest management strategy for shelf-life extension of fresh tomato (Solanum lycopersicum L.): An overview." *Journal of Food Science* 87 (6):2256–2290.

Yadav, Srasti, GK Mehrotra, and PK Dutta. 2021. "Chitosan based ZnO nanoparticles loaded gallic-acid films for active food packaging." *Food Chemistry* 334:127605.

Yin, Nan, Renpeng Du, Fangkun Zhao, Ye Han, and Zhijiang Zhou. 2020. "Characterization of antibacterial bacterial cellulose composite membranes modified with chitosan or chitooligosaccharide." *Carbohydrate Polymers* 229:115520.

Zhang, Huiyun., Ying Liang, Xinling Li, and Huaibin Kang. 2020. "Effect of chitosan-gelatin coating containing nano-encapsulated tarragon essential oil on the preservation of pork slices." *Meat Science* 166:108137.

Zhang, Ning, Yang Hong, Qiming Qin, and Lu Liu. 2013. "VSDI: A visible and shortwave infrared drought index for monitoring soil and vegetation moisture based on optical remote sensing." *International Journal of Remote Sensing* 34 (13):4585–4609.

4 Gene Delivery Methods
Advances and Opportunities of Nanoparticles-Based Techniques

Venus Borgohain, Swati, Reetu, and Vinay Kumar
Central University of Punjab

CONTENTS

4.1 INTRODUCTION

Increasing population globally has resulted in increased demand for sustainable agriculture and, subsequently, crop production (Chugh, Siddique, and Solaiman 2021). Genetic engineering can produce recombinant products with multiple improved traits, including better growth and improved yield related traits, broad and specific disease resistance, protection against biotic and abiotic stress and traits associated with improving the nutritional values of diverse commercial agricultural products (Zhang et al. 2021). The genetic engineering approach facilitates the delivery of desired genetic/DNA fragments in host cells and produces transgenic plants with successful expression of the incorporated DNA fragments (Nandy, Mitra, and Maity 2020). The whole concept of transgenic plants is based on the incorporation of foreign

DOI: 10.1201/9781003397861-4

DNA fragments into the genome of the plant of interest. Gene delivery is defined as the process of stable introduction of a foreign genetic fragment (either DNA or RNA) (Luiz et al. 2022). However, the naked genetic material does not directly enter into the targeted cells due to its large size and hydrophobic nature. Thus, many diverse approaches have been utilized for the delivery of genetic fragments in various types of plant systems (Ameen Rashid and Dhahir Lateef 2016).

Nowadays, newly emerging technologies have been adopted with diverse perspectives for improving the productivity of agricultural crops. Nanotechnology has been extensively utilized at diverse fronts in agriculture for crop improvement (Bhardwaj, Lata, and Garg 2022). The unique properties of different organic and inorganic materials (nanopesticides, nanofertilizers and nanostimulants) at the nanoscale make them suitable parameters for developing different tools for crop improvement. The development of biosensors for different biochemical analyses is another emerging area for which nanotechnology has great potential in the coming days (Lata, Bhardwaj, and Garg 2022). Another emerging area in which nanotechnology has been given much attention is gene delivery/transformation systems, which are directly associated with transgenic development (Dizaj, Jafari, and Khosroushahi 2014).

In plants, genetic transformation techniques enable the integration of foreign genes into crop plants for creation of new traits in genetically modified organisms. So, genetic engineering contributes significantly to an overall increase in crop productivity (Nandy, Mitra, and Maity 2020). In the present scenario, both viral and non-viral systems have been exploited to be used for the delivery of genetic materials. However, cellular and extracellular barriers are major concerns for stable and successful genetic transformation (Demirer et al. 2019). Utilization of nanoparticles for delivery of desired genetic fragments has been reported recently and offers opportunities to overcome many challenges associated with conventional gene delivery methods in plant transformation protocols (Scherer et al. 2002). This chapter is written to enlist all methods of genetic transformation, the utilization of different nanoparticles in genetic engineering and the associated limitations and advantages in the present scenario. In fact, this work highlights the strength of nanotechnology in genetic transformation and subsequently in crop improvement.

4.2 ROLES OF NANOPARTICLES IN GENETIC TRANSFORMATION

In the present scenario, the diverse applications of nano-sized particles, having unique physiochemical properties with a high surface area and diverse particle morphology, in agriculture have been indicated. So, nanoparticles are classified based upon their sources as carbon-based nanoparticles, metal-based nanoparticles and metal oxide nanoparticles (Garg et al. 2022). Nanoparticles act as a bullet that encapsulates the packing materials (DNA, RNA or protein, as well as chemicals such as fertilizers, herbicides and pesticides) or attaches the material to its surface and facilitates their entry of into host cells (Bi et al. 2020). Thus, nanotechnology has enabled the researcher to improve the efficiency of genetic engineering without causing much damage to the host plant and enhance crop growth and development (Chugh, Siddique, and Solaiman 2021). However, the series of steps that are necessary for the entry of nanoparticles into plant cells are mentioned in a sequential manner.

The detailed description of the interaction of the cell wall, cell membrane and cellular organelles with nanoparticles is also mentioned here.

4.3 BARRIERS FOR ENTRY OF NANOPARTICLES INTO PLANTS

4.3.1 Cell Wall, Unique Feature of Plant Cell and Major Barrier of Any Entry

The plant cell wall acts as a first line of defense against the entry of any type of nanoparticle. However, the nature of the barrier and extent of resistance of the cell wall depend on the nature of the plant species as well as environmental conditions, including the presence of cuticles, casparian strips, tightly bound strips and tightly bound cell wall material for adaptation to minimize water loss in drought-tolerant plant species that resist the uptake of nanoparticles (Schwab et al. 2016). The cell wall thickenings, suberin, and lignin act as key resistant materials for the entry of nanoparticles. It was observed that suberin acts as a stimulant for cell wall formation in epidermal as well as endodermal cells and influences the passage of nanoparticles. However, it was also found that plant cell walls act as semi-permeable barriers and allow limited uptake of specific NPs (Kurepa et al. 2010). The diameter of the cell wall pore (5–20 nm) is a key determinant for the introduction of a range of different sizes of nanoparticles (Zulfiqar et al. 2019). The nanoparticles with a diameter of 50 mm have been observed to localize in the cytoplasm (Cañas et al. 2008). Fleischer, O'Neill, and Ehwald (1999) documented that monocot plants, especially the Gramineae family, showed resistance to the uptake of nanoparticles as compared to dicots. Pores can also be created as a result of the interaction of synthesized nanoparticles. Generally, nanoparticles enter an unbroken cell membrane via apertures or holes present in the membrane (Ma et al. 2010). In leguminous plants, symbiosis with N2-fixing bacteria facilitates pore formation across the cell wall of root hairs and improves the uptake of nanoparticles (Mateos et al. 2001).

4.3.2 Cell Membrane Acts as the Second Line of Defense

The second line of defense is documented by the cell membrane, which regulates the movement of nanoparticles. The transport of nanoparticles through the cell membrane depends on many factors, including size, shape and surface charge (Schwab et al. 2016). The different arrangements of the cell membrane, including endocytosis, ion channel and carrier proteins, contribute significantly to the internalization of NPs at subcellular level. The receptor-mediated clathrin-dependent fluid phase endocytosis (CEPE) pathway has been observed for the entry of positively charged nanoparticles (Onelli et al. 2008). In contrast, the entry of neutrally charged and polyethylene glycol-layered quantum dots (ODs) has been facilitated by the clathrin-independent FPE pathway (Etxeberria, Gonzalez, and Pozueta 2009). However, uptake of gold NPs (Au) has been documented with both clathrin-dependent and clathrin-independent endocytosis processes in Nicotiana tabaccum (Onelli et al. 2008). Positively charged gold nanoparticles internalize spontaneously after attaching to negatively charged cell membranes, while weak interactions between negatively charged gold nanoparticles and cell membrane have been documented and are irregularly absorbed. Banerjee et al. (2019) documented the smooth movement of oxidized single-walled carbon nanotubes (SWCNTs) of length less than 500 nm through the cell membrane of *N. tobaccum*. This observation provided evidence that gold nanoparticles of size less than 120 nm are easily internalized by cells through the CFPE mechanism, while nanoparticles of size 200–500 nm are internalized by an unspecific FPE mechanism. The non-endocytic pathways include the involvement of aquaporins for the uptake of nanoparticles (Fitzpatrick and Reid 2009). The aquaporins have also been documented for manipulation of nanoparticle transport (Taylor et al. 2014). The nanoparticles block the apoplast movement at the cellular level and are responsible for decreasing water transport, which results in the shape and size of aquaporins (Lü et al. 2010).

4.3.3 Plant Different Tissues/Organs Also Act as Entry Point for Nanoparticles

Different unique cellular modifications, including stomata and bases of trichomes, as well as wounds or eroded cuticles, also act as entry points for the uptake of nanoparticles (Uzu et al. 2010). Stomata are well-known modified cellular structures documented on both the abaxial and adaxial surfaces of leaves and used for the exchange of gases. The stomatal density is highly correlated with the uptake frequency of nanoparticles by the plant (Eichert and Goldbach 2008). However, nanoparticles block the stomata/guard cells and drastically reduce the transpiration and photosynthesis rates, as documented earlier (Farmer 1993), and subsequently impact the canopy temperature and inhibit plant growth. Wounds caused by different plant pathogens also provide the option for the direct entry of nanoparticles into the cytoplasm of plant cells (Dietz and Herth 2011). Root hairs are also specialized extensions of root cells with a thin, porous cuticle as compared to normal plant cells, and a number of reports have been documented of the entrapment of nanoparticles by root hairs that subsequently reduces their development (Miralles, Church, and Harris 2012). Overall, nanoparticles are absorbed by the active transport system of the cell membrane. The exposed concentration of nanoparticles in the plant system is the rate-limiting factor for uptake and accumulation of nanoparticles in the plant system (Gottschalk and Nowack 2011). Nanoparticles accumulate in the xylem of the vascular bundle after entering through the plant cell wall and membrane of the root epidermis and are further transported to different parts of the plant, including the leaves. Similarly, phloem has also been documented for the translocation of nanoparticles from the shoot to the root portion of plants (Liu et al. 2021).

4.4 GENE DELIVERY BASED UPON NANOPARTICLES

Magnetofection is a magnetic nanoparticle (MNP)-based transfection that is used to deliver the nucleic acids into the living cell using magnets. In this process, genetic materials are conjugated to an attachment that is already attached to the surface of nanoparticles (Bi et al. 2020). In vitro, magnetofection has been carried out by placing the genetic material-MNP complex in the appropriate medium, and further facilitation of the process in cell culture has been achieved using magnets. The conjugation process has been known to increase the sedimentation value of the complex,

TABLE 4.1
Comparison between Different Modes of Gene Delivery

	Physical Mode of Gene Delivery	Chemical Mode of Gene Delivery	Biological Mode of Gene Delivery	Nanoparticle-Based Gene Delivery
Efficiency	Low to high	Low to High	High	High
hosts range	Broad range of hosts	Generally broad range of hosts	Narrow range of hosts/hosts specific	Broad range of hosts
Damage to host tissue	High damage to hosts tissues	Generally no damage to hosts tissues	No damage to hosts tissue	No damage to hosts tissues
Toxicity	No toxicity in host plants	May or may not cause toxicity in host plants	May or may not cause cytotoxicity in host plants	No toxicity in host plants
Plant regeneration efficiency	Low to high	Low	High	High
Integration of transgene	Random	Random	Random	Target/site specific

increase the contact time with the target cells, and facilitate the speed of transfection (Dobson 2006) (Table 4.1). In addition, this method has been documented with a minimum requirement of nucleic acids for transfection as compared to other methods (Huth et al. 2004). The magnetic property of nanoparticles is a major factor that can be easily manipulated by altering their sizes. The other factors that affect the magnitude of force are magnetic field strength and magnetic field gradient. Scherer et al. (2002) documented that a particular size range (200–1,000 nm) under a static magnetic field of 250 mT is best used for magnetofection with an incubation period (5–20 minutes).

The superparamagnetic iron oxide nanoparticles have been reported to be extensively used for magnetofection. The pulsed magnetic field and oscillating magnetic field have been reported to increase the efficiency of transfection (Fus-Kujawa et al. 2021). Furthermore, the transfer of the DNA-MNP complex into the target cells has been reported to be tracked with confocal fluorescent microscopy and transmission electron microscopy (TEM). The internalization of MNP-DNA complexes has been confirmed by energy dispersive spectroscopy. The higher magnetofection rate has been correlated with a circular expression vector as compared to a linear plasmid (Zhao et al. 2017). The correct ratio of DNA to MNPs has been estimated for complete bonding of nucleic acids with the MNPs. The MNPs have been used for the delivery of DNA or genes into the cells conjugated with polymer-coated magnetic particles. Polyethyleneimine (PEI) has been reported to be attached by covalent bonding or through electrostatic interaction between PEI-coated MNPs and plasmid DNA (Nguyen et al. 2018). Paysen et al. (2020) also documented that the cellular mechanism of uptake of MNPs is endocytosis in nature. The delivery of DNA has been achieved using magnetic particles after mixing DNA and plasmid in an optimum ratio. The concentration of DNA has been much higher than the concentration of MNPs. After proper formation of DNA-MNP complexes, these complexes have been placed into a medium containing the cell culture that needs to be transfected. Depending on the type of target cells, different natures and compositions of the medium have been used. After successful transfection, the gene of interest has been found to be integrated into the genome of the tested organism to form a transgenic organism. The confirmation has been carried out using different techniques, including PCR, dot blot and RT-PCR.

4.4.1 Pollen Magnetofection

The magnetofection process in plants is a very difficult process due to the presence of the cell wall, which is the first line of defense against the delivery of foreign DNA. However, pollen grains have a unique structure with surface apertures and a thinner pollen wall. Thus, pollen grains are suitable targets for the entry of foreign DNA through surface apertures using MNPs (Wang et al. 2022). Pollen grains contain one vegetative cell and one generative cell, which further divide into two male gametes, or sperm cells. Zhang et al. (2021) reported that the uninucleate stage of pollen is an appropriate stage for the introduction of exogenous DNA. Further in vitro fertilization techniques could be used for the development of transgenic plants. In pollen magnetofection, some chemicals are required, including sucrose, boric acid (H_3BO_3), potassium nitrate (KNO_3), calcium nitrate ($CaNO_3$), manganese sulfate ($MnSO_4$), magnesium sulfate ($MgSO_4{\cdot}7H_2O$) and gibberelic acid (GA_3). For this purpose, pollen grains should be collected before dehiscence. The magnetofected pollens are cultured on pollen solid medium and could be examined by fluorescent microscopy and the Gus assay to determine the expression of foreign genes (Kim, Shin, and Park 2007). Magnetofected pollen grains are artificially pollinated by the emasculated flowers for the generation of transgenic seeds, which are subsequently screened the putative transgenic seeds using selectable markers (Figure 4.1).

4.4.2 Carbon Nanotube-Mediated DNA Delivery

Carbon nanotubes (CNTs) have also been exploited for the delivery of genes and other biomolecules across the plasma membrane of cells without using external force (Liang et al. 2022). CNTs are also known as being capable of traversing the chloroplast membrane, which allows for modifying the chloroplast genome to regulate the expression pattern of genes of interest. Generally, CNTs are subdivided into multi-walled carbon nanotubes (MWCNTs) and single-walled carbon nanotubes (SWCNTs or SWNTs) (Schwab et al. 2016). SWNTs are 700 times more efficient than MWCNTs for delivering plasmid DNA in plant systems (Demirer et al. 2019). The basic reasons for good efficiency are a high aspect ratio, excellent tensile strength and the good protection ability of biomolecules from degradation. In addition, single-walled CNTs have high intrinsic near-infrared (NIR) fluorescence, which facilitates efficient tracking of CNTs carrying foreign molecules (Zheng et al. 2003). Oligonucleotides/DNA fragments could be conjugated with SWNTS using sonication via non-covalent pi-pi interaction. However, for large double-stranded DNA, this interaction is not very efficient. Secondly, large DNA molecules are also prone to shearing during sonication. The DNA loading on the surface of SWNTs is further improved by coating them with polyethylenimine (PEI) (Figure 4.2). Both carboxylated SWNTs and PEI are

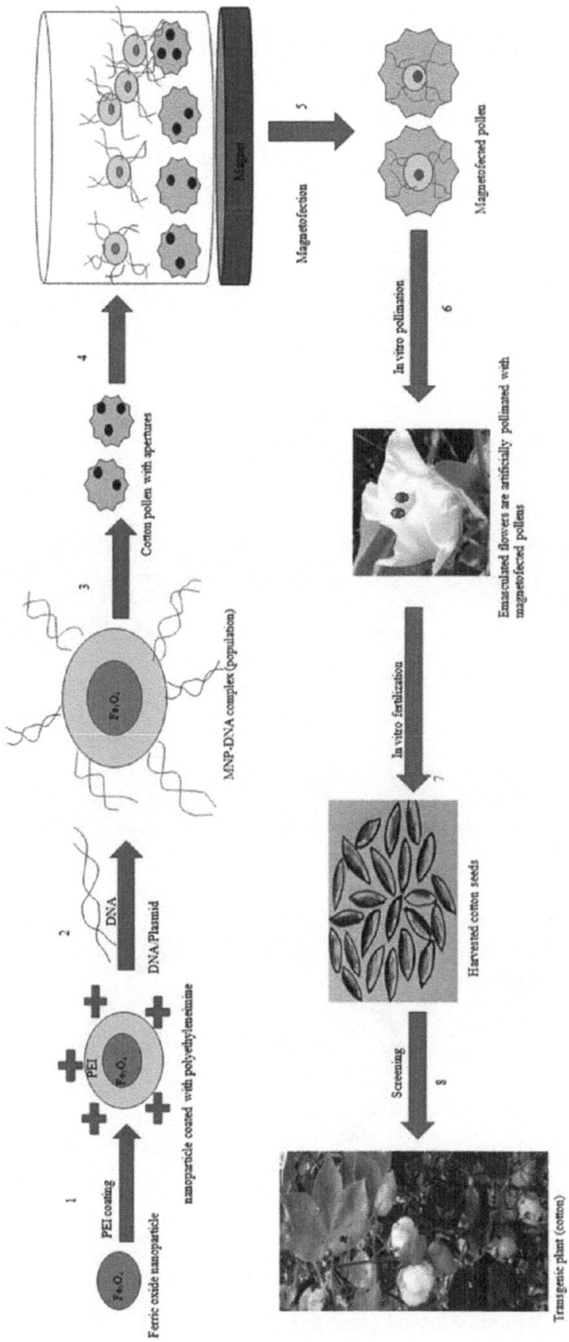

FIGURE 4.1 Pollen magnetofection.

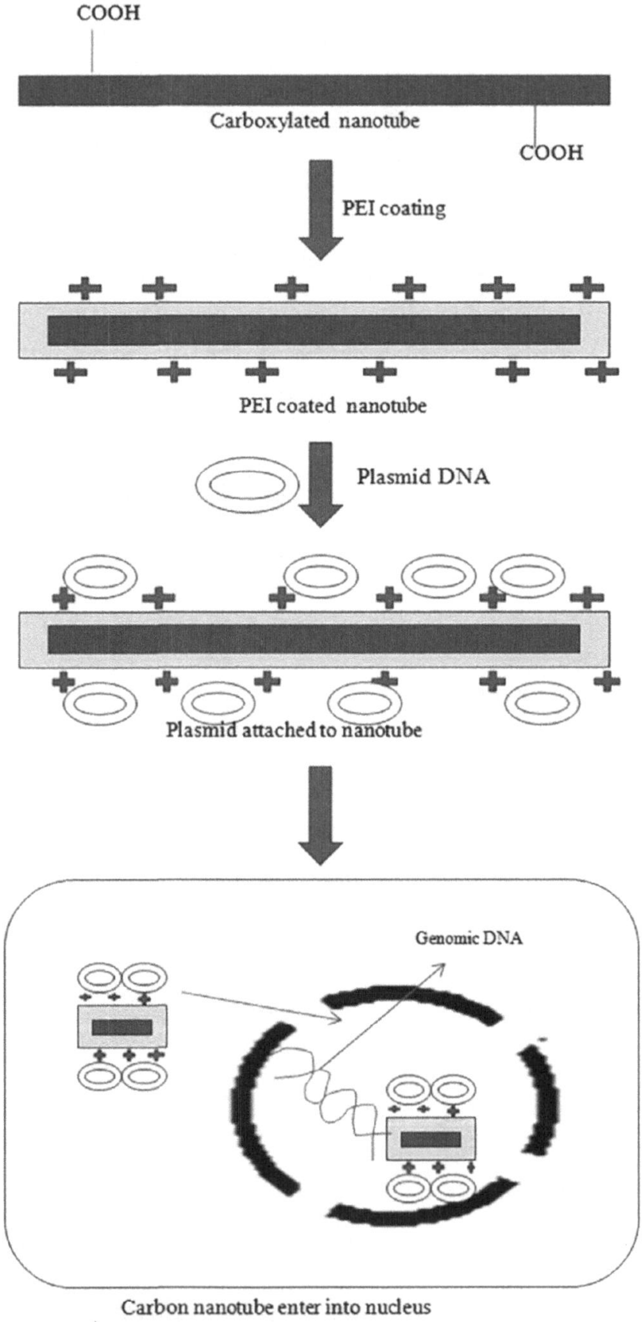

FIGURE 4.2 Carbon nanotube-based gene delivery.

incubated at high temperatures to form a covalent bond between the carbohydroxyl group (COOH) on SWNT and the primary amine (NH3) on PEI (Liang et al. 2022).

4.4.3 Entry of Carbon Nanotubes into Intact Plants

After an appropriate incubation period, the DNA-PEI-COOH solution can be injected into the abaxial surface of leaves. Internalization of these DNA-coated nanotubes can be confirmed using Cy3-tagged DNA–CNTs and high-resolution confocal imaging. Internalization of DNA-SWNTs into intact leaves can also be monitored by using TEM and NIR imaging (Schwab et al. 2016). Gene expression analysis using qPCR and confocal microscopy has been used to determine the activity of fluorescent proteins. qPCR and western blotting have also been used for checking the activity of non-fluorescent proteins. qPCR and western blotting can be used (Demirer et al. 2019).

SWCNT-based transgene delivery also enables the delivery of multiple genes. This approach helps to incorporate the different biosynthetic pathways that can be used for producing recombinant molecules. Since the transgene (s) delivered by SWNT has been reported as an independent unit and not integrated into the host genome (Schwab et al. 2016). Thus, SWNTs could be used for delivering nuclease-based genome editing proteins such as zinc finger nucleases, transcription activator-like effector nucleases and Cas (CRISPR-associated) nucleases for transient expression of these nucleases, which will be useful in efficient genome modification (Demirer et al. 2019). SWNT-based delivery of plasmids encoding CRISPR could also be used for GMO labeling. SWNTs are also reported to offer the transfer of foreign DNA into the chloroplast, enabling its genome transformation. SWNTs have also been reported for the delivery of siRNA and sgRNAs (Liang et al. 2022).

4.5 ADVANTAGES AND LIMITATIONS

Agrobacterium-mediated gene transfer and biolistic methods are the most common conventional methods for the introduction of foreign DNA into plant cells, but these techniques are not free from limitations (Schwab et al. 2016). The gene delivery via *Agrobacterium* is host-specific and results in random and uncontrolled transgene integration. This method is not suitable for chloroplast genome transformation. The gene gun/biolistic gun uses external force for DNA delivery, which may damage host tissue and decrease transformation efficiency. SWNT-based DNA delivery, a non-toxic method, is not a host-specific mode of gene delivery (Liang et al. 2022). Toxicity of SWNTs could be measured with the help of qPCR. It could be used for gene delivery in both model and crop plants (monocots and dicots), and hence can be used for crop improvement. This approach does not allow transgene integration into the host genome and avoids tagging GMOs. This approach is also cost- and time-efficient. In addition, since it does not use any external force for DNA delivery, it does not cause any damage to the host tissue (Demirer et al. 2019). The major limitation associated with SWNT-based DNA delivery is that this approach is not optimized for all diverse crops. In addition, optimization of the chemistry associated with SWNTs is also required.

Nanoparticle-based gene delivery is an efficient way of transferring genes in plants. It delivers the desired foreign gene in a specific targeted region and prevents random integration of foreign genes in the host genome (Dizaj, Jafari, and Khosroushahi 2014). It is a highly efficient gene delivery system. Unlike gene guns, nanoparticle-based gene delivery does not cause damage to host tissues. It is a non-toxic way of transferring genes. This mode of gene delivery is not host-specific, i.e., nanoparticles can deliver desired genes to different plants that need to be transformed. Magnetofection can increase plant transformation efficiency by providing direction to nanoparticles for targeted integration of transgenes in the host plant (Bi et al. 2020).

4.6 CONCLUSION

Crop improvement, a result of genetic engineering, has made it possible to meet the needs of the increasing population with high yields and increased resistance to pests and diseases. Genetically engineered plants show more tolerance to stress. Conventional plant genetic transformation methods have certain disadvantages. *Agrobacterium*-mediated gene transfer is host-specific and also results in uncontrolled foreign gene integration. The gene gun method uses external force, which may damage the host tissues and have low efficiency. These drawbacks can be overcome by the utilization of nanoparticles for the delivery of foreign DNA into plants. Nanoparticle-based plant transformation does not use external force for the delivery of genes and is non-toxic to host cells. Nanoparticles allow controlled delivery of DNA, protect it from degradation, and enable tracking their delivery inside host cells by using fluorescent proteins. This method is cost-efficient and scalable. Nanoparticles can be used in genome editing because of their high efficiency and targeted gene delivery. Pollen magnetofection has emerged as a new technique to deliver foreign genes into pollen grains via small apertures using magnets and by in vitro fertilization, new transgenic plants can be generated. This method provides direction for gene transfer, reduces time, and increases efficiency. CNTs are also used for delivering transgenes into plants. The transgene does not get integrated into the host genome. Single-walled CNTs are more efficient compared to multiwalled. Nanoparticle-based gene delivery may aid high-throughput screening, which allows for the indentification of genotypes of the desired phenotypes.

However, gene delivery via nanoparticles needs to be explored more in different commercially important plants. Optimization of the surface of nanoparticles and in vitro conditions are required for efficient and targeted delivery of foreign DNA in a wide range of hosts. The biosafety level of crop varieties improved by nanoparticle-based gene delivery is still under debate. These doubts may be resolved by the green synthesis of nanoparticles, which will eliminate the use of toxic chemicals. There are limited guidelines and a legislative framework supporting the use of nanoparticles in agriculture and the food industry. Efforts must be made to assess the risk factors related to nanoparticles if they are present, and priority should be given to making the nanoparticle-based products healthy and eco-friendly. The permissible level of nanoparticles in crops should also be determined in order to avoid health risks. Nanoparticle-based improved crop varieties may result in the production of large quantities of nutrition-rich food in a limited area and will contribute to sustainable development.

ACKNOWLEDGMENT

The authors would like to express sincere thanks to the Vice Chancellor of the Central University of Punjab for providing the basic facilities for conducting the work in the laboratory.

REFERENCES

Ameen Rashid, Ahmad Hama, and Djshwar Dhahir Lateef. 2016. "Novel Techniques for Gene Delivery into Plants and Its Applications for Disease Resistance in Crops." *American Journal of Plant Sciences* 07 (01): 181–93.

Banerjee, Koushik, Pragati Pramanik, Aniruddha Maity, D. C. Joshi, Shabir Hussain Wani, and P. Krishnan. 2019. Methods of Using Nanomaterials to Plant Systems and Their Delivery to Plants (Mode of Entry, Uptake, Translocation, Accumulation, Biotransformation and Barriers). *Advances in Phytonanotechnology: From Synthesis to Application*, 123–152. Elsevier Inc.

Bhardwaj, Sweta, Suman Lata, and Rajni Garg. 2022. "Application of Nanotechnology for Preventing Postharvest Losses of Agriproducts." *The Journal of Horticultural Science and Biotechnology.* 98: 1–14.

Bi, Qunjie, Xu Song, Ao Hu, Tianying Luo, Rongrong Jin, Hua Ai, and Yu Nie. 2020. "Magnetofection: Magic Magnetic Nanoparticles for Efficient Gene Delivery." *Chinese Chemical Letters* 31 (12): 3041–46.

Cañas, Jaclyn E., Monique Long, Shawna Nations, Rodica Vadan, Lenore Dai, Mingxiang Luo, Ramya Ambikapathi, E. Henry Lee, and David Olszyk. 2008. "Effects of Functionalized and Nonfunctionalized Single-Walled Carbon Nanotubes on Root Elongation of Select Crop Species." *Environmental Toxicology and Chemistry* 27 (9): 1922–31.

Chugh, Gaurav, Kadambot H.M. Siddique, and Zakaria M. Solaiman. 2021. "Nanobiotechnology for Agriculture: Smart Technology for Combating Nutrient Deficiencies with Nanotoxicity Challenges." *Sustainability* 13 (4): 1–20.

Demirer, Gozde S., Huan Zhang, Natalie S. Goh, Eduardo González-Grandío, and Markita P. Landry. 2019. "Carbon Nanotube-Mediated DNA Delivery without Transgene Integration in Intact Plants." *Nature Protocols* 14 (10): 2954–71.

Dietz, Karl Josef, and Simone Herth. 2011. "Plant Nanotoxicology." *Trends in Plant Science* 16 (11): 582–89. doi.10.1016/j.tplants.2011.08.003.

Dizaj, Solmaz Maleki, Samira Jafari, and Ahmad Yari Khosroushahi. 2014. "A Sight on the Current Nanoparticle-Based Gene Delivery Vectors." *Nanoscale Research Letters* 9 (1): 1–9.

Dobson, Jon 2006. "Gene Therapy Progress and Prospects: Magnetic Nanoparticle-Based Gene Delivery." *Gene Therapy* 13 (4): 283–87.

Eichert, Thomas, and Heiner E. Goldbach. 2008. "Equivalent Pore Radii of Hydrophilic Foliar Uptake Routes in Stomatous and Astomatous Leaf Surfaces—Further Evidence for a Stomatal Pathway." *Physiologia Plantarum* 132 (4): 491–502.

Etxeberria, Ed, Pedro Gonzalez, and Javier Pozueta. 2009. "Evidence for Two Endocytic Transport Pathways in Plant Cells." *Plant Science* 177 (4): 341–48.

Farmer, Andrew M. 1993. "The Effects of Dust on Vegetation—A Review." *Environmental Pollution* 79 (1): 63–75.

Fitzpatrick, Kate L., and Rob J. Reid. 2009. "The Involvement of Aquaglyceroporins in Transport of Boron in Barley Roots." *Plant, Cell and Environment* 32 (10): 1357–65.

Fleischer, Axel, Malcolm A. O'Neill, and Rudolf Ehwald. 1999. "The Pore Size of Non-Graminaceous Plant Cell Walls Is Rapidly Decreased by Borate Ester Cross-Linking of the Pectic Polysaccharide Rhamnogalacturonan II." *Plant Physiology* 121 (3): 829–38.

Fus-Kujawa, Agnieszka, Pawel Prus, Karolina Bajdak-Rusinek, Paulina Teper, Katarzyna Gawron, Agnieszka Kowalczuk, and Aleksander L Sieron. 2021. "An Overview of Methods and Tools for Transfection of Eukaryotic Cells in Vitro." *Frontiers in Bioengineering and Biotechnology* 9 (July): 1–15.

Garg, Rajni, Priya Rani, Rishav Garg, Mohammad Amir, Nadeem Ahmad, Afzal Husain, and Juliana Heloisa. 2022. "Biomedical and Catalytic Applications of Agri-Based Biosynthesized Silver Nanoparticles." *Environmental Pollution* 310 (June): 119830.

Gottschalk, Fadri, and Bernd Nowack. 2011. "The Release of Engineered Nanomaterials to the Environment." *Journal of Environmental Monitoring* 13 (5): 1145–55.

Huth, Stephanie, James Lausier, Soeren W. Gersting, Carsten Rudolph, Christian Plank, Ulrich Welsch, and Joseph Rosenecker. 2004. "Insights into the Mechanism of Magnetofection Using PEI-Based Magnetofectins for Gene Transfer." *Journal of Gene Medicine* 6 (8): 923–36.

Kim, Su Sung, Dong Il Shin, and Hee Sung Park. 2007. "Transient β-Glucuronidase Expression in Lily (Lilium Longflorum L.) Pollen via Wounding-Assisted Agrobacterium-Mediated Transformation." *Biotechnology Letters* 29 (6): 965–69.

Kurepa, Jasmina, Tatjana Paunesku, Stefan Vogt, Hans Arora, Bryan M. Rabatic, Jinju Lu, M. Beau Wanzer, Gayle E. Woloschak, and Jan A. Smalle. 2010. "Uptake and Distribution of Ultrasmall Anatase TiO_2 Alizarin Red s Nanoconjugates in Arabidopsis Thaliana." *Nano Letters* 10 (7): 2296–2302.

Lata, Suman, Sweta Bhardwaj, and Rajni Garg. 2022. "Nanomaterials for Sensing and Biosensing: Applications in Agri-Food Diagnostics." *International Journal of Environmental Analytical Chemistry*: 1–18. doi.10.1080/03067319.2022.2115895.

Liang, Weibing, Ming Chen, Lin Li, Liqiang Yan, Xiuli Wang, Xiongzhi Wu, and Chenghong Lei. 2022. "Polyethyleneimine-Functionalized Carbon Nanotubes Enabling Potent Antimycotic Activity of Lyticase." *Polymers* 14 (5):959.

Liu, Cai, Yanchong Yu, Hanzhu Liu, and Hua Xin. 2021. "Effect of Different Copper Oxide Particles on Cell Division and Related Genes of Soybean Roots." *Plant Physiology and Biochemistry* 163 (June): 205–14.

Lü, Peitao, Jinping Cao, Shenggen He, Jiping Liu, Hongmei Li, Guiping Cheng, Yuelian Ding, and Daryl C. Joyce. 2010. "Nano-Silver Pulse Treatments Improve Water Relations of Cut Rose Cv. Movie Star Flowers." *Postharvest Biology and Technology* 57 (3): 196–202.

Luiz, Marcela Tavares, Jessyca Aparecida Paes Dutra, Larissa Bueno Tofani, Jennifer Thayanne Cavalcante de Araújo, Leonardo Delello Di Filippo, Juliana Maldonado Marchetti, and Marlus Chorilli. 2022. "Targeted Liposomes: A Nonviral Gene Delivery System for Cancer Therapy." *Pharmaceutics* 14 (4): 821.

Ma, Xingmao, Jane Geiser-Lee, Yang Deng, and Andrei Kolmakov. 2010. "Interactions between Engineered Nanoparticles (ENPs) and Plants: Phytotoxicity, Uptake and Accumulation." *Science of the Total Environment* 408 (16): 3053–61.

Mateos, Pedro F., David L. Baker, Maureen Petersen, Encarna Velázquez, Jose I. Jiménez-Zurdo, Eustoquio Martínez-Molina, Andrea Squartini, et al. 2001. "Erosion of Root Epidermal Cell Walls by Rhizobium Polysaccharide-Degrading Enzymes as Related to Primary Host Infection in the Rhizobium-Legume Symbiosis." *Canadian Journal of Microbiology* 47 (6): 475–87.

Miralles, Pola, Tamara L. Church, and Andrew T. Harris. 2012. "Toxicity, Uptake, and Translocation of Engineered Nanomaterials in Vascular Plants." *Environmental Science and Technology* 46 (17): 9224–39.

Nandy, Debdyuti, Arup Kumar Mitra, and Amrita Maity. 2020. "Review Target-Specific Gene Delivery in Plant Systems and Their Expression : Insights into Recent Developments." *Journal of Bioscience* 45: 30.

Nguyen, Anh H., Gaser N. Abdelrasoul, Donghai Lin, Hamid Maadi, Junfeng Tong, Grace Chen, Richard Wang, et al. 2018. "Polyethylenimine-Coated Iron Oxide Magnetic Nanoparticles for High Efficient Gene Delivery." *Applied Nanoscience* 8 (4): 811–21.

Onelli, Elisabetta, Cristina Prescianotto-Baschong, Marco Caccianiga, and Alessandra Moscatelli. 2008. "Clathrin-Dependent and Independent Endocytic Pathways in Tobacco Protoplasts Revealed by Labelling with Charged Nanogold." *Journal of Experimental Botany* 59 (11): 3051–68.

Paysen, Hendrik, Norbert Loewa, Anke Stach, James Wells, Olaf Kosch, Shailey Twamley, Marcus R. Makowski, Tobias Schaeffter, Antje Ludwig, and Frank Wiekhorst. 2020. "Cellular Uptake of Magnetic Nanoparticles Imaged and Quantified by Magnetic Particle Imaging." *Scientific Reports* 10: 1–8.

Scherer, F., M. Anton, U. Schillinger, J. Henke, C. Bergemann, A. Krüger, B. Gänsbacher, and C. Plank. 2002. "Magnetofection: Enhancing and Targeting Gene Delivery by Magnetic Force in Vitro and in Vivo." *Gene Therapy* 9 (2): 102–9.

Schwab, Fabienne, Guangshu Zhai, Meaghan Kern, Amalia Turner, Jerald L. Schnoor, and Mark R. Wiesner. 2016. "Barriers, Pathways and Processes for Uptake, Translocation and Accumulation of Nanomaterials in Plants—*Critical Review.*" *Nanotoxicology* 10 (3): 257–78..

Taylor, Andrew F., Elizabeth L. Rylott, Christopher W.N. Anderson, and Neil C. Bruce. 2014. "Investigating the Toxicity, Uptake, Nanoparticle Formation and Genetic Response of Plants to Gold." *PLoS One* 9 (4).

Uzu, Gaëlle, Sophie Sobanska, Géraldine Sarret, Manuel Muñoz, and Camille Dumat. 2010. "Foliar Lead Uptake by Lettuce Exposed to Atmospheric Fallouts." *Environmental Science and Technology* 44 (3): 1036–42.

Wang, Zuo-Ping, Zhong-Bao Zhang, Deng-Yu Zheng, Tong-Tong Zhang, Xiang-Long Li, Chun Zhang, Rong Yu, Jian-Hua Wei, and Zhong-Yi Wu. 2022. "Efficient and Genotype Independent Maize Transformation Using Pollen Transfected by DNA-Coated Magnetic Nanoparticles." *Journal of Integrative Plant Biology* 64 (6): 1145–56.

Zhang, Meng, Hengling Wei, Pengbo Hao, Aimin Wu, Qiang Ma, Jingjing Zhang, Hantao Wang, et al. 2021. "GhGPAT12/25 Are Essential for the Formation of Anther Cuticle and Pollen Exine in Cotton (Gossypium Hirsutum L.)." *Frontiers in Plant Science* 12 (May). doi. 10.3389/fpls.2021.667739.

Zhao, Xiang, Zhigang Meng, Yan Wang, Wenjie Chen, Changjiao Sun, Bo Cui, Jinhui Cui, et al. 2017. "Pollen Magnetofection for Genetic Modification with Magnetic Nanoparticles as Gene Carriers." *Nature Plants* 3 (12): 956–64.

Zheng, Ming, Anand Jagota, Michael S. Strano, Adelina P. Santos, Paul Barone, S. Grace Chou, Bruce A. Diner, et al. 2003. "Structure-Based Carbon Nanotube Sorting by Sequence-Dependent DNA Assembly." *Science* 302 (5650): 1545–48.

Zulfiqar, Faisal, Míriam Navarro, Muhammad Ashraf, Nudrat Aisha Akram, and Sergi Munné-Bosch. 2019. "Nanofertilizer Use for Sustainable Agriculture: Advantages and Limitations." *Plant Science* 289: 110270.

5 Effect of Nanomaterials on Plant Physiology and Function

Monika Mahajan
Punjab Agricultural University

CONTENTS

5.1 INTRODUCTION

Nanotechnology is a promising field of research that offers sustainable remedies to the challenges faced by modern agriculture by means of nanomaterials (NMs). These NMs have a small size, ranging from 1 to 100 nm. This imparts unique characteristics to them, like a large surface-to-volume ratio, promising optical properties and remarkable physiochemical properties in comparison to their bulk counterparts (Ijaz et al. 2020). The large surface area of NMs helps them interact effectively with the target sites. Because of its versatile physiochemical properties, it is showing fantastic results in different fields like life sciences, chemical engineering and electronics. From last so many years, NMs are gaining interest in plant sciences especially for target specific agrochemicals release and biomolecule delivery in plants. More recently, NMs have shown great potential for addressing the challenges in the agriculture sector. This will help in attaining sustainable food production and security (Khan et al. 2017). Numerous applications of NMs are known in agriculture sector that varies from improved seed germination, crop yield, increased plant growth and protection to either abiotic or biotic stress (Servin et al. 2015).

DOI: 10.1201/9781003397861-5

The development of nanosensors for observing environmental stress is one of the biggest innovations in nanotechnology for increasing the fighting potential of plants. With the help of nanosensors, soil conditions, penetration of agrochemicals, crop growth, diseases and environmental pollution can be measured and monitored for sustainable agriculture and ecosystems. It plays an important role in resolving the numerous and unique challenges, like reduced crop yield under various biotic and abiotic stresses (Usman et al. 2020). The agronomic application of nanotechnology has resulted in the release of nanofertilizers in a controlled manner. This has ensured the minimum use of fertilizers through site-directed delivery of nutrients (DeRosa et al. 2010). Today, for genetic modifications or crop improvement, delivery of CRISPR/Cas 9 mRNA and small guide RNA (sgRNA) is being done using engineered NMs (Glass et al. 2018). Besides having tremendous applications, the potentiality and benefits of NMs in the plant sciences and agriculture sectors, their interaction with plant responses, uptake, localization, and mobilization need to be fully explored. The potential risks of their accumulation in plants and contaminating the environment are of major concern to the researchers. So, there is a need for the safe design of NMs with the minimum influence on plant growth and development. The present book chapter has briefly summarized different NMs used in the plant kingdom for sustainable crop production and their interaction with plant responses. The positive and negative impacts of these NMs on physiology and function have been discussed.

5.2 NMs IN PLANT KINGDOM

Various types of nanoparticles (NPs) and NMs are used in the plant kingdom, ranging from carbon nanotubes (CNTs), fullerene and graphene to metals (Ag, Cu and Au), metal oxides (ZnO, CuO, TiO_2 and Fe_3O_4), silicon-based quantum dots, nanogels, etc. (Figure 5.1). Metal NPs such as gold, silver and copper used in plant sciences are quite hazardous (Khan et al. 2022). However, green approach, such as the use of plant extracts, is being developed. Oxidized NPs such as ZnO, TiO_2, CaO and MgO, which have superior catalytic, electrical and light absorption properties, are also in use (Jahan et al. 2018). Nanocomposite materials based on titanium have been synthesized for the removal of arsenic from polluted water (Ashraf et al. 2019). Carbon-based NMs are used for the removal of pollutants from different environmental compartments. They are also very effective biochemical sensors for detecting very low chemical substances in different environments, which has an important role in precision agriculture (Mauter and Elimelech 2008).

Core/Shell NPs with a combination of organic/inorganic, organic/organic and organic/inorganic materials and polymeric shells to improve the biocompatibility of NPs are being manufactured (Chaudhuri and Paria 2012). Mesoporous silica NPs (MSNs) with a high functional surface area have also been synthesized (Torney et al. 2007). These MSNs are used for the delivery of DNA, proteins and other factors. NMs are also used for intracellular labelling and imaging by circumventing biological barriers like the cell wall and cell membrane, which results in the perfect target site localization of imaging agents. Quantum dots are another remarkable innovation in nanotechnology that can be excited by a single light source. The uptake and translocation of NPs can be traced through *in vivo* imaging (Koo et al. 2015).

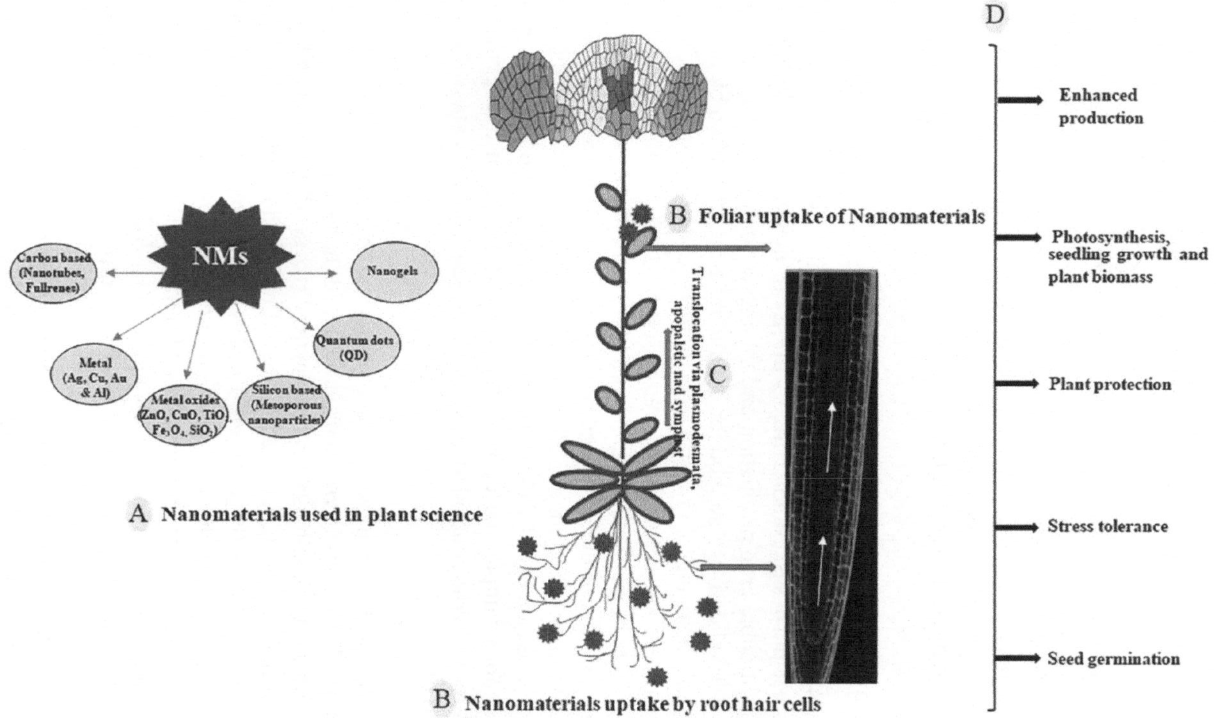

FIGURE 5.1 (a) Nanomaterials (NMs) used in plant science as divided into different categories: carbon-based NMs (CNMs) like graphene, fullerenes and CNTs, which include single-walled carbon nanotubes (SWCNTs) or multi-walled carbon nanotubes (MWCNTs); metallic NPs, like silver (Ag), copper (Cu), gold (Au) and aluminium (Al); metal oxides such as ZnO, CuO, TiO_2, Fe_2O_3, SiO_2, etc.; quantum dots (QDs) and nanogels. (b) Schematic view of NMs delivery methods in plants, i.e., either through foliar or root systems. (c) Once NMs enter the external layers, they either follow an apoplastic or symplastic path. (d) Application of NMs in plant sciences by affecting their physiology and functions.

Nanogels are nano-synthesized ionic or non-ionic hydrogels composed of synthetic or natural polymeric chains. They have excellent physiochemical, bioconjugation and encapsulation properties and are of great interest to researchers (Neamtu et al. 2017).

5.3 APPLICATIONS OF NANOMATERIALS IN AGRICULTURE

Exposure to different NMs in plants led to various morpho-physiological changes. These variations depend on several parameters such as NMs properties, their size, dosage, surface coating, NMs interaction with host plants and exposure time. Applications of NMs promote earlier plant germination, enhance the rate of photosynthesis and increase the plant production (Figure 5.1). Some of the potential effects of different types of NMs on plant physiology and functions are compiled in Table 5.1 and have been discussed as follows.

5.3.1 Nanofertilizers for Enhanced Production

Nanotechnology is being utilized by nanoagriculture for improving plant yields for food, fuel and other purposes. Nanofertilizers are a big innovation in the field of agriculture for sustainable crop production and the protection of agricultural products. It proposes a novel delivery method for enhancing crop production by making less use of chemical pesticides and fertilizers, thereby reducing the cost of cultivation. These fertilizers are encapsulated and delivered in the form of nanosized emulsions. This results in a controlled, slow and highly regulated release of active ingredients for a long time. It increases nutrient use efficiency, reduces soil toxicity and minimizes environmental pollution (Sambangi et al. 2022). Various nanosized formulations have been prepared from synthetic fertilizers, ammonia, urea, peat, and plant wastes. For example, mesoporous NPs loaded with urease, enzymes or urease inhibitors result in the controlled release of nitrogen from urea via hydrolysis (Dimkpa et al. 2020).

Nanofertilizers of metallic oxides like ZnO, TiO_2 and CuO have improved yields in many plant species such as mung bean, cucumber, rape and peanuts (Prasad et al. 2012; Tarafdar et al. 2014). Fertilizers incorporated with nanocarbon have resulted in early seed germination and enhanced root growth in rice seedlings (Liu and Tong 2008). Nanoencapsulated phosphorous has increased the yield of maize, soybeans, wheat and vegetables by 10%–20% (Liu and Lal 2015). Another important finding was reported in wheat, where a 40-day reduction in crop cycle was caused by NPs loaded with NPK (Abdel-Aziz et al. 2016). In addition, foliar application of nano-chitosan has boosted tomato and pea yields by up to 20% (Walker et al. 2004).

A similar growth-promoting effect was also observed in beetroot, capsicum and cucumber. Zhao et al. (2013) studied the effect of CeO_2 NPs on the cucumber life cycle at different concentrations of 0, 400 and 800 mg/kg and revealed that at low concentrations it has not affected plant growth, but at higher concentrations it has reduced the yield by 31.6%. Earlier studies have reported that the application of Si NPs in wheat has reduced Cd uptake and translocation in its aerial parts (Hussain et al. 2019). Thus, enhance the wheat's growth by reducing the oxidative stress, which was found to be directly dependent on the amount of Si NPs used.

TABLE 5.1
Impact of Different Nanomaterials on Plant Physiology and Functions

Functions	Nanomaterials (NMs)	Plant Species	Beneficial Impact	References
Plant production	Urease loaded Mesoporous nanoparticles	Crop plant spp.	Led to control release of nitrogen from urea	Dimkpa et al. (2020)
	ZnO, TiO_2 and CuO	Mung, bean, cucumber, rape, peanuts	Higher yield	Tarafdar et al. (2014)
	Nanocarbon	Rice seedlings	Enhanced germination of seeds and root growth	Liu and Tong (2008)
	Nanoencapsulated phosphorous	Maize, soybean, wheat and vegetables	Increased the yield by 10%–20%	Liu and Lal (2015)
	NPK-loaded nanoparticles	Wheat	Reduction in crop cycle by 40 days	Abdel-Aziz et al. (2016)
	Nano-chitosan	Tomato and pea	Boost up the yield up to 20%	Walker et al. (2004)
	CeO_2	Cucumber	Reduction in yield at high concentration	Zhao et al. (2013)
	Si NPs	Wheat	Reduced cadmium stress	Hussain et al. (2019)
Seed germination	Carbon nanotubes (CNTs), C_{60}, graphene	Sorghum, switchgrass, tomato, *Glycine max*, *Zea mays*, *Hordeum vulgare*	Enhanced growth and sed germination	Pandey et al. (2018), Lahiani et al. (2013)
	AgNPs	*Brassica juncea*	Root length and vigour at lower dose but inhibition at higher dose	Sharma et al. (2012)
	MWCNTs, $nSiO_2$ and $nTiO_2$	*Capsicum annum* L. seedlings	Enhanced growth	Yang et al. (2007)
Photosynthesis	$nCeO_2$, citric acid (CA) + $nCeO_2$	Tomato	Increased the photosynthesis	Barrios et al. (2016)
	nano-TiO_2	Spinach	Increased the photosynthetic rate	Zheng et al. (2005)
	Alumina NPs	*Lemna minor* (duckweed)	Enhanced photosynthesis	Juhel et al. (2011)
	Zinc oxide nanoparticles carrying phycomolecule ligands	Cotton	Increased antioxidant enzyme activities	Venkatachalam et al. (2017)
	ZnO NPs	Tomato	Increased antioxidant activity	Singh et al. (2016)

(Continued)

TABLE 5.1 (*Continued*)
Impact of Different Nanomaterials on Plant Physiology and Functions

Functions	Nanomaterials (NMs)	Plant Species	Beneficial Impact	References
	Fe/Fe_3O_4, Cu/CuO and FeSO4·7H_2O, $CuSO_4$· H_2O	Lettuce	Enhanced chlorophyll content, plant growth, and catalase (CAT)	Trujillo-Reyes et al. (2014)
	Mesoporous nanoparticle	Wheat and lupin	Increased the growth and biomass	Ma et al. (2016)
Stress tolerance	Carbon NPs	*Arabidopsis*	Effect photo morphogenesis and flowering time control	Kumar et al. (2018)
	CeO_2	Corn	Increased antioxidant enzyme activity	Ma et al. (2016)
	Se-NPs	Strawberry	Increased antioxidant enzyme activities	Zahedi et al. (2019)
	Silicon dioxide, zinc oxide, selenium graphene, graphene nanoribbons	Sugarcane	Chilling stress tolerance	Elsheery et al. (2020)
	TiO_2 NPs	Chickpea	Resistance to cold stress	Hasanpour et al. (2015)
	Silicon nanoparticles	Rice seedlings	Reducing the Zn-induced stress	Song et al. (2014)
	Nano-SiO_2	*Cucurbita pepo* L.	Tolerance against salt stress	Siddiqui et al. (2014)
Plant protection	Silica nanosphere	–	Protection against aphids (chewing or sucking)	Li et al. (2007)
	AuNPs	–	Protection against damaging insects	Patil et al. (2016)
	ZnO	Crop plant spp.	Protection against *Fusarium spp. Penicillium, Alternaria and other fungal species*	Servin et al. (2015)
	CuO	*Eichhornia*	*Against fungus Aspergillus niger and Fusarium culmorum*	Vanathi et al. (2016)
	Nano-Cu formulation	Tomato	*Phytophthora infestans*	Giannouri et al. (2013)
	SiO_2	Weeds	Reduce the growth of weeds	Sharifi-Rad et al. (2016)
	Pectin nanoparticles loaded with metsulfuron-methyl	Weeds	Cytotoxic to *Chenopodium album* weed	Kumar et al. (2017)

5.3.2 Seed Germination

Application of NMs has shown an earlier and enhanced rate of seed germination in many plant species. Earlier studies have shown that carbon NMs of diverse dimensionalities, such as CNTs, graphene and C_{60} have increased the seed germination, shoot and root length of sorghum, rice and switchgrass seedlings by acting as enhancers (Pandey et al. 2018). Similar results are also reported in tomatoes and other valuable crops like *Glycine max*, *Zea mays* and *Hordeum vulgar* grown in media supplemented with CNTs and graphene (Lahiani et al. 2013).

Many studies have reported plant responses as enhancement or inhibition of growth depending on the dosage of AgNPs. Sharma et al. (2012) reported a 326% and 133% increase in root length and vigour index, respectively, in *Brassica juncea* seedlings treated with 50 ppm of AgNPs, whereas inhibition was found in seedlings treated with a higher dose. Similarly, multi-walled carbon nanotubes (MWCNTs), $nSiO_2$ and $nTiO_2$ exposure in *Capsicum annum* L. seedlings showed increased plant growth at low concentrations and decreased growth with higher concentrations (Yang et al. 2007). A recent report has shown that the green nanoemulsions synthesized from turmeric oleoresin and citrus seed oil have increased the seed germination and seedling emergence of two onion cultivars when incubated with them. This nanopriming treatment altered the metabolism of onion seedlings, which enhanced seed performance and plant growth (Acharya et al. 2020).

5.3.3 Photosynthetic Rate, Seedling Growth and Plant Biomass

The other study was conducted by Barrios et al. where tomato plants were grown for 7 months in soil supplemented with coated and uncoated cerium oxide NPs ($nCeO_2$). The authors have found that treatment CA + $nCeO_2$ has enhanced the chlorophyll Chl-a and Chl-b and plant growth at 250 mg/kg dosage, compared to the other treatments and their bulk counterparts (Barrios et al. 2016). This treatment has also reduced C_e uptake by the roots and its translocation to the above-ground parts. Another study has shown that the exposure of $nTiO_2$ increased germination, chlorophyll content, dry weight, photosynthetic rate and ribulose bisphosphate (RuBP) carboxylase/oxygenase activity in spinach. This has enhanced nitrogen metabolism and improved vegetative growth in spinach (Zheng et al. 2005). Application of alumina NPs resulted in enhanced numbers of fronds, root length, biomass and photosynthesis in duckweed, *Lemna minor* (Juhel et al. 2011). Similarly, enhancement in root length was observed in radish and rape after treatment with aluminium NPs (Lin and Xing 2007).

Venkatachalam et al. have reported that the application of plant growth promoters, i.e., phycomolecule ligands coated with zinc oxide nanoparticles (ZnONPs), increased the growth and biomass of cotton. The plants have also shown increased antioxidant enzyme activities and reduced MDA and free radical content (Venkatachalam et al. 2017). Similar studies have reported the stimulatory effect of ZnO NPs on seed germination, vigour, antioxidant activity of tomatoes at lower doses but their inhibitory effect at higher doses (Singh et al 2016). The enhanced chlorophyll content, plant growth, and catalase (CAT) activity were observed in lettuce grown hydroponically with nanocore-shell nanomaterials (Fe/Fe_3O_4, Cu/CuO) and $FeSO_4{\cdot}7H_2O$ and

$CuSO_4 \cdot H_2O$ at 10 and 20 mg/L, but these NMs showed negative effects at higher concentrations (Trujillo-Reyes et al. 2014). On the other side, Kumar et al. have observed that plant processes such as photomorphogenesis and the time of flowering are also affected by carbon nanoparticles (CNPs). The group has reported increased seed germination rate, large cotyledon area, elongated hypocotyl length, increased chlorophyll content and early flowering (Kumar et al. 2018).

5.3.4 Stress Tolerance

Many studies have reported the tolerance potential of plants to various abiotic stresses when treated with diverse NPs. For example, the study conducted on corn plants grown in CeO_2 (400 or 800 mg/kg) treated soil has shown decreased lipid peroxidation, cell death and increased antioxidant enzyme activities such as catalase and ascorbate peroxidase, thus suggesting that induced antioxidant machinery protects the plants against oxidative stress (Ma et al. 2016). Another similar study shows the positive impact of selenium SeNPs foliar application on fruit quality improvement under stress conditions (Zahedi et al. 2019). One of the recent reports has shown the chilling tolerance of sugarcane exposed to the foliar application of different NPs such as silicon dioxide, zinc oxide, selenium graphene, and graphene nanoribbons (Elsheery et al. 2020). Similarly, tolerance responses of chickpea to cold stress were observed after $nTiO_2$ application by decreasing H_2O_2 content and by enhancing the expression of photosynthetic genes *CaL Rubisco*, *CaS Rubisco* and *Ca chlorophyll a/b*-genes at the transcription levels (Hasanpour et al. 2015). Similarly, silicon NPs help reduce Zn-induced stress in hydroponically grown rice seedlings and salt stress in *Cucurbita pepo* L. cv. by increasing the photosynthetic rate, total chlorophyll, proline, carbonic anhydrase activity, water use efficiency, transpiration rate and stomatal conductance (Siddiqui et al. 2014; Rastogi et al., 2019).

5.3.5 Plant Protection

The role of nanotechnology in plant protection has resulted in an exponential increase in crop production. It is better than conventional practices where fungicides, pesticides and herbicides are used on a large scale. Nanotechnology has initiated the use of nanoformulations or encapsulated pesticides with useful pesticidal properties (Haq and Ijaz 2019). For instance, silica nanosphere-based formulations consisting of pesticides have controlled chewing or sucking aphids by decreasing their penetration into the plant cell sap (Li et al. 2007). Similarly, Patil et al. (2016) observed decreased catalytic potential of protease and trypsin in damaging insects when exposed to bioactive AuNPs fabricated with latex. ZnO NPs have shown protection against *many fungal species* (Servin et al. 2015) and CuO NPs synthesized from *Eichhornia crassipes* by the green approach showed protection against *Fusarium culmorum* and *Aspergillus niger* (Vanathi et al. 2016). The application of the Nano-Cu formulation in tomatoes has shown protection against *Phytophthora infestans* (Giannousi et al. 2013). Nanotechnology has shown a potential role in making nanoweedicides. For example, SiO_2 NPs have been reported to reduce germination, shoot and root length, photosynthetic pigments, and fresh and dry weights in weeds (Sharifi-Rad et al. 2016).

5.3.6 NP Phytotoxicity

The development of nanotechnology has increased NP-induced environmental pollution. Its growing amounts in the environment have made it necessary to access their toxicity levels. Many studies have shown their accumulation in the plant leaf vasculature leads to phytotoxicity (Paramo et al. 2020). NMs possible toxicity to plants and their transfer to the food chain are still unknown and need to be fully explored. The uptake of NPs and NMs differs and depends upon their size, dosage, chemical composition and type of plant species exposed to them. Most of the studies in tomato, onion and wheat have shown that excessive doses of metal-based NMs lead to oxidative bursts in plants by interfering with detoxification machinery and the electron transport chain (ETC) (Jampílek and Kráľová 2019). A report published that the overall growth, nutrient content and macromolecule conformation are significantly affected by silver NPs in radish sprouts (Zuverza et al. 2016).

A throughput analysis of crop plants at the genetic and metabolic levels is needed to see the response of various genes and metabolites affected by NM exposure. An earlier study revealed the phytotoxicity of CuO NPs on Arabidopsis and rice seedlings by decreasing the growth of roots and increasing the generation of free radicals (Wang et al. 2015). Yanik and Vardar have revealed through their study that Al_2O_3 NP exposure to wheat plants has induced morphological, molecular and cellular alterations in the roots. The reduced root elongation was found to be due to the deposition of callose, lignification and damaged root cortex cells. Besides that, DNA fragmentation due to Al_2O_3 NP toxicity was also observed and has become an important marker of programmed cell death (Yanik and Vardar 2015). Hao et al. (2018) observed that carbon nanomaterials (CNMs), like fullerene (C_{60}), reduced graphene oxide (rGO) regardless of their doses, reduced shoot and root growth and resulted in environmental toxicity to rice. Graphene-exposed tomatoes, cabbage and spinach showed oxidative stress necrosis due to increased free radical generation and cell death at higher doses (Begum et al. 2011). Stampoulis et al. exposed *zucchini* to nAg or its bulk powder and observed about a 50% reduction in their rates at higher concentrations. Also, *zucchini* accumulated 4.7 times greater concentrations in their stems as compared to the plants from their corresponding bulk solutions (Stampoulis et al. 2009).

5.4 CONCLUSION

Nanotechnology has given a lot to agriculture by improving farming systems through enhanced crop production. The development of engineered NMs for sustainable agriculture will help to ensure food security. Although NMs have given us a new way to boost farming, more assurance is still needed to protect the environment from their pollution. Therefore, there is a need to design NMs with all necessary properties like reasonable concentrations showing high impact, stability, controlled release, target specificity, solubility, and less environmental pollution or plant toxicity. Hence, the need of the hour is to investigate the potential toxicity and interaction of various NMs with biomolecules present in living systems and the environment before considering them safe for utilization.

REFERENCES

Abdel-Aziz Heba Mahmoud Mohammad, Hasaneen Mohammed Nagib Abdelghany, Omer Aya Mohammad. 2016. "Nano chitosan-NPK fertilizer enhances the growth and productivity of wheat plants grown in sandy soil". *Spanish Journal of Agricultural Research*, 14: e0902.

Acharya Pratibha, Jayaprakasha Guddadarangavvanahally K., Semper James, Patil Bhiman S. 2020. "1H Nuclear magnetic resonance and liquid chromatography coupled with mass spectrometry-based metabolomics reveal enhancement of growth-promoting metabolites in onion seedlings treated with green-synthesized nanomaterials". *Journal of Agricultural and Food Chemistry*, 68: 13206–13220.

Ashraf Sana, Ali Qasim, Zahir, Zahir Ahmad, Ashraf Sobia, Asghar Hafiz Naeem. 2019. "Phytoremediation: environmentally sustainable way for reclamation of heavy metal polluted soils". *Ecotoxicology and Environmental Safety*, 174: 714–727.

Barrios Ana Cecilia, Rico Cyren M., Trujillo-Reyes Jesica, Medina-Velo Illya A., Peralta-Videa Jose R., Gardea-Torresdey Jorge L. 2016. "Effects of uncoated and citric acid coated cerium oxide nanoparticles, bulk cerium oxide, cerium acetate, and citric acid on tomato plants". *Science of Total Environment*, 563: 956–964.

Begum Parvin, Ikhtiari Refi, Fugetsu Bunshi. 2011. "Graphene phytotoxicity in the seedling stage of cabbage, tomato, red spinach, and lettuce". *Carbon*, 49: 3907–3919.

Chaudhuri Rajib Ghosh, Paria Santanu. 2012. "Core/shell nanoparticles: classes, properties, synthesis mechanisms, characterization, and applications". *Chemical Reviews*, 112: 2373–2433.

DeRosa Maria C., Monreal Carlos, Schnitzer Morris, Walsh Ryan, Sultan Yasir. 2010. "Nanotechnology in fertilizers". *Nature Nanotechnology*, 5: 91.

Dimkpa Christian O., Fugice Job, Singh Upendra, Lewis Timothy D. 2020. "Development of fertilizers for enhanced nitrogen use efficiency: trends and perspectives". *Science of Total Environment*, 731: 139113.

Elsheery Nabel I., Sebastian John Sunoj Valiaparambil, Wen Yin, Zhu J.J., Muralidharan G., Cao Kun-Fang. 2020. "Foliar application of nanoparticles mitigates the chilling effect on photosynthesis and photoprotection in sugarcane". *Plant Physiology and Biochemistry*, 149: 50–60.

Giannousi Kleoniki, Avramidis Ioannis, Dendrinou-Samara Catherine. 2013. "Synthesis, characterization and evaluation of copper-based nanoparticles as agrochemicals against *Phytophthora infestans*". *RCS Advances*, 3: 21743–21752.

Glass Zachary, Lee Matthew, Li Yamin, Xu Qiaobing. 2018. "Engineering the delivery system for CRISPR-based genome editing". *Trends in Biotechnology*, 36: 173–185.

Hao Yi, Ma Chuanxin, Zhang Zetian, Song Youhong, Cao Weidong, Guo Jing, Zhou Guopeng, Rui Yukui, Liu Liming, Xing Baoshan. 2018. "Carbon nanomaterials alter plant physiology and soil bacterial community composition in a rice-soil-bacterial ecosystem". *Environmental Pollution*, 232: 123–136.

Haq Imran Ul, Ijaz Siddra. 2019. "Use of metallic nanoparticles and nanoformulations as nanofungicides for sustainable disease management in plants". In: Prasad R., Kumar V., Kumar M., Choudhary D., editors. *Nanobiotechnology in Bioformulations*. Springer; Cham, Switzerland, pp. 289–316.

Hasanpour H., Maali-Amiri Reza, Zeinali Hassan. 2015. "Effect of TiO_2 nanoparticles on metabolic limitations to photosynthesis under cold in chickpea. *Russian Journal of Plant Physiology*, 62: 779–787.

Hussain Shahid, Rengel Zed, Qaswar Muhammad, Amir Mamoona, Zafar-ul-Hye Muhammad. 2019. "Arsenic and heavy metal (cadmium, lead, mercury and nickel) contamination in plant-based foods". *Plant and Human Health*, 2: 447–490.

Ijaz Irfan, Gilani Ezaz, Nazir Ammora, Bukhari Aysha. 2020. "Detail review on chemical, physical and green synthesis, classification, characterizations and applications of nanoparticles". *Green Chemistry Letters and Reviews*, 13: 223–245.

Jahan Shanaz, Alias Yatimah Binti, Bakar Ahmad Farid Bin Abu, Yusoff Ismail Bin. 2018. "Toxicity evaluation of ZnO and TiO_2 nanomaterials in hydroponic red bean (Vigna angularis) plant: physiology, biochemistry and kinetic transport". *Journal of Environmental Science*, 72: 140–152.

Jampílek Josef, Králová Katarína. 2019. "Impact of nanoparticles on photosynthesizing organisms and their use in hybrid structures with some components of photosynthetic apparatus". In: Prasad R., ed. *Plant Nanobionics*. Springer; Cham, pp. 255–332.

Juhel Guillaume, Batisse Emeline, Hugues Quitterie, Daly Deirdre, van Pelt Frank N.A.M., O'Halloran John, Jansen Marcel A.K. 2011. "Alumina nanoparticles enhance growth of Lemna minor". *Aquatic Toxicology*, 105: 328–336.

Khan M. Nasir, Mobin M., Abbas Zahid Khorshid, AlMutairi Khalid A., Siddiqui Zahid H. 2017. "Role of nanomaterials in plants under challenging environments". *Plant Physiology and Biochemistry*, 110: 194–209.

Khan M. Nasir, Mobin M., Abbas Zahid Khorshid, AlMutairi Khalid A., Siddiqui Zahid H. 2022. "Classification and properties of nanoparticles," In Rangappa S. M., Parameswaranpillai J., Yashas Gowda, T.G., Siengchin S., Seydibeyoglu M.O., editors *Nanoparticle-Based Polymer Composites*. Woodhead Publishing, Cambridge. pp. 15–54.

Koo Yeonjong, Wang Jing, Zhang Qingbo, Zhu Huiguang, Chehab E. Wassim, Colvin Vicki L., Alvarez Pedro J. J., Braam Janet. 2015. "Fluorescence reports intact quantum dot uptake into roots and translocation to leaves of arabidopsis thaliana and subsequent ingestion by insect herbivores". *Environmental Science and Technology*, 49: 626–632.

Kumar Abhishek, Singh Anamika, Panigrahy Madhusmita, Kumar Sahoo Pratap, Panigrahi Kishore C.S. 2018. "Carbon nanoparticles influence photomorphogenesis and flowering time in Arabidosis thaliana". *Plant Cell Reports*, 37: 901–912.

Kumar, Sandeep, Bhanjana, Gaurav, Sharma, Amit., Dilbaghi, Neeraj., Sidhu MC, and Kim Ki-Hyun. 2017. "Development of nanoformulation approaches for the control of weeds". Science of Total Environment, 86: 1272–1278.

Lahiani Mohamed H., Dervishi Enkeleda, Chen Jihua, Nima Zeid, Gaume Alain, Biris, Alexandru S., Khodakovskaya Mariya. 2013. "Impact of carbon nanotube exposure to seeds of valuable crops". *ACS Applied Materials & Interfaces*, 5: 7965–7973.

Li Zhu-Zhu, Chen Jian-Feng, Liu Fan, Liu An-Qi, Wang Qing, Sun Hai-Yan, Wen Li-Xiong. 2007. "Study of UV-shielding properties of novel porous hollow silica nanoparticle carriers for avermectin". *Pest Management Science*, 63: 241–246.

Lin Daohui, Xing Baoshan. 2007. "Phytotoxicity of nanoparticles: inhibition of seed germination and root growth". *Environmental Pollution*, 150: 243–250.

Liu Ruiqiang, Lal Rattan. 2015. "Potentials of engineered nanoparticles as fertilizers for increasing agronomic productions". *Science of Total Environment*, 514: 131–139.

Liu Ying, Tong Zhen, Prud'homme Robert K. 2008. "Stabilized polymeric nanoparticles for controlled and efficient release of bifenthrin". *Pest Management Science*, 64: 808–812.

Ma Xingmao, Wang Qiang, Rossi Lorenzo, Zhang Weilan. 2016. "Cerium oxide nanoparticles and bulk cerium oxide leading to different physiological and biochemical responses in Brassica rapa". *Environmental Science and Technology*, 50: 6793–6802.

Mauter Meagan S., Elimelech Menachem. 2008. "Environmental applications of carbon-based nanomaterials". *Environmental Science & Technology*, 42: 5843–5859.

Neamtu Iordana, Rusu Alina Gabriela, Diaconu Alina, Nita Loredana Elena, Chiriac Aurica P. 2017. "Basic concepts and recent advances in nanogels as carriers for medical applications". *Drug Delivery*, 24: 539–557.

Pandey Kamal, Lahiani Mohamed H., Hicks Victoria K., Hudson M. Keith, Green Micah J., Khodakovskaya Mariya. 2018. "Effects of carbon-based nanomaterials on seed germination, biomass accumulation and salt stress response of bioenergy crops". *PLoS one*, 13: e0202274.

Paramo Luis A., Feregrino-Pérez Ana A., Guevara Ramón, Mendoza Sandra, Esquivel Karen. 2020. "Nanoparticles in agroindustry: applications, toxicity, challenges, and trends". *Nanomaterials*, 10: 1654.

Patil Chandrashekhar D., Borase Hemant P., Suryawanshi Rahul K., Patil Satish V. 2016. "Trypsin inactivation by latex fabricated gold nanoparticles: a new strategy towards insect control". *Enzyme and Microbial Technology*, 92: 18–25.

Prasad Tollamadugu N.V.K.V., Sudhakar P., Sreenivasulu Yasodam, Latha P., Munuswamy V., Raja Reddy K., Sreeprasad T.S., Sajanlal P.R., Pradeep Thalappil. 2012. "Effect of nanoscale zinc oxide particles on the germination, growth and yield of peanut." *Journal of Plant Nutrition* 35 (7): 905–927.

Rastogi Anshu, Tripathi Durgesh Kumar, Yadav Saurabh, Chauhan Devendra Kumar, Živčák Marek, Ghorbanpour Mansour, El-Sheery Nabil Ibrahim, Brestic Marian. 2019. "Application of silicon nanoparticles in agriculture." *3 Biotech* 9 (2): 90.

Sambangi Pratyusha, Gopalakrishnan Subramaniam, Pebam Monika, Rengan Aravind Kumar. 2022. "Nanobiofertilizers on soil health, chemistry, and microbial community: benefits and risks." *Proceedings of Indian National Science Academy* 88: 1–12.

Servin Alia, Elmer Wade, Mukherjee Arnab, Torre-Roche Roberto De la, Hamdi Helmi, White Jason C., Bindraban Prem, Dimkpa Christian. 2015. "A review of the use of engineered nanomaterials to suppress plant disease and enhance crop yield." *Journal of Nanoparticle Research* 17 (2): 92.

Sharifi-Rad J., Sharifi-Rad M., da Silva J.A.T. 2016. "Morphological, physiological and biochemical responses of crops (Zea mays L., Phaseolus vulgaris L.), medicinal plants (Hyssopus officinalis L., Nigella sativa L.), and weeds (Amaranthus retroflexus L., Taraxacum officinale F. H. Wigg) exposed to SiO_2 nanoparticles." *Journal of Agricultural and Science Technology* 18: 1027–1040.

Sharma Priyadarshini, Bhatt Deepesh, Zaidi M. G. H., Saradhi P. Pardha, Khanna P. K., Arora Sandeep. 2012. "Silver nanoparticle-mediated enhancement in growth and antioxidant status of Brassica juncea." *Applied Biochemistry and Biotechnology* 167: 2225–2233.

Siddiqui, Manzer H., Al-Whaibi Mohamed H., Faisal Mohammad, and Al Sahli Abdulaziz A. 2014 "Nano-silicon dioxide mitigates the adverse effects of salt stress on Cucurbita pepo L." *Environmental Toxicology and Chemistry* 33 (11): 2429–2437.

Singh, Ajey, Singh N.B., Hussain Imtiyaz, Singh Himani, Yadav Vijaya, and Singh S.C. 2016. "Green synthesis of nano zinc oxide and evaluation of its impact on germination and metabolic activity of Solanum lycopersicum." *Journal of Biotechnology* 233: 84–94.

Song, Alin, Li Ping, Fan Fenliang, Li Zhaojun, and Liang Yongchao. 2014. "The effect of silicon on photosynthesis and expression of its relevant genes in rice (Oryza sativa L.) under high zinc stress." *PLoS One* 9: e113782.

Stampoulis, Dimitrios, Sinha Saion K., and White Jason C. 2009 "Assay-dependent phytotoxicity of nanoparticles to plants." *Environmental Science Technology* 43 (24): 9473–9479.

Tarafdar, J.C., Raliya Ramesh, Mahawar Himanshu, and Rathore Indira. 2014. "Development of zinc nanofertilizer to enhance crop production in pearl millet (Pennisetum americanum)." *Agricultural Research* 3: 257–262.

Torney, François, Trewyn Brian G., Lin Victor S.-Y., and Wang Kan. 2007. "Mesoporous silica nanoparticles deliver DNA and chemicals into plants." *Nature Nanotechnology* 2: 295–300.

Trujillo-Reyes, J., Majumdar S., Botez C.E., Peralta-Videa J.R., and Gardea-Torresdey J.L. 2014 "Exposure studies of core-shell Fe/Fe3O4 and Cu/CuO NPs to lettuce (Lactuca sativa) plants: are they a potential physiological and nutritional hazard?" *Journal of Hazardous Materials* 267: 255–263.

Usman, Muhammad, Farooq Muhammad, Wakeel Abdul, Nawaz Ahmad, Cheema Sardar Alam, Rehman Hafeez ur, Ashraf Imran, and Sanaullah Muhammad. 2020. "Nanotechnology in agriculture: current status, challenges and future opportunities." *Science of Total Environment* 721: 137778.

Vanathi, P., Rajiv P., Sivaraj Rajeshwari 2016. "Synthesis and characterization of eichhornia-mediated copper oxide nanoparticles and assessing their antifungal activity against plant pathogens." *Bulletin of Materials Science* 39 (5): 1165–1170.

Venkatachalam, P., Priyanka N., Manikandan K., Ganeshbabu I., Indiraarulselvi P., Geetha N., Muralikrishna K., Bhattacharya R. C., Tiwari M., Sharma N., and Sahi S. V. 2017. "Enhanced plant growth promoting role of phycomolecules coated zinc oxide nanoparticles with P supplementation in cotton (Gossypium hirsutum L.)." *Plant Physiology and Biochemistry* 110: 118–127.

Walker, Racha, Morris Suzanne, Brown Phil, and Gracie Alistair 2004. "Evaluation of Potential for Chitosan to Enhance Plant Defense." Australian Government, RIRDC Report 4, p. 49.

Wang, Xiaoping, Yang Xiyu, Chen Siyu, Li Qianqian, Wang Wei, Hou Chunjiang, Gao Xiao, Wang Li, and Wang, Shucai 2015. "Zinc oxide nanoparticles affect biomass accumulation and photosynthesis in Arabidopsis." *Frontiers in Plant Science* 6: 1243.

Yang, Fan, Liu Chao, Gao Fengqing, Su Mingyu, Wu Xiao, Zheng Lei, Hong Fashui, and Yang Ping 2007. "The improvement of spinach growth by nano-anatase TiO_2 treatment is related to nitrogen photoreduction." *Biological Trace Element Research* 119 (1): 77–88.

Yanık, Fatma and Vardar Filiz 2015. "Toxic effects of aluminum oxide (Al_2O_3) nanoparticles on root growth and development in Triticum aestivum." *Water Air Soil Pollution* 226 (8): 296.

Zahedi, Seyed Morteza, Abdelrahman Mostfa, Hosseini, Marjan Sadat, Hoveizeh, Narjes Fahadi, and Tran, Lam-Son Phan 2019. "Alleviation of the effect of salinity on growth and yield of strawberry by foliar spray of selenium-nanoparticles." *Environmental Pollution* 253: 246–258.

Zhao, Lijuan, Sun Youping, Hernandez-Viezcas Jose A., Servin Alia D., Hong Jie, Niu Genhua, Peralta-Videa Jose R., Duarte-Gardea Maria, and Gardea-Torresdey Jorge L. 2013. "Influence of CeO_2 and ZnO nanoparticles on cucumber physiological markers and bioaccumulation of Ce and Zn: a life cycle study." *Journal of Agricultural and Food Chemistry* 61 (48): 11945–11951.

Zheng, Lei, Hong Fashui, Lu Shipeng, and Liu Chao 2005. "Effect of nano-TiO_2 on strength of naturally aged seeds and growth of spinach." *Biological Trace Element Research* 104 (2): 83–91.

Zuverza-Mena, Nubia, Armendariz, Raul, Peralta-Videa, Jose R., and Gardea-Torresdey, Jorge L. 2016. "Effects of silver nanoparticles on radish sprouts: root growth reduction and modifications in the nutritional value". *Frontiers in Plant Science*, 7: 90.

6 Emerging Trends in Pest Management in Agriculture

Future Perspectives

Rameswar Singh Rattan
Mahindra Summit Agriscience Ltd

Vinay Kumar
Central University of Punjab

CONTENTS

6.1 INTRODUCTION

Pests are an important biotic stress for decreasing agricultural productivity. Today, the major problem in agricultural pest management (Deguine et al. 2021, Dara 2019) is the development of suitable agents that can kill insects physically. With the onset of the green revolution (Boethling 1996, De Vito 1996, Garlett 1996), the use of chemicals such as herbicides, pesticides and fertilizers has positively impacted crop yields. However, indiscriminate use of these agricultural inputs has also led to increased insect resistance, resurgence, biomagnification and loss of yield. The reduced yield warrants the importation of food in addition to bringing more land under farming (Smith et al., 2008). Crop yields are not increasing in proportion to world growth; hence, there is a need for novel approaches to meet the challenges. With restricted land and the availability of agricultural inputs, novel technologies such as biotechnology, green chemistry and nanotechnology are required to meet the food demands of the growing population of the world. With better systems to monitor environmental impact and the delivery of nutrients or insecticides, new technologies will enhance

DOI: 10.1201/9781003397861-6

our knowledge of the biology of different crops, leading to increases in yields or nutritional values in addition to protecting crops from insect pests and diseases.

Agricultural biotechnological techniques (Alemu 2020) have shown promising potential for agricultural sustainability and the production of safe foods (van den Bergh and Holley, 2002). One can argue that the application of modern agricultural biotechnology might stimulate food production (Vitousek et al., 1986). The technologies have generated considerable debate, with some supporting the view that they can sustain the growing population with the high potential of transgenic crops in reducing the severity of human hunger. However, adoption and acceptance of these technologies are critical to providing choice and accurate information to consumers. The present article addresses the prospects of these technologies in terms of their adoption.

6.2 TRENDS IN CROP PROTECTION

6.2.1 Modern Crop Protection Chemicals

The population of the world is increasing, and so is the demand for food. Therefore, it has become imperative to find sustainable and environmentally safe solutions through modern chemistry by evaluating current shortcomings as well as probable future scenarios. Organic chemistry has played an important role in insect pest management in agriculture (Smith et al., 2008) and will continue to guide the discovery of novel molecules through modern synthetic chemistry methods and technological interventions targeting biological efficacy as well as sustainable environmental protections (Anastas and Kirchhoff, 2002). The principles of mechanistic toxicology allow molecular designs that are safe for humans and animals, more selective and less persistent in the environment (Dow AgroSciences, 2000). In the future, sustainable crop protection (Kalogiannidis et al., 2022, Rodham 2000) by novel molecules will play an important role through a targeted and holistic approach by combining various screens with high throughput.

Smith et al. (2008) explained that high-throughput methods should focus on screening commercial targets, analogue chemistry with known activity, phytochemicals (cues from nature) and designing methodologies targeting the biochemical process at a molecular level to discover modern crop protection compounds. Remarkable achievements have been made in insecticide discovery and development of successful commercial crop protection products by using holistic methodologies, e.g., avermectins (fermentation products of *Streptomyces avermitilis*), emamectin, pyrethroids, nicotinoids, spinosad (fermentation product of *Saccharopolyspora spinosa*), milbemectin (isolated from *Streptomyces hygroscopicus*), etc. (Copping and Menn, 2000; Thompson et al., 2000). It has become vital that new technologies used in the design of novel crop protection compounds not only focus on superior efficacy but also on human health and environmental safety.

6.2.2 GM Crops and Insect Pest Management

In recent years, biotechnology has emerged as a successful alternative (not a replacement) to synthetic chemicals. Genetic engineering enables the addition of desired traits by transferring target genes from one organism to another or modifying genes in an organism. There is a great advantage to this technology embedded in the seed as per conventional local farming practices, as it provides biological and economic advantages (Duke, 2005; Gianess, 2005). Its application may be considered as an appropriate tool to solve agricultural problems. It reduces the usage of pesticides and helps to increase yields by providing better protection against insect pests and diseases (Alhassan, 2003; James, 2004). It may help to enhance the nutritional value and taste of the food by using appropriate techniques and materials.

The decreased use of pesticides can be beneficial for improving diverse and beneficial insect populations, in addition to environmental benefits such as reduced leaching of pesticides in water, less disposal waste and a lower fuel requirement for pesticide application. Moreover, crop yields can be improved by using its application in a more efficient way for better productivity (Bridges et al., 2003). Vaughan (2003) discussed that insect resistant GM crops can act as partial substitutions of *Bt* genes or alternative genes for synthetic pesticides and highlighted the need to create an insect protein gene pool with different modes of action. A major use of biotechnological applications in agriculture is to develop herbicide-tolerant, insect-pest-resistant and disease-resistant crops (Altieri, 1998). This can be achieved by inserting monogenic traits and could be a profitable opportunity for corporates. An added benefit of transgenic crops could be farming on marginal lands by developing crops with polygenic traits that could grow on salty, dry or acidic soils. However, it is a challenging and slow process due to multiple genes. Despite their numerous benefits, there have been apprehensions regarding transgenic crops (Qaim 2009). The case of the *Bt* (*Bacillus thuringiensis*) gene that produces an insect-killing toxin. Transgenic crops produced by using *Bt* genes and insect pests against which such modifications provide protection are tabulated in Table 6.1. These crops have already reached the end user.

Several other crops, such as black gram, chickpea, banana, wheat, papaya, soybean, squash and tobacco, are also being modified with the *Bt* gene or alternative gene for insect tolerance and other traits. Public apprehension regarding *Bt* corn is huge, as studies showed high mortality of monarch butterflies because of genetically modified corn pollen (Losey et al., 1999). Another study showed that transgenic maize (Cry 1Ab and Cry3Bb1) does not affect *Chrysoperla carnea* or ladybird beetles (Li et al., 2008). Environmentalists have mentioned the monarch butterfly as a prime example of the ecological risks associated with transgenic crops. However, one should not haste in declining genetically modified crops, as detailed research is required for conclusive evidence of negative effects. Moreover, research has shown that the comprehensive effect in the case of the monarch butterfly is not perturbing (MacKenzie, 1999).

6.2.3 Nanotechnology and Nanopesticides Formulations

Nanotechnology and nanoparticle engineering have become important technological interventions in the research and development of pharmaceuticals since the mid-seventies (Rosen and Abribat, 2005). Implementation of nanotechnology in precision farming was initiated around the 1980s (Rickman et al., 1999) to augment production and reduce the use of farm inputs. The most exciting examples are in the development of

TABLE 6.1
Transgenic Crops, Gene Used and Protection Against Various Insect Pests

Crops	Gene	Product Name/ Status	Protection (Against)	Scientific Names	Company
Bt Cotton	Cry 1Ac	Bollgard, InGard	Cotton ballworm complex	*Heliothis virescence*, *Helicoverpa zea*, *Pectinophora gossypiella*, *Helicoverpa armigera*, *Earias vittella* and *E. insulana*	Mahyco-Monsanto, Monsanto
	Cry 1Ac Cry 2Ab2	Bollgard II	Cotton ballworm complex and bud worm	*Helicoverpa armigera*, *Earias vittella* and *E. insulana*, *Spodoptera exigua* and *S. frugiperda*	Mahyco-Monsanto
	Cry1Ac & Cry1F	–	Cotton ballworm complex	*Helicoverpa armigera*, *Earias vittella* and *E. insulana*	Dow AgroSciences
	Cry1Ac and Cry1EC	–	Cotton ballworm complex	*Helicoverpa armigera*, *Earias vittella* and *E. insulana*	JK Agrigenetics Ltd.
	Fusion of Cry 1Ab, Cry 1Ac	–	Cotton ballworm complex	*Helicoverpa armigera*, *Earias vittella* and *E. insulana*	Nath Seeds
Brinjal	Cry1Ac	Ninja	Fruit borer, fruit and shoot borer	*Helicoverpa armigera*, *Leucinodes arbonalis*	Mahyco
Canola	Cry1AC	UGA188-20B	Diamondback moth and	*Plutella xylostella*, *Trichoplusia ni*	
Cauliflower	Cry 1AC	Na (Level 1 trial)	Diamondback moth	*Plutella xylostella*	Sungro Seeds Research Ltd
Bt corn (Maize)	Cry1Ab	Mon810, Bt 11, Bt176 Campa, YieldGard, MaisGard, NatureGard, Knockout	European stem borer	*Ostrinia nubilalis*	Monsanto, Novartis, Mycogen, Syngenta,
	Cry 1Ac	Bt-Xtra	European stem borer	*Ostrinia nubilalis*	Monsanto
	Cry 9Ca	Libertylink, Starlink	European stem borer	*Ostrinia nubilalis*	AgrEvo
	Cry 3Bb	–	Corn rootworms	*Diabrotica* spp	–
Rice	Cry1Ac	–	–	–	Mahyco
	Cry1Ab Cry1Ac	–	–	–	Bayer Cropsciences
Pigeon pea	Cry1Ab	–	Pod Borer	*Helicoverpa armigera*	ICRISAT
Potato	Cry3Ab, Cry3Aa	NewLeaf	Colorado potato beetle	*Leptinotarsa decemlineata*	Monsanto

nanopesticides (Kannan et al., 2022). Subsequently, this technology was used in agrochemical formulations and delivery systems. Nanotechnology promises the development of pesticides with improved activity and their use in targeted applications. Researchers believe that applications of nanopesticides will lead to decreased pesticide use.

The reformulation of the active ingredients of patented pesticides into nano-scale formulations for agri-chemicals may help companies extend the validity of their patents (Friends of the Earth, 2008). Another way is to encapsulate pesticides into nano-capsules, as they offer better control over the release of pesticides. Many nano-agrochemicals were initially developed by reprocessing existing pesticides, fungicides and seed treatment chemicals by either minimizing the particle size of chemical formulations to nanosize or nanoencapsulating active ingredients (ETC Group, 2004; Green and Beestman, 2007; Joseph and Morrison, 2006, Mulqueen 2003). Such capsules break open in response to specific triggers such as heat, sunlight or an alkaline environment, thus making them more potent (ETC Group, 2004). Nanoencapsulation provides inbuilt pesticides for crops as well as triggers to control the release and availability of pesticides (Kumari et al., 2010). The released toxin is then readily absorbed by the plant and is not easily run off in water or soil due to irrigation and rain.

Nanopesticides (Deka et al., 2021; Konappa et al., 2021; Chaudhary 2023) are now widely used in agricultural formulations containing 100–250 nm nanoparticles for commercial purposes. Small particles enhance its water solubility, thereby improving its action potential. Some nanoemulsions are water- or oil-based and contain uniform suspensions of 200–400 nm pesticide nanoparticles (Rickman et al., 1999; Joseph and Morrison, 2006). Likewise, nanosilica mediated gene delivery in plants and pest control have gained more emphasis. Some of the recently studied pesticidal nanoformulations are tabulated in Table 6.2.

It has become possible to control various pests such as white fly (*Trialeurodes vaporium*), mustard beetle (*Phaedon cochleariae*), coconut mite (*Dermanyssus gallinae*) and granary weevil (*Sitophilus granaries*) by electrostatically spraying amorphous silica (Ulrichs et al., 2006, 2009). Nanosilica did not show any adverse effect on plant, fish or silkworm physiology. The nanostructured silica can kill arthropod pests based on the physico-sorption of lipid from the insect cuticle. Admixture experiments with silica and different synthetic pesticides show that some of them have synergistic effects and can be applied in combination in pest management programme. Immunity in beneficial insects can be developed using nanosilica. This was observed in the case of *Bombyx mori*, where grasserie disease caused by nuclear polyhedrosis virus (NPV) could be controlled with the help of lipophilic amorphous silica nanoparticles (LASN) and live BmNPV conjugate as a drug (Barik et al., 2008). The IARC (International Agency for Research on Cancer) has rated amorphous silica dust as non-carcinogenic. The USDA has also considered amorphous silica as safe for human.

6.3 ENVIRONMENTAL AND SOCIO-ECONOMIC IMPACT OF EMERGING TECHNOLOGIES

Certainly, these emerging technologies may facilitate and enhance agricultural production. However, environmental concerns are also related to these technologies and their applications. The production of crop protection molecules needs to be revamped as the process mitigates environmental and health benefits.

TABLE 6.2
Different Pesticidal Nanoformulations and Their Properties

Type of Nanoparticle	Size (nm)	Pesticide	Encapsulatin Efficiency (%)	Release	Advantage	Activity	Reference
Poly(ε-caprolactone)	181–407	Essential oil	75.8–84.4	Sustained release	–	103.53 and 112.64 μL/L	Ahsaei et al. (2020) Boehm et al. (2000)
Cellulose nanocrystals	8.9 ± 1.7 nm in width and 78.4 ± 19.1 nm in length	Emamectin benzoate	87.5	–	–	0.31 μg/mL	Elabasy et al. (2019)
Porous hollow silica nanoparticles	100	Avermectin	58.3	Sustained release pattern	–	–	Wen et al. (2005)
	60–100	Avermectin	62.5	Sustained release	Photostability of avermectin	–	Li et al. (2007)
Diacyl hydrazine							
	42–138	–	–	–	Higher stability	0.015 mg L^{-1}	Pandey et al. (2020)
Polyvinylpyridine (PVPy) and polyvinylpyridine-co-styrene	100–250	Tebuconazole and chlorothalonil		Initial fast followed by slow release	–	*G. trabeum*	Li et al. (2001)
Ag/TiO_2	100	Chlorfenapyr	–	Nanoformulation had strong virulence than nanoformulation	Stable in darkness	The effect of preventing from cabbage *Spodoptera litura* was better than chlorfenapyr emulsifiable concentrate	Yan et al. (2006)
Chitosan and sodium alginate	–	Imidacloprid	81.57	Slow release	Thermodynamically stable after encapsulation	Toxicity was higher against *Martianus dermestoides* adults than normal imidacloprid	Guan et al. (2008)

The development of new molecules through modern chemistry for sustainable crop protection may offer efficient and safe production of molecules through improved processes (Donlar Corporation, 1996; Sparks and Lorsbach, 2017). With a substantial degree of chemical application, new commercial crop varieties exhibiting protection against insect pests and diseases and improved yields are being bred and developed. The adoption of gene-based technology has been unprecedented, with the success of some traits in a few crops (James, 2004). Likewise, nanoparticle-based pesticide applications, with their improved absorption into plants and bioavailability of the active ingredients of the compounds, will have more efficacy compared to traditional formulations. As such, they may not be washed off as readily, thereby increasing their effectiveness. However, at the same time, they may pose new challenges and hazards to consumers and farm workers (Colvin, 2003; Scrinis, 2006; Scrinis and Lyons, 2007).

The smaller size and improved solubility of nanopesticides may have better efficacy; however, they may contaminate soils, natural water resources and food chains across a larger area in different climates. Additionally, toxins from nanopesticides can be released into living organisms as well as the environment. The Environment Protection Agency of the USA reiterated that silver ions from laundry released in washing machines are equivalent to pesticides, though not very hazardous, and earlier stated that a major federal pesticide law must not be applied to washing machines as they are considered household appliances. Similarly, FAO conceded to establishing a connection between food safety and the use of nanotechnologies in the food and agriculture sectors through scientific studies on existing and emerging applications of nanotechnologies in the food and agriculture sectors (FAO/WHO, 2009). However, little has been done to identify the possible suspected hazards of nanomaterials to workers and consumers (Nel et al., 2006; Mueller and Nowack, 2008). Indeed, creating an effective oversight system for the synthesis of new products, biotech crops and nanotechnology requires policymakers to focus on appropriate risk assessment and management assumptions with respect to existing health and ecological regulations that will translate from the macro to the nano world.

The new technological developments have opened doors to the agrochemical industry in the last decade and have experienced tremendous growth in the global market (James, 1999, 2002; Roco 2003; Aitken et al., 2006). The economic impacts of biotech, nanotech and modern chemistry developments will affect farmers differently, based on the size of their land holdings and the amount of capital invested. This will affect farmers in the long run in the developing world through intensification in agriculture and low input prices with high productivity. Biotech crops have a significant role in increasing global production of crops such as soybeans, corn, cotton, canola, etc. Biotechnology is beneficial to farmers and consumers, with advantageous effects for the environment and human health. It will be a large-scale and capital-intensive farming operation that will reap early profits (ETC Group, 2004; Smith et al., 2008). However, companies must consider equally the economic, environmental and social components of sustainability (Reinert, 2001) while working on new technologies and work simultaneously with governments in framing policies.

6.4 OUTPUTS: SUSTAINABLE AGRICULTURE

Sustainable agricultural growth is pivotal for developing countries. It is important to adopt a holistic approach to sustain the growth that biotechnology, modern synthetic chemistry and nanotechnology research can offer. A diligent system is required to deliver innovations based on newer technologies to the final product development stage and ensure their reach to the end users of the agriculture value chain (Sastry et al., 2007), which can help alleviate poverty in the developing world at large. The prospective applications of nanotechnology are extensive and broad, but their current usage in the development of pesticide formulations is limited. However, nanotechnology-based products and their applications in agriculture will continue to grow as these technological interventions have a very high probability of solving current challenges faced by industry. In contrast, increased efficacy may provide positive validation, as in the case of the success of *Bt* crops.

Shelton (2010) argued that *Bt* brinjal provided superior protection against fruit and shoot borer, but other arthropod pests, *viz.* mites, jassids, mealybugs, whiteflies and leaf-eating beetles, still need to be managed (Brookes and Barfoot, 2009). However, pesticide usage has been reduced by 80% against fruit and shoot borer since the introduction of the *Bt* concept (Krishna and Qaim, 2008). Through the convergence of nanotechnology and biotechnology, it may be possible to strengthen the delivery of new traits to crop plants through genetic engineering and breeding programmes. In addition, researchers are trying to introduce foreign DNA and chemicals into cells, e.g., silica nanoparticles (Torney et al., 2007; Friends of the Earth, 2008).

6.5 CONCLUSION

Recent developments in biotechnology, molecular biology and metabolic engineering have provided ways to alter endogenous biochemical pathways to develop GM crops for specific purposes (Kinney, 1998, Mandel 2004). Additionally, molecular biology and metabolic engineering can contribute to the development of future pesticides with novel modes of action. Hence, there is a dire need to understand the gap and map unknown pathways and processes with respect to their impact on the environment and human health. Although the consequences of the spread of biotechnology are not yet known, commentators identify the failure to anticipate the debate surrounding genetically modified organisms as a mistake that the nanotechnology industry should avoid. Academic and research organizations around the world are racing to commercialize these technologies. However, there is sufficient evidence that the new materials may have significant health, safety and environmental risks as well as significant social, economic and ethical challenges that the new technology product constitutes, thereby requiring procedures and a feedback system to avoid the similar pitfalls of past novel materials and technologies.

REFERENCES

Ahsaei, S.M., K. Talebi-Jahromi, and G. Amoabediny. 2022. "Insecticidal activity of polycaprolactone nanoparticles decorated with chitosan containing two essential oils against Tribolium confusum." *International Journal of Pest Management* 68, no. 3: 237–245.

Aitken, R.J., M.Q. Chaudhry, A.B.A. Boxall, and M. Hull. 2006. "Manufacture and use of nanomaterials: current status in the UK and global trends." *Occupational Medicine* 56, no. 5: 300–306.

Alemu, M. 2020. "Trend of biotechnology applications in pest management: a review." *International Journal of Applied Sciences and Biotechnology* 8, no. 2: 108–131.

Alhassan, W.S. 2003. "*Agrobiotechnology Application in West and Central Africa (2002 Survey Outcome)*". IITA: Ibadan.

Altieri, M.A. 1998. "The Environmental Risks of Transgenic Crops: An Agroecological Assessment. https://www. pmac. net/miguel. htm/. Accessed 18 September 1999.

Anastas, P.T., and M.M. Kirchhoff. 2002. "Origins, current status, and future challenges of green chemistry." *Accounts of Chemical Research* 35, no. 9: 686–694.

Barik, T.K., B. Sahu, and V. Swain. 2008. "Nanosilica —from medicine to pest control." *Parasitology Research* 103: 253–258.

Boehm, A.-L.L.R., R. Zerrouk, and H. Fessi. 2000. "Poly epsilon-caprolactone nanoparticles containing a poorly soluble pesticide: formulation and stability study." *Journal of Microencapsulation* 17, no. 2 : 195–205.

Boethling, R.S. 1996. "Designing biodegradable chemicals." In: *Designing Safer Chemicals: Green Chemistry for Pollution Prevention:* DeVito, S.C., Garlett, R.L; Eds: Chapter 8. American Chemical Society, Washington, DC.

Bridges, A., R. Kimberly, M. Magin, and J.W. Stave. 2003. "Agricultural Biotechnology (GMO). Methods of Analysis." In: *Food Analysis*. 3rd Edition. Kluwer Academic/ Plenum publishers, New York, 301–311.

Brookes, G., and P. Barfoot. 2009. "Global impact of biotech crops: income and production effects, 1996–2007". *AgBioForum* 12: 184–208.

Chaudhary, A. 2023. "Nanopesticides: a promising nanoagrochemicals in crop protection." *Chemical Science Review and Letters*, ISSN 2278-6783.

Colvin, V. 2003. "The potential environmental impact of engineered nanomaterials". *Nature Biotechnology* 21(10): 1167–1170.

Copping, L.G., and J.J. Menn. 2000. "Biopesticides: a review of their action, applications and efficacy." *Pest Management Science: Formerly Pesticide Science* 56, no. 8 (2000): 651–676.

Dara, S.K. 2019. "The new integrated pest management paradigm for the modern age." *Journal of Integrated Pest Management* 10, no. 1: 12.

Deguine, J.P., J.N. Aubertot, R.J. Flor, F. Lescourret, K.A.G. Wyckhuys, and A. Ratnadass. 2021. "Integrated pest management: good intentions, hard realities. A review." *Agronomy for Sustainable Development* 41, no. 3: 38.

Deka, B., A. Babu, C. Baruah, and M. Barthakur. 2021. "Nanopesticides: a systematic review of their prospects with special reference to tea pest management." *Frontiers in Nutrition* 8: 686131.

DeVito, S.C. 1996. "General principles for the design of safer chemicals: toxicological considerations for chemists" In: *Designing Safer Chemicals: Green Chemistry for Pollution Prevention:* DeVito, S.C., Garlett, R.L., Eds: Chapter 2. American Chemical Society, Washington, DC.

Donlar Corporation. 1996. "Production and Use of Thermal Polyaspartic Acid. In: The Presidential Green Chemistry Challenge Award Program: Summary of 1999 Award entries and recipients. EPA744-K-96-001. US Environmental Protection Agency, Office of Pollution Prevention and Toxins, Washington, DC.

Dow AgroSciences LLC. 2000. "Spinosad: A New Natural Product for Insect Control. In: The Presidential Green Chemistry Challenge Award Program: Summary of 1999 award entries and recipients. EPA744-R-00-001. US Environmental Protection Agency, Office of Pollution Prevention and Toxins, Washington, DC.

Duke, S.O. 2005. "Taking stock of herbicide resistant crops ten years after introduction". *Pest Management Science* 61: 211–218.

Elabasy, A., A. Shoaib, M. Waqas, M. Jiang, and Z. Shi. 2019. "Synthesis, characterization, and pesticidal activity of emamectin benzoate nanoformulations against Phenacoccus solenopsis Tinsley (Hemiptera: Pseudococcidae)." *Molecules* 24, no. 15: 2801.

ETC Group. 2004. "Down on the farm: the impact of nano-scale technologies on food and agriculture."

FAO/ WHO. 2009. "Food and Agriculture Organization of the United Nations/World Health Organization. FAO/WHO Expert Meeting on the Application of Nanotechnologies in the Food and Agriculture Sectors: Potential Food Safety Implications: Meeting Report. Rome. 104.

Friends of the Earth. 2008. "Nanotechnology in Food and Agriculture." *Friends of the Earth Australia.*

Garlett, R.L. 1996. "Pollution prevention, green chemistry and the design of safer chemicals. In: *Designing safer Chemicals: Green Chemistry for Pollution Prevention:* DeVito, S.C., Garlett, R. L., Eds: Chapter 1. American Chemical Society, Washington, DC.

Gianess, L.P. 2005. "Economic and herbicide use impacts of glyphosate resistant crops". *Pest Management Science* 61: 241–245.

Green, J.M., and G.B. Beestman. 2007. "Recently patented and commercialized formulation and adjuvant technology". *Crop Protection* 26: 320–327.

Guan, H., D. Chi, J. Yu, X. Li. 2008. "A novel photodegradable insecticide: preparation, characterization and properties evaluation of nano-imidacloprid." *Pesticide Biochemistry and Physiology* 92: 83–91.

James, C. 1999. "*Preview Global Review of Commercialized Transgenic Crops.*" ISAAA Briefs: Makati City.

James, C. 2002. "*Preview: Global Status of Commercialized Transgenic Crops.*" ISAAA Briefs No. 27. ISAAA: Ithaca, NY.

James, C. 2004. "*Preview: Global Status of Commercialized Biotech and GM Crops.*" ISAAA Brief 32.Joseph, T., and M. Morrison. 2006. "Nanotechnology in Agriculture and Food. Nanoforum Report." https://www.nanoforum.org/dateien/temp/nanotechnology%20in%20agriculture%20and%20food.pdf?08122006200524. Accessed January 2010.

Kalogiannidis, S., D. Kalfas, F. Chatzitheodoridis, and O. Papaevangelou. 2022. "Role of crop-protection technologies in sustainable agricultural productivity and management." *Land* 11, no. 10: 1680.

Kannan, M., Mohan, M., Elango, K., Govindaraju, K., and Mani, M. 2022. "Principles and application of nanotechnology in pest management." In: *Trends in Horticultural Entomology:* Mani, M. Ed., 49–79. Springer, Singapore.

Kinney, A.L. and Knowlton, S. 1998. "Designer oils: The high oleic acid soybean." In: *Genetic Modification in the Food Industry*: Roller, S., Harlander, S. Eds. Blackie, London, 193–213.

Konappa, N., S. Krishnamurthy, U.C. Arakere, S. Chowdappa, R. Akbarbasha, and N.S. Ramachandrappa 2021. "Nanofertilizers and nanopesticides: recent trends, future prospects in agriculture." In: *Advances in Nano-Fertilizers and Nano-Pesticides in Agriculture*, 281–330. Woodhead Publishing, Cambridge.

Krishna, V.V., and M. Qaim. 2008. "Potential impacts of Bt eggplant on economic surplus and farmers' health in India." *Agricultural Economics* 38: 167–180.

Kumari, A., S.K. Yadav, and S.C. Yadav. 2010. "Biodegradable polymeric nanoparticles based drug delivery systems." *Colloids and Surfaces B: Biointerfaces* 75: 1–18.

Li, Y., M. Meissle, and J. Romeis. 2008. "Consumption of Bt maize pollen expressing Cry1Ab or Cry3Bb1 does not harm adult green lacewings, Chrysoperla carnea (Neuroptera: Chrysopidae)." *PLoS One* 3, no. 8: e2909.

Li, Y., L. Yan, P. Heiden, and P. Laks. 2001. "Use of nanparticles for controlled release of biocides in solid wood." *Journal of Applied Polymer Science* 79: 458–465.

Li, Z.Z., J.F. Chen, F. Liu, A.Q. Liu, Q. Wang, H.Y. Sun, and L.X. Wen. 2007. "Study of UV-shielding properties of novel porous hollow silica nanoparticle carriers for avermectin." *Pest Management Science* 63, no.3: 241–246.

Losey, John E., Linda S. Rayor, and Maureen E. Carter. 1999. "Transgenic pollen harms monarch larvae." *Nature* 399, no. 6733: 214–214.

MacKenzie, D. 1999. "Unpalatable truths." *New Scientist* 162, no. 2182: 18–19.

Mandel, G.N. 2004. "Gaps inexperience, inconsistencies, and overlaps: crisis in the regulation of genetically modified plants and animals". *William and Mary Law Review* 45: 2167, 2175.

Mueller, N.C., and B. Nowack. 2008. "Exposure modeling of engineered nanoparticles in the environment." *Environmental Science & Technology* 42, no. 12: 4447–4453.

Mulqueen, M. 2003. "Recent advances in agrochemical formulations." *Advances in Colloid and Interface Science* 106: 83–107.

Nel, A., T. Xia, L. Madler, and N. Li. 2006. "Toxic potential of materials at the nanolevel." *Science* 311, no. 5761: 622–627.

Pandey, A., S. Srivastava, N. Aggarwal, C. Srivastava, A. Adholeya, and M. Kochar. 2020. "Assessment of the pesticidal behaviour of diacyl hydrazine-based ready-to-use nanoformulations." *Chemical and Biological Technologies in Agriculture* 7, no. 1: 1–15.

Qaim, M. 2009. "The economics of genetically modified crops." *Annual Review of Resource Economics* 1: 665–693.

Reinert, K.H. 2001. "Integrating green chemistry and sustainability into a specialty chemical company." *Pure and Applied Chemistry* 73, no. 8: 1269–1272.

Rickman, D., J.C. Luvall, J. Shaw, P. Mask, D. Kissel, and D. Sullivan. 1999. "Precision Agriculture: Changing the Face of Farming.". www.ghcc.msfc.nasa.gov/precisionag/. Accessed March 27, 2023.

Roco, M.C. 2003. "Broader societal issues of nanotechnology." *Journal of Nanoparticle Research* 5: 181–189.

Rodham, D.K. 2000. "Colloid and interface science in formulation research for crop protection products." *Current Opinion in Colloid & Interface Science* 5, no. 3–4: 280–287.

Rosen, H., and T. Abribat. 2005. "The rise and rise of drug delivery." *Nature Reviews Drug Discovery* 4, no. 5: 381–385.

Sastry, R.K., N.H. Rao, R. Cahoon, and T. Tucker. 2007. "Can Nanotechnology Provide the Innovations for a Second Green Revolution in Indian Agriculture?" Paper presented at the NSF Nanoscale Science and Engineering Grantees Conference, December 3–6.

Scrinis, G. 2006. "Nanotechnology and the environment: the nano-atomic reconstruction of nature." *Chain Reaction* 97: 23–26.

Scrinis, Gyorgy, and Kristen Lyons. 2007. "The emerging nano-corporate paradigm: nanotechnology and the transformation of nature, food and agri-food systems." *The International Journal of Sociology of Agriculture and Food* 15, no. 2: 22–44.

Shelton, A.M. 2010. "The long road to commercialization of Bt brinjal (eggplant) in India." *Crop Protection* 29: 412–414.

Smith, K., D.A. Evans, G.A. El-Hiti. 2008. "Role of modern chemistry in sustainable arable crop protection." *Philosophical Transactions Royal Society. B: Biological Sciences* 363: 623–637.

Sparks, T.C., and B.A. Lorsbach. 2017. "Agrochemical discovery: Building the Next Generation of Insect Control Agents." In: Gross et al., Advances in Agrochemicals: Ion Channels and G Protein-Coupled Receptors (GPCRs) as Targets for Pest Control. ACS Symposium Series; American Chemical Society: Washington, DC.

Thompson, G.D., R. Dutton, and T.C. Parks. 2000. "Spinosad—a case study: an Example from a natural product discovery program." *Pest Management Science* 56: 696–702.

Torney, F., B. Trewyn, V.S.Y. Lin, and K. Wang. 2007. "Mesoporous silica nanoparticles deliver DNA and chemicals into plants." *Nature Nanotechnology* 2: 295–300.

Ulrichs, C., F. Krause, T. Rocksch, A. Goswami, and I. Mewis. 2006. "Electrostatic application of inert silica dust based insecticides onto plant surfaces." *Communications in Agricultural and Applied Biological Sciences* 71(2): 171–178.

Ulrichs, C., T. Mucha-Pelzer, C. Büttner, I. Mewis, E. Scobel, and E. Bauer. 2009. "New approaches in the control of insect pests in tomato." *Acta Horticulturae (ISHS)* 821: 189–200.

van den Bergh, J.C.J.M., and J.M. Holley. 2002. "An environmental–economic assessment of genetic modification of agricultural crops." *Futures* 34: 807–822.

Vaughan, G.M. 2003. "Plants and protection against insects: alternative strategies based on gene technology." *Acta Agriculturae Scandinavica Section B: Plant Soil Science* 53 (4): 34–40.

Vitousek, P.M., P.R. Ehrlich, A.H. Ehrlich, and P.A. Matson. 1986. "Human appropriation of the products of photosynthesis." *Bioscience* 34: 368–73.

Wen, L.X., Z.Z. Li, H.K. Zou, A.Q. Liu, and J.F. Chen. 2005. "Controlled release of avermectin from porous hollow silica nanoparticles." *Pest Management Science* 61: 583–590.

Yan, J., K. Huang, Y. Wang, and S. Liu. 2006. "Application of anti-pollution nano pesticides preparations for chlorfenapyr." *Journal of Chemical Industry and Engineering* 57: 91–96.

7 Nanosensors for the Detection of Contaminants in Aqueous Water

Gamze Ersan and Mahmut Selim Ersan
Arizona State University

CONTENTS

7.1 INTRODUCTION

Due to anthropogenic activities, more than 10 million tons per year of toxic chemicals, including pharmaceuticals, heavy metals, pathogens, pesticides, and per- and polyfluoroalkyl substances (PFAS), have been released into aquatic systems. With increasing urbanization and industrialization, new groups of chemicals are constantly being produced, and the quantification of these toxic chemicals is an urgent need for understanding their fate in the environment. The main water contaminant issues and their global challenges have been listed in Table 7.1. Contaminant types have been changing depending on wastewater utility providers and source types. Thus, water quality detection/monitoring have covered various issues/challenges.

The standard testing methods for these contaminants using chromatography (Fang et al., 2015), atomic adsorption spectroscopy (Richardson and Ternes, 2015), ICP-MS (Richardson and Ternes, 2015), electrophoresis (Knob et al., 2012), or an ion-selective electrode (Cheng et al., 2016) are in general expensive, time-consuming, complex sample preparation procedures, and not convenient for on-site application. Therefore, robust technologies are being developed for the identification and quantification of contaminants in water. For site studies, several sensors have been implemented to help prescreen potential contaminants rapidly (Whitesides, 2006; Cheng et al., 2016;

DOI: 10.1201/9781003397861-7

TABLE 7.1
Current Contaminants in Aqueous Water, Their Water Quality Issue, and Challenges

Contaminants	Sources	Water Quality Issue	Challenges
Persistent organic contaminants (PAHs, PCBs etc.)	Agriculture, wastewater disposal sites, fuel combustion, sewage, and sludge	Long biologic/ environmental persistent, bioaccumulations to the food web, slow breakdown over time, toxic, adverse health impacts	Long-lasting chemical, persistent in environment, break down very slowly
Inorganic contaminants – heavy metals (i.e., Cr, Pb, As, etc.)	Agriculture, mining	Contaminant private wells through groundwater and surface water seepage and run off, chronic toxic, and adverse health impacts (e.g., high blood pressure and cancer)	Alternate sources of potable water including deep aquifers or rainwater, environmentally responsible mining techniques, and eco-friendly methods for removing impurities are also possible
Per- and polyfluoroalkyl substances (PFAS)	Fire frightening foam, food storages, landfills, waste disposal facilities, sewage, waste treatment plants, military bases, and airports	Long biologic/ environmental persistent, never break down over time, toxic, adverse health impacts	Long-lasting chemical, persistent in environment
Pesticides/fertilizers	Agriculture	Adverse health impacts (i.e., endocrine distributions)	Preventing the flow of pesticides from agricultural land
Pharmaceuticals (e.g., antibiotics, painkillers, antidepressant, etc.)	Industry, residential, sewage and urban wastewater,	Effects on rivers' ecotoxicology and public health	Advanced wastewater polishing treatment method using reverse osmosis/nanofiltration/ activated carbon
Viruses (i.e., SARS-CoV-2, COVID, Influenza) and microbiological pathogens	Urban wastewater	Adverse health effects	Drinking water disinfection techniques
Diverse contaminants (i.e., non-biodegradable plastics, oil, radioactive substances)	Hazardous waste, loss from storage facilities, spillage during transport	Long-term contamination of drinking water resources	Monitoring of mitigation processes such as natural attenuation

Source: Data obtained from Florescu et al. (2010), Schwarzenbach et al. (2010), Wambu and Ho (2016) and Ersan et al. (2023).

Fang et al. 2016). However, due to the complexity of water sources and the detection thresholds of contaminants that must be identified, creating a sensor is a difficult undertaking. Besides, water quality parameters including temperature, pH, dissolved organic matter, existing competing ions, and concentration of contaminants can have an influence on the identification and quantification of sensors. Physical, chemical, and biological factors may need to be taken into account for water quality monitoring with sensing technology, including temperature, color, density, conductivity, turbidity, total dissolved solids, disinfectants, phosphate and nitrogen contents, micronutrients, and other existing metal contaminants (Kruse, 2018). Thus, high demand exists for sensitive, selectivity, easy-to-use, cost-effective, environmentally friendly, rapidly operating, and portable operating sensors for real-time monitoring of contaminants in water. Current sensing technologies are colorimetric, electrochemical, and optic sensors, which have been commonly used for the detection of contaminants in water (Duke et al., 2010). A promising approach to water contamination detection is the combination of nanotechnology and cutting-edge water techniques in the manufacturing of sensors. So far, as presented in Figure 7.1, the researchers have been focused on several topics related to sensing technology for contaminant detection in aqueous water. Most researchers have worked on sensitivity and linearity, low detection limits, linear range issues, stability, and reproducibility of sensors.

7.2 NANOSENSORS FOR THE DETECTION OF CONTAMINANTS

Nanomaterials can be used in sensor technology for the detection of contaminants in aqueous water. Due to their high surface area to volume ratio, unique optical properties, and surface functionalization ability, nanomaterials can be a good candidate for nanosensor technology (Willner and Vikesland, 2018). So far, several nanomaterials such as semiconductor quantum dots (QTs), metal (silver/gold) nanomaterials, and carbon-based nanomaterials (graphene) have been created with distinct morphologies and structures for sensing anionic (F^-, CN^-, AsO_3^{-3}/AsO_4^{-3}, and PFOS/PFOA) and cationic (dyes and pesticides) contaminant detections in an aqueous environment (Griffin et al. 2009; Li et al., 2010; Wang et al. 2009; Eigler et al. 2013; Mosconi et al. 2015). Water quality nanosensors employed for sensing, such as colorimetric, electrochemical, and optical, are discussed in this chapter.

A large detecting area, an electrical connection, mechanical properties, conductivity, power consumption, and compactness are only a few of the performance criteria. The following less significant factors were overlooked: (i) operation at room temperature; (ii) fabrication techniques (i.e., careful substrate preparation to take advantage of the bulk properties); (iii) the degree of alignment of carbon nanotubes; (iv) distinction between pristine single-multiwalled carbon nanotubes, graphene and their various derivatives, including graphene oxide, reduced graphene oxides, and nitrogen-doped graphene; and (v) energy conservation and manufacturing costs, which are not anticipated to continue to be obstacles to the adoption of novel sensors in commercial applications for very long.

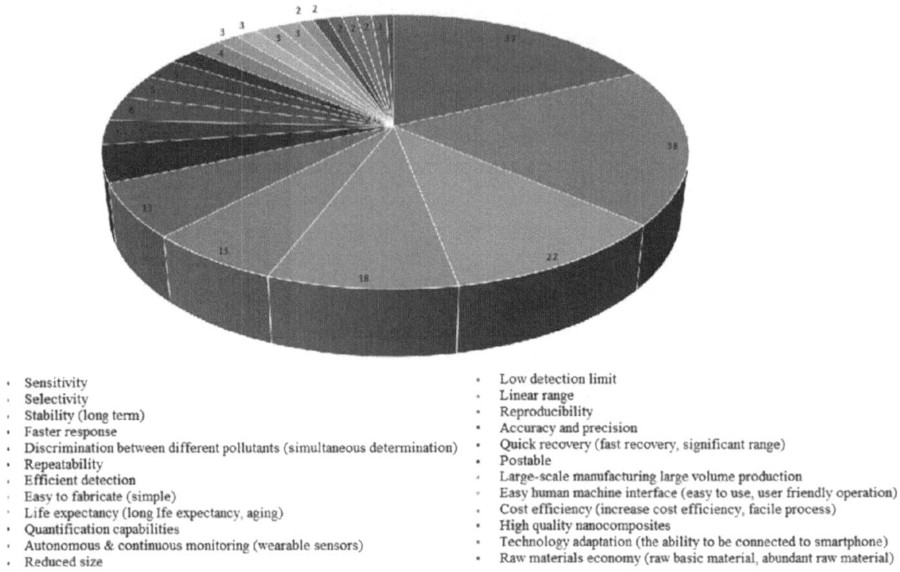

FIGURE 7.1 Number of published papers in the current literature for the topic of sensing technology.
*Published articles distribution between 2015 and 2022.

7.2.1 Nanomaterial-Based Optical Sensors for Colorimetric Detection

Colorimetric detection is the signal recognition method for the contaminants, which can be turned into a bare-eye sensitive color change. Because of its ease of use, rapid analysis, and cost-effectiveness, this approach to chemical and biological analysis is quite appealing (Wang et al., 2012). Due to their distinctive localized surface plasmon resonance, nanomaterials (particularly those made of gold and silver metals) have molar extinction coefficients up to five orders of magnitude lower than those of typical organic dyes in the visible spectrum (Willets and Van Duyne, 2007). The surface plasmon resonance is the most important optical property for each nanomaterial, which depends on the composition, particle size, hydrophobicity, aggregation, surface functionalization, interparticle distance, and dielectric environment (Kelly et al. 2003; Ghosh and Pal, 2007; Anh et al. 2022). The plasmon resonance frequency can also affect the color change of aqueous media. For instance, gold metal-based nanomaterials with a wine-red hue can have a high absorption band at 520 nm, and during their aggregation stages, the surface plasmon band underwent a considerable redshift, changing the color of the solution from red to blue (Zhang et al., 2014). Besides, the morphological differences of dispersed versus aggregated silver metal-based nanomaterials are related to the color change from yellow to brown (Vilela et al. 2012; Anh et al. 2022). On the other side, two strong absorption bands in the visible range may be seen in gold metal-based nanomaterials, and their colors are closely related to changes in the length-to-diameter ratio (Link and El-Sayed, 1999).

The aggregation stage of metal- or carbon-based nanomaterials can be triggered through crosslinking or non-crosslinking strategies. For the crosslinking approach, nanomaterials functionalized with ligand can act as a recognition probe for the detection contaminant, which can bind to the target contaminant. Due to the trigger's aggregation of nanomaterials, the interparticle distance is decreased (Anh et al., 2022). On the other side, the non-crosslinking method (also called de-protection) is based on the removal of stabilizing agents on the nanomaterial surface because of the existing of target contaminant. In this approach, metal-based nanomaterials (such as gold) are well protected by stabilizing agents to prevent nanomaterial aggregation. The only limitation of the non-crosslinking approach is losing coating agents of nanomaterials due to the higher affinity of target contaminants than linkers (or stabilizing agents), and thus their interaction prevents the binding of the linker and ligand on the functionalized nanomaterials. Consequently, this approach inhibits the aggregation of those nanomaterials (Mao et al., 2021). Overall, with only several steps of preparation and testing, the colorimetric sensing approach based on the aggregation of nanomaterials is a fast and simple strategy. The apparent color change can be monitored with the bare eye without any complicated instruments. However, the main limitation is that the aggregation of nanomaterials depends on several factors, including temperature, pH, existing ions, and their forces, which can have negative impacts on the detection of contaminants in water. Therefore, current researchers have been focusing more on electrochemical and fluorescent sensors for contaminant detection.

7.2.2 Nanomaterial-Based Electrochemical Sensor

Electrochemical sensors based on nanomaterials can measure changes in voltage, current, conductance, or impedance. These sensors also use nanomaterials as recognition elements to convert the signal into digital information. Basically, electrochemical sensors consist of three main electrodes including, working, reference, and counter (auxiliary) electrodes. The working electrodes can be the material surface where the reaction or process of interest happens. The electrochemical sensing approach happens at the working electrode and is initiated by the transfer of electrons between the electrode surface and the adsorbed contaminant on the surface of the electrode. So far, different electrochemical techniques such as potentiometry, amperometry, cyclic voltammetry, impedance spectroscopy, and conductometry (capacitive and resistive) have been used for electrochemical sensing technology due to the different detection mechanisms involved in electrochemical reactions (Wei et al., 2009). Electrochemical sensing techniques can deliver sensitive and trustworthy results for pollutant detection in waterways because of their cost-effectiveness, detection limitations, rapid real-time response, and ease of use. There are still certain gaps in knowledge regarding reproductivity, reusability, ease of fabrication, etc., even though this technique is a strong contender for in-process monitoring and detection of trace quantities of water contaminants in real time.

Besides, the geometric configuration, design and backbone materials of the working electrode can play a crucial role in electrochemical sensing technology. Nanotechnology and electrochemical sensing can be used to create durable and dependable electrical equipment for water pollution detection (Tajik et al., 2021). To simulate portability for an on-field sensing technique, the working electrode is scaled down to mimic portability for an on-site sensing technique. The conductivity, sensitivity, ability to reduce biofouling, and surface area to detect target contaminants are all increased by nanoelectrodes or working electrodes that have been modified employing nanomaterials and surface modifications (Vecitis et al. 2011; Chen et al. 2013; Maleki et al. 2019).

7.2.3 Nanomaterial-Based Optical Sensor

So far, for the detection of organics (such as dissolved organic carbon, DOC, chemical oxygen demands or COD, and biological oxygen demand or BOD), residual chlorine, disinfection by products (such as trihalomethane, THM), methyl mercury (MeHg) concentration, UV_{254} absorbance, turbidity equivalent, and in situ fluorescence-dissolved organic matter (FDOM) sensors have been used in a variety of environments to monitor water quality (Bergamaschi, et al., 2013). The researchers have been recently focusing on fluorescence sensors for the detection of emerging contaminants (such as PFAS, pesticides, antibiotics, and pharmaceuticals) and heavy metals in water. However, Fluorescence probes continued to experience a number of unavoidable issues, such as photobleaching, background noise, ground substance disturbance, weak analyte labeling degrees, and diminished accuracy. To address these concerns, alternative nanosensor can start to be developed using lanthanide-doped up conversion nanomaterials (Yin et al., 2020) or gold nanoparticle-based ones (Anh et al., 2022). So far, different nanoparticles have been used to improve sensitivity and reliability. Fluorescence sensors constructed from organic luminogens (Yin et al., 2020), carbon dots (Jariah et al., 2022), nanowires (Allen et al., 2008; Luo et al., 2009), carbon-based nanomaterials (Forzani et al., 2006; Garcia-Aliaro, 2010; So et al., 2008; Villamizar et al., 2008), graphene nanosheets (Liu et al., 2012, 2011; Zhan et al., 2014), and MoS_2 nanosheets (Luo et al., 2009) have been developed for contaminant detection due to their characteristics (i.e., surface plasmon bon localization in visible spectrum, easy to synthesis, and various functionalization ability). Therefore, nanomaterial-based sensors usually exhibit high sensitivity and a quick reaction to water contaminants due to their considerable carrier mobility and high sensitivity to electrical disturbances. With different types of toxic contaminants, such as metal cations, several anions, and other compounds (pesticides, antibiotics, and PFAS), a nanomaterial-based sensor shows the detectability of toxic contaminants in the water environment with different detection limits ranging from ppb to ppm. Herein, it is now possible to quickly identify water contaminants and overcome the shortcomings of traditional sensing methods. However, numerous factors, including concentration of carbon nanomaterials, medium composition, characteristics of target contaminants, wavelength and intensity of the excitation light, and spectrometer parameter settings, can impact the determined PL spectral intensity (Mandal et al., 2019; Tran et al., 2023).

The primary contributions to the high affinity interaction that powers the sensing process are the dominating hydrogen bond interactions between the probes and contaminants with various chain lengths/benzene rings of contaminants and conformations, which create variable fluorescence response signals. Additionally, the low contamination concentration of these probes contributed to their sensitive fluorescence responses (up to 40 nM). With 92.9% accuracy, several contaminants in various background waters (surface water, groundwater, ocean, etc.) could be distinguished. Excellent identification capabilities of the suggested sensor array made them essential for hastening the development of highly effective pollutants discriminating in rapid environmental monitoring.

7.3 CONCLUSION

Water pollutants, such as organic and inorganic chemicals and microbes, are important indicators of water quality because they have a big impact on aqueous water and living organisms. The scope and focus of this chapter are the fundamentals of sensors for rapid detection of contaminants in water. Based on the visible color change, varying types of dyes have been developed for anionic and cationic sensors. The fundamental concept is to convert physical phenomena into optical signals for the detection of contaminants. Since synthesized organic dyes are dissolved in organic solvents, their stability and sensitivity limitations restrict restricted their application for on-site studies. Besides, the electrochemical-based approach also has been also utilized to detect contaminants in aqueous water; however, real-time monitoring is nevertheless hampered by low electrochemical activity, decreased sensitivity, and significant background interferences. Due to its excellent optimization and spatial sampling capabilities, the nano-enabled sensor with infrared may be the ideal option for contaminant detection when there is a shortage of timely and reliable contaminant monitoring. The monitoring of possible contamination in aquatic waterways therefore requires quick and effective fluorescence detection methods.

REFERENCES

Ainun Jariah, Muhandis Shiddiq, Bidayatul Armyna, Tahir Dahlang. 2022. "Sensor heavy metal from natural resources for a green environment: a review relation between synthesis method and luminescence properties of carbon dots", *Luminescence*, 37 (8), 1246–1258,

Behrooz Maleki, Mehdi Baghayeri, Masoud Ghanei-Motlagh, Farrokhzad Mohammadi Zonoz, Amirhasssan Amiri, Fatemeh Hajizadeh, AbduRahman Hosseinifar, Ehsan Esmaeiln Zhad. 2019. "Polyamidoamine dendrimer functionalized iron oxide nanoparticles for simultaneous electrochemical detection of Pb^{+2} and Cd^{+2} ions in environmental waters", *Measurment*, 140, 81–88.

Beibei Zhan, Chen Li, Jun Yang, Gareth Jenkins, Wei Huang, Xiaochen Dong. 2014. "Graphene field-effect transistor and its application for electronic sensing", *Small*, 10, 4042.

B.A. Bergamaschi, J.A. Fleck, B.D. Downing, E. Boss, B.A. Pellerin, N.K. Ganju, D.H. Schoellhamer, A.A. Byington, W.A. Heim, M. Stephenson and R. Fujii. 2011. "Methyl mercury dynamics in a tidal wetland quantified using in situ optical measurements." Limnology and Oceanography, 56(4), 1355–1371.

Cheng Fang, Mallavarapu Megharaj, Ravendra Naidu. 2015. "Chemical oxidization of some AFFFs leads to the formation of 6:2FTS and 8:2FTS", *Environmental Toxicology and Chemistry*, 34, 2625–2628.

Cheng Fang, Zuliang Chen, Mallavarapu Megharaj, Ravendra Naidu. 2016. "Potentiometric detection of AFFFs based on molecular imprinting polymer", *Environmental Technology and Innovation*, 5, 52–59.

Cristina García-Aljaro, Lakshmi N. Cella, Dhamanand J. Shirale, Miso Park, Francisco Javier Muñoz, Marylynn V. Yates, Ashok Mulchandani. 2010. "Carbon nanotubes-based chemiresistive biosensors for detection of microorganisms", *Biosensors and Bioelectronics*, 26, 1437–1441.

Dario Mosconi, Daniela Mazzier, Simone Silvestrini, Alberto Privitera, Carla Marega, Lorenzo Franco, Alessandro Moretto. 2015. "Synthesis and photochemical applications of processable polymers enclosing photo luminescent carbon quantum dots", *ACS Nano*, 9, 4156–4164.

Di Wei, Marc J. A. Bailey, Piers Andrew, Tapani Ryhänen. 2009. "Electrochemical biosensors at the nanoscale", *Lab on a Chip*, 9 (15), 2123–2131.

Diana Vilela, María Cristina González, Alberto Escarpa. 2012. "Sensing colorimetric approaches based on gold and silver nanoparticles aggregation: chemical creativity behind the assay. A review", *Analytical Chemical Acta*, 751, 24–43.

Dingbin Liu, Zhuo Wang, Xingyu Jiang. 2011. "Gold nanoparticles for the colorimetric and fluorescent detection of ions and small organic molecules", *Nanoscale*, 3, 1421.

Enos W. Wambu, Yuh-Shan Ho. 2016. "A bibliometric analysis of drinking water research in Africa", *Water Research Commission*, 42, 4.

Erica S. Forzani, Xiulan Li, Peiming Zhang, Nongjian Tao, Ruth Zhang, Islamshah Amlani, Raymond Tsui, Larry A. Nagahara. 2006. "Tuning the chemical selectivity of SWNT-FETs for detection of heavy-metal ions", *Small*, 2, 1283.

Feng Wang, Wei Niu, Jiantao Guo, Peter G. Schultz. 2012. "Unnatural amino acid mutagenesis of fluorescent proteins", *Angewandte Chemie International Edition*, 51, 10132–10135.

Gamze Ersan, Gabriel Antonio Cerrón-Calle, Mahmut S. Ersan, Sergi Garcia-Segura. 2023. "Opportunities for in situ electro-regeneration of PFAS-laden carbons", *Water Research*, 232, 119718.

George M. Whitesides. 2006. "The origins and the future of microfluidics", *Nature*, 442, 368–373.

Hye-Mi So, Dong-Won Park, Eun-Kyoung Jeon, Yo-Han Kim, Beom Soo Kim, Chong-Kyo Lee, Sun Young Choi, Sung Chun Kim, Hyunju Chang, Jeong-O Lee. 2008. "Detection and titer estimation of Escherichia coli using aptamer-functionalized single-walled carbon-nanotube field-effect transistor", *Small*, 4, 197.

Jelani Griffin, Anant Kumar Singh, Dulal Senapati, Eumin Lee, Kevin Gaylor, Jeanette Jones-Boone, Paresh Chandra Ray. 2009. "Sequence-specific HCV RNA quantification using the size-dependent nonlinear optical properties of gold nanoparticles", *Small*, 5, 839–845,

Ji Chen, Bowen Yao, Chun Li, Gaoquan Shi. 2013. "An improved hummers method for eco-friendly synthesis of graphene oxide", *Carbon* 64, 225–229.

K. Lance Kelly, Eduardo Coronado, Lin Lin Zhao, George C. Schatz. 2003. "The optical properties of metal nanoparticles: the influence of size, shape, and dielectric environment", *Journal of Physical Chemistry. B*, 107, 668–677.

Katherine A. Willets, Richard P. van Duyne. 2007. "Localized surface plasmon resonance spectroscopy and sensing", *Annual Review of Physical Chemistry*, 58, 267–297.

Lihui Mao, Qianqian Wang, Yuhao Luo, Yongping Gao. 2021. "Detection of Ag+ ions via an anti-aggregation mechanism using unmodified gold nanoparticles in the presence of thiamazole", *Talanta*, 222, 121506.

Linbao Luo, Jiansheng Jie, Wenfeng Zhang, Zhubing He, Jianxiong Wang, Guodong Yuan, Wenjun Zhang, Lawrence Chi Man Wu1, Shuit-Tong Lee. 2009. "Silicon nanowire sensors for Hg2+ and Cd2+ ions", *Applied Physics Letters*, 94, 193101.

Marjorie R. Willner, Peter J. Vikesland. 2018. "Nanomaterial enabled sensors for environmental contaminants" *Journal of Nanobiotechnology*, 16, 95.

Mark Allen, Erwin M. Sabio, Xiubin Qi, Bokuba Nwengela, M. Saif Islam, Frank E. Osterloh. 2008. "Metallic LiMo3Se3 nanowire film sensors for electrical detection of metal ions in water", *Langmuir*, 24 (13), 7031–7037.

Nam-Anh Tran, Nguyen Thi Hien, Nguyen Minh Hoang, Hai-Linh Thi Dang, Do Quang Huy, Tran Van Quy, Nguyen Thi Hanh, Ngoc Hung Vu, Van-Duong Dao. 2023. "Carbon dots in environmental treatment and protection applications", *Desalination*, 548, 116285.

Nguyen Ha Anh, Mai Quan Doan, Ngo Xuan Dinh, Tran Quang Huy, Doan Quang Tri, Le Thi Ngo Loan, Bui Van Hao, Anh-Tuan Le. 2022. "Gold nanoparticles-based SERS nanosensor for thiram and chloramphenicol monitoring in food samples: insight into effects of analyte molecular structure on their sensing performance and signal enhancement" *Applied Surface Science*, 584, 152555.

Paulami Mandal, Dibakar Sahoo, Priyatosh Sarkar, Kaushik Chakraborty, Sukhen Das. 2019. "Fluorescence turn-on and turn-off sensing of pesticides by carbon dot-based sensor", *New Journal Chemistry*, 43 (30), 12137–12151.

Peter Kruse. 2018. "Review on water quality sensors", *Journal of Physics D: Applied Physics*, 51, 203002.

Radim Knob, Vítězslav Maier, Jan Petr, Václav Ranc, Juraj Sevčík. 2012. "On-line preconcentration of perfluorooctanoic acid and perfluorooctanesulfonic acid by nonaqueous capillary electrophoresis", *Electrophoresis*, 33, 2159–2166.

Raquel A. Villamizar, Alicia Maroto, F. Xavier Rius, Isabel Inza, Maria J. Figuera. 2008. "Fast detection of Salmonella Infantis with carbon nanotube field effect transistors", *Biosensors and Bioelectrontronics*, 24 (2), 279–283.

Rebecca M. Duke, Emma B. Veale, Frederick M. Pfeffer, Paul E. Kruger and Thorfinnur Gunnlaugsson. 2010. "Colorimetric and fluorescent anion sensors: an overview of recent developments in the use of 1, 8-naphthalimide-based chemosensors", *Chemical Society Review*, 39, 3936–3953.

René P. Schwarzenbach, Thomas Egli, Thomas B. Hofstetter, Urs von Gunten, Bernhard Wehrli. 2010. "Global water pollution and human health", *Annual Review of Environment and Resource*, 35, 109–136.

Rui Yin, Shuaiyuan Yang, Qianming Li, Shuaidi Zhang, Hu Liu, Jian Han, Chuntai Liu, Changyu Shen. 2020. "Flexible conductive Ag nanowire/cellulose nanofibril hybrid nanopaper for strain and temperature sensing applications", *Science Bulletin*, 65 (11), 899–908.

Siegfried Eigler, Michael Enzelberger-Heim, Stefan Grimm, Philipp Hofmann, Wolfgang Kroener, Andreas Geworski, Christoph Dotzer, Michael Röckert, Jie Xiao, Christian Papp, Ole Lytken, Hans-Peter Steinrück, Paul Müller, Andreas Hirsch. 2013. "Wet chemical synthesis of graphene", *Advanced Materials*, 25, 3583–3587.

Somayeh Tajik, Hadi Beitollahi, Fariba Garkani Nejad, Zahra Dourandish, Mohammad A. Khalilzadeh, Ho Won Jang, Richard A. Venditti, Rajender S. Varma, Mohammadreza Shokouhimehr. 2021. "Recent developments in polymer nanocomposite-based electrochemical sensors for detecting environmental pollutants", *Industrial & Engineering Chemistry Research*, 60 (3), 1112–1136.

Stephan Link and Mostafa A. El-Sayed. 1999. "Spectral properties and relaxation dynamics of surface plasmon electronic oscillations in gold and silver nanodots and nanorods", *Journal of Physical Chemistry B*, 103, 8410–8426.

Sujit Kumar Ghosh, Tarasankar Pal. 2007. "Interparticle coupling effect on the surface plasmon resonance of gold nanoparticles: from theory to applications", *Chemical Review*, 107, 4797–4862.

Susan D. Richardson, Thomas A. Ternes. 2015. "Water analysis: emerging contaminants and current issues", *Analytical Chemistry*, 77, 3807–3838.

Chad D. Vecitis, Guandao Gao, and Han Liu. 2011. "Electrochemical carbon nanotube filter for adsorption, desorption, and oxidation of aqueous dyes and anions." *The Journal of Physical Chemistry C*, 115 (9), 3621–3629.

Xiaohui Wang, Yan Li, Yumin Du. 2009. "Novel water-soluble chitosan derivatives/quantum dots nanocomposite: synthesis, characterization and photoluminescence properties", *Journal of Nanoscience and Nanotechnology,* 12, 6866–6875.

Yifan Zhang, Jaesang Lee, Stephen R. Forrest. 2014. "Tenfold increase in the lifetime of blue phosphorescent organic light-emitting diodes", *Natural Communication*, 5, 5008.

Yuxin Liu, Xiaochem Dong, Peng Chen. 2012. "Biological and chemical sensors based on graphene materials", *Chemical Society Review*, 41, 2283.

Zhihua Li, Yanwei Wang, Qianqian Yu. 2010. "Significant parameters in the optimization of synthesis of silver nanoparticles by chemical reduction method", *Journal of Materials Engineering and Performance*, 19, 252–256.

8 Packaging Nanomaterials in the Food Industry

Rafael Vargas-Bernal
Universidad de Guanajuato

CONTENTS

8.1 INTRODUCTION

The food industry is a multi-trillion-dollar business worldwide, so it is always improving food production safely and efficiently (Basavegowda et al. 2020). By 2050, the production of packaging materials is expected to reach 250 million metric tons (Chen et al. 2022). Worldwide, this sector represents the third largest industry dedicated to ensuring distribution and storage, protecting food from its environment, and extending its shelf life (Basavegowda et al. 2020). Utilizing functional food packaging aims to decrease food waste, raising both the value chain's efficiency and the ultimate cost of the food product. Because food can experience mechanical, biological, or chemical damage, it must be packaged using polymer-based basic protection to withstand mechanical shock and resist microorganisms or chemical attacks, as shown in Figure 8.1a (Kim et al. 2019). Polymers for food packaging such as poly(ethylene terephthalate) (PET), polyethylene (PE) and polypropylene (PP), being permeable to oxygen, fail to extend the life of food, so to overcome this deficiency, it is necessary to add fillers. But even these boosters are harmful from a health and environmental standpoint. Metals and metal oxides reduce transparency because they obstruct the passage of visible light and radio frequency signals.

Through food packaging, it is possible to artificially produce the physicochemical conditions necessary for food to be durable, safe, and of high quality (Basavegowda et al. 2020). The use of traditional plastic packaging implies the consumption of

DOI: 10.1201/9781003397861-8

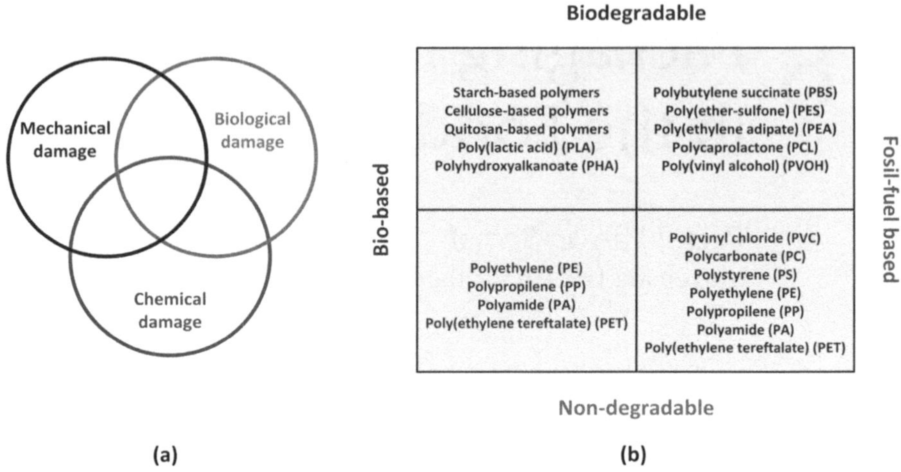

FIGURE 8.1 Role of food packaging: (a) Types of damage in food products. (b) Examples of polymeric materials with possible uses in food packaging.

fossil fuel, which implies environmental contamination during its synthesis, processing, and disposal. The conventional petroleum-derived plastics are the most commonly materials used nowadays in packaging (Chen et al. 2022). Unfortunately, these are not sustainable, renewable, or degradable. Biopolymers are non-toxic and biodegradable, but they have poor mechanical properties and are susceptible to water (Chaudhary et al. 2020; Deepika et al. 2023). The ultimate change is associated with the use of green materials extracted from plants, biodegradable edible materials, and biodegradable polymer-based nanocomposites. The classification of polymer types for various engineering applications is outlined in Figure 8.1b.

Food during its lifetime is susceptible to putrefaction since it is not necessarily consumed immediately after being removed from the orchard, cut from the trees, extracted from the animal, or processed (Ashfaq et al. 2022). Also, they must maintain their nutrients as well as their color, flavor, smell, and texture, which are known as organoleptic properties. One of the strategies to maintain and manage the quality of food properly is packaging. Conventional food packaging offers communication, convenience, containment, and preservation as passive functions. These basic functions must be optimized to design active, smart, and improved packaging. In addition to traditional properties, novel packaging should offer properties such as reduced antimicrobial activity, an efficient oxygen barrier, optimized ultraviolet impermeability, etc. (Sirviö et al. 2020).

The biopolymers used in multifunctional nanocomposites are starch, cellulose, and chitosan within the group of polysaccharides, and poly(vinyl alcohol) (PVA), poly(lactic acid) (PLA), polyhydroxyalkanoates (PHAs), and polybutylene succinate (PBS) within the group of synthetic biopolymers (Chaudhary et al. 2020; Chen et al. 2022). Both their mechanical strength and their barrier properties against water and oxygen in biopolymers can be improved through strengthening and processing.

Biopolymers are modified through coating and/or functionalization mechanisms to modify their physical cross-linking and surface properties, which are optimized to improve mechanical resistance and/or rigidity as well as barrier properties.

Nanomaterials have at least one of their lengths between 1 and 100 nm and present novel physical and chemical properties concerning their macro-sized counterparts (Chaudhary et al. 2020). In addition, they can modify the bulk properties of polymers, such as flexibility, recyclability, durability, barrier properties, and flame resistance. Nanomaterials have applications in the electrical, magnetic, optical, biological, agricultural, construction, medical, defense, energy, aerospace, and food sectors, among others (Siddiqui et al. 2022). Nanomaterials can provide better preservation and maintain food quality by modifying their physical and chemical properties, such as durability, surface reactivity, water and oxygen barrier, specific surface area, strength, flexibility, antimicrobial properties, and reusability (Siddiqui et al. 2022). Because each nanomaterial has both different physical properties and a unique chemical structure and composition, their use results in different practical applications within food packaging. Nanomaterials allow the implementation of sensors that report on the safety and quality of food (McClements 2020). Successful packaging design implies mastering the structure of materials and the interaction of different types of components, which optimizes multifunctionality, and, due to the diversity of options, a single design is fortunately not the only possibility.

Two interaction mechanisms between nanomaterials used in nanomaterial-based packaging have been identified so far (Amin et al. 2020). The first involves evenly distributing the nanomaterials in a matrix across a high interfacial area, which leads to improved physical properties. The second consists of the active interaction among food, nanomaterials, and the environment. Toxic microbes derived from food are the main reason for diseases of the digestive system that can end in hospitalization or even death in humans (Amin et al. 2020). Pathogens in food appear on its surface or in its packaging. These microbes produce the loss of products reflected in the millions of dollars of investment paid by the food industry.

The content of the chapter was organized as follows: In Section 8.2, a brief description is given of the types of food packaging that, thanks to nanomaterials, have been, are being, and will be used in the market. Subsequently, the different types of nanomaterials and the properties that they incorporate into the packaging are discussed in Section 8.3. In Section 8.4, some key examples of applying nanomaterials to food packaging are introduced. The future perspectives that must be carried out to achieve the massive incorporation of nanomaterials in a safe and controlled manner are presented in Section 8.5. Finally, the conclusions about the use of nanomaterials in food packaging are given in Section 8.6.

8.2 PACKAGING TYPES

Packaging materials, including nanomaterials, can be synthesized by adding nanoparticles to plastic films and containers or manufacturing a multilayer nanocomposite (Ashfaq et al. 2022). Both the mechanical and physical properties are improved through improved packaging. The active characteristics of the packaging

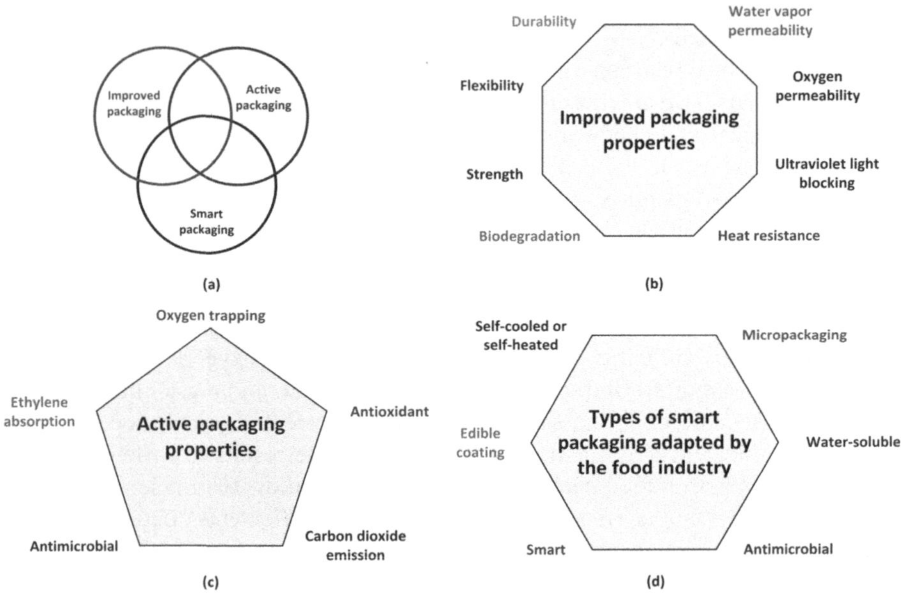

FIGURE 8.2 Food packaging. (a) Types of packaging obtained using nanomaterials. (b) Main properties of improved food packaging. (c) Main properties of active food packaging. (d) Smart packaging technologies adapted by the food industry.

derive from its antibacterial, antioxidative, and ultraviolet light-absorbing properties. Smart packaging properties originate from the ability to monitor and/or control the conditions of food. A summary of the possible types of packaging obtained using nanomaterials is illustrated in Figure 8.2a.

8.2.1 Improved Packaging

This type of packaging uses nanoparticles to tune physical and mechanical properties (Ashfaq et al. 2022). Properties that can be improved include durability, flexibility, strength, biodegradation, heat resistance, ultraviolet light blocking, and oxygen and water vapor permeability, as shown in Figure 8.2b. Packaging seeks to optimize gas exchange, produce constant humidity, and maintain a constant temperature to increase the duration of the food. The inclusion of metal oxides (SiO_2, ZnO, and TiO_2) in the biopolymers optimizes the oxygen barrier, mechanical properties, and light permeability. Metal oxides react photocatalytically to produce reactive oxygen species (ROS) that oxidize the cytoplasm of bacterial cells, which subsequently leads to cell death (Chaudhary et al. 2020). Metallic materials such as magnesium (Mg), zinc (Zn), titanium (Ti), and silver (Ag) embedded in synthetic biopolymers increase their resistance to water (Babu et al. 2021). Chitosan-based nanocomposites and either silver (Ag) or gold (Au) nanoparticles exhibit antimicrobial properties.

8.2.2 Active Packaging

This packaging increases the shelf life of food by preventing the invasion of microbes, humidity, oxidation, overripeness, etc. (Ashfaq et al. 2022). Active food packaging materials are designed to reduce oxygen or water permeability or to protect their exposure to light and heat (Amin et al. 2020; Deepika et al. 2023). The additives used in this type of packaging must fulfill the functions of trapping oxygen, absorbing ethylene, acting as antioxidants, emitting carbon dioxide, and containing antimicrobial agents, as depicted in Figure 8.2c. Antimicrobial agents have the function of prohibiting the growth of microbes through the destruction of their molecular structure or by avoiding the metabolic pathway. Nanomaterials can generate reactive oxygen species (ROS), which prevent DNA replication and ATP production, causing damage or death to the cell. Poly(vinyl alcohol), polyethylene (PE), or kaolin-based nanopowders combined with silver (Ag), titanium (Ti), and zinc (Zn) nanoparticles have been used to protect fruits from microbes such as *Escherichia coli* and *Staphylococcus aureus* (Babu et al. 2021). Avoiding contamination by pathogens such as *Listeria monocytogenes* and *Salmonella enterica* found in sewage or polluted water is feasible with gelatin-based nanocomposites of fish skin with bimetallic copper and silver (Cu-Ag) nanoparticles. The silver nanoparticles damage the DNA of the bacteria, preventing ATP synthesis, which leads to the death of the cell. However, zinc oxide (ZnO) nanoparticles exhibit better antimicrobial activity compared to silver (Ag) nanoparticles for bacterial species such as *Escherichia coli*, *Staphylococcus*, *Salmonella*, and *Bacillus* (Yang et al. 2021).

8.2.3 Smart Packaging

Through this type of packaging, it is possible to monitor the environment and condition of the food (Ashfaq et al. 2022). Smart food packaging materials act as sensing elements for food and give information about possible contamination or spoilage (Amin et al. 2020). This type of packaging takes advantage of technologies such as sensors, radio frequency identification tags, and barcodes to perform the entire logistics. This logistics implies communicating, monitoring, sensing, storing, and tracking the food from its origin to its destination. When food is intelligently packaged, the temperature, acidity (pH), and oxygen levels will be monitored (Babu et al. 2021). Gold nanoparticles can be used to measure temperature through the changing colors of the packaging. Metal oxide nanomaterials are used to monitor gas levels released by pathogens in food. Nanomaterials based on titanium dioxide (TiO_2), magnesium oxide (MgO), cadmium telluride (CdTe), cadmium sulfide (CdS), and carbon dots prevent the growth of biofilms formed by bacteria. Biopolymers such as chitosan and poly(vinyl alcohol) (PVA), together with silica, can form films to reduce the oxygen absorptivity and moisture content of fruits such as cherries. Silver (Ag) and/or copper (Cu) nanofilms have been used in electronic noses to detect trimethylamine vapor in spoiled meat. Nanomaterials also have the function of making the packaging strong, light, and less susceptible to oxygen (Riahi et al. 2022). Metal oxides, metal, graphene, and mesoporous silica have contributed to improving the antibacterial properties of active and intelligent packaging. Through these properties, harmful bacteria are prevented from causing food deterioration, and these are made safe and fresh when they reach consumers.

Nanosensors are used in conjunction with packaging materials to sense chemical moieties associated with food spoilage through the detection and measurement of physical variables related to the acidity, color, and temperature of packaged foods (Basavegowda et al. 2020). Even the presence of pesticides, herbicides, and fungicides in food can be determined through the nanosensors placed on the packaging. Nanomaterials can be used in biological and chemical sensors to determine the presence of allergens, pathogens, and/or toxins, as well as the freshness of food, in search of safety and a friendly environmental effect (Mustafa et al. 2020).

So far, six technologies have been distinguished to develop intelligent packaging adapted by the food industry: self-cooled or self-heated packaging, micropackaging, edible coating, smart packaging, antimicrobial packaging, and water-soluble packaging (see Figure 8.2d) (Babu et al. 2021). Self-heating can facilitate heating up to 140°C thanks to the materials of the packaging itself. Self-cooling allows a package to be cooled through the adsorption of carbon dioxide from the environment, which causes a decrease in the temperature of the food. Micropackaging implies that the use of clay nanoparticles prevents the combination of liquids and gases from breaking down food products. Smart packaging allows the use of an electronic nose that includes different types of sensors to know the state of the food and whether it is in good condition or not. Edible packaging based on lipids, proteins, polysaccharides, and surfactants in the form of composite films protects the food for a defined time. Antimicrobial packaging shields food from pathogens, for example, by absorbing oxygen that makes it putrefied. When a package is water soluble, it dissolves in water at room temperature and is often used in foods such as sauces, soups, cereals, powders, and beverages.

8.3 NANOMATERIALS FOR PACKAGING

Nanomaterials offer unique physicochemical properties for food packaging, including antimicrobial, therapeutic, increasing shelf life, and maintaining color and flavor for food products (Babu et al. 2021). These properties are obtained thanks to the possibility of functionalizing nanomaterials with different types of biomolecules. Nanomaterials improve the performance of conventional and non-biodegradable food packaging in bioavailability, texture, taste, flavor, antimicrobial properties and overall shelf life. To achieve this, layer-by-layer hybrid components with opposite electrical charges are alternately arranged to obtain electrostatic interactions (Kim et al. 2019).

Functional properties from organic and inorganic nanomaterials are incorporated into biopolymers (Ashfaq et al. 2022). Among the properties that are intended to be improved with nanomaterials are thermal stability, tensile strength, antibacterial and antifungal activity, oxygen barrier, water vapor absorption, solubility, fracture deformation, swelling, degradation, resistance to ultraviolet radiation, elongation, and shelf life, as depicted in Figure 8.3a. Amidst the organic fillers that can be incorporated are nanocellulose, nanostarch, protein nanoparticles, and chitosan nanoparticles. The main inorganic nanomaterials for the food packaging industry are silver (Ag),

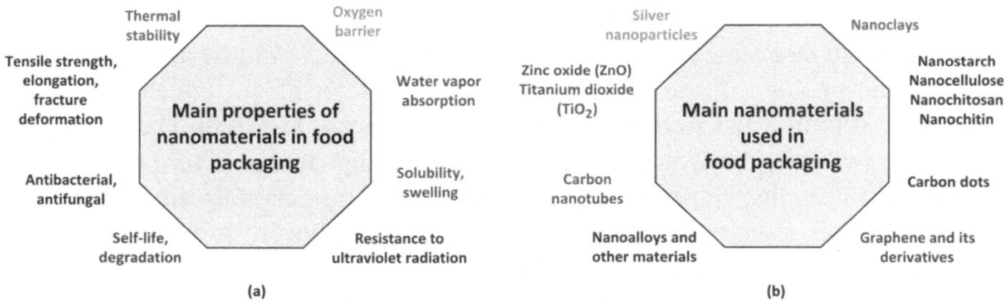

FIGURE 8.3 Nanomaterials in food packaging. (a) Main properties that nanomaterials contribute to food packaging. (b) Main nanomaterials used in food packaging.

gold (Au), iron (Fe), silver oxide (AgO), titanium dioxide (TiO_2), iron oxide (FeO), zinc oxide (ZnO), silicon dioxide (SiO_2), carbon nanotubes, and nanoclay (Wang et al. 2019; Babu et al. 2021). The interaction mechanism between nanomaterials and food can be either by direct contact or by gradual migration and reaction with organic molecules in food. A summary of the main nanomaterials used in food packaging is illustrated in Figure 8.3b.

Silver nanoparticles have antifungal, anti-yeast, antimicrobial, and antiviral properties that can be used for sterilization (Chaudhary et al. 2020; Yang et al. 2021). The disadvantage of these nanoparticles is that they bind to the cell through their surface, so lipopolysaccharides in food are degraded through a hole in the cell. To avoid this problem, the silver ions must be embedded in clay, and later this material can be incorporated into a polymeric matrix. Zinc oxide nanoparticles have antibacterial properties against *Salmonella aureus*, *Escherichia coli*, and *Bacillus atrophaeus*, are commercially available, and have low toxicity (Chaudhary et al. 2020). When ultraviolet light is applied to zinc oxide (ZnO) in the air, it generates hydrogen peroxide, which leads to oxidative stress that kills bacterial cells, making the material antibacterial in nature. Titanium dioxide (TiO_2) nanoparticles provide properties of thermal stability, biodegradability, low toxicity, antibacterial activity, and photocatalysis (Chaudhary et al. 2020). The antibacterial activity of these nanoparticles in the air occurs when sunlight or ultraviolet light gives rise to reactive oxygen species (ROS) through oxide anions and hydroxyl radicals in the hydrogen peroxide produced in the interaction, which damage the microbial cell walls.

Nanoclays have gas and water barrier characteristics thanks to their efficient cation exchange, large surface area, and ease of swelling (Chaudhary et al. 2020). Since clays form clusters of platelets in nature, these must be exfoliated to be broken up into individual flakes, which are then evenly distributed in the polymer matrix. When nanoclays are incorporated into biopolymers, barrier properties, hydrophobicity, mechanical strength, and creep resistance are added. Starch is a non-toxic, biodegradable, and renewable natural polysaccharide used as a low-cost, abundant, and environmentally friendly packaging material (Chaudhary et al. 2020). At the macromolecular level, the starch is made up of granules containing branched amylopectin and linear amylose. To produce nanostarch, the constituents must be separated to extract the amylose in the form of nanoscale platelets. This nanomaterial has barrier properties, high mechanical properties, and is hydrophilic. Cellulose at the nanoscale can have morphology in the form of fibers, whiskers, or fibrils, which alternately consist of crystalline and amorphous chains (Chaudhary et al. 2020). The crystalline part due to hydrogen bonding provides a stable, axially rigid structure to regularly function as fibers that reinforce polymers. Nanocellulose crystals are mixed with carrageenan, poly(lactic acid) (PLA), pectin, and guar gum to produce packaging with good mechanical and barrier properties (Chen et al. 2022).

Carbon nanotubes (CNTs), classified as single-walled or multi-layered, where each layer is one atom thick, are used as smart sensors and antimicrobial elements and have high elastic moduli and tensile strength that can be exploited in biopolymers (Chaudhary et al. 2020). *Escherichia coli* bacteria are toxic to single-walled carbon nanotubes. In addition, nanotubes can be used to design gas sensors to detect the decomposition of putrefied meat or food. Carbon quantum dots offer properties such as functionalization capacity, nanosize, non-toxic, antibacterial, biocompatibility, antioxidant, UV barrier, photostability, photoluminescence, tensile strength, ease of synthesis, and low cost (Deepika et al. 2023). They can provide properties such as freshness, quality, safety, and preservation to the following foods: fruits, honey, butter, seafood, leaves, seeds, and peels. Graphene is a carbon-based material with atomic thickness, a high area-to-volume ratio, and high sensitivity to changes in environmental conditions, making it useful for sensing applications in smart packaging (Hashim et al. 2022). In addition, the use of functionalized graphene in food packaging aims to determine the freshness and presence of contaminants. This guarantees that liquid foods such as dairy products, fruit drinks, and vines, do not lose their quality.

Both metallic nanoparticles and their oxides, as they possess antimicrobial and antioxidant properties, are being used together with nanobiosensors to establish follow-up and monitoring of the condition of food (Basavegowda et al. 2020). Multimetallic nanoparticles have a wider range of activity, chemical and/or physical stability, higher selectivity, surface-to-volume ratio, size, and shape than their monometallic counterparts. Bi- or trimetallic nanoparticles are formed when two or three metals are combined to form nanoalloys. In addition, it is possible to form multimetallic ceramic oxides when they are oxidized. Both types of nanoparticles improve mechanical properties and shield against moisture and gases to reduce food waste. Furthermore, the above nanocomposites can be enriched with essential oils to improve their operation. These oils are obtained from secondary metabolites derived from plant oils through thermal processes involving chemical reactions. The relevant properties that these oils contribute to food packaging are insecticidal, antioxidant, antifungal, antiviral, and/or antibacterial.

Due to their highly volatile nature and poor aqueous solubility, essential oils must be encapsulated in nanocarriers to better exploit their unbeatable properties for use in food packaging (Kumar A. et al. 2022). Potential nanocarriers include lipid nanoparticles, emulsions, nanoclays, biopolymeric materials, and inclusion complexes. Essential oils can come from oranges, lemons, eucalyptus, mint, tea tree, basil, thyme, etc. These have larvicidal, antifungal, and antimicrobial properties. Essential oils are volatile, aromatic, oily liquids extracted from leaves, herbs, flowers, roots, bark, buds, seeds, and fruits. These have been used for the preservation of cucumbers and strawberries, using an aqueous phase and a cosurfactant to dilute them in a material used for food packaging. Until now, the most widely used nanomaterials in food packaging are zinc oxide (ZnO) nanoparticles, nano clay (montmorillonite and/or kaolinite), silver (Ag) nanoparticles, and titanium dioxide (TiO_2) nanoparticles (Chaudhary et al. 2020). These materials have barrier properties against carbon dioxide, oxygen, and flavor compounds, have antimicrobial activity, and are temperature-tolerant.

Seven major aspects of nanomaterials that need to be highlighted have called for their use in food packaging applications, namely antimicrobial, mechanical stability, biosensors, food preservation, food protection, food safety, and eco-friendly (Babu et al. 2021). To improve the mechanical stability of the packings, nanomaterials improve the tensile strength, elasticity, and stiffness of the composite material. Nanomaterials can be used as biosensors for the detection of pathogens, allergens, pollutants, and pesticides. In food preservation, nanomaterials have the function of increasing the shelf life of food as well as the release or absorption of antioxidants depending on the type of food or medium in which it was packaged (Biswas et al. 2022). From the point of view of food protection, nanomaterials have the function of protecting the polymer against ultraviolet rays, acting as efficient gas barriers, being resistant to water, and behaving in a photostable manner. In food safety, nanomaterials have the function of producing non-toxic and chemically inert packaging. For food packaging to be eco-friendly, nanomaterials are used to make it biodegradable and biocompatible.

8.4 APPLICATION OF NANOMATERIALS TO FOOD PACKAGING

In this section, a set of possible materials that have been proposed for use in food packaging is provided to highlight the relevance of nanomaterials in this globally relevant economic sector.

The composite material poly(vinyl alcohol) and carbon dots was synthesized to produce a transparent, photoluminescent material with ultraviolet ray-blocking activity (Alaş et al. 2022). This material presented good antimicrobial and antioxidant activity, emitting green and red colors, and was applied for the conservation of strawberries, inhibiting the growth of fungi. Edible food packaging is based on food-grade biopolymers, proteins, lipids, and polysaccharides (Kumar L. et al. 2022). The films that are formed have antioxidant, antimicrobial, sensory, and mechanical properties. Through this type of packaging, it is possible to protect vegetables, fruits, meat, and dairy products. Positively charged chitin nanowhiskers and negatively charged cellulose nanofibers have been used to reinforce poly(ethylene terephthalate) (PET) as a food packaging material (Kim et al. 2019). This material is flexible, transparent to light and radio frequency, has oxygen barrier properties, is antibacterial, has bending stress resistance, is renewable, and is hydrophilic (Sirviö et al. 2020).

Starch nanomaterials with different nanoscale architectures have been synthesized from native starches smaller than 1,000 nm in size (Alves et al. 2021). Edible starch-based films present a water vapor barrier, antioxidant and antimicrobial properties, biodegradability, thermal reinforcement, and mechanical properties.

These nanomaterials can be coated with metallic silver nanoparticles and hybridized with graphene oxide to enhance their antimicrobial activity against different bacteria (Yang et al. 2021). These materials are used to protect tubers/roots, fruits, grains, and seeds. A poly(vinyl alcohol)/pluronic/ZnO nanoparticle-based composite nanomaterial film has been proposed for active packaging (Amin et al. 2020). This nanocomposite presented an opaque improved antimicrobial and antifungal activity with a higher percentage of ZnO nanoparticles. The nanocomposite has satisfactory antimicrobial, barrier, thermal, biodegradable, and morphological properties in food packaging. Cellulose nanofibrils or nanocrystals mixed with poly(lactic acid) and a rosin emulsion to produce hybrid and multifunctional materials capable of offering water vapor and oxygen barrier properties (Le Gars et al. 2020; Sirviö et al. 2020). Rosin is a low-cost, antimicrobial, abundant, renewable, antioxidant, and hydrophobic blend of natural compounds.

A nanocomposite based on natural building blocks of cellulose nanofibrils and zein protein nanoparticles exhibited thermal stability, tensile strength, stiffness, biocompatibility, hydrophobicity, sustainability, environmentally friendliness, and antibacterial properties for food packaging applications (Li et al. 2020). To reinforce the properties of cellulose nanofibrils, it is necessary to add zinc oxide (ZnO) nanoparticles, chitin nanocrystals, carbon nanotubes, graphene oxide, metal nanoparticles, and polyaniline nanoparticles. These nanomaterials provide electrical, photocatalytic, and antibacterial properties that can be used for interesting applications. Zein is a safe, renewable, antimicrobial, biocompatibility, antioxidant, and water-insoluble prolamin protein. Traditional packaging tends to use opaque and metallic foil, which is not beneficial to consumers as they want multifunctional packaging with a transparent window to check the food before purchasing (Chen et al. 2022). The use of cellulose nanocrystals, modified with cationic trivalent metal ions such as aluminum and iron, was proposed to develop materials that are transparent, resistant to oxygen and water permeation, absorbent to ultraviolet light, and flexible. Aminated nanocrystalline cellulose has been conjugated with a natural rose bengal-based photosensitizer to develop an antibacterial booster against *Salmonella enterica* and *Listeria monocytogenes* by applying light irradiation to produce reactive oxygen species (ROS), which destroy bacterial cell walls (Koshani et al. 2021). To reduce and/or eliminate bacteria using light, a photoactive molecule, light, and oxygen must interact by absorbing light to form reactive oxygen species (ROS) to kill bacteria by damaging biomolecules such as proteins, lipids, or DNA. Since fresh natural products have short lifespans before decomposing, they need to be packaged against oxygen (Chiang et al. 2021). Edible chitosan and pectin-based nanocoatings have been proposed to provide a gas barrier and high transparency. These can be used for slow-ripening bananas and/or browning apple slices.

Nanocomposites using carrageenan and natural titanium dioxide (TiO_2) nanotubes and/or doped with copper oxide (CuO) have been used to provide hydrophobicity, mechanical resistance, barrier properties against water vapor, antibacterial properties, as well as blocking ultraviolet light (Ezati et al. 2021). These were used to pack bananas for 12 days, with amazing results. The relevance of the use of TiO_2 nanotubes is their photocatalytic activity, biocompatibility, stability, and antimicrobial properties under ultraviolet light. Antimicrobial activity is obtained through reactive oxygen species that form electron-hole pairs that destroy microbial cell walls and their DNA, and thereby kill the cells. The use of copper oxide (CuO) consists of extending the catalytic activity now to the visible range because the natural range of TiO_2 covers only the ultraviolet range. A nanocomposite based on semolina flour, kaolin, and ZnO nanoparticles has been proposed as a material to produce active food packaging films (Jafarzadeh et al. 2019). This material presents antimicrobial properties at low temperatures to monitor the properties of mozzarella cheese, such as acidity, fat content, and moisture, as well as color, texture, taste, and overall acceptability.

Selenium nanoparticles have reduced biocidal and toxicological effects, intense biological action, and bioavailability in biopolymers (Ndwandwe et al. 2021). In addition, they exhibit antioxidant, antimicrobial, and antifungal properties in polymers with chemical stabilizers. Nanocomposites can be used to package meat, fish, and fresh foods. Reinforcing agents such as silver and zinc oxide nanoparticles can be added. Red meat packaging is challenging to retain color, slow weight loss, improve tenderness, avoid contamination, and eliminate waste (Nile et al. 2020). The lifetime of these products is extended by creating an atmosphere with formulated gas mixtures coupled with the use of nanomaterials with the properties required for each type of food. Gelatin and/or agar-based nanocomposites have been fortified with sulfur quantum dots to produce functional films for food packaging (Priyadarshi et al. 2021). The resulting properties of the material are antimicrobial and antioxidant, as well as barrier properties to ultraviolet rays.

Carbon nanostructures such as carbon or graphene quantum dots, graphene, carbon nanotubes, and natural or reduced graphene oxide can enhance the shelf life of foods without losing their nutritional value (Raul et al. 2022). These materials present environmentally friendly, food preservation, antibacterial, and filling properties for active food packaging. In addition, these materials are applied to the analysis, safety, preservation, biodegradable, environmental, and economic packaging of foods. These have low toxicity and versatile activities. Films based on cellulose nanofibers reinforced with titanium dioxide (TiO_2) nanotubes and copper oxide (Cu_2O) nanoparticles have been used for tomato packaging (Riahi et al. 2022). These materials are photocatalytically active and present visible and UV light shielding properties, are flexible, biocompatible, mechanically resistant, slightly opaque, have water barrier properties, and have antibacterial properties. When tomatoes are packaged, ethylene can be generated, giving rise to detrimental effects on the vegetables such as discoloring, reducing their weight, and softening them, and this can be reduced or avoided thanks to the use of this packaging. Gelatin and/or carrageenan-based films and carbon dots have been proposed for active packaging (Roy et al. 2021). Carbon dots have been derived from Enoki mushrooms and provide ultraviolet light absorption, transparency, photoluminescence, and antioxidant properties to food packaging.

8.5 FUTURE PERSPECTIVES

Undoubtedly, the main concern before extensively using nanomaterials in food packaging is their toxicity (Ashfaq et al. 2022). For nanomaterials to be accepted in the food industry, rigorous standards and tests must be established to ensure their safety and safe use in food packaging to minimize food waste (Babu et al. 2021). The toxicity of nanomaterials is influenced by their concentration, exposure time, sensitivity to the medium, type of material, and composition. Up until now, there has been insufficient data on the toxicity of nanomaterials and their dangerous effects on humans. Nanomaterials can migrate among the different media, and in packaging, it is no exception since a microenvironment is generated between the packaging and the food. Nanomaterials by nature are cytotoxic for the food and biomedical sectors, so many researchers around the world continue to investigate their effects on humans, but a universal theory has not yet been established. To achieve the establishment of regulations and standards associated with the safety and technology involved with food packaging, it is necessary to develop a greater amount of multidisciplinary research.

8.6 CONCLUSION

The development of food packaging that is much more reliable and effective will continue to be a worldwide demand in future decades due to climate change issues. Nanomaterials, as in many other fields of science and engineering, will continue to contribute to offering solutions to optimize the quality and level of specialization required. The multifunctional materials guarantee the lifetime of the food, so this area of research will continue to be developed to exploit its potential much more widely. These materials protect food from oxygen, water, ultraviolet rays, dust, bacteria, etc. An analysis of the possible migration trajectories to produce toxicity by nanomaterials must still be studied and reviewed because the side effects that nanomaterials would have when ingested by humans are unknown.

In a general way, without describing specific application examples, it can be pointed out that each nanomaterial has a particular effect on one or more properties, so the choice of the nanomaterial(s) for multifunctional food packaging is not as simple as it could be considered because different nanomaterials can affect the same property. Therefore, a lot of research still needs to be done to fully take advantage of nanoscale materials in packaging intended to protect food.

REFERENCES

Alaş, Melis Özge, Gamze Doğan, Mustafa Serkan Yalcin, Sadin Ozdemir, and Rükan Genç. 2022. "Multicolor Emitting Carbon Dot-Reinforced PVA Composites as Edible Food Packaging Films and Coatings with Antimicrobial and UV-Blocking Properties." *ACS Omega* 7 (34): 29967–29983.

Alves, Maria Jaízia dos Santos, Wilson Daniel Caicedo Chacon, Talita Ribeiro Gagliardi, Ana C. Agudelo Henao, Alcilene Rodrigues Monteiro, and Germán Ayala Valencia. 2021. "Food Applications of Starch Nanomaterials: A Review." *Starch* 73 (11–12): 2100046.

Amin, Khaled M., Abir M. Partila, Hassan A. Abd El-Rehim, and Noha M. Deghiedy. 2020. "Antimicrobial ZnO Nanoparticle-Doped Polyvinyl Alcohol/Pluronic Blends as Active Food Packaging Films." *Particle & Particle Systems Characterization* 37 (4): 2000006.

Ashfaq, Alweera, Nazia Khursheed, Samra Fatima, Zayeema Anjum, and Kaiser Younis. 2022. "Application of Nanotechnology in Food Packaging: Pros and Cons." *Journal of Agriculture and Food Research* 7: 100270.

Babu, Punuri Jayasekhar, Akriti Tirkey, and Tingirikari Jagan Mohan Rao. 2021. "A Review on Recent Technologies Adopted by Food Industries and Intervention of 2D-Inorganic Nanoparticles in Food Packaging Applications." *European Food Research and Technology* 247: 2899–2914.

Basavegowda, Nagaraj, Tapas K. Mandal, and Kwang-Hyun Baek. 2020. "Bimetallic and Trimetallic Nanoparticles for Active Food Packaging Applications: A Review." *Food and Bioprocess Technology* 13: 30–44.

Biswas, Rahul, Mahabub Alam, Animesh Sarkar, Md Ismail Haque, Md. Moinul Hasan, and Mominul Hoque. 2022. "Application of Nanotechnology in Food: Processing, Preservation, Packaging and Safety Assessment." *Heliyon* 8 (11): e11795.

Chaudhary, Pallavi, Faria Fatima, and Ankur Kumar. 2020. "Relevance of Nanomaterials in Food Packaging and Its Advanced Future Prospects." *Journal of Inorganic and Organometallic Polymers and Materials* 30: 5180–5192.

Chen, Cong, Wenjing Sun, Lu Wang, Mehdi Tajvidi, Jinwu Wang, and Douglas J. Gardner. 2022. "Transparent Multifunctional Cellulose Nanocrystal Films Prepared Using Trivalent Metal Ion Exchange for Food Packaging." *ACS Sustainable Chemistry & Engineering* 10 (29): 9419–9430.

Chiang, Hsu-Cheng, Bailey Eberle, Drew Carlton, Thomas J. Kolibaba, and Jaime C. Grunlan. 2021. "Edible Polyelectrolyte Complex Nanocoating for Protection of Perishable Produce." *ACS Food Science & Technology* 1 (4): 495–499.

Deepika, Lokesh Kumar, and Kirtiraj K. Gaikwad. 2023. "Carbon Dots for Food Packaging Applications." *Sustainable Food Technology* 1 (2): 185–199.

Ezati, Parya, Zohreh Riahi, and Jong-Whan Rhim. 2021. "Carrageenan-Based Functional Films Integrated with CuO-Doped Titanium Nanotubes for Active Food-Packaging Applications." *ACS Sustainable Chemistry & Engineering* 9 (28): 9300–9307.

Hashim, Norazlina, Suhaila Abdullah, and Kamal Yusoh. 2022. "Graphene Nanomaterials in the Food Industries: Quality Control in Promising Food Safety to Consumers." *Graphene and 2D Materials Technologies* 7: 1–29.

Jafarzadeh, Shima, Jong-Whan Rhim, Abd Karim Alias, Fazilah Ariffin, and Shahrom Mahmud. 2019. "Application of Antimicrobial Active Packaging Film made of Semolina Flour, Nano Zinc Oxide and Nano-Kaolin to Maintain the Quality of Low-Moisture Mozzarella Cheese During Low-Temperature Storage." *Journal of the Science of Food and Agriculture* 99 (6): 2716–2725.

Kim, Taehyung, Thang Hong Tran, Sung Yeon Hwang, Jeyoung Park, Dongyeop X. Oh, and Byeong-Su Kim. 2019. "Crab-on-a-Tree: All Biorenewable, Optical and Radio Frequency Transparent Barrier Nanocoating for Food Packaging." *ACS Nano* 13 (4): 3796–3805.

Koshani, Roya, Jingbin Zhang, Theo G. M. van de Ven, Xiaonan Lu, and Yixiang Wang. 2021. "Modified Hairy Nanocrystalline Cellulose as Photobactericidal Nanofillers for Food Packaging Application." *ACS Sustainable Chemistry & Engineering* 9 (31): 10513–10523.

Kumar, Amit, Rohini Kanwar, and Surinder K. Mehta. 2022. "Recent Development in Essential Oil-based Nanocarriers for Eco-Friendly and Sustainable Agri-Food Applications: A Review." *ACS Agricultural Science & Technology* 2 (5): 823–837.

Kumar, Lokesh, Dakuri Ramakanth, Konala Akhila, and Kirtiraj K. Gaikwad. 2022. "Edible Films and Coatings for Food Packaging Applications: A Review." *Environmental Chemistry Letters* 20: 875–900.

Le Gars, Manon, Benjamin Dhuiège, Aurore Delvart, Mohamed N. Belgacem, Karim Missoum, and Julien Bras. 2020. "High-Barrier and Antioxidant Poly(lactic acid)/Nanocellulose Multilayered Materials for Packaging." *ACS Omega* 5 (36): 22816–22826.

Li, Qing, Ruohang Gao, Liying Wang, Mengyue Xu, Yang Yuan, Lulu Ma, Zhili Wan, and Xiaoquan Yang. 2020. "Nanocomposites of Bacterial Cellulose Nanofibrils and Zein Nanoparticles for Food Packaging." *ACS Applied Nano Materials* 3 (3): 2899–2910.

McClements, David Julian. 2020. "Nanotechnology Approaches for Improving the Healthiness and Sustainability of the Modern Food Supply." *ACS Omega* 5 (46): 29623–29630.

Mustafa, Fatima, and Silvana Andreescu. 2020. "Nanotechnology-Based Approaches for Food Sensing and Packaging Applications." *RSC Advances* 10 (33): 19309–19336.

Ndwandwe, Bongekile K., Soraya P. Malinga, Eugénie Kayitesi, and Bhekisisa C. Dlamini. 2021. "Advances in Green Synthesis of Selenium Nanoparticles and Their Application in Food Packaging." *International Journal of Food Science & Technology* 56 (6): 2640–2650.

Nile, Shivraj Hariram, Venkidasamy Baskar, Dhivya Selvaraj, Arti Nile, Jianbo Xiao, and Guoyin Kai. 2020. "Nanotechnologies in Food Science: Applications, Recent Trends, and Future Perspectives." *Nano-Micro Letters* 12: 45.

Priyadarshi, Ruchir, Zohreh Riahi, Jong-Whan Rhim, Sanghee Han, and Seok-Geun Lee. 2021. "Sulfur Quantum Dots as Fillers in Gelatin/Agar-Based Functional Food Packaging Films." *ACS Applied Nano Materials* 4 (12): 14292–14302.

Raul, Prasanta K., Angshuman Thakuria, Bodhaditya Das, Rashmi R. Devi, Gaurav Tiwari, Chidugundi Yellappa, and Dev Vrat Kamboj. 2022. "Carbon Nanostructures as Antibacterials and Active Food-Packaging Materials: A Review." *ACS Omega* 7 (14): 11555–11559.

Riahi, Zohreh, Parya Ezati, Jong-Whan Rhim, Reza Bagheri, and Gholamreza Pircheraghi. 2022. "Cellulose Nanofiber-Based Ethylene Scavenging Antimicrobial Films Incorporated with Various Types of Titanium Dioxide Nanoparticles to Extend the Shelf Life of Fruits." *ACS Applied Polymer Materials* 4 (7): 4765–4773.

Roy, Swarup, Parya Ezati, and John-Whan Rhim. 2021. "Gelatin/Carrageenan-Based Functional Films with Carbon Dots from Enoki Mushroom for Active Food Packaging Applications." *ACS Applied Polymer Materials* 3 (12): 6437–6445.

Siddiqui, Junaid, Mahtab Taheri, Arif Ul Alam, and M. Jamal Deen. 2022. "Nanomaterials in Smart Packaging Applications: A Review." *Small* 18 (1): 2101171.

Sirviö, Juho Antti, Mostafa Y. Ismail, Kaitao Zhang, Mysore V. Tejesvi, and Ari Ämmälä. 2020. "Transparent Lignin-Containing Wood Nanofiber Films with UV-blocking, Oxygen Barrier, and Antimicrobial Properties." *Journal of Materials Chemistry A* 8 (16): 7935–7946.

Wang, Pei-Long, Lin-Hua Xie, Elizabeth A. Joseph, Jian-Rong Li, Xiao-Ou Su, and Hong-Cai Zhou. 2019. "Metal-Organic Frameworks for Food Safety." *Chemical Reviews* 119 (18): 10638–10690.

Yang, Tianxi, Teena Paulose, Benjamin W. Redan, James C. Mabon, and Timothy V. Duncan, 2021. "Food and Beverage Ingredients Induce the Formation of Silver Nanoparticles in Products Stored within Nanotechnology-Enabled Packaging." *ACS Applied Materials & Interfaces* 13 (1): 1398–1412.

9 Nanotechnology for Enhancement of Food Value

Barbara Sawicka
University of Life Sciences

Piotr Barbaś
Plant Breeding and Acclimatization Institute – National Research Institute

Dominika Skiba
University of Life Sciences in Lublin

Piotr Pszczółkowski
Research Centre for Cultivar Testing

Barbara Krochmal-Marczak
Carpathian State College

CONTENTS

9.1 INTRODUCTION

The term "nanotechnology" was first used in 1974 in Japan at the University of Tokyo regarding the possibility of designing materials precisely at the nanometer level (Rathien and Read 2005). The science combines selected areas of solid-state physics, chemistry, materials science and molecular biology. Nanotechnology uses matter at the atomic, molecular and supramolecular scales (Awan et al. 2016).

DOI: 10.1201/9781003397861-9

Campbell (2015) defines nanotechnology as a branch of materials science. He states that under this umbrella term is a set of techniques and methods for creating a variety of structures of nanometric size (from 10 to 1,000 nm), i.e., at the level of single molecules. According to Buriro et al. (2023), nanoparticles are structures below 100 nm in size that have only one dimension. They have special, rare features due to their large surface area compared to their mass, which increases their reactivity. The most well-known nanoparticles are iron oxide, manganese oxide, or zinc oxide, especially noble metal nanoparticles.

The use of nanotechnology in products seems limitless. Nanoparticle materials will be widely used in medicine, pharmacy, food processing, food storage, and the chemical and cosmetics industries (Shoseyov and Levy 2008; Pulizzi 2016; Mohammad et al. 2022). From nanoproducts, there are nanocapsules containing drugs that reach a specific place, creams containing nanoparticles that absorb UV radiation and protect against sunburn, dressings that prevent infection and accelerate wound healing, polymer packaging for food storage and equipment covered with coatings containing nanoparticles with antimicrobial activity, among others (Pulizzi 2016; Pateiro et al. 2021). The use of nanocompounds is support for the protection, preservation and increase of suitability of food for consumption. Metallic nanoparticles (MNP) are part of antibacterial products that extend the shelf life of food products. Biopolymers/polymers are easily combined with various MNPs, which enables the combination of organic chemicals with metal nanocompounds, which in turn act as transporters for many materials (Buriro et al. 2023). However, a risk assessment is required before using nanoparticles in food packaging. Quality and safety are desirable in food production technology, including MNP, especially in its preservation. Nanotechnology creates various proposals for solutions for the food industry, from food production through its processing, preservation and storage to packaging. Nanomaterials offer an opportunity to improve the quality and safety of food and, at the same time, generate its health benefits (Dasgupta et al. 2015).

In food production, nanotechnology is used for primary production (e.g., crops and feed), processing of agricultural products, their storage, and packaging. For the production of nanomaterials, so-called techniques are used top-down (reduction of particle sizes) or bottom-up (building new structures from existing nanoparticles). It is also possible to use nanotechnology in food monitoring and control. Through the use of nanomaterials, nanobiosensors, nanobiorobots and nanobioreceptors, among others, in health care, agriculture, food processing, electronics, energy and transport, it has practically differentiated every industry (Anonymous 2023). Therefore, there is a possibility for participants in various industries to take advantage of the growing market, thanks to the income from commercialization technology (Buriro et al. 2023). Nanotechnology itself and its products are cheaper, more efficient and safer than earlier technologies. As the world's natural resources are shrinking, nanotechnology is an open door to the future (Anonymous 2022; Malik et al. 2023). The recent activities of scientists in the field of nanotechnology provide a significant impetus for the industry to deal with all aspects of the use of nanotechnology in food science.

9.2 NANOTECHNOLOGY IN FOOD PROCESSING

Nanotechnology is used in food processing to improve the quality and efficiency of production as well as introduce functionality into its products. This field has provided and still provides many alternative approaches in this industry, both in terms of improving the physicochemical, organoleptic, sensory and rheological properties as well as increasing the stability of nutrients and their bioavailability (Malik et al. 2023).

Approaches to the synthesis of nanoparticles depend on the size of the starting material and are divided into top-bottom and bottom-up due to the possibility of use in the production of functional food or nutraceuticals, i.e., food additives. Therefore, nanobiotechnology will affect the food production process (from farm to fork), as well as the preservation of products, the improvement of their organoleptic, sensory and rheological characteristics, and the proper creation of food packaging. Manufactured nanomaterials are also used in the food industry (Table 9.1). In particular, nanoparticles can occur in so-called functional foods, which thanks to them may have interesting properties, e.g., a salty taste with low sodium content (Leśkiewicz 2013). Food nanoencapsulation is a technological process that consists of creating microscopic capsules of nutrients, additives or flavors that are then added to food to improve its taste, smell or nutritional value. These capsules are typically a few to several hundred nanometers in size and consist of materials such as polysaccharides, lipids or proteins. They can be produced using various techniques such as emulsification, encapsulation or co-preservation. Food nanocapsules are currently widely used in the food industry; however, there are some concerns about their health safety (Pulizzi 2016; Shukla et al. 2017; Petkova et al. 2022). Microencapsulation and nanoencapsulation technologies (ME) enable the production of microcapsules or nanocapsules that have different sizes and internal structures. The distinction between them is based on their size, shape and structure. Microcapsules are spheres or spheres with a diameter of 1–800 μm that contain an internal phase such as a liquid, particulate or gas. Microcapsules can be made of various materials such as polysaccharides, proteins, lipids and synthetic polymers (Awan et al. 2016; Shukla et al. 2017; Patra et al. 2018; Umachandran et al. 2018). Nanocapsules, on the other hand, are spheres or spheres less than 1 μm in diameter that also contain an internal phase such as a liquid, solid particle or gas. They are typically made of materials such as polysaccharides, proteins, lipids or synthetic polymers (Pateiro et al. 2021). In the case of micro- and nanocapsules, the internal structure can be controlled during production, which allows you to control the release of substances contained in the capsules. For example, the release of the substance can be controlled by the composition and wall thickness of the capsule, which allows for sustained release of the active ingredients in the body. Microencapsulation or nanoencapsulation technology requires a precise manufacturing process that should be planned and controlled using appropriate software, such as Microsoft Process Manufacturing (Patra et al. 2018; Mohammad et al. 2022; Buriro et al. 2023; Husain et al. 2023; Martins et al. 2023) (Table 9.1).

The classification and division of nanotechnologies were presented by Mohammad et al. (2022) (Figure 9.1). The use of nanomaterials in the food sector is divided into different categories based on the type of nanomaterials used and their intended purpose.

TABLE 9.1
Examples of Nanotechnology of Encapsulation and Transport of Functional Food Ingredients

Nanotechnologies	Feature	Examples	References
Edible coatings	To preserve the quality	Gelatin-based edible coatings containing cellulose nanocrystal Chitosan/nanosylica	Pulizzi (2016), Patra et al. (2018), El-Naggar et al. (2022), Mohammad et al. (2022), Buriro et al. (2023)
Hydrogels	Can be easy placed into capsules, protects drugs from extreme environments	Protein hydrogels	Buriro et al. (2023)
Polymeric micelles	Solubilize water-insolublecompounds	PEO-b-PCL polyethylene glycol block-poly(caprolactone) Methoxy polyethylene glycol) palmitate polymeric micelles	Shukla et al. (2017)
Nanoemulsions	Stabillity to dropled agregation	Beta-carotene based on Nano emulsion	Pulizzi (2016), Pateiro et al. (2021)
Liposomes	Since liposome surrounds an aqueous inside a hydrophobic membrane	Cationic lipid	Shukla et al. (2017)
Inorganic NPs	They display good encapsulation capability and their rigid surfaces allow controlled functionalization	Mesoporus sillica nanopracticles	Awan et al. (2016)

Nanomaterials used in the food sector can be in the form of powders, liquid solutions or suspensions and can be added to foods to improve their quality, safety and shelf life. For example, nanoscale particles of certain metals, such as silver, can be used as antimicrobial agents to inhibit the growth of bacteria and fungi in food. Nanocapsules and nanoliposomes can be used as delivery systems for bioactive compounds, vitamins and flavors, allowing for better absorption and increased bioavailability (Panyala et al. 2008; Langauer-Lewowicka et al. 2014; Silver 2003; Milea et al. 2020; Mohammad et al. 2022). However, the use of nanomaterials in the food industry also raises concerns regarding their safety and potential adverse health effects. As a result, regulatory agencies around the world have implemented guidelines and regulations regarding the use of nanomaterials in food and food packaging to ensure their safety and minimize potential risks (Malik et al. 2023).

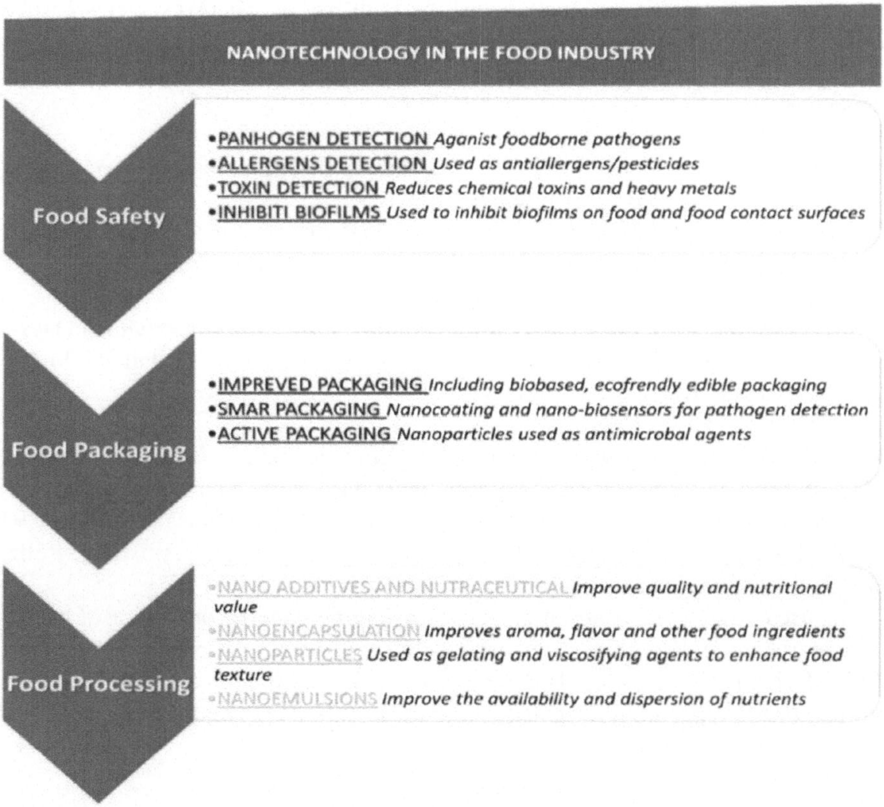

FIGURE 9.1 Classification of nanotechnology in food.

9.3 NANOTECHNOLOGY IN FOOD PACKAGING

The following are currently used for the production of packaging with unique properties:

1. Nanocomposites—whose task is to extend the durability of food, e.g., bottles, various types of barrier films.
2. Bactericidal and anti-allergic textiles covered with an impregnating layer of silica powder with an admixture of metallic silver nanocultures.
3. Materials detecting bacterial contamination.
4. Silica nanopowders for catalytic and analytical applications, phyllosilicates as fillers for polymer composites (Silver 2003; Pulizzi 2016).

Production of packaging with carbon tubes built into their structure. This type of packaging has better mechanical strength, and additionally, nanotubes destroy bacterial cells, thus protecting the product against their adverse effects (Waszkiewicz-Robak and Świderski 2008; Pulizzi 2016; Umachandran et al. 2018).

The following are currently used for the production of packaging with unique properties:

1. Nanocomposites—whose task is to extend the durability of food, e.g., bottles and various types of barrier films.
2. Bactericidal and anti-allergic textiles covered with an impregnating layer of silica powder with an admixture of metallic silver nanocultures.
3. Materials detecting bacterial contamination.
4. Silica nanopowders for catalytic and analytical applications, phyllosilicates as fillers for polymer composites (Echegoyen and Nerín 2013; Pulizzi 2016; Maliket al. 2023).

Production of packaging with carbon tubes built into their structure. This type of packaging has better mechanical strength, and additionally, nanotubes destroy bacterial cells, thus protecting the product against their adverse effects (See Figure 9.2) (Jakubiak and Foltynowicz 2004; Waszkiewicz-Robak and Świderski 2008; Pulizzi 2016; Umachandran et al. 2018).

The use of nanocomposites as an active packaging material or coating of packaging materials can also be used to improve food packaging (Echegoyen and Nerín 2013; Mackevica et al. 2016). The agro-food processing industry is interested in the antimicrobial properties of organic compounds such as essential oils, organic acids

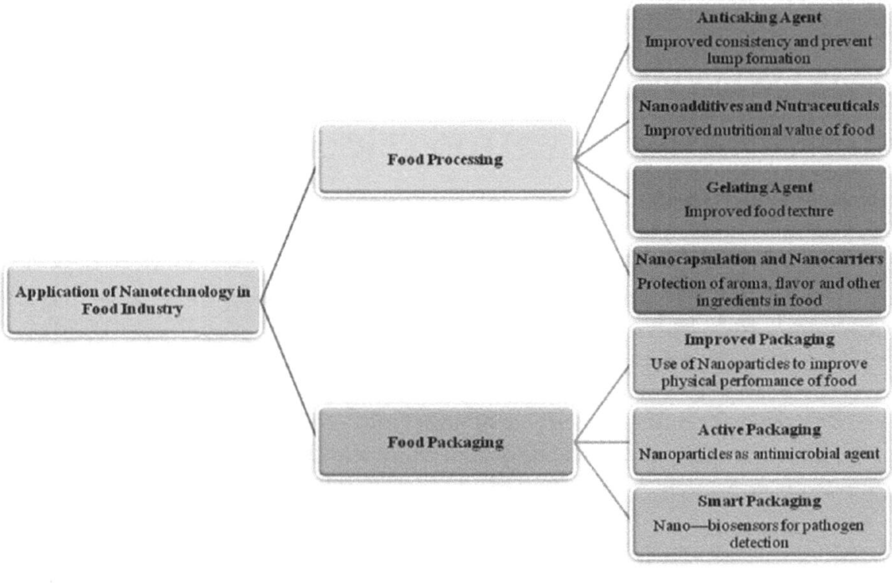

FIGURE 9.2 The role of nanotechnology in various aspects of the food sector.

and bacteriocins and their use in polymer matrices as safe antimicrobial packaging (Echegoyen and Nerín 2013; Umachandran et al. 2018).

Innovation at the intersection of food packaging and food safety is extremely important, as packaging plays a key role in maintaining food quality and safety. They protect food from contamination, oxidation, moisture, light and other factors that can affect food quality and safety. Modern packaging technologies allow for longer food storage, reduce food waste and prevent the loss of nutritional value (Teymouri and Shekarchizadeh 2022).

1. Innovations at the intersection of food packaging and food safety include:
2. biodegradable and ecological packaging—packaging that does not pose a threat to the natural environment after use, e.g., packaging made of cellulose, starch, algae, etc. (El-Naggar et al. 2022);
3. intelligent packaging equipped with sensors that allow you to monitor the condition of food, e.g., temperature, humidity and pressure, and also indicate the expiration date (Mihindukulasuriya and Lim 2014);
4. antimicrobial packaging, controlled release packaging and biosensors;
5. packaging that increases the shelf life of food—packaging with additives that extend the shelf life of food, e.g., oxygen absorbers, antimicrobiotics, etc. (Lin et al. 2014);
6. packaging safe for health—packaging that does not contain harmful substances, such as Bisphenol A, phthalates and polychlorinated biphenyls (PCBs) (Lin et al. 2014; Echegoyen and Nerín 2013; Mackevica et al. 2016; FDA, 2023; Malik et al. 2023).

Many nanoparticles, such as copper, chitosan, gold and silver, as well as metal oxide nanoparticles, such as titanium oxide and zinc oxide, have antibacterial properties. Lin et al. (2014) proved the migration of Ti from nano-(TiO_2)-polyethylene composite packaging to food simulators.

9.4 NANOTECHNOLOGY IN FOOD SAFETY

Nanoparticles are characterized by specific physicochemical properties compared to macroparticles of the same composition, so they can interact with all living organisms, causing toxicity. Recently, liposomal nanoparticles have been recognized as innovative in the food and medical industries (Singh et al. 2017; Shukla et al. 2017; Mohammad et al. 2022). Liposomal particles have a vesicular structure and consist of one or more lipid membranes that surround an aqueous interior. Thanks to this, they have many applications in the food industry. For example, they are used to deliver drugs directly to cells, which increases their effectiveness and reduces the risk of side effects (Shukla et al. 2017).

In the food industry, liposomes can be used as carriers of substances that can be added to food products, such as vitamins, antioxidants or anti-inflammatory compounds. Liposomes can also be used to protect against UV and ionizing radiation, as lipid membranes can act as a protective barrier for cells. Liposomes are also used as carriers for DNA or RNA that can be introduced into cells. This allows the treatment of genetic diseases and other diseases associated with genetic defects (Mohammad et al. 2022). Liposomal particles have a vesicular structure and consist of one or more lipid membranes that surround an aqueous interior. Thanks to this, they have many applications in the food industry (Chen et al. 2006). For example, they are used to deliver drugs directly to cells, which increases their effectiveness and reduces the risk of side effects (Shukla et al. 2017).

However, despite technical and technological progress in the storage and preservation of food, controlled and monitored sanitary conditions and restrictive regulations in this area, food safety is still a serious problem for civil society around the world (Shukla et al. 2017; Mahommed et al. 2022). Both pathogens and their toxins can be transmitted through food and cause various diseases. This is a real threat both for human and animal health (Singh et al. 2017; Umachandran et al. 2018; Buriro et al. 2023).

The most important part of all the main segments, given the major changes taking place in European countries that are further accelerating the growth of the European market, is nanotechnology. The applications section also allows the use of nanotechnology in electronics, energy technology, security system protection, environmental protection and transport. The nanotechnology market can be divided into nanomaterials, nanotools and nanodevices (EFSA 2020; Malik et al. 2023). Within the EU, five main countries can be distinguished, including Germany, France, Austria, the UK and Spain, which are at the top of the EU in terms of the number of patents in nanotechnology. This is also confirmed by the profiles of market players such as ASML, STMicroelectronics, Bruker Nano Analytics, etc. (EFSA 2023). The usual traditional methods for detecting contaminants in food or animal feed are inaccurate or self-contaminating during processing, while classical molecular techniques used to identify toxins or pathogenic microorganisms are very expensive (EFSA 2023). Nanotechnology in general and antimicrobial nanomaterial coatings in particular should potentially ensure food safety in the food industry related to surface contamination by preventing microbial adhesion and biofilm formation (Czyż et al. 2011; Shukla et al. 2017; Wu et al. 2023).

9.5 NANOTECHNOLOGY IN ANIMAL FOOD PRODUCTION

Nanosensor technology for the production and testing of (manufactured) food and nano-layer antimicrobial coatings for food industry apparatus. According to Główd et al. (2014), depending on the degree of safety of nanofoods, they are divided into three groups: foods of least risk, foods of moderate risk and foods of special risk. This division is based on the way the human body digests them, the way they move through the human digestive tract, and whether food additives are persistent, insoluble, or not digested by the digestive tract at all. So-called nanofillers (e.g., fullerenes or nanotubes) have found their way into food production and have been used as carriers of bioactive substances—so-called controlled dosage reservoirs of bioactive substances, i.e., carriers of flavor, aroma and active substances of medicines and as carriers of appropriately dispersed active substances in food, e.g., in functional foods and dietary supplements (Waszkiewicz-Robak and Świderski 2008; Langauer-Lewowicka and Pawlas 2014; Głód et al. 2014). Many chemical companies taking advantage of nanotechnology advances are working on or have already introduced potentially more effective pesticides, plant growth regulators and chemical fertilizers. One US agrochemical company has developed a method to further reduce the size of particles and chemical emulsions to the nanoscale and has designed nanocapsules that are designed to burst under the right conditions and the substance contained therein to activate under the influence of sunlight on the applied plant or object (Perez-de-Loue 2009). The advantage of nanocapsules containing chemical or biological plant protection products and growth promoters is their slow mode of action, gradually releasing these active ingredients. Preparations containing nanoparticles against pests are also available. An example is the production of pheromones by IBC for aphid control.

Nanotechnology has also found application in animal production and the related problems of environmental pollution through the emission of unpleasant odors and gases that contribute to global warming. The use of coating solutions seems to be more effective in this respect, with a catalytic effect. Myczko (2006) reports that the use of catalytic oxides containing silver nanoparticles (Ag) and photocatalysts with titanium nanoparticles (TiO_2) is used in animal husbandry, primarily for sanitation and deodorization of livestock buildings and the reduction of gas emissions, mainly ammonia (by about 50%) and nitrous oxide (from 68% to 75%). In addition, titanium oxide nanoparticles have the ability to neutralize odors, which can be used in the modern construction of large-scale livestock facilities, mainly piggeries and chicken coops. It is emphasized that the interactions between preparations containing nanoparticles and animals, insects and microorganisms in livestock buildings and their impact on products of animal origin are not fully understood (Liu et al. 2014).

In animal production, based on nanotechnology, there are, e.g., preparations of organic origin (viruses, fullerenes, carbon nanotubes, dendrimers) and inorganic: Ag, Au and Cu metals; metal oxides; TiO_2, ZnO and CuO; oxides of indium (In), tin (Sn), cerium (Ce), aluminum (Al) and other metals; quantum dots; ionic ceramics for decontamination and disinfection of livestock facilities, equipment, clothing and air-conditioning installations (Silver 2003; Langauer-Lewowicka and Pawlas 2014; Liu et al. 2014; Shukla et al. 2017). The properties of nanocoatings are mainly used to keep air-conditioning filters and filter-ventilation systems clean. They contain active nanoforms of elements or chemical compounds, such as silver, copper, or silica with a strong antimicrobial effect. Their advantage is the possibility of using them in the presence of people and the very low concentration of these elements in the preparations (below 10 ppm) (Czyż et al. 2011; Patra et al. 2018). The strong bactericidal and fungicidal effects of copper and silver have been known since ancient times. A good example was the storage of water, milk and other drinks in silver vessels by Sumerian sailors and the Phoenicians. Beverages stored in this way were fit for consumption for a long time (Silver 2003).

Nanoparticles in the animal feed industry can be a carrier for feed as a feed additive, or a specific feed or its specific ingredient can be fragmented to the size of nanostructures. Because nanoparticles have the ability to penetrate cells and tissues, they can be used, for example, in the production of feedstuffs. Therefore, they are used to create capsules or matrices that protect the active feed ingredient from degradation in the body (e.g., in the rumen) until it reaches its destination. An example may be nano-liposomes, which, by encapsulating the feed particle, protect it against oxidation and digestive enzymes of rumen microorganisms, thus increasing the bioavailability of certain active ingredients, such as antioxidants, B vitamins and vitamin K (Shukla et al. 2017). Moreover, there are applications of liposomal nanoparticles in food (Chen et al. 2006; Couch et al. 2016; Shukla et al. 2017; Sawicka et al. 2019).

9.6 SAFETY OF NANOPRODUCTS

The safety of nanoproducts should be regulated by human safety regulations. Such regulations are formulated and legally sanctioned in the USA; in the European Union, their functioning and observance are worse (Buriro et al. 2023). Therefore, it is necessary to analyze the ability of the environment to degrade materials for human safety before they are used in food and packaging (Macêdo et al. 2020). The emission of graphene-based carbon nanotubes from polymilk nanocomposite films was reported by Kotsilkov et al. (2018). A major environmental problem is the impact of pollution and the emission of nanocompounds from the polymer film. With PLA sheets and the migration of nanoparticles from them, as calculated, the migration limit is from 0.028 to 0.053 mg/cm^2. These values are below the total EU level (0.10 mg/cm^2). According to Dimitrijevic et al. (2015), the quality of nanomaterials is determined by their impact on humans. Although the exact mechanism is unknown, it is assumed that oxidative stress is the basis of the toxicity of nanomaterials and nanoproducts. The key problem is the rules for the use of nanomaterials in food. Indeed, they fall within the scope of the Novel Food Regulation (258/97), which applies to "foods and food ingredients containing novel or intentionally modified primary molecular structures" and is subject to a pre-authorization risk assessment procedure for authorized products when used as the main ingredient (e.g., nanoemulsions). They must be previously entered in the EU register and use this entry if they are used as food additives, in accordance with a separate article in relation to methods (Regulation 1333/2008) (Gallocchio et al. 2015). This aspect is very important to underline because, for example, Switzerland is the only country where nanoparticle rules are incorporated into existing legislation; in comparison, in other European countries, nanoparticle management policy is based only on information from the industry (Buriro et al. 2023).

The most important problems related to nanotechnology include safety aspects related to nanomaterials and the possibility of nanoparticle migration (Buriro et al. 2023). A new tool for genome alteration is the bacteriophage and plasmid resistance system of prokaryotes (CRISPR/Cas9) (Rozov et al. 2019; Ahmad et al. 2020). This technology has emerged as a new way and possibility of application in science in various research studies used to develop new varieties of crops with improved traits and in many other fields of science and industry. This method is known by other names, such as "DNA scissors" or "molecular scissors." It allows for very precise DNA cuts for insertion, repair, or gene editing (Witkowska et al. 2017). Currently, a big problem is the regulation of crops resulting from this method. For example, the United States deregulates CEO-edited organisms under its laws, while the EU seeks to regulate all crops produced by this method as genetically modified (GM) crops. More importance should be given to the regulation of products and processes in genetically modified (GM) crops, as genome editing rules are also based on existing legislation. Similarly, the results of targeted nucleases (e.g., ZFN, TALEN, CRISPR/Cas) are the same as those of conventional mutagenesis techniques, although public opinion is divided on how to regulate them (Rozov et al. 2019).

Witkowska et al. (2017), Ahmad et al. (2021) proved that each nanomaterial has its own individual properties, so toxicity can probably be determined on a case-by-case basis. Therefore, the legal regulation must also apply to standards for commercial products to ensure their high quality, as well as to prepare appropriate occupational health and safety regulations and ensure environmental protection (Waszkiewicz-Robak and Świderski 2008; Suwała and Pietruszka 2017; Shukla et al. 2017).

9.7 TOWARD THE FUTURE

Nanotechnology will be the next step in the generation of technologies (10), which will make the modernization of the industry and investments in the field of nanotechnology necessary in the coming years. Nanotechnology will be the next step in the generation of technologies (Malik 2023), which will make the modernization of the industry and investments in the field of nanotechnology necessary in the coming years.

The "Europe Nanotechnology Market Outlook 2027" presented the results of research on the market divided into segments of the nanotechnology industry and interpreted the trends and market changes that are taking place and the opportunities to accelerate the growth of the industries. The European nanotechnology market is seeking growth with private government R&D support, cross-country partnerships and strategic alliances, and an increase in commitment to other more powerful devices is expected. Currently, the health service, the agricultural sector and the agri-food industry are the main sectors in Europe where nanotechnology is part of an important breakthrough in the service of agriculture in the diagnosis and treatment of chronic diseases, e.g., diseases of the digestive, circulatory and neurodegenerative systems (Azadfar et al. 2022). Therefore, the nanotechnology market should be analyzed by segment (by type, region and application) to gain insight into the source of the market scenario as well as forecasts for higher quality segments until 2027. It is allowed to use nanotechnology in other branches of the economy, in environmental protection and in transport. Particular sectors of the European and global markets affect the division of the nanotechnology market into nanomaterials, nanotools and nanodevices. The greatest development of nanotechnology in the European market is observed in the US and EU countries, both in terms of research and development, analysis of nanotechnology patents, financing and regulations, as well as in-depth monitoring of investments and recent research and development in the field of nanotechnology. The applications of nanotechnology are vast, yet the risks associated with unlimited applications remain unclear and unarticulated. There is therefore a need for measures such as education on pollution and health problems. By combining the aspect of sustainable development with the development of nanotechnology, a successful and optimal future for nanotechnology can only be ensured.

9.8 CONCLUSIONS

Nanotechnology has many applications in food production, both in the production process and in the food products themselves. This brings with it an improvement in food quality and a number of amenities, such as:

1. Increasing food durability by using nanoparticles of silver, zinc oxide and other nanomaterials in packaging or adding them directly to food to improve shelf life and delay oxidation and spoilage processes;
2. Improving the nutritional value by increasing the bioavailability of nutrients, such as vitamins and minerals, which allows the body to absorb them better;
3. Improving the texture and taste of food products thanks to the use of nanoproducts, which allow for better quality and greater variety;
4. Streamlining production processes allows for faster and more efficient production processes;
5. Improves emulsion stability and solubility of ingredients in food products such as margarines, sauces and powdered products;
6. Increasing health security: nanotechnology improves the safety of food of both plant and animal origin;
7. Nanometals in ionic or non-ionic form are components of nanosols with antistatic and antibacterial properties;
8. Modern applications of metallic nanoparticles can be used to detect and treat cancer in medicine and regulate the growth of cancer cells. Organic nanoparticles are used in the treatment of malignant deposits. In turn, biogenic AgNPs act as strong antiviral agents with a broad spectrum of activity and also as antidiabetic and anti-inflammatory agents.

REFERENCES

Ahmad, Aftab, Zubair Ghouri Ghouri, Nayla Munawar, Muhammad Ismail, Sidra Ashraf, and Syed Ovais Aftab. 2021. "Regulatory, Ethical, and Social Aspects of CRISPR Crops." In *CRISPR Crops*, edited by Ahmad Aftab, Shahid Hameed Khan, and Zahoor Ahmad Khan, 129–146. Singapore: Springer.

Anonymous. "Europe Nanotechnology Market Outlook Report 2022–2027: Lucrative Opportunities for Industry Participants to Tap the Fast-Growing Market." PR Newswire. Last Modified 2022. https://www.prnewswire.com/news-releases/europe-nanotechnology-market-outlook-report-2022-2027-lucrative-opportunities-for-industry-participants-to-tap-the-fast-growing-market-301648784.html.

Anonymous. "Europe Nanotechnology Market Outlook Report 2022–2027: Lucrative Opportunities for Industry Participants to Tap the Fast-Growing Market." PR Newswire. Last Modified 2023. https://www.prnewswire.com/news-releases/europe-nanotechnology-market-outlook-report-2022-2027-lucrative-opportunities-for-industry-participants-to-tap-the-fast-growing-market-301648784.html.

Awan, Iqra Zubair, S.B. Hussain, Anwar Ul Haq, and Abdul Qadeer Khan. 2016. "Wondrous Nanotechnology." *Journal of the Chemical Society of Pakistan* 38 (06): 1026–1054.

Azadfar, M., F. Lashgarara, S.J. Farajollah Hosseini, M. Omidi Najafabadi, and H. Afshari. 2022 . "Using Nanotechnology in Food Production: Ensuring Food Security Development in Iran." *Journal of Agricultural Science and Technology* 24 (6): 1331–1343.

Buriro, Tahreem, Ghous Bakhsh Buriro, Ehtisham Taj, Hamza Tahir, Iqbal Jan, Muhammada Armughana Saghir, Mubeen Asad, Sabahat Umar, and Nayab Bano. 2023. "A Study on Applications and Usage of Metal-Based Nanoparticles in Foods." *Journal of Xi'an Shiyou University, Natural Science Edition* 19 (2): 645–651. ISSN: 1673-064X.

Campbell N., C. Deane, and P. Murphy. 2015. Adversing nanotechnology: Imaging the invisible. *Science technology and Human Values* 40 (6): 965–997.

Chen, H., J. Weiss, and F. Shahidi. 2006. "Nanotechnology in nutraceuticals and functional foods." *Journal of Food Technology* 60(3): 30–36.

Couch, Lynn Monahan, Michelle Wien, J. Lynne Brown, and Patricia Davidson. 2016. "Food Nanotechnology: Proposed Applications, Safety Issues." *Agro Food Industry Hi Tech* 27 (1): 36–39.

Czyż, Katarzyna, Zbigniew Dobrzański, Bożena Patkowska-Sokoła, and Magdalena Zabłocka. 2011. "Rozwój i zastosowania nanotechnologii." *Przegląd Hodowlany* 10:32–35.

Dasgupta, N., S. Ranjan, D. Mundekkad, C. Ramalingam, R. Shanker, and A. Kumar. 2015. "Nanotechnology in agro-food: From field to plate." *Food Research International*, 69: 381–400.

Dimitrijevic, Mirjana, Nedjeljko Karabasil, Marija Boskovic, Vlado Teodorovic, Dragan Vasilev, Vesna Djordjevic, Natasa Kilibarda, and Nikola Cobanovic. 2015. "Safety aspects of nanotechnology applications in food packaging." In *58th Meat Industry Conference "Meat Safety and Quality: Where it goes?"*, 57–60. Procedia Food Science.

Echegoyen, Yolanda, and Cristin Nerín. 2013. "Nanoparticle Release from Nano-Silver Antimicrobial Food Containers." *Food and Chemical Toxicology* 62: 16–22.

EFSA. "The European Union One Health 2020, Zoonoses Report." EFSA Journal. Last Modified 2020. https://www.efsa.europa.eu/en/efsajournal/pub/6971.

EFSA. "European Food Safety Authority." European Union. https://european-union.europa.eu/institutions-law-budget/institutions-and-bodies/institutions-and-bodies-profiles/efsa_pl. Accessed March 2023.

El-Naggar, Mehrez E., Hasanin Mohamed, and Hashem Amr H. 2022. "Eco-Friendly Synthesis of Superhydrophobic Antimicrobial Film Based on Cellulose Acetate/Polycaprolactone Loaded with the Green Biosynthesized Copper Nanoparticles for Food Packaging Application." *Journal of Polymers and the Environment* 30 (5): 1820–1832.

FDA. 2023. "FDA U.S. Food and Drug." https://www.fda.gov/. Accessed February 28, 2023).

Gallocchio, Federica, Simone Belluco, and Antonia Ricci. 2015. "Nanotechnology and Food: Brief Overview of the Current Scenario." *Procedia Food Science*, 5: 85–88.

Głód, Dagmara, Marek Adamczak, and Włodzimierz Bednarski 2014. "Wybrane Aspekty Zastosowania Nanotechnologii W Produkcji Żywności." Żywność. Nauka. Technologia. Jakość 5 (96): 36–52.

Husain, Shaheen, Aditya Nandi, Faizan Zarreen Simnani, Utsa Saha, Aishee Ghosh, Adrija Sinha, Aarya Sahay, Shailesh Kumar Samal, Pritam Kumar Panda, and Suresh K. Verma. 2023. "Emerging Trends in Advanced Translational Applications of Silver Nanoparticles: A Progressing Dawn of Nanotechnology." *Journal of Functional Biomaterials* 14: 47.

Kotsilkov, Stanislav, and Nikolay K. Vitanov. 2014. "Risk and nanotechnology: On risks connected to presence of nanoparticles in food and food packages." *Conference: Business Education and Market Dynamics. Proceedings of an International Conference*, 76–83. May 16–18, VarnaAt: Varna, Bulgaria.

Langauer-Lewowicka, Henryka, and Krystyna Pawlas 2014. "Nanocząstki, Nanotechnologia: Potencjalne Zagrożenia Środowiskowe I Zawodowe/Nanoparticles, Nanotechnology: Potential Environmental and Occupational Hazards." *Environmental Medicine* 17 (2): 7–14.

Lin, Qin Bao, Li On, Zhong Huai Ning, Zhao Qoan, Xiao Da-Hui, and Wang Zhi-Wei 2014. "Migration of Ti from Nano-TiO2-Polyethylene Composite Packaging into Food Simulants." *Food Additives & Contaminants Part A: Chemistry, Analysis, Control, Exposure & Risk Assessment* 31 (7): 1284–1290.

Liu, J., Y. Huang, A. Kumar, A. Tan, S. Jin, A. Mozhi, and X. J. Liang. 2014. "pH-sensitive nano-systems for drug delivery in cancer therapy." *Biotechnology Advances* 32(4): 693–710.

Macêdo, Daniel Charles Dos Santos, Cavalcanti, Iago Dillion Lima, Medeiros, Sandrelli Meridiana de Fátima Ramos Dos Santos, Souza, Jaqueline Barbosa de, Nogueira, Mariane Cajubá de Britto Lira, & Cavalcanti, Isabella Macário Ferro. (2022). "Nanotechnology and tuberculosis: An old disease with new treatment strategies." *Tuberculosis (Edinb)*, 135, 102208.

Mackevica, Aiga, Mikaela Emil Olssona, and Steffen Fossa Hansen. 2016. "Silver Nanoparticle Release from Commercially Available Plastic Food Containers into Food Simulants." *Journal of Nanoparticle Research* 18 (5). doi.10.1007/s11051-015-3313-x.

Martins, Valtera F.R., Manuela E. Pintado, Rui MSC Morais, and Alcina M.M.B. Morais. 2023. "Valorisation of Micro/Nanoencapsulated Bioactive Compounds from Plant Sources for Food Applications Towards Sustainability." *Foods* 12 (1): 32.

Malik, Sziza, Muhammad Khalid, and Waheed Yasir. 2023. "Nanotechnology: A Revolution in Modern Industry." *Molecules* 28 (2): 661.

Mihindukulasuriya, S.D.F, and Lim, L.-T. 2014. "Nanotechnology Development in Food Packaging: A Review." *Trends in Food Science & Technology* 40 (2): 149–167. ISSN 0924-2244.

Milea, Stefania Adelina, Vasile Mihaela Aida, Oana Crăciunescu, Ana-Maria Prelipcean, Gabriela Elena Bahrim, Gabriela Râpeanu, Anka Oancea, and Nicoleta Stănciuc. 2020. "Co-Microencapsulation of Flavonoids from Yellow Onion Skins and Lactic Acid Bacteria Lead to Multifunctional Ingredient for Nutraceutical and Pharmaceutics Applications." *Pharmaceutics* 12 (11): 1053.

Mohammad Zahra H., Faizan Ahmad, Salam A. Ibrahim, and Sadaf Zaidi. 2022. "Application of Nanotechnology in Different Aspects of the Food Industry." *Discover Food* 2 (12). doi.10.1007/s44187-022-00013-9.

Myczko, Andrzej. 2006. "Zastosowanie Nanotechnologii w Praktyce Rolniczej" (The Application of Nanotechnology to the Agricultural Practice). *Inżynieria Rolnicza* 2: 45–50.

Panyala, Nagender Reddy, Eladia M. Pena-Méndez, and Jozef Havel. 2008. "Silver or Silver Nanoparticles: A Hazardous Threat to the Environment and Human Health." *Journal of Applied Biomedicine* 6: 117–129.

Pateiro, Mirian, Belén Gómez, Paulo Eduardo Munekata, Francisco J. Barba, Predrag Putnik, Danijela Bursać Kovacevic, and José Manuel Lorenzo. 2021. "Nanoencapsulation of Promising Bioactive Compounds to Improve Their Absorption, Stability, Functionality and the Appearance of the Final Food Products." *Molecules* 26: 1547.

Patra, Jayanta Kumar, Han Seung Shin, and Spiros Paramithiotis. 2018. "Editorial: Application of Nanotechnology in Food Science and Food Microbiology." *Frontiers in Microbiology* 9: 714.

Perez de Luque, Alejandro. 2009. "Nanotechnology in Protection against Parasitic Plants." *Pest Management Science* 65: 540–545.

Petkova, Dorina, Dasza Mihaylova, and Ivelina Desseva. 2022. "Microencapsulation in Food Industry—An Overview." BIO Web of Conferences vol. 45, 02005, 68th Scientific Conference with International Participation "Food Science, Engineering and Technology—2021".

Pulizzi, Fabio. 2016. "Nanotechnology in Food: Silver-Lined Packaging." *Nature Nanotechnology* 18:5.

Rathien, D., and L. Read. 2005. Enabling Technologies for Australien Nanotechnology Innovative Industries. Prime Minister's Science, Engineering and Innovation Council ((PMSEIC), 11 March, 2005, 39 pp.

Rozov, Sergiej M., Natalia V. Permyakova, and Elena V. Deineko. 2019. "The problem of the low rates of CRISPR/Cas9-mediated knock-ins in plants: approaches and solutions." *International Journal of Molecular Sciences* 920 (13):3371.

Sawicka, Barbara, Honorata Danilcenko, Elvyra Jariene, Anna Kiełtyka-Dadasiewicz, and Dominika Skiba. 2019. "Importance of Nanotechnology in Agriculture in the 21st Century." Paper presented at the International Conference 'Rural Developments', September 19–21, 2019, Kaunas.

Shoseyov, O., and I. Levy. 2008. *Nano BioTechnology, BioInspired Devices and Materials of the Future*. 303–327. New Jersey: Humana Press.

Shukla, Shruti, Yuvaraj Haldorai, Seung Kyu Hwang, Vivek K Bajpai, Yun Suk Huh, and Young-Kyu Han. 2017. "Current Demands for Food-Approved Liposome Nanoparticles in the Food and Safety Sector." *Frontiers in Microbiology, Enclosed Food Microbiology* 8. doi.10.3389/fmicb.2017.02398.

Silver, Simon. 2003. "Bacterial Silver Resistance: Molecular Biology and Uses and Misuses of Silver Compounds." *FEMS Microbiology Reviews* 27(2–3):341–353.

Singh, Trepti, Shruti Shukla, Pradeepa Kumara, Verinder Wahla, Vivek K Bajpai, and Irfan A Rather. 2017. "The Application of Nanotechnology to Food Science: Perception and Review." *Frontiers in Microbiology, Section Food Microbiology* 8. doi.10.3389/fmicb.2017.01501.

Suwała, Grzegorz, and Jarosław Pietruszka. 2017. "Consumers' Awareness of the Application of Nanotechnology in Food Production." In *Food Quality and Safety Analyses, Trends, Challenges*, edited by Polskie Towarzystwo Towaroznawcze, 169–184. Kraków: Polskie Towarzystwo Towaroznawcze.

Teymouri, Zahra, and Hajar Shekarchizadeh. 2022. "A Colorimetric Indicator Based on Copper Nanoparticles for Volatile Sulfur Compounds to Monitor Fish Spoilage in Intelligent Packaging." *Food Packaging and Shelf Life* 33:100884, ISSN 2214-2894.

Umachandran, Krishnan, Barbara Sawicka, Arifullah Mohammed, Abdul-Nabi Nasir, and Antonella Pasqualone. 2018. "Relevance of Nanotechnology in Food Processing Industries." *International Journal of Agriculture Sciences* 10 (7):5730–5733.

Waszkiewicz-Robak, Bolesław, and Franciszek Świderski. 2008. "Health and Safety Aspects of Nanotechnology." *Bromatologia i Chemia Toksykologiczna* 41 (3):202–208.

Witkowska, Aleksandra, Agnieszka Kierepa, Arleta Kowala-Piaskowska, and Iwona Mozer-Lisewska. 2017. "Crispr/Cas9 Jako Nowa Metoda W Walce Z Hiv – Perspektywy I Historia Rozwoju." *Postępy Biologii Komórki* 44 (4):371–380

Wu, Tongjiao, Santudprom Phacharapan, Natsuki Inoue, Mitsuki M. Sakamoto, and Yoshinori Kamitani. 2023. "Antibacterial and Cleaning Efficacy of Alkaline Electrolytic Silver Ionized Water on E. coli Planktonic Cells, Biofilms and Sweet Potato Stains on Food Contact Surfaces." *Food Control* 149:109717. ISSN 0956-7135.

10 Resource Recovery from Biomass for Biofuel Generation

Nnabuk Okon Eddy
University of Nigeria

Rajni Garg
Galgotias College of Engineering & Technology

Richard Alexis Ukpe
Federal University Otuoke

Rishav Garg
Galgotias College of Engineering & Technology

Anduang Odiongenyi
Akwa Ibom State University

Eze Ifeanyi Samson
University of Nigeria

CONTENTS

DOI: 10.1201/9781003397861-10

10.1 INTRODUCTION

Evaluation of the global records on the consumption of fuels reveals that fossil fuels are inherently demanded by almost all nations of the world. Unfortunately, fossil fuel has environmental challenges such as the release of gaseous pollutants into the environment (Höök and Tang 2013), the tendency to cause global warming (Johnsson, Kjärstad, and Rootzén 2019), the fact that it is not renewable (Holechek et al. 2022) and the fact that the raw materials are natural and not evenly distributed around the globe (Herndon 2017). Besides, unpredicted events have continued to increase the global energy crisis, and the surge for a solution has limited avenues. The COVID-19 incident and the Russian invasion, for example, have markedly driven some countries to higher inflation (Allam et al. 2022). If the consideration of the listed facts is properly analysed, then the need for an alternative source of energy becomes necessary because of the following concerns:

1. Renewability: Unlike fossil fuel, renewable fuel will guarantee the ability to have the product without fear of depletion or interruption in production or supply.
2. Green environment: Greener or environmentally friendly fuel will ensure a lower contribution to environmental contamination or pollution.
3. Accessibility: An accessible source of fuel indicates that the raw materials and the production technology are relatively cheap and that the majority of consumers can have access to the product.

Renewable sources of energy include hydroelectric, solar, wind, biomass and geothermal power (Salvarli and Salvarli 2016). Huge successes have been recorded and sustained in the utilization of hydropower for the generation of electricity by several countries around the world (Ebhota and Tabakov 2018). However, the major challenges that are associated with hydropower stations are (i) a drop in water level during the dry season (Wei et al. 2020) and (ii) the source is most useful when it is naturally designed and is not common in most parts of the world. Wind power is sensitive to climate and weather changes and is therefore limited to some zones of the globe (Luthfi et al. 2023). Geothermal energy is also sensitive to climate and weather conditions and cannot be accessible to all (Salhein et al. 2022). Energy from biomass has been approved as the most accessible, feasible and relatively cleaner source (Reid et al. 2020). The major energy products that can be derived from biomass are biofuels such as biohydrogen, bioethanol, biodiesel, etc. (Jeswani et al. 2020). In addition to the previously highlighted advantages, biofuels have flexible feedstocks for their production and are less hazardous because of their low carbon emission capacity (Neupane 2022).

10.2 DIFFERENT CLASSES OF BIOFUELS

Biofuels can be classified based on different concepts, the source of raw materials and types of end products or feedstocks, among other considerations. However, one of the most acceptable and common classification systems is based on progress in the developmental trend of the source of feedstock over time. Consequently, we have first-, second- and third-generation biofuels (Awogbemi et al. 2021).

The first-generation biofuels mainly consider bioethanol and biodiesel and include those produced from edible plant feedstock, including starch-based tubers (such as cassava, potato, etc.), cereals (such as maize, wheat, etc.) and sugar-based plant materials (e.g., sugar cane). (Dahman et al. 2019). The second-generation (olive green) biofuels use lignocellulose and carbohydrates as the major feedstocks. The unique distinction between first- and second-generation biofuels is that the latter are synthesized from non-edible plants such as food waste, wood, waste vegetable oil, industrial wastes, grasses, agricultural wastes and some other biomass (Kumar et al. 2022). Third-generation biofuels are produced from algal feedstocks, including macroalgae and microalgae (Abbasi et al. 2021). The fourth-generation biofuels are products of enhanced oil-rich feedstocks produced from genetic engineering manipulations (Kumar et al. 2022). The major advantage of the fourth-generation fuel over the third-generation biofuels is the enhancement of the lipid content and biomass yield due to enhanced CO_2 sequestering capacity (Mizik and Gyarmati 2022).

The consideration of the generational evolution of biofuels reveals that second-generation biofuels have received intense promotion, diversification and applications compared to others. Therefore, this review shall concentrate on second-generation biofuels, with emphasis on the utilization of feedstock and catalyst from waste.

10.3 BIOFUEL SYNTHESIS

The initial step required in the production of biofuels is the acquisition of the required feedstocks. In consideration of second-generation biofuels, the feedstock is waste biomass. The feed stock must be converted to another form, as discussed ahead.

10.3.1 Bioethanol

In the production of ethanol, the feedstock may be waste food materials that are rich in cellulose or sugar. The conversion can be achieved by fermentation. The structural formula of the cellulose unit (which is the most abundant organic polymer) is shown in Figure 10.1. It consists of a beta-1,4 chain joined to glucose units. Cellulose

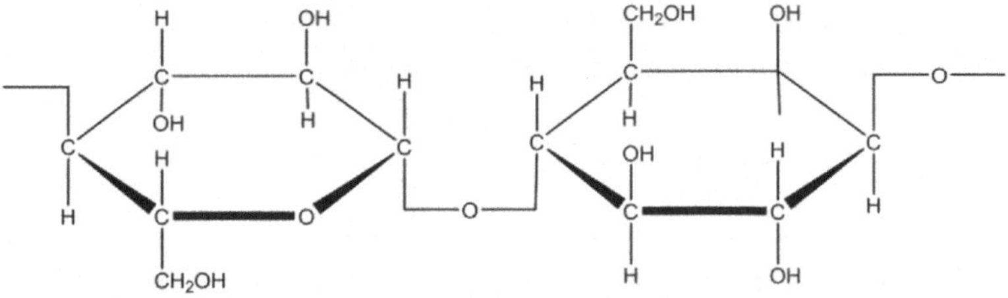

(1*S*,2*S*,3*S*,4*R*,5*S*,6*S*)-6-((((2*S*,3*R*,4*S*,5*S*,6*S*)-5-ethyl-3,4-dihydroxy-6-(hydroxymethyl)tetrahydro-2*H*-pyran-2-yl)methoxy)methyl)-5-(hydroxymethyl)-3-(methoxymethyl)-4-methylcyclohexane-1,2,4-triol

FIGURE 10.1 Structural formula of the cellulose unit.

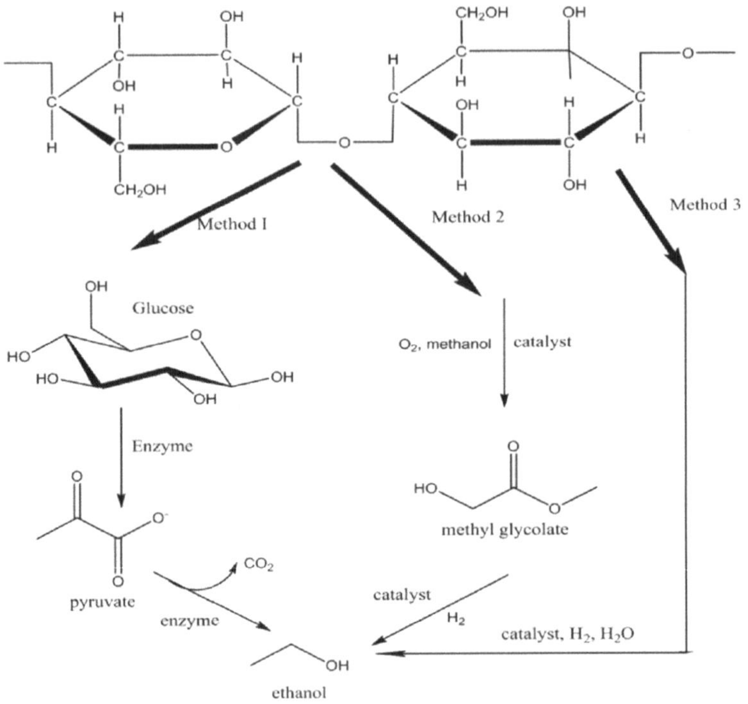

FIGURE 10.2 Scheme 1: Synthesis of bioethanol using cellulose as the starting material.

is present in higher plants, bacteria and some lower plants (Barahona et al. 2020). Three methods of conversion of cellulose to bioethanol are outlined in Scheme 1 (Figure 10.2). Song et al. (2019) have described a direct synthetic method for the production of bioethanol from cellulose that started with the hydrolysis of the cellulose to glucose, followed by two steps of H_2WO_4-catalysed retro aldol reactions, and finally, the catalytic reduction of ethane-1,2-diol (formed from glycoaldehyde) to ethanol using Pt/ZrO_2 mixed catalysed under the atmosphere of hydrogen gas. The reported synthesis is shown in Scheme 2 (Figure 10.3). Schemes 1 and 2 described above are non-enzymatic pathways. However, enzymatic pathways rely on the use of enzymes for catalysis such as yeast, etc. (Vasić et al. 2021).

10.3.2 Biodiesel Synthesis

The basic process in the synthesis of biodiesel is the catalysed transesterification of feedstock (basically lipids) with low molecular mass alcohol (Pasha et al. 2021). Consequently, the type of feedstock (lipid source) and catalysts are essential steps that can be controlled towards an optimum and efficient production process. Feedstock can be differentiated into first-, second-, third- and fourth-generation lipids, respectively. Also, catalysis can be enzyme-, acid-, alkali-, supercritical- and heterogenous-based.

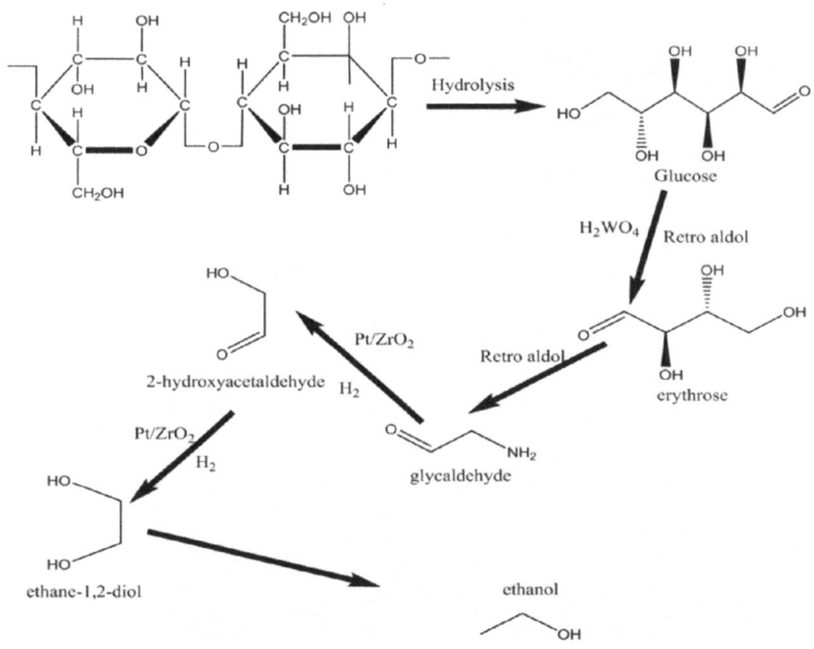

FIGURE 10.3 Scheme 2: Examples of synthetic routes for direct conversion of cellulose to bioethanol fuel.

The final product of the transesterification process is glycerol. At the end of the production, the biodiesel must be separated from the glycerol. Separation can be achieved using methods such as decantation, centrifugation (Abdul Rahman et al. 2022), membrane (Gomes, Pereira, and Barros 2010), electrostatic methods (Ribeiro et al. 2020), etc. Figure 10.4 shows the various steps involved in the biosynthesis of biodiesel, covering the conversion of different oils that are suitable feedstocks, different types of catalysts and the products of the transesterification process (biodiesel and glycerin), as well as some methods that can be used to separate biodiesel from the moisture.

10.3.3 Biohydrogen Gas

Biohydrogen is a useful gas for the synthesis of biogas, biodiesel, bioethanol and other products (Saratale et al. 2019). Bio-H_2 can be produced in several ways, including the following:

i. Production using microalgae, either as feedstock for fermentation by microbes or by direct production. The production route involving algae has extra advantages such as the absence of lignin, low cultivation cost, rapid growth rate and high biomass value (Dawood et al. 2021). Microalgae lack root stems and leaves but use sunlight, oxygen and water to enrich

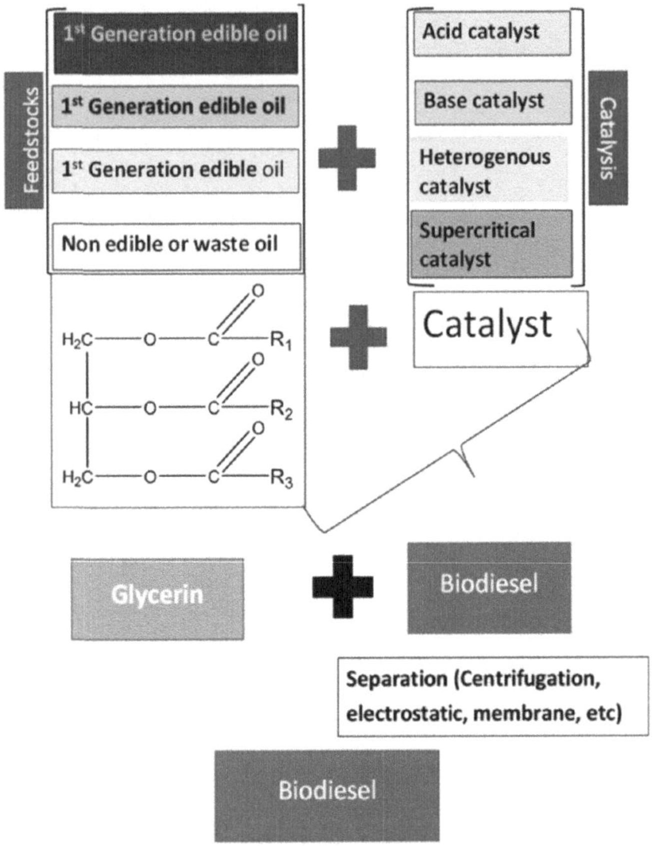

FIGURE 10.4 Demonstration of steps involved in the synthesis of biodiesel.

their biomass content, which can be harvested, concentrated, pretreated and applied as feedstock for the production of biohydrogen as outlined in Figure 10.5.

ii. Photolysis of water in the presence of cyanobacteria and green algae can produce biohydrogen from light using nitrogenase or hydrogenation, as demonstrated in Figure 10.6 (Kim et al. 2011).
iii. Lignocellulose-rich biomass has also been accepted as a good feedstock for the production of biofuel (Singh et al. 2015). Under this approach, optimum conditions would be defined by biomass that has a high carbohydrate content that can easily be processed or pretreated, is accessible and contains contents that can be easily fermented into biohydrogen.

An extension of the listed methods has been reported in terms of some stages of modification, such as pretreatment technology and hybrid technology such as photo and dark reactions, as well as direct and indirect photosynthesis (Chen et al. 2021).

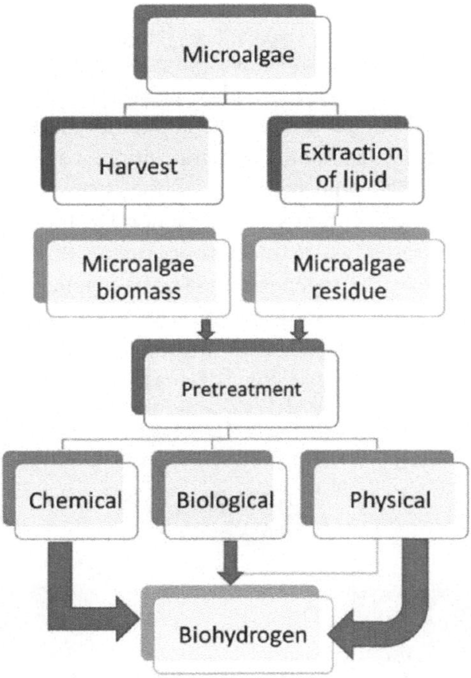

FIGURE 10.5 Steps towards the production of biohydrogen from microalgae.

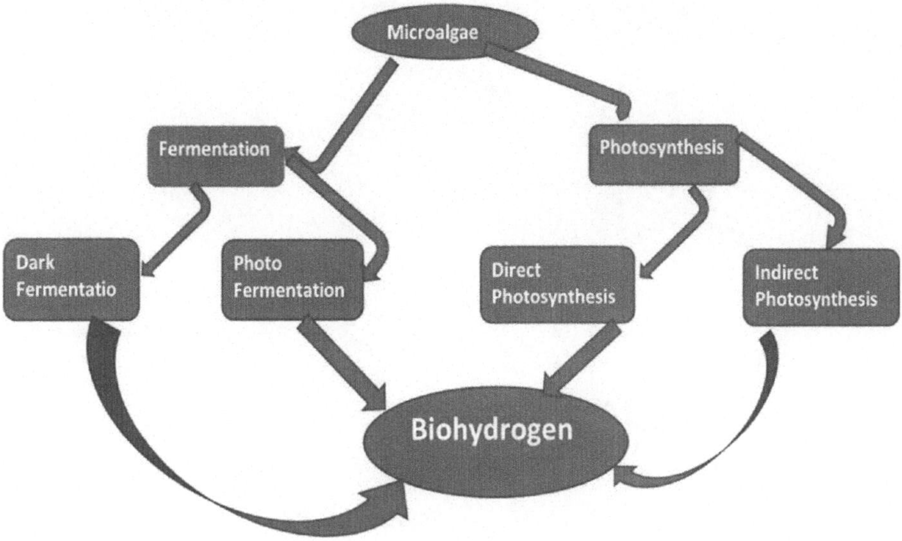

FIGURE 10.6 Biohydrogen synthetic route in the presence of microalgae as the feedstock.

10.3.4 Biomethane

The operation of domestic and industrial cooking systems requires natural gas to be replaced with biomethane (Onu and Mbohwa 2021). Other possible applications of biomethane include electricity generation and as fuel for some stationary and mobile engines (Ferreira et al. 2019). Biomethane composition is dominated by carbon, constituting about 50% to >70%, and hydrogen, which may constitute a maximum of 50% and a minimum of 20% (Adams 2017). Generally, biomethane can be synthesized by anaerobic digestion of waste by microorganisms through four stages, including hydrolysis, acidogenesis, acetogenesis and methanogenesis stages (Cakmak et al. 2021), as shown in Figure 10.7. Some biomass feedstocks that are useful for biomethane production have been reported to include agricultural waste, animal dung, etc. The major challenge to the efficiency of anaerobic digestion is the removal of by-products, which is about 30%–40% (Li et al. 2017).

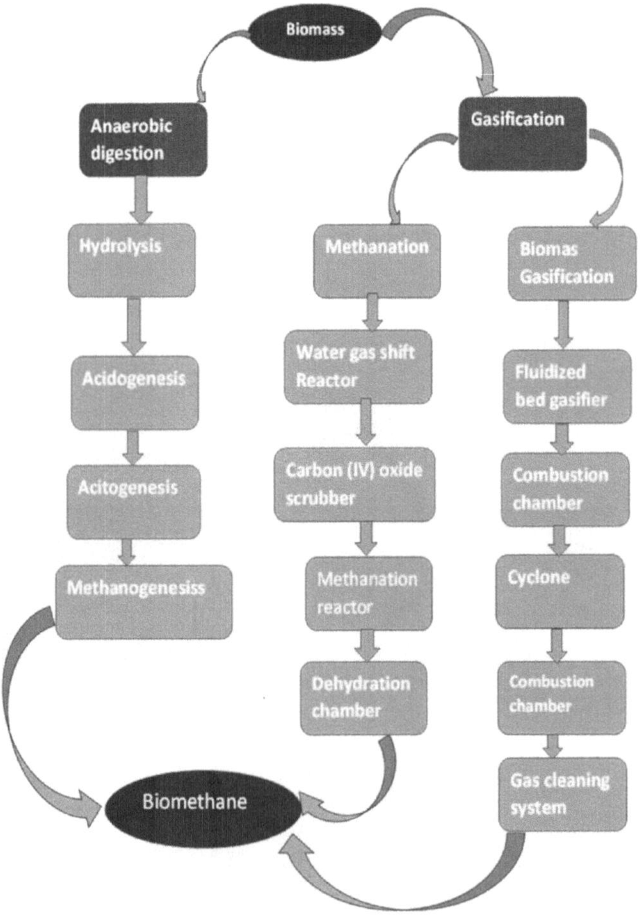

FIGURE 10.7 Methods of synthesizing biomethane.

10.4 RECENT DEVELOPMENTS IN BIOFUEL SYNTHESIS

The major component of the work reported by Selaimia et al. (2015) was the synthesis of biodiesel using olive residue and waste vegetable oil. The results generated from the investigation showed a pour point value of −18, a flash point of 75°C, a density at 15°C of 827.1 kg/m^3 and a viscosity of 5.12 mm/s. The physiochemical characteristics of the methyl ester from the raw materials were also provided by the authors. They concluded that the product aligned with the standard expected for quality biodiesel.

Recovery of catalyst from eggshell wastes was applied to catalyse the production of biodiesel from waste cooking oil by Primadi et al. (2021). More than 80% optimum yield was observed, and the reaction was almost unfeasible in the absence of the catalyst. Their study shed light on the replacement of edible materials with waste biomass for biodiesel production. A similar investigation was also undertaken by Gandhi et al. (2021), who also applied an egg shell as the catalyst for the manufacturing of biodiesel from waste oil and obtained an optimum yield of 95.5% after the application of the Deluca and Termini fuzzy entropy weighted indexes. Amesho et al. (2022) observed that oil from *Jatropha curcas* is a good feedstock for biodiesel production. An optimal production efficiency of 92% was observed under various experimentally controlled conditions.

Sivamani et al. (2022) were able to synthesize biodiesel with an 86% yield using sunflower oil and $Ca(OH)_2$ as the feedstock and catalyst, respectively. This optimum level of biodiesel production was achieved when the concentration of the catalyst was 23 g+0.3 L of methanol, the period was 24 hours and the agitation speed was 100 rpm. The accelerating function of activated carbon (prepared from beach tree biomass) in the catalysis of biodiesel production from vegetable oils was observed by Narowska et al. (2020). The biodiesel yield was 99% after 4 hours of preparation.

Elkady et al. (2015) reported their work on the production of biodiesel from waste vegetables using a K mixer reactor, such that a yield as high as 97% was recorded while the vapourization point of the biodiesel compared to the oil was much reduced (300°C–120°C). The presence of methyl ester in the final product was confirmed. Their study also covered the investigation of variables such as catalyst concentration (NaOH), methanol-to-oil ratio, mixed solvent (i.e., tetrahydrofuran and methanol) and reaction temperature. Waste cooking oil was also employed by Muppaneni et al. (2015) for the supercritical production of biodiesel, which met ASTM standards. The oil and the produced fuel were characterized. At ethanol-oil and hexane-oil ratios of 40:1 and 2:100, a biodiesel yield of 88% was observed at an operational temperature of 300°C for 20 minutes. Kavitha and Murugavelh (2021) have also reported high-quality biodiesel production from an equal ratio blend of waste cooking oil and *Sterculia foetida* oil. The mixed oil showed palmitic and linolenic acid contents of 18% and 21% respectively while the 95% product was recovered. Oil from *Croton gratissimus* was analysed by Jiyane et al. (2021) and applied in the manufacturing of biodiesel under a Zr/KOH-catalysed reaction condition. The optimum product yield of up to 85% was obtained, and the influence of several factors was investigated. The results were informative in highlighting the contribution of catalyst, time, mixing ratio and other factors. The research work by Sarno and Iuliano (2019) revealed *Thermomyces lanuginosus* (TL) lipase as an excellent precursor for the synthesis of

90% biodiesel using nanoparticles based on Fe_3O_4/Au composites. The yield efficiency averaging 90% was observed at a catalyst load of 2%, a temperature of 45°C, a substrate concentration of 20% and a methanol-to-oil ratio of 1:6.

Under varying experimental conditions, palm oil was used for biodiesel production by Jalaluddin et al. (2022) and a catalyst synthesized by calcination of kapok fruit peel waste. Varying conditions such as THF/methanol mixed, oil/methanol mixed, temperature and catalyst loads were investigated, and the optimum yield was above 90%. Also, low-quality palm oil was efficient in producing 85% biodiesel using 4% oil weight and CaO-ZnO composites as catalysts. Wahyono et al. (2022) produced biofuels from different blends of five oils mixed at different ratios and obtained efficiencies ranging from 92.99% to 99.58%. Other reported works on biodiesel production are presented in Table 10.1, including utilization of the non-edible and waste oils.

Nishida et al. (2023) reported the production of bioethanol from soybean hulls with a reported 79% yield, while Olawale et al. (2021) synthesized bioethanol by co-culture of *Aspergillus niger* and *Saccharomyces cerevisiae* using watermelon peels as substrate and obtained a yield exceeding 90%. A 91% yield of bioethanol has been reported for *Abrus seed flour* hydrolysate after 90 minutes (Obeta et al. 2021). Up to 90% yield was recorded for bioethanol obtained by Verhe et al. (2022) using an organic fraction of municipal solid waste and refuse-derived fuel. Waste paper was used as a feedstock for bioethanol production by Zimife et al. (2022), and the reported efficiency was up to 80%. The wastes were converted to glucose by treatments with H_2SO_4 and NaOH, and the bioethanol was obtained after the fermentation action of *Saccharomyces cerevisiae* ceased. The fermentation functions of *Aspergillus terreus* and *Trichoderma harzanium* were taken into advantage for the production of biodiesel from rice husk, and the yield obtained was favourably high, although *Aspergillus terreus* gave a better yield (Devi and Munjam 2022). According to Zhan et al. (2022), Bamboo stem was observed to yield 67% of bioethanol; however, improvement in the

TABLE 10.1
Review of Published Work on Biodiesel Production

Feedstock	Operating Condition/ Property	Catalyst	Optimum yield (%)	References
Rubber seed oil	Time = 60 minutes Temperature = 60°C	Natural zeolite supported PbO, ZnO and ZrO_2	86.22	Pulungan et al. (2021)
Microalgae oil	Oil:methanol = 1:15 Time = 50 minute Temperature = 70°C	Metal organic framework	92	Javed et al. (2023)
Soybean oil	Temperature = 50°C Stirlling speed = 320 rpm	NaOH	71.2	Hossain et al. (2010)

(Continued)

TABLE 10.1 (*Continued*)
Review of Published Work on Biodiesel Production

Feedstock	Operating Condition/ Property	Catalyst	Optimum yield (%)	References
Palm oil	Temperature = 50°C Time = 120 minutes	Calcined *Moringa* leaf	92.87	Kolakoti et al. (2022)
Sunflower oil	Temperature = 70°C	NaOH	92.94	Boucheta et al. (2023)
Brachychiton populneus seed oil	Oil: methanol = 1:9 Temperature = 87°C	Nickel oxide	97.5	Dawood et al. (2021)
Pongamia pinnata *Carthamus oxyacantha* *Citrullus colocynthis* *Sinapis arvensis* *Ricinus communis*	Time = 120 minutes	Nano magnetic Fe/SnO	97.4 98.1 98.5 97.0 97.0	Hanif et al. (2022)
Jatropha oil	Viscosity = 4.982 mm^2/s Density = 0.887 g/cm^3 Pour point = 8°C	H_2SO_4	98.00	Humphrey et al. (2017)
Waste vegetable oil	Viscosity = 4.706 mm^2s Densoty = 0.850 g/cm^3 Pour point = 4°C	H_2SO_4	95.00	Humphrey et al. (2017)
Castor oil	Viscosity = 3.867 mm^2/s Density = 0.860 g/cm^3 Pour point = 2°C	H_2SO_4	90.00	Humphrey et al. (2017)
Madhuca indica	Viscosity (cSt) = 4.81 Pour point = 9°C %S = nill Flash point (°C) = 124 Cloud point (°C) = 12	KOH/NaOH	74.36	Ahmad et al. (2011)
Jatropha curcas	Viscosity (cSt) = 5.67 Pour point = −9°C %S = 0.0008 Flash point (°C) = 96 Cloud point (°C) = −6	KOH/NaOH	70.00	Ahmad et al. (2011)
Pongamia pinnata	Viscosity (cSt) = 5.5 Pour point = 3°C %S = 0.008 Flash point (°C) = 90 Cloud point (°C) = 6	KOH/NaOH	77.16	Ahmad et al. (2011)
	Viscosity (cSt) = 3.83 Pour point = 6°C %S = nill Flash point (°C) = 120 Cloud point (°C) = −2.5	KOH/NaOH	70.00	Ahmad et al. (2011)
Palm oil	Time = 45 minutes	NaOH	96	Marwan et al. (2015)

pretreatment of the sample led to a yield approaching 80%, which clearly defined that the pretreatment step can affect the yield for bioethanol.

Based on the work reported by Amru et al. (2023), biohydrogen was produced by the fermentation of palm oil biomass carbon sources, and the yield was 1.55 mol H_2/mol of sugar. Ethanol, acetate and nutylate were recognized as the most suitable metabolites in the process. Mutsvene et al. (2023) reported the success behind the conversion of brewery waste to biohydrogen using an anaerobic baffled reactor. A production rate of 16.17 ml/h was observed, and it was concluded that the method is cost-effective. Based on the study conducted by Bao et al. (2012) on the application of the dark fermentation of starch by *Bacillus* sp. and *Brevumdimonas* sp., biohydrogen average concentration of 1.04 mol of biohydrogen per mole of glucose was obtained. The volume was two times higher than that obtained when either of the enzymes was used singly, which led to the conclusion that a combination of enzymes can increase the yield of biohydrogen. The production of 51% biomethane from fish waste has been reported by Ingabire et al. (2023). The C:N ratio was reported to be low for the synthesized biomethane. Literature has shown that the quality and yield of biomethane can be significantly affected by C/N and the types and levels of pretreatment. Consequently, Budiyono et al. (2023) reported that different pretreatments gave rise to different production efficiency of biomethane using coffee grounds and rich husk as feedstocks. Enhanced efficiency was observed when the C/N was low.

10.5 CHALLENGES AND FUTURE PROSPECT

The major challenges associated with the production of biofuels in their current state include the existence of appropriate technology, raw materials, environmental consequences, cost and equipment. From the analysed information, the production of biofuels requires less technology compared to fuels from petroleum refineries. Diverse forms of raw materials are useful for biofuel synthesis. Current efforts are targeting the application of waste biomass, non-edible or non-useful plant and animal biomass, microalgae, etc. Catalyst synthesized from waste materials such as egg shells, periwinkle and oysters has also been found to be suitable for the catalysis of biofuel production. The application of catalysts associated with particles in the nanodimension has also proven to enhance efficiency more than conventional ones. As shown in the foregoing analysis, biofuel has fewer environmental hazards compared to fossil fuels.

10.6 CONCLUSION

Biofuels such as bioethanol, biodiesel, biohydrogen and biomethane have significant potential for replacing fossil fuels. The resources required are abundant and cover a wide variety of materials that guarantee availability in almost all countries in the world. Biofuels are classified into four major classes based on generational uniqueness. The current generational focus on biofuel is also embracing the adoption of natural and artificial microalgae, waste biomass and other materials that will render other applications less competitive. The production requirements for biofuels differ for different types of biofuels but generally involve raw material acquisition, pretreatment, synthesis, product purification and storage. Waste biomass of various grades, sources and quality has found its usefulness in biodiesel, bioethanol, biohydrogen and biomethane production. Consequently, their application contributes to reducing the volume of waste in our environment through resource recovery.

REFERENCES

Abbasi, Mostafa, Mir Saman Pishvaee, and Shayan Mohseni. 2021. "Third-Generation Biofuel Supply Chain: A Comprehensive Review and Future Research Directions." *Journal of Cleaner Production* 323: 129100.

Abdul Raman, Aziz Abdul , Hooi W. Tan, and Archina Buthiyappan. 2022. "Two-Step Purification of Glycerol as a Value Added by Product from the Biodiesel Production Process." 7 (November 2019): 1–9.

Adams, Paul. 2017. "GHG Emissions from Biomethane Gas-to-Grid Injection via Anaerobic Digestion". In *Greenhouse Gas Balances of Bioenergy Systems*, edited by Patricia Thornley, Paul Adams, 141–58. Cambridge: Academic Press.

Ahmad, Mushtaq, Mir Ajab, Muhammad Zafar, and Shazia Sult. 2011. "Biodiesel from Non Edible Oil Seeds: A Renewable Source of Bioenergy." In *Economic Effects of Biofuel Production*, edited by Marco Aurélio dos Santos Bernardes. Rijeka: IntechOpen.

Allam, Zaheer, Simon Elias Bibri, and Samantha A. Sharpe. 2022. "The Rising Impacts of the COVID-19 Pandemic and the Russia-Ukraine War: Energy Transition, Climate Justice, Global Inequality, and Supply Chain Disruption." *Resources* 11 (11):99.

Amesho, Kassian T.T., Yuan Chung Lin, Chin En Chen, Pei Cheng Cheng, and Vinoth Kumar Ponnusamy. 2022. "Optimization and Kinetics Studies of Biodiesel Synthesis from Jatropha Curcas Oil under the Application of Eco-Friendly Microwave Heating Technique: An Environmentally Benign and Sustainable Bio-Waste Management Approach." *Sustainable Environment Research* 32 (4).

Awogbemi, Omojola, Daramy Vandi Von Kallon, Emmanuel Idoko Onuh, and Victor Sunday Aigbodion. 2021. "An Overview of the Classification, Production and Utilization of Biofuels for Internal Combustion Engine Applications." *Energies* 14 (18): 1–42.

Bao, Meidan, Haijia Su, and Tianwei Tan. 2012. "Biohydrogen Production by Dark Fermentation of Starch Using Mixed Bacterial Cultures of Bacillus Sp. and Brevumdimonas Sp." *Energy and Fuels* 26 (9): 5872–78.

Barahona, Patricia Portero, Bernardo Bastidas Mayorga, Jesús Martín-Gil, Pablo Martín-Ramos, and Enrique Javier Carvajal Barriga. 2020. "Cellulosic Ethanol: Improving Cost Efficiency by Coupling Semi-Continuous Fermentation and Simultaneous Saccharification Strategies." *Processes* 8 (11): 1459.

Boucheta, Abbes, Razika Bared, Soufi Kacimi, and Article History. 2023. "Synthesis of Ni/Bentonite Catalysts and Its Application for Transesterification of Sunflower Oil into Biodiesel." *Algerian Journal of Environmental Science* 9 (2): 3157–61.

Budiyono, Budiyono, Hashfi Hawali Abdul Matin, Ihzani Yulistra Yasmin, and Iwang Septo Priogo. 2023. "Effect of Pretreatment and C/N Ratio in Anaerobic Digestion on Biogas Production from Coffee Grounds and Rice Husk Mixtures." *International Journal of Renewable Energy Development* 12 (1): 209–15.

Cakmak, Ece Kendir, Merve Atasoy, Isaac Owusu-Agyeman, Kasra Khatami, and Zeynep Cetecioglu. 2021. "Circular City Concept for Future Biorefineries." In *Clean Energy and Resource Recovery: Wastewater Treatment Plants as Biorefineries*, Volume 2, edited by Alicia An, Vinay Tyagi, Manish Kumar, and Zeynep Cetecioglu, 335–52. Elsevier.

Chen, Wei Hsin, Wasif Farooq, Muhammad Shahbaz, Salman Raza Naqvi, Imtiaz Ali, Tareq Al-Ansari, and Nor Aishah Saidina Amin. 2021. "Current Status of Biohydrogen Production from Lignocellulosic Biomass, Technical Challenges and Commercial Potential through Pyrolysis Process." *Energy* 226:120433.

Dahman, Yaser, Kashif Syed, Sarkar Begum, Pallavi Roy, and Banafsheh Mohtasebi. 2019. Biofuels: Their Characteristics and Analysis. In *Biomass, Biopolymer-Based Materials, and Bioenergy: Construction, Biomedical, and Other Industrial Applications*, edited by Deepak Verma, Elena Fortunati, Siddharth Jain, and Xiaolei Zhang, 277–325. Sawston: Wood Head Publishing.

Dawood, Sumreen, Apurav Krishna Koyande, Mushtaq Ahmad, Muhammad Mubashir, Saira Asif, Jiří Jaromír Klemeš, Awais Bokhari, et al. 2021. "Synthesis of Biodiesel from Non-Edible (Brachychiton Populneus) Oil in the Presence of Nickel Oxide Nanocatalyst: Parametric and Optimisation Studies." *Chemosphere* 278:130469.

Devi, Manubolu Harika, and Srinivas Munjam. 2022. "Bioethanol Production from Rice Straw and Cellulose Degradation Using Aspergillus Terreus and Trichoderma Harzanium." *Biosciences Biotechnology Research Asia* 19 (September): 699–711. https://www.bio-tech-asia.org/vol19no3/bioethanol-production-from-rice-straw-and-cellulose-degradation-using-aspergillus-terreus-and-trichoderma-harzanium/.

Ebhota, William S., and Pavel Y. Tabakov. 2018. "The Place of Small Hydropower Electrification Scheme in Socioeconomic Stimulation of Nigeria." *International Journal of Low-Carbon Technologies* 13 (4): 311–19.

Elkady, Marwa F., Ahmed Zaatout, and Ola Balbaa. 2015. "Production of Biodiesel from Waste Vegetable Oil via KM Micromixer." *Journal of Chemistry* 2015. Article ID 630168.

Ferreira, Aline S., Antonébia A. Mota, Aldayr M. Oliveira, Francisco I.L. Rodrigues, Sara N. Pacífico, José E. Da Silva, Bruno T.O. Abagaro, et al. 2019. "Equilibrium and Kinetic Modelling of Adsorption: Evaluating the Performance of Adsorbent in Softening Water for Irrigation and Animal Consumption." *Revista Virtual de Quimica* 11 (6): 1752–66.

Gandhi, Chander Prakash, Rajni Garg, and Nnabuk Okon Eddy. 2021. "Application of Biosynthesized Nano-Catalyst for Biodiesel Synthesis and Impact Assessment of Factors Influencing the Yield." *Nanosystems: Physics, Chemistry, Mathematics* 12 (6): 808–17.

Gomes, Maria Carolina Sérgi, Nehemias Curvelo Pereira, and Sueli Teresa Davantel de Barros. 2010. "Separation of Biodiesel and Glycerol Using Ceramic Membranes." *Journal of Membrane Science* 352 (1–2): 271–76.

Hanif, Maryam, Ijaz Ahmad Bhatti, Muhammad Zahid, and Muhammad Shahid. 2022. "Production of Biodiesel from Non-Edible Feedstocks Using Environment Friendly Nano-Magnetic Fe/SnO Catalyst." *Scientific Reports* 12 (1): 1–14.

Herndon, J. Marvin. 2017. "New Concept on the Origin of Petroleum and Natural Gas Deposits." *Journal of Petroleum Exploration and Production Technology* 7 (2): 341–52.

Holechek, Jerry L., Hatim M.E. Geli, Mohammed N. Sawalhah, and Raul Valdez. 2022. "A Global Assessment: Can Renewable Energy Replace Fossil Fuels by 2050?" *Sustainability* 14 (8): 1–22.

Höök, Mikael, and Xu Tang. 2013. "Depletion of Fossil Fuels and Anthropogenic Climate Change: A Review." *Energy Policy* 52: 797–809.

Hossain, A.B.M. Sharif, Boyce Amru Nasrulhaq, Aishah Salleh, and Somasundram Chandran. 2010. "Biodiesel Production from Waste Soybean Oil Biomass as Renewable Energy and Environmental Recycled Process." *African Journal of Biotechnology* 9 (27): 4233–40.

Humphrey, Ibifubara, Nsikan Ime Obot, and Michael A. C. Chendo. 2017. "Utilization of Some Non-Edible Oil for Biodiesel Production." *Nigeria Journal of Pure and Applied Physics*. Vol. 7. www.njpap.futa.edu.ng.

Ingabire, Hortence, Boniface Ntambara, and Ezgad Mazimpaka. 2023. "Characterization and Analysis of Fish Waste as Feedstock for Biogas Production." *International Journal of Low-Carbon Technologies* 18:212–17.

Jalaluddin Jalaluddin, Zainuddin Ginting, Syariful Maliki, Arif Setiawan, and Zulfa Zulfa. 2022. "Biodiesel Production from Crude Palm Oil Using Kapok Skin KOH (Ceiba Pentandra) Catalyst as Solid Green Catalyst." *Journal of Ecological Engineering* 23 (5): 286–92.

Javed, Fahed, William B. Zimmerman, Tahir Fazal, Ainy Hafeez, Maria Mustafa, Naim Rashid, and Fahad Rehman. 2023. Green Synthesis of Biodiesel from Microalgae Cultivated in Industrial Wastewater via Microbubble Induced Esterification Using Bio-MOF-Based Heterogeneous Catalyst. *Chemical Engineering Research and Design*. Vol. 189:707–20.

Jeswani, Harish K., Andrew Chilvers, and Adisa Azapagic. 2020. “Environmental Sustainability of Biofuels: A Review: Environmental Sustainability of Biofuels.” *Proceedings of the Royal Society A: Mathematical, Physical and Engineering Sciences* 476.doi.10.1098/rspa.2020.0351.

Jiyane, Phiwe Charles, Kaniki Tumba, and Paul Musonge. 2021. “Production of Biodiesel From Croton Gratissimus Oil Using Sulfated Zirconia and KOH as Catalysts.” *Frontiers in Energy Research* 9 (August): 1–16.

Johnsson, Filip, Jan Kjärstad, and Johan Rootzén. 2019. “The Threat to Climate Change Mitigation Posed by the Abundance of Fossil Fuels.” *Climate Policy* 19 (2): 258–74.

Kavitha, Munisamy Sambasivam, and Somasundaram Murugavelh. 2021. “Biodiesel Production from Reactive Extraction of Sterculia and Waste Cooking Oil Blend Using an Acid Catalyst.” *International Journal of Ambient Energy* 42 (12): 1435–40.

Kim, M. S., D. H. Kim, and J. K. Lee. 2011. “Biohydrogen.” In *Comprehensive Biotechnology, Second Edition*, edited by Murray Moo-Young, 3:115–25. Burlington: Academic Press.

Kolakoti, Aditya, Muji Setiyo, and Muhammad Latifur Rochman. 2022. “A Green Heterogeneous Catalyst Production and Characterization for Biodiesel Production Using RSM and ANN Approach.” *International Journal of Renewable Energy Development* 11 (3): 703–12.

Kumar, Ramesh, Alak Kumar Ghosh, Rashmi Dhurandhar, and Sankha Chakrabortty. 2022. “Downstream Process: Toward Cost/Energy Effectiveness.” In *Handbook of Biofuels*, edited by Sanjay Sahay, 249–60. Burlington: Academic Press.

Li, Hailong, Daheem Mehmood, Eva Thorin, and Zhixin Yu. 2017. “Biomethane Production Via Anaerobic Digestion and Biomass Gasification.” *Energy Procedia* 105: 1172–77.

Luthfi, Abdullah Amru Indera, Peer Mohamed Abdul, Jamaliah Md Jahim, Nurul Sakinah Engliman, Nur Syakina Jamali, Jian Ping Tan, Shareena Fairuz Abdul Manaf, Mohd Shaiful Sajab, and Nurul Adela Bukhari. 2023. “Isolation and Characterization of Biohydrogen-Producing Bacteria for Biohydrogen Fermentation Using Oil Palm Biomass-Based Carbon Source.” *Applied Sciences* 13 (1):656.

Marwan, Suhendrayatna, and Eti Indarti. 2015. “Preparation of Biodiesel from Microalgae and Palm Oil by Direct Transesterification in a Batch Microwave Reactor.” *Journal of Physics: Conference Series* 622 (1) doi.10.1088/1742-6596/622/1/012040.

Mizik, Tamás, and Gábor Gyarmati. 2022. “Three Pillars of Advanced Biofuels’ Sustainability.” *Fuels* 3 (4): 607–26.

Muppaneni, Tapaswy, Harvind K. Reddy, and Shuguang Deng. 2015. “Supercritical Synthesis of Ethyl Esters via Transesterification from Waste Cooking Oil Using a Co-Solvent.” *Journal of Environmental Protection* 6(9): 986–94.

Mutsvene, Boldwin, Manimagalay Chetty, Sheena Kumari, and Faizal Bux. 2023. “Biohydrogen Production from Brewery Wastewater in an Anaerobic Baffled Reactor. A Preliminary Techno-Economic Evaluation.” *South African Journal of Chemical Engineering* 43: 9–23.

Narowska, Beata Edyta, Marek Kułażyński, and Marcin Łukaszewicz. 2020. “Application of Activated Carbon to Obtain Biodiesel from Vegetable Oils.” *Catalysts* 10(9): 1–14.

Neupane, Dhurba. 2022. “Biofuels from Renewable Sources, a Potential Option for Biodiesel Production.” *Bioengineering* 10 (1): 29.

Nishida, Verônica Sayuri, Adenise Lorenci Woiciechowski, Kim Kley Valladares-Diestra, Luis Alberto Zevallos Torres, Luciana Porto de Souza Vandenberghe, Arion Zandoná Filho, and Carlos Ricardo Soccol. 2023. “Second Generation Bioethanol Production from Soybean Hulls Pretreated with Imidazole as a New Solvent.” *Fermentation* 9 (2): 93.

Obeta, Joseph Chikodili, Emmanuel Chekwube Ossai, and Obioma Uzoma Njoku. 2021. "Optimization and Characterization of Bioethanol Production from Abrus Seed Flour." *International Journal of Energy Research* 45 (3): 3883–98.

Olawale, Bashir Ismail, Mahmud Yerima Iliyasu, Bishir Musa, Ahmad Abdulrahman, and Ahmed Faruk Umar. 2021. "Production of Bioethanol by Co-Culture of Aspergillus Niger and Saccharomyces Cerevisiae Using Watermelon Peels as Substrate." *Path of Science: International Electronic Scientific Journal.* 7(10): 6001–6011.

Onu, Peter, and Charles Mbohwa. 2021. "New Approach and Prospects of Agrowaste Resources Conversion for Energy Systems Performance and Development." In *Agricultural Waste Diversity and Sustainability Issues: Sub-Saharan Africa as a Case Study*, edited by Peter Onu and Charles Mbohwa, 97–118. Elsevier.

Pasha, Mustafa Kamal, Lingmei Dai, Dehua Liu, Miao Guo, and Wei Du. 2021. "An Overview to Process Design, Simulation and Sustainability Evaluation of Biodiesel Production." *Biotechnology for Biofuels* 14:1–24.

Primadi, Thutug Rahardiant, Fauziatul Fajaroh, Aman Santoso, and Olvi Dyah Fernanda. 2021. "Biodiesel Production from Cooking Oil Waste (Jelantah) Using Egg Shells Waste Catalyst CaO/CoFe2O4." *AIP Conference Proceedings* 2384 (1): 50004.

Pulungan, Ahmad Nasir, Agus Kembaren, Nurfajriani Nurfajriani, Feri Andi Syuhada, Junifa Layla Sihombing, Muhammad Yusuf, and Rahayu Rahayu. 2021. "Biodiesel Production from Rubber Seed Oil Using Natural Zeolite Supported Metal Oxide Catalysts". *Polish Journal of Environmental Studies* 30 (6): 5681–89.

Reid, Walter V., Mariam K. Ali, and Christopher B. Field. 2020. "The Future of Bioenergy." *Global Change Biology* 26(1): 274–86.

Ribeiro, Kelly Cristina, Khalil Kashe, Troner Assenheimer, Cristine Carretoni, Márcio Portilho, José Carlos Pinto, and Márcio Nele. 2020. "Biodiesel / Water / Glycerol Emulsion Separation Using Electrocoalescence : Model and Process Emulsions." *Energy Fuels* 34(2):1737–43.

Salhein, Khaled, C. J. Kobus, and Mohamed Zohdy. 2022. "Forecasting Installation Capacity for the Top 10 Countries Utilizing Geothermal Energy by 2030." *Thermo* 2(4): 334–51.

Salvarli, Mustafa Seckin, and Huseyin Salvarli. 2016. "For Sustainable Development : Future Trends in Renewable Energy and Enabling Technologies." In *Renewable Energy Resources, Challenges and Applications*, edited by Mansour Al Qubeissi, Ahmad El-kharouf, Hakan Serhad Soyhan, 1–15. London: Intechopen.

Saratale, Ganesh Dattatraya, Rijuta Ganesh Saratale, J. Rajesh Banu, and Jo-shu Chang. 2019. Biohydrogen Production from Renewable Biomass Resources. In *Biohydrogen*, edited by Ashok Pandey, S. Venkata Mohan, Jo-Shu Chang, Patrick C. Hallenbeck, Christian Larroche, 247–77. Elsevier.

Sarno, Maria, and Mariagrazia Iuliano. 2019. "Biodiesel Production from Waste Cooking Oil." *Green Processing and Synthesis* 8 (1): 828–36.

Selaimia, Radia, Abdelsalem Beghiel, and Rabah Oumeddour. 2015. "The Synthesis of Biodiesel from Vegetable Oil." *Procedia: Social and Behavioral Sciences* 195: 1633–38.

Singh, Anoop, Surajbhan Sevda, Ibrahim M. Abu Reesh, and Karolien Vanbroekhoven. 2015. "Biohydrogen Production from Lignocellulosic Biomass : Technology and Sustainability." *Energies* 8(11):13062–80.

Sivamani, Selvaraju, Marwan Ahmed, Sulieman Al, Aseela Musalem, Awad Anthroon, Azeezah Mohammed, Masoud Kashoob, et al. 2022. "Sciences Heterogeneous Catalyzed Synthesis of Biodiesel from Crude Sunflower Oil." *Journal of the Nigerian Society of Physical Sciences* 4(1): 16–19.

Song, Haiyan, Pan Wang, Shi Li, Weiping Deng, Yanyun Li, Qinghong Zhang, and Ye Wang. 2019. "Direct Conversion of Cellulose into Ethanol Catalysed by a Combination of Tungstic Acid and Zirconia-Supported Pt Nanoparticles." *Chemical Communications* 55 (30): 4303–6.

Vasić, Katja, Željko Knez, and Maja Leitgeb. 2021. "Bioethanol Production by Enzymatic Hydrolysis from Different Lignocellulosic Sources." *Molecules* 26 (3): 753.

Verhe, Roland, Saju Varghese, Johan M. Thevelein, Jasmijn Hazegh Nikroo, Margaux Lambrecht, Emile Redant, and Gilles De Clercq. 2022. "Production of Bio-Ethanol from the Organic Fraction of Municipal Solid Waste and Refuse-Derived Fuel." *Biomass* 2 (4): 224–36.

Wahyono, Yoyon, Hadiyanto, Mochamad Arief Budihardjo, Yogi Hariyono, and Rifqi Ahmad Baihaqi. 2022. "Multifeedstock Biodiesel Production from a Blend of Five Oils through Transesterification with Variation of Moles Ratio of Oil: Methanol." *International Journal of Technology* 13 (3): 606–18.

Wei, Li, Li Jiheng, Guo Junhong, Bao Zhe, Fu Lingbo, and Hou Baodeng. 2020. "The Effect of Precipitation on Hydropower Generation Capacity: A Perspective of Climate Change." *Frontiers in Earth Science* 8 (September): 1–13.

Zhan, Yawei, Meixin Wang, Tengfei Ma, and Zhiqiang Li. 2022. "Enhancing the Potential Production of Bioethanol with Bamboo by γ-Valerolactone/Water Pretreatment." *RSC Advances* 12 (26): 16942–54.

Zimife, M. Egbe, C. Nwinyi Obinna, F. Ahuekwe Eze, and Z. George Dango. 2022. "Bioethanol Production from Waste Paper: An Alternative Energy Source." *IOP Conference Series: Earth and Environmental Science* 1054 (1):012002.

11 Synthesis of Value-Added Nanomaterials from Agricultural Wastes

Wardishan Kassem Alrifai and Birsen Korkmaz
Erciyes University

Ebubekir Yılmaz
Nuh Naci Yazgan University

Ismail Ocsoy
Erciyes University

CONTENTS

DOI: 10.1201/9781003397861-11

11.1 INTRODUCTION

Nanotechnology can be summed up as the manipulation of atoms and molecules within the nanoscale, with one or more dimensions of about 100 nm where a structure, tool, or system can be designed, manufactured, and exploited. Nanotechnology has evolved into an interdisciplinary science with tremendous endeavors to convert early promises into reality. Through the two production approaches, Top-Down and Bottom-Up, as Figure 11.1 reveals, various nanomaterials have been generated and integrated with larger systems, providing the ability to significantly alter pharmacology, medicine, food, and agricultural aspects.

The green synthesis of nanomaterials has increasingly drawn significant interest from numerous researchers due to its ease of use, low cost, and environmental friendliness. The value-added nanoscaled output originates from diverse wastes comprising industrial effluents, crop residues and food wastes, thereby proposing viable choices to combat the issues arising from topical and universal pollution. Agricultural production has dramatically increased with the increase in the human population, and it is considered one of the key sectors producing an enormous amount of waste. It has been recorded that about 24 million tons of food are produced globally in the agricultural sector. Although food is essential for human persistence, the agricultural influence on the environment is quite obvious (saac Oluseun and Olufemi Adebukola 2020).

Agricultural waste is undesirable or unsalable products that come entirely from agricultural activities—growing crops or raising livestock with the intention of making a profit or supporting one's livelihood (He, Zhang, and Zeng 2019). Generally, agricultural wastes come in a variety of forms: (i) Green wastes made up of decomposing organic plant resources, such as leaves and crop residues; (ii) Pesticides and fecal matter, which have a history of seeping into the soil, causing damage to its fertility in addition to habitats and water bodies; and (iii) Gas emissions generated from factories and agricultural equipment that are able to harm the atmosphere. Both incidental and intentional disposal of these materials can occur, in which the generated wastes can either be exploited as raw materials for the bio-economy or left to pile inappropriately and pose a threat to public health and food safety (saac Oluseun and Olufemi Adebukola 2020).

Tons of agricultural waste, including straws, wood, and forest leftovers, are burned in fields as a form of waste disposal in many parts of the world. This is typically the

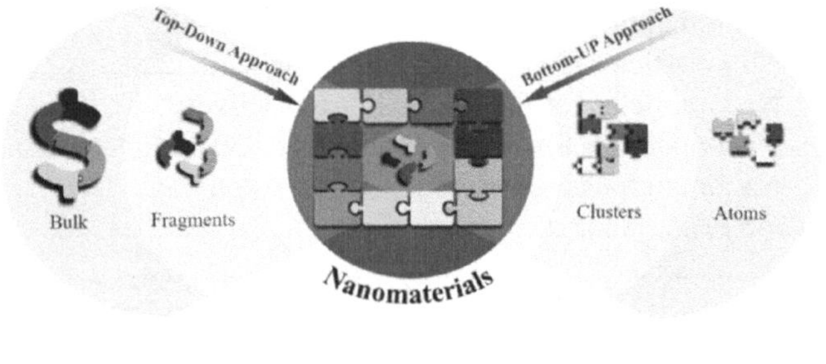

FIGURE 11.1 Nanomaterial top-down and bottom-up synthetic approaches.

case due to the large amount of residue created having no other viable, cost-effective uses. However, for reasons related to the environment, ecology, and public health, burning agricultural waste is becoming less and less popular, and it is acknowledged that agricultural wastes now have concealed economic value. Therefore, utilization of agricultural wastes as raw materials for providing value-added nanomaterials is necessary to reduce the detrimental effects of agriculture on the environment, aquatic life, and human health.

This chapter provides an overview of the plenty types of biogenic nanoparticle that have been developed using plant extracts, microorganisms, and other biomolecules, along with their manipulation in the biomedical industry, particularly in dentistry.

11.2 SYNTHESIS OF BIOGENIC NANOPARTICLES

Nanobiotechnology is considered the nexus of both nanotechnology and biotechnology since it is a promising discipline devoted to the development, advancement, and implementation of nanoscale structures for a progressive biotechnology. One significant research area in this domain is the production of nanoparticles (NPs) with a variety of chemical compositions, forms and sizes (Akhtar, Panwar, and Yun 2013). In general, a variety of chemical and physical techniques are utilized in NP synthesis. Meanwhile, these techniques use hazardous chemicals such as toxic capping agents used for stabilizing NPs (Karaagac et al. 2021) and require a significant amount of energy and money, ruling them out for use in biomedicine. In order to synthesize NPs, it is therefore necessary to design procedures that are safe, biocompatible, reliable and benign. For this purpose, scientists are increasingly considering biological systems as prospective environmentally friendly "nano-factories" (Jain et al. 2011), where numerous biological systems have been utilized in recently published findings for the synthesis of NMs. These biomolecules mostly consist of proteins, enzymes, nucleic acids (DNA, RNA), peptides, and plant extracts, along with some microorganisms such as bacteria, viruses, algae and fungi (Figure 11.2).

11.2.1 MICROORGANISM-DIRECTED NANOPARTICLES

One effective strategy for the green synthesis of NPs is the exploitation of microorganisms. These uni- or multicellular organisms depict an enormous structural variety, simple cultivability and quick growth, rendering them potential biofactories for green NPs (Purohit, Chattopadhyay, and Singh 2019). Despite the fact that current chemical processes can rapidly and in huge quantities synthesize NPs of a specific size and shape, they are labor-intensive, expensive, and generate hazardous wastes that are dangerous for both the environment and human health. Enzymatic activities carried out by microorganisms are able to eliminate the employment of costly chemicals and offer an eco-friendly and less energy-intensive green approach that is able to be performed in a simple cell culture medium at ambient temperatures and pressures (Li et al. 2011; Jeffryes, Agathos, and Rorrer 2015).

Numerous prokaryotic or complex eukaryotic microorganisms such as viruses, bacteria and microalgae have lately been examined for the synthesis of metal, metal oxide, and other NPs. As an analogy, the first case of microbial synthesis of AuNPs

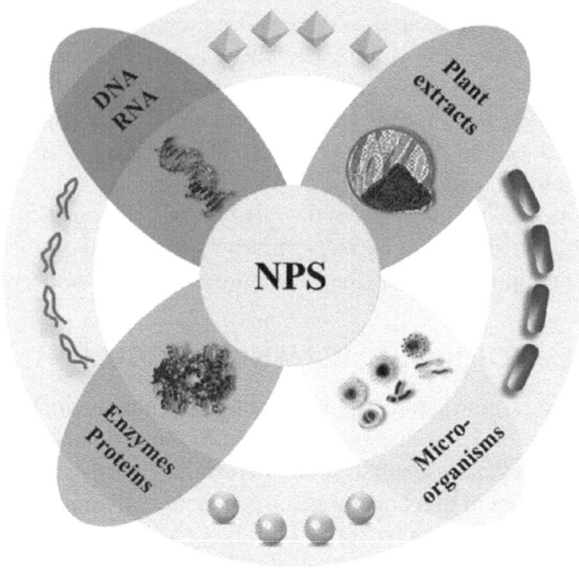

FIGURE 11.2 Biogenic nanomaterials synthesized starting from various biomolecules.

was described in *Bacillus subtilis* 168, revealing the existence of 5–25 nm octahedral NPs within the cell wall. Spherical AuNPs with a size scope of 10–20 nm have been reported in *Rhodopseudomonas capsulata* bacteria at lower concentrations, and networks of nanowires have been found at higher concentrations (Shedbalkar et al. 2014). Compared to bacteria, fungi have a better scope due to their strong resistance to metal, potent ability to adhere to walls, and potential for metal bioaccumulation. For example, *Penicillium brevicompactum*, *A. clayatus*, and *F. oxysporum* were employed to produce silver nanoparticles (Ag NPs) outside of cells through the reduction of an aqueous silver ion into water. Additionally, ZnO nanoparticles whose sizes remained less than 100 nm even after three months were produced from Mycelia of *Aspergillus furnigatus*, with functional groups such as ≡C-O-R and =C=C= from the heterocyclic compounds serving as capping ligands. Furthermore, *Asperigillus niger* was able to synthesize magnetic iron nanoparticles (Fe NPs) intracellularly after incubating precursor salt with homogenized mycelia for six days without the direct or indirect use of organic solvents in the synthesis (Keshu et al. 2021). Likewise, microalgal cultures are also showing promise as platforms for the environmentally friendly synthesis of NPs, including metallic nanoparticles, nanostructured, nanocrystalline biopolymers, and hierarchically structured NPs that can allow the manufacture of biomaterials like scaffolds for tissue engineering or drug delivery, supercapacitors, and optoelectronic materials. For instance, *Cyclotella* and *Thalassiosira,* two genera of diatoms, are recognized to produce chitin nanofibrils in stationary phase cultures, with *Thalassiosira* excreting highly crystalline, long aspect ratio (30 nm diameter, 10s of microns in length) fibrils that do not require subsequent processing or harsh chemicals. Nitrogen-doped carbon materials have also been templated using nanocrystalline chitin fibrils in super-capacitor electrodes. Meanwhile, *Thalassiosira weissflogii* crystalline β-chitin fibrils have been employed as templates for organic-inorganic silica hybrid nanocomposites, the formation of nanostructured crystals with heavy metal cations such as Gd^{3+}, Pb^{+} and UO^{2+} and as a substance to aid in the accumulation of human pathogens (Jeffryes, Agathos, and Rorrer 2015).

11.2.2 Biomolecule-Directed Nanoparticles

11.2.2.1 DNA-Based NMs

Due to its controllable structure and adaptable features, DNA has increasingly been extensively employed as a flexible tool in the production of different NPs. DNA-directed nanostructures have been discovered to have potential applications in a broad range of scientific and technical areas, including nanomedicine, where they can be used to create antimicrobial and anticancer agents as well as mini-sensing devices and electronic components (Ocsoy, Tasdemir, Mazicioglu, Celik, et al. 2018). DNA metallization has the special ability to function as a scaffold for the production of nanostructures and nanoassemblies. Thymine (T), adenine (A), cytosine (C) and guanine (G) are the four nucleobases that make up DNA, along with the phosphate backbone. Each component contains a unique chemical group that facilitates the binding of metal ions (Some et al. 2018). One strategy for the building-up process is *in situ reduction* and growth, and the other is *electrostatic interaction*, often known as molecular hybridization. The first method typically involves DNA template binding by a metal precursor in two steps, followed by in situ reduction with DNA-directed NP growth. Conversely, in the second method, DNA is utilized to control the electrostatic interaction or molecular hybridization that results in the assembly of pre-synthesized NPs into desired nanostructures (Huang et al. 2015). One example is the bare colloidal metallic (CM) NPs (copper, gold, silver, and alloys) and the CM NPs obtained on the surface of graphene oxide (GO) in aqueous solution that were both synthesized by Ocsoy et al. in a single pot using a common method. The CM NPs-GO nanocomposites were utilized as an efficient antibacterial agent to inactivate bacteria that cause plant pathogens and as a biosensor for detecting dopamine (Ocsoy, Tasdemir, Mazicioglu, Celik, et al. 2018).

11.2.2.2 Protein- and Enzyme-Based NMs

Rather than DNA, proteins and enzymes function as nanoreactors in order to develop biocompatible NPs. Metal nanoparticles, metal oxide, and quantum dots are a few of the types of NPs that have so far been efficiently manufactured using proteins. To form primary nanocrystals or seeds, metal ions can interact with functional groups (thiol, carboxyl or amine) in proteins; subsequently, the presence or addition of reducing agents leads to the production of NPs. It is worth noting that, in the same manner as proteins, which act as templates or stabilizing agents for the fabrication of NPs, enzymes also play an important stabilizing role by producing hydrogen peroxide (H_2O_2) and superoxide anion (O^{2-}) via enzymatic reactions that function as reducing agents to generate biogenic NPs (Demirbas et al. 2020). For instance, Xie and his colleagues used a plasma protein, bovine serum albumin (BSA), to manufacture anisotropic Au nanoplates, revealing the protein's dual roles in reducing Au (III) and controlling the formation of Au(0) into plate-like structures. According to this study, the BSA carboxyl side groups were what caused the plate-like Au nanostructures to develop. Moreover, gelatin can be employed as a reducing and stabilizing agent to produce spherical Au NPs via the metal ion-induced oxidation of amine into nitrite, as reported by Zhu and coworkers (Nasaruddin et al. 2021). On the other hand, distinct enzymes with unique mechanisms of action can serve as reducing and capping agents for the formation of several NPs with diverse morphologies, particularly Ag NPs. For example, to reduce Ag^+ ions for the development of Ag NPs in an in vitro system, macerating enzymes were used. In another study, a silver enhancement kit (EnzMetTM), and horseradish peroxidase for catalyzing the reduction of Ag^+ ions were also added simultaneously, leading to the fast generation of Ag NPs (Some et al. 2018).

11.2.3 Plant Extract-Directed Nanoparticles

For thousands of years, people have employed plants for a variety of therapeutic purposes, including in Ayurveda and traditional Chinese medicine. Lately, investigations combining nanotechnology and biotechnology have been pursued thanks to the unique features of phytochemicals found in plant extracts, the most essential component of plant-mediated NP synthesis (Vincent et al. 2022). The use of biomolecule-directed NPs has grown in order to provide a remedy for the issues caused by conventional, i.e., physical and chemical synthesis techniques. Synthesis of biogenic NPs from plant extracts has been extensively used as a reducing agent and/or capping agent through their metabolite and biomolecule content (Ocsoy, Tasdemir, Mazicioglu, Celik, et al. 2018; Ullah and Lim 2022). Plant-mediated NPs are readily accessible, simple to manufacture, non-toxic, cost-efficient, and have good repeatability and reproducibility compared to microorganism-based NPs, which have low dispersion and a slow rate of production and require multistep processes such as microbial isolation, culturing, and maintenance (Karaagac et al. 2021; Mondal, Anweshan, and Purkait 2020; Akhtar, Panwar, and Yun 2013).

Due to the abundance of polyphenols, terpenes, flavones, ketones, amides, alkaloids, enzymes, polypeptides, and proteins that serve as reducing or stabilizing agents, many different types of NMs have been developed to date using plant extracts, including red cabbage, blueberries, and blackberries (Karaagac et al. 2021; Vincent et al. 2022). Especially noteworthy is that extracts can be prepared from a variety of plant sections, such as leaves, fruits, peels, flowers, barks, roots, and seeds, based on the maximum constituent availability (Ullah and Lim 2022) in which each part should be dried, downsized and then extracted via boiling and heating (Vincent et al. 2022). For instance, in order to reduce metallic moieties into a nanoscale form, several metallic NMs, such as iron, zinc, silver, and gold, have been produced from various plant parts using their constituents' reducing, capping and stabilizing properties along with their antioxidant capabilities (Ullah and Lim 2022). Many scientists have developed diverse kinds of NPs through plant-mediated synthesis that can be used for a vast range of purposes, such as antibacterials, anticancer agents, antioxidants, catalysts, biosensors, carriers, fuel cells, and photocatalysts (Ocsoy, Tasdemir, Mazicioglu, and Tan 2018). To give an idea, Ghodake et al. used dried clove buds (*Syzygium aromaticum*) in solution to examine the quick and extracellular synthesis of gold nanocrystals, and they established the water-soluble flavonoids of clove buds as a bioreductant and stabilizing agent in nanoproduction (Sood and Chopra 2018). An additional example is the generation of eco-friendly and antimicrobial Ag NPs starting from *Matricaria chamomilla* medicinal flower extract by Dogru and her colleagues (Dogru et al. 2017). Also metal oxide nanoparticles can be synthesized starting from plants; for example, the green biosynthesis of CuONPs was performed by Koca et al. from *Punica Granatum* peel (Koca et al. 2018) with satisfying sustainability and from citrus lemon fruits by Zikalala et al. (2018).

11.3 AGRICULTURAL WASTE-DIRECTED NANOMATERIALS

Over the years, the rapid intensification of agricultural production due to the increase in the world population has increased agricultural waste, thereby causing extreme environmental problems. Nanotechnology has become a multi-purpose platform that might offer practical, ecologically friendly, and economically advantageous solutions for society's overall sustainability (Diallo et al. 2013). Recycling agricultural waste into usable materials encourages efficient, eco-friendly utilization of plant waste and biomass (Devadiga et al. 2015). Agricultural waste is a valuable resource that can be collected separately and processed to produce NPs (Abd-Elsalam et al. 2021). Several NPs have been effectively synthesized by using different agricultural wastes, as shown in Figure 11.3. This section is focused on the synthesis of various NPs derived from different types of agricultural waste.

11.3.1 Metal Oxide Nanoparticles (MO NPs)

Numerous agricultural wastes are produced and dumped carelessly into the environment, posing environmental challenges. Researchers developed an alternative use of some agricultural waste as a potential source for the synthesis of MO NPs. Agricultural wastes are widely known for their accessibility, affordability, and ecological acceptability (Kumar et al. 2020).

11.3.1.1 Iron Oxide Nanoparticles (IO NPs)

IO NPs have been used in medicine and pharmaceutical applications (Siddique and Chow 2020). Researchers have been working on developing simple, dependable, and efficient green chemical methods for the production of IO NPs. In recent research, agricultural wastes were utilized for the IO NP synthesis. To remove chromium ions from tannery effluent, magnetic IO NP was made using sugarcane bagasse (Jabasing et al. 2018). Pomegranate, carob, oak, green tea, and *Ruellia tuberosa* were used for the biosynthesis of IO NP (Vasanthraj et al. 2019; Machado et al. 2013). The leaf extract of grape and garlic vines is used for IO NP synthesis through a green method (Luo et al. 2014; Prasad 2016).

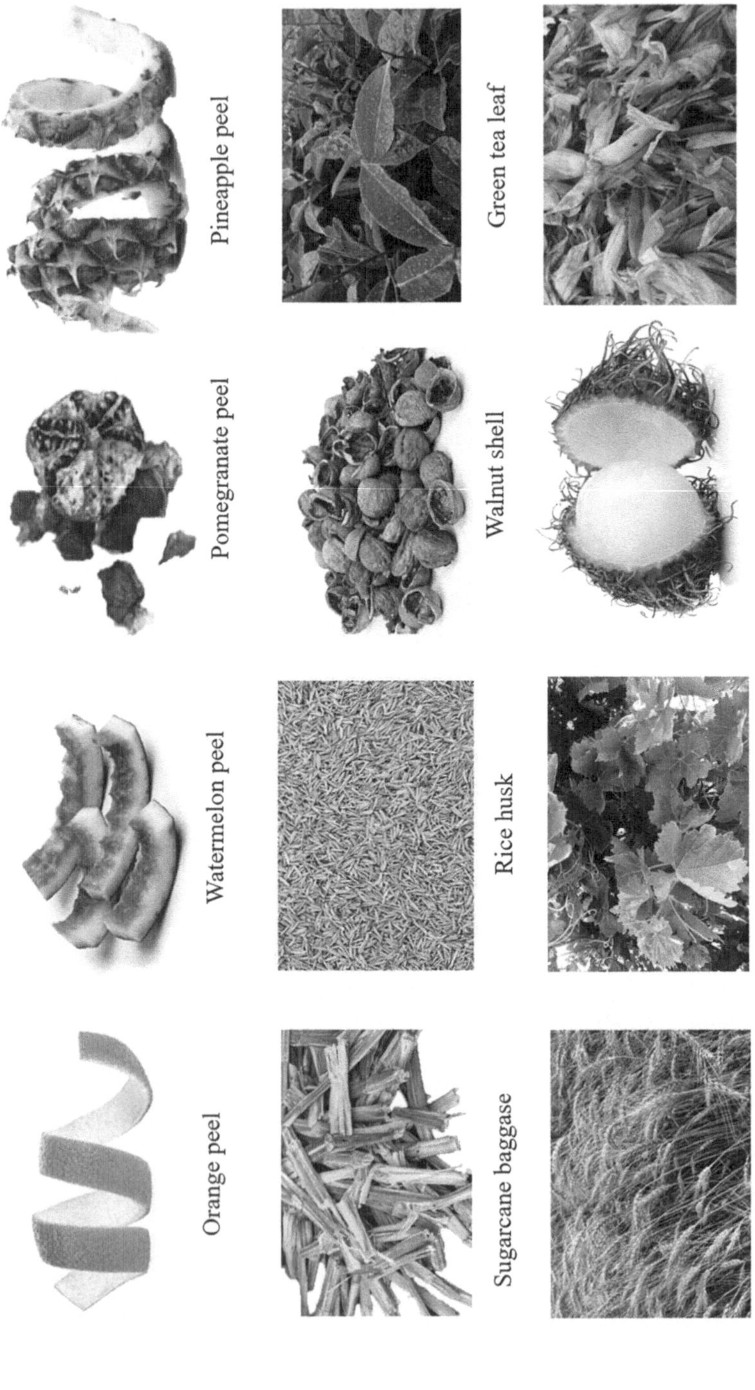

FIGURE 11.3 Commonly used agricultural wastes for the synthesis of NPs.

11.3.1.2 Titanium Oxide Nanoparticles (TiO_2 NPs)

TiO_2 NPs have been investigated since they have unique properties (Lai et al. 2015). Recent technological advancements have the potential to form TiO_2 NPs from agricultural waste. It has been reported that the leaf extract of *Calotropis gigantea* (*L.*) is promising for TiO_2 NP fabrication (Marimuthu et al. 2013). Additionally, the leaves of *Syzygium cumini* have been used to create TiO_2 NPs using a non-toxic, economical, and environmentally friendly production technique (Sethy et al. 2020).

11.3.1.3 Zinc Oxide Nanoparticles (ZnO NPs)

ZnO NPs can be utilized in a variety of applications due to their antibacterial, antioxidant, and anticancer properties, as well as their usage in photocatalytic degradation (Abdol-Aziz et al. 2019). Agricultural wastes and leaf extracts are often utilized for the green biosynthesis of ZnO NPs (Hassan et al. 2015). In order to synthesize pure ZnO NPs, the significance of using green sources has been reported (Ali et al. 2018).

11.3.1.4 Copper Oxide Nanoparticles (CuO NPs)

Recently, CuO NPs have drawn much attention because of their properties as antioxidants, antimicrobials, and anticancer agents (Krishnan and Mahalingam 2017). Agricultural wastes are utilized for the production of CuO NPs. However, there have been some difficulties in the production of CuO NPs. For the green biosynthesis of CuO NPs, aloe vera, walnut shells, dried husk from *Zea mays* L., and date stones have been utilized (Abtahi et al 2021; Nwanya et al. 2019).

11.3.2 Hybrid Nanoparticles

Hybrid NPs typically contain two or more different components that are combined through specific interactions that enhance their functional capabilities in a synergistic way. The combination of different components and structures with different types of interactions leads to a virtually infinite variety of unique task-specific materials. A variety of agricultural waste products are utilized for the production of hybrid NPs and nanocomposites (NCs) for technical and biomedical applications (Abd-Elsalam 2020). Hybrid NPs have innovative potential in agricultural applications. Additionally, they are utilized in the application of nanosensors and nanobiosensors for the detection of plant pathogens and the transport of genes or DNA in plants (Singh et al. 2021; Abd-Elsalam 2020). Hybrid NPs are therefore particularly beneficial for biological and agricultural applications.

Researchers may synthesize complex hybrid NPs with multimodal functions and design nanomaterials with outstanding biocompatibility and safety by implementing "green chemistry" methodologies and employing sustainable resources.

11.4 NANOMATERIAL-BASED DENTISTRY APPLICATIONS

Nanomaterials (NMs) have attracted great interest in many areas of dentistry due to their small size, large surface area/volume ratio, antimicrobial effect, and chemical functionality. Various nanotools can be utilized in a variety of dental applications.

11.4.1 Nanomaterials and Tissue Engineering

The periodontal ligament (PDL) regeneration is the greatest challenge in the regeneration process (Nyman et al. 1987). Many biomaterials are being investigated for PDL regeneration. The PDL that serves to connect teeth to the alveolar bone is essential to physiological function, and their interactions with NPs are the subject of much research (Yu et al. 2013). Numerous NPs are suggested as alternatives for the regeneration of PDL. Two- and three-layered membranes have been developed to replicate the intricate structure of periodontal tissue. A hybrid, three-layered superparamagnetic scaffold containing a magnetic component that can enhance osteogenic processes was recently produced (Sprio et al. 2018).

The NPs show considerable potential for the regeneration or recovery of alveolar bone defects. Hydroxyapatite and collagen are the two main components of bone tissue. nHA and nanoscale collagen have been widely used together to reproduce the composition and microstructure of the natural bone. According to Wang et al. (2016), the most successful alveolar bone regeneration ability was demonstrated with a nHA/collagen ratio of 7/3. Liu et al. (2011) found that the rhBMP-2-loaded nHA/collagen/polylactic acid were effective in a manner similar to autogenous bone.

11.4.2 Nanoparticles to Improve Oral Hygiene

Nanorobotic dentifrice (dentifrobots) is a toothpaste or mouth rinse that can be used to identify and destroy bacterial pathogens in the plaque or elsewhere in the oral cavity. These nanodevices (1–10 μm) may reach areas that are challenging to clean with a toothbrush or floss. Dentifrobots with the proper configuration might recognize and eliminate harmful bacteria found in plaque and elsewhere while enabling the 500 species of beneficial oral microflora to thrive in a healthy ecosystem (Grumezescu 2016).

11.4.3 Nanoparticles for Periodontal Drug Delivery

Local drug delivery systems (LDDS) increase drug concentration at the infection site and minimize side effects; clinicians prefer them over systemic antibiotics, particularly in situations with mild to moderately localized periodontitis. LDDS are available in these forms: irrigation systems, fibers, gels, strips, films, microparticles, and NPs. The use of a NPs drug delivery technology, which selectively deposits a regulated amount of drug near the area of interest, reduces dosage-related adverse effects and facilitates the drug's reaching the subgingival area (Joshi et al. 2016). These NPs are either directly delivered at the site of action or loaded with active drugs for sustained and controlled release. Triclosan has served as an excellent example of drug delivery through the incorporation of NPs in the treatment of periodontal disease. Triclosan-loaded NPs were produced and identified by the emulsification-diffusion process as a new delivery mechanism suitable for the treatment of periodontal disease (Piñón-Segundo et al. 2005).

11.4.4 Nanomaterials for Dental Implants and Osseointegration

Dental implants are one of the most popular and effective options for re-establishing occlusion after teeth are lost. The perspectives on implant design and surface modification have evolved to mimic nature as a result of the development of NPs-based nanotechnology in this century (Ding et al. 2021).

Until now, the surface of dental implants has been reported to be modified by innovative nanotechnologies, such as coating with nanomaterials (Hu et al. 2019; Xia et al. 2018). The most recommended nanocoating for dental implants is nHA because it resembles the inorganic parts of alveolar bone. Dental implants with a nHA surface have been shown to promote improved alveolar bone development. For instance, it has been demonstrated that nHA coating on Ti-6Al-4V implants with increased crystallinity improves osteogenesis and angiogenesis as well as osseointegration in rat bone marrow-derived stem cells (Hu et al. 2019; Xia et al. 2018). Particle size is also a key modulator of the bone-implant interface. For instance, smaller particles of nanodiamonds, alumina nanofibers, and calcium phosphate nanocrystals have been shown to improve osseointegration in vivo and have superior osteoblast proliferation (Hu et al. 2019; Padovani et al. 2015).

11.4.5 Nanomaterials in Prosthodontics

Acrylic resins and silicone elastomers are also widely utilized in prosthodontics. These materials have acceptable esthetics and biocompatibility, but they also exhibit fatigue failure in the oral cavity, poor tensile and tearing strength, and polymerization shrinkage. Metals such as Ti or cobalt alloys, which are utilized for skeletal prosthetics, have poor corrosion resistance and uncertain biocompatibility. The ductility and brittleness of ceramics (ZrO_2 or Al_2O_3) used in the production of crowns and bridges are low (Wang et al. 2015). Nanotechnology has been incorporated into these materials to improve clinical satisfaction.

Nanofillers have been added to dental impression materials to improve flow and enhance precision details. TiO_2 or Al_2O_3 can be used as fillers in polymethyl methacrylate to reduce the porosity of dentures and bacterial attachment (Abiodun-Solanke et al. 2014).

11.5 CONCLUSION

There is a broad range of plant species found in nature that can be ideal candidates for building synthesis areas in nanotechnology that have significant effects on how nanoscience develops. Hence, plant-mediated NP synthesis has emerged in the field of nanobiotechnology to produce eco-friendly, inexpensive, easily producible, and accessible innovative materials that can be safely applied in biomedical and bioanalytical applications. Utilizing agricultural waste has several benefits since it is the most effective method of waste management and contributes to the creation of high-quality products from low-cost materials as well. These value-added nanomaterials can be effectively utilized in biomedical sectors because of their biocompatibility.

REFERENCES

Abd-Elsalam Kamel Ahmed. 2020. *Multifunctional Hybrid Nanomaterials for Sustainable Agri-Food and Ecosystems: A Note from the Editor.* Amsterdam: Elsevier Inc.

Abd-Elsalam Kamel A., Periakaruppan Rajiv, Rajeshkumar S. 2021. *Agri-Waste and Microbes for Production of Sustainable Nanomaterials*. Amsterdam: Elsevier.

Abdol-Aziz Rabiatul A., Abd Karim Siti Fatma, Rosli Nor Asikin. 2019. "The Effect of Ph on Zinc Oxide Nanoparticles Characteristics Synthesized from Banana Peel Extract." *Key Engineering Materials* 797:271–279.

Abiodun-Solanke Iyabode M. Funmilayo, Ajayi DM, Arigbede Abiodun. 2014. "Nanotechnology and Its Application in Dentistry." *Annals of Medical and Health Sciences Research* 4 (Suppl. 3):171–177.

Abtahi Froushani Seyyed Meysam, Zamani Asghar, Abbasi Ardeshir. 2021. "Cytotoxic Effects of Combined Walnut Shell-Extracted Copper Oxide: Nanoparticles and Hydroalcoholic Extracts of Aloe Vera against Human Leukemic Cell Line." *Zahedan Journal of Research in Medical Sciences* 23(1):E94572.

Adejumo Isaac Oluseun, and Adebiyi Olufemi Adebukola. 2020. Agricultural solid wastes: Causes, effects, and effective management." In Saleh Hossam A. (eds) *"Strategies of Sustainable Solid Waste Management. 8*. London: IntechOpen.

Akhtar Mohd Sayeed, Panwar Jitendra, and Yun Yeoung-Sang. 2013. "Biogenic Synthesis of Metallic Nanoparticles by Plant Extracts." *ACS Sustainable Chemistry & Engineering* 1 (6):591–602.

Ali Attarad, Phull Abdul-Rehman, Zia Muhammed. 2018. "Elemental Zinc to Zinc Nanoparticles: Is ZnO NPs Crucial for Life? Synthesis, Toxicological, and Environmental Concerns. " *Nanotechnology Reviews* 7(5):413–41.

Demirbas Ayse, Karaytuğ Tuna, Arabaci Nihan, Yilmaz Ebru Sebnem, and Ocsoy Ismail. 2020. "Synthesis of Metallic and Metal Oxide Nanomaterials." InNeha Srivastava, Manish Srivastava, P. K. Mishra, Vijai Kumar Gupta (eds.) *Green Synthesis of Nanomaterials for Bioenergy Applications,* 99–123. John Wiley & Sons.

Devadiga Aishwarya, Shetty K. Vidya, Saidutta M.B. 2015. "Timber Industry Wasteteak (Tectona Grandis Linn.) Leaf Extract Mediated Synthesis of Antibacterial Silver Nanoparticles." *International Nano Letters* (5):205–214.

Diallo Mamadou S., Fromer Neil A., Jhon Myung S. 2013. "Nanotechnology for Sustainable Development: Retrospective and Outlook." *Journal of Nanopartical Research* 15:2044.

Ding Qinfeng, Cui Jinjie, Shen Hangqi, He Chuanglong, Wang Xudong, Shen Steve G.F., Lin Kaili. 2021. "Advances of nanomaterial applications in oral and maxillofacial tissue regeneration and disease treatment." *Wiley Interdisciplinary Review: Nanomedicine and Nanobiotechnology* 13:1669.

Dogru Esra, Demirbas Ayse, Altinsoy Berrak, Duman Fatih, and Ocsoy Ismail. 2017. "Formation of Matricaria chamomilla Extract-Incorporated Ag Nanoparticles and Size-Dependent Enhanced Antimicrobial Property." *Journal of Photochemistry and Photobiology B* 174:78–83.

Grumezescu Alexandru 2016. *Nanobiomaterials in Dentistry: Applications of Nanobiomaterials*. William Andrew: Norwich, NY.

Hassan Saad S. M., El Azab Waleed I. M., Ali Hager R., Mansour Mona S. M. 2015. "Green Synthesis and Characterization of ZnO Nanoparticles for Photocatalytic Degradation of Anthracene." *Advances in Natural Sciences: Nanoscience and Nanotechnology* 6(4):045012.

He Ke, Zhang Junbiao, and Zeng Yangmei. 2019. "Knowledge Domain and Emerging Trends of Agricultural Waste Management in the Field of Social Science: A Scientometric Review." *Science of the Total Environment* 670:236–244.

Hu Zhekai, Wang Xiuhui, Xia Wei, Wang Zongliang, Zhang Peibiao, Xia Lunguo, Lin Kaili, Zhu Min. 2019. "Nano-Structure Designing Promotion Osseointegration of Hydroxyapatite Coated Ti-6Al-4V Alloy Implants in Diabetic Model." *Journal of Biomedical Nanotechnology* 15(8):1701–1713.

Huang Jiale, Lin Liqin, Sun Daohua, Chen Huimei, Yang Dapeng, and Li Qingbiao. 2015. "Bio-inspired Synthesis of Metal Nanomaterials and Applications." *Chemical Society Reviews* 44 (17):6330–74.

Isaac Oluseun, Adejumo, and Adebiyi Olufemi Adebukola. 2020. "Agricultural Solid Wastes: Causes, Effects, and Effective Management." In *Strategies of Sustainable Solid Waste Management*, edited by M. Saleh Hosam, Ch. 10. Rijeka: IntechOpen.

Jabasingh S. Anuradha, Belachew Habtamu, Yimam Abubeker. 2018. "Iron Oxide Induced Bagasse Nanoparticles for the Sequestration of Cr6+ Ions from Tannery Effluent Using a Modified Batch Reactor." *Journal of Applied Polymer Science* 135(36):46683.

Jain Navin, Bhargava Arpit, Majumdar Sonali, Tarafdar J. C., and Panwar Jitendra. 2011. "Extracellular Biosynthesis and Characterization of Silver Nanoparticles Using Aspergillus flavus NJP08: A Mechanism Perspective." *Nanoscale* 3 (2):635–41.

Jeffryes Clayton, Agathos Spiros N., and Rorrer Gregory. 2015. "Biogenic Nanomaterials from Photosynthetic Microorganisms." *Current Opinion in Biotechnology* 33:23–31.

Joshi Deeksha, Garg Tarun, Goyal Amit K., Rath Goutam. 2016. "Advanced Drug Delivery Approaches against Periodontitis." *Drug Delivery* 23:363–377.

Karaagac Zehra, Dadi Seyma, Ekrikaya Semiha, and Ocsoy Ismail. 2021. "Plant Extract Guided Nanomaterials and Plant Disease Management." In *Nanotechnology in Sustainable Agriculture*, 271–290. Taylor & Francis Group.

Keshu, Rani Manviri, Yadav Jyoti, Meenu, Chaudhary Sudha and Shanker Uma. 2021. "An Updated Review on Synthetic Approaches of Green Nanomaterials and Their Application for Removal of Water Pollutants: Current Challenges, Assessment and Future Perspectives." *Journal of Environmental Chemical Engineering* 9 (6):106763.

Koca Fatih Doğan, Demirezen Yilmaz Dilek, Duman Fatih, and Ocsoy Ismail. 2018. "Comparison of Phytotoxic Effects of Bio-Synthesised Copper Oxide Nanoparticle and Ionic Copper on Elodea canadensis." *Chemistry and Ecology* 34 (9):839–853.

Krishnan Bama, Mahalingam Sundrarajan. 2017. "Improved Surface Morphology of Silver/Copper Oxide/Bentonite Nanocomposite Using Aliphatic Ammonium Based Ionic Liquid for Enhanced Biological Activities." Journal of Molecular Liquids 241:1044–58.

Kumar Harish, Gehlaut Avneesh Kumar, Gupta Himanshu, Gaur Ankur, Park Jin-Won. 2020. "Development of Copper Loaded Nanoparticles Hydrogel Made from Waste Biomass (Sugarcane Bagasse) for Special Medical Application." *Key Engineering Materials* 847:102–7.

Lai Yuekun, Wang Luning, Liu Dawei, Chen Zhong, Lin Changjian. 2015. "TiO2-Based Nanomaterials: Design, Synthesis, and Applications." *Journal of Nanomaterials* Article ID 250632.

Li Xiangqian, Xu Huizhong, Chen Zhe-Sheng, and Chen Guofang. 2011. "Biosynthesis of Nanoparticles by Microorganisms and Their Applications." *Journal of Nanomaterials* 2011:1–16.

Liu Hong-Chen, Ling-Ling E., Wang Dong-Sheng, Su Fang, Wu Xia, Shi Zhan-Ping, Lv Yan, Wang Jia-Zhu. 2011. "Reconstruction of Alveolar Bone Defects Using Bone Morphogenetic Protein 2 Mediated Rabbit Dental Pulp Stem Cells Seeded on Nano-Hydroxyapatite/Collagen/poly(L-lactide)." Tissue Engineering Part A 17(19–20):2417–2433.

Luo Fang, Chen Zuliang, Megharaj Mallavarapu, Naidu Ravendra. 2014. "Biomolecules in Grape Leaf Extract Involved in One-Step Synthesis of Iron-Based Nanoparticles." RSC Advances 4(96):53467–53474.

Machado S., Pinto S.L., Grosso J.P., Nouws H.P.A., Albergaria J.T., Delerue-Matos C. 2013. "Green Production of Zero-Valent Iron Nanoparticles Using Tree Leaf Extracts." Science of the Total Environment 445–446:1–8.

Marimuthu Sampath, Rahuman Abdul Abdul, Jayaseelan Chidambaram, Kirthi Arivasan Vishnu, Santhoshkumar Thirunavukkarasu, Velayutham Kanayairam, et al. 2013. "Acaricidal Activity of Synthesized Titanium Dioxide Nanoparticles Using Calotropis Gigantea against Rhipicephalus Microplus and Haemaphysalis Bispinosa." *Asian Pacific Journal of Tropical Medicine* 6(9):682–688.

Mondal Piyal, Anweshan A., and Purkait Mihir Kumar. 2020. "Green Synthesis and Environmental Application of Iron-Based Nanomaterials and Nanocomposite: A review." *Chemosphere* 259:127509.

Nasaruddin Ricca Rahman, Chen Tiankai, Yao Qiaofeng, Zang Shuangquan, and Xie Jianping. 2021. "Toward Greener Synthesis of Gold Nanomaterials: From Biological to Biomimetic Synthesis." *Coordination Chemistry Reviews* 426:213540.

Nwanya Assumpta Chinwe, Razanamahandry Lovasoa Christine, Bashir A.K.H., Ikpo Chinwe O., Nwanya Stephen C., Botha Subelia, et al. 2019. "Industrial Textile Effluent Treatment and Antibacterial Effectiveness of Zea Mays L. Dry Husk Mediated Bio-Synthesized Copper Oxide Nanoparticles." *Journal of Hazardous Materials* 375:281–289.

Nyman Sture, Gottlow Jan, Lindhe Jan, Karring Thorkild, Jan Wennstrom 1987. "New Attachment Formation by Guided Tissue Regeneration." *Journal of Periodontal Research* 22:252–254.

Ocsoy Ismail, Tasdemir Didar, Mazicioglu Sumeyye, and Tan Weihong. 2018. "Nanotechnology in Plants." *Advances in Biochemical Engineering and Biotechnology* 164:263–275.

Ocsoy Ismail, Tasdemir Didar, Mazicioglu Sumeyye, Celik Cagla, Katı Ahmet, and Ulgen Funda. 2018. "Biomolecules Incorporated Metallic Nanoparticles Synthesis and Their Biomedical Applications." *Materials Letters* 212:45–50.

Padovani Gislaine C., Feitosa Victor P., Sauro Salvatore, Tay Franklin R., Duran Gabriela, Paula Amauri J., Duran Nelson. 2015. "Advances in Dental Materials through Nanotechnology-Facts, Perspectives and Toxicological Aspects." *Trends in Biotechnology* 33(11):621–636.

Piñón-Segundo E., Ganem-Quintanar A., Alonso-Perez D., Quintanar-Guerrero D. 2005. "Preparation and Characterization of Triclosan Nanoparticles for Periodontal Treatment." *International Journal of Pharmaceutics* 294:217–232.

Prasad Arun S. 2016. "Iron Oxide Nanoparticles Synthesized by Controlled Bio-Precipitation Using Leaf Extract of Garlic Vine (Mansoa Alliacea)." Materials Science in Semiconductor Processing 53:79–83.

Purohit, Jyotika, Chattopadhyay Anirudha, and Singh Nirbhay K.. 2019. "Green Synthesis of Microbial Nanoparticle: Approaches to Application." In Prasad, R. (eds) *Microbial Nanobionics. Nanotechnology in the Life Sciences. Springer: Cham,* 35–60.

Sethy Naresh Kumar, Arif Zeenat, Mishra Pradeep Kumar, Kumar Pradeep. 2020. "Green Synthesis of TiO_2 Nanoparticles from Syzygium Cumini Extract for Photo-Catalytic Removal of Lead (Pb) in Explosive Industrial Wastewater." *Green Processing and Synthesis* 9(1):171–81.

Shedbalkar Utkarsha, Singh Richa, Wadhwani Sweety, Gaidhani Sharvari, and Chopade B.A. 2014. "Microbial Synthesis of Gold Nanoparticles: Current Status and Future Prospects." *Advances in Colloid and Interface Science* 209:40–8.

Siddique Sarkar, Chow James C.L. 2020. "Application of Nanomaterials in Biomedical Imaging and Cancer Therapy." Nanomaterials 10(9):1700.

Singh Sushila, Sangwan Seema, Sharma Pankaj, Devi Parvesh, Moond Monika. 2021. "Nanotechnology for Sustainable Agriculture: An Emerging Perspective." *Journal of Nanoscience and Nanotechnology* 21:3453–3465.

Some Sudip, Sen Ipsita Kumar, Mandal Amitava, Aslan Tugrul, Ustun Yakup,Yilmaz Ebru Şebnem, Katı Ahmet, Demirbas Ayse, Mandal Amit Kumar, and Ocsoy Ismail. 2018. "Biosynthesis of Silver Nanoparticles and Their Versatile Antimicrobial Properties." *Materials Research Express* 6 (1):012001.

Sood Richa, and Chopra Dimple Sethi. 2018. "Metal-Plant Frameworks in Nanotechnology: An Overview." *Phytomedicine* 50:148–156.

Sprio Simone, Campodoni Elisabetta, Sandri Monica, Preti Lorenzo, Keppler Tobias, Müller Frank A., Pugno Nicola M., and Tampieri Anna. 2018. "A Graded Multifunctional Hybrid Scaffold with Superparamagnetic Ability for Periodontal Regeneration." *International Journal of Molecular Science* 19(11):3604.

Ullah Aziz, and Lim Sung In. 2022. "Plant Extract-Based Synthesis of Metallic Nanomaterials, Their Applications, and Safety Concerns." *Biotechnology and Bioengineering* 119 (9):2273–2304.

Vasantharaj Seerangaraj, Sathiyavimal Selvam, Senthilkumar Palanisamy, LewisOscar Felix, and Pugazhendhi Arivagalan. 2019. "Biosynthesis of Iron Oxide Nanoparticles Using Leaf Extract of Ruellia tuberosa: Antimicrobial Properties and Their Applications in Photocatalytic Degradation." *Journal of Photochemistry and Photobiology B: Biology* 192:74–82.

Vincent Jei, Lau Kam Sheng, Evyan Yang Chia-Yan, Chin Siew Xian, Sillanpaa Mika, and Chia Chin Hua. 2022. "Biogenic Synthesis of Copper-Based Nanomaterials Using Plant Extracts and Their Applications: Current and Future Directions." *Nanomaterials* 12(19):3312.

Wang Yan-Fu, Wang Cheng-Yue, Wan Peng, Wang Shao-Gang, and Wang Xiu-Mei. 2016. "Comparison of Bone Regeneration in Alveolar Bone of Dogs on Mineralized Collagen Grafts with Two Composition Ratios of Nano-Hydroxyapatite and Collagen." *Regenerative Biomaterials* 3(1):33–40.

Wang Wei, Liao Susan, Zhu Yuhe, Liu Ming, Zhao Qian, and Fu Yating. 2015. "Recent Applications of Nanomaterials in Prosthodontics." *Journal of Nanomaterials* 15(2): 1–11.

Xia Lunguo, Youtao Xie, Bing Fang, Xiuhui Wang, and Kaili Lin. 2018. "In Situ Modulation of Crystallinity and Nano-Structures to Enhance the Stability and Osseointegration of Hydroxyapatite Coatings on Ti-6Al-4V implants." *Chemical Engineering Journal* 347:711–720.

Yu Na, Prodanov Ljupcho, Riet Joost te, Yang Fang, Walboomers X Frank, and Jansen John A. 2013. "Regulation of Periodontal Ligament Cell Behavior by Cyclic Mechanical Loading and Substrate Nanotexture." *Journal of Periodontology* 84(10):1504–1513.

Zikalala Nkosingiphile, Matshetshe Kabo, Parani Sundararajan, and Oluwafemi Oluwatobi S. 2018. "Biosynthesis Protocols for Colloidal Metal Oxide Nanoparticles." *Nano-Structures & Nano-Objects* 16:288–299.

12 Advances and Applications of Nanomaterial-Assisted Phytoremediation

Israt Jahan
Mardin Artuklu University

CONTENTS

12.1 INTRODUCTION

Urbanization, industrialization and other anthropogenic activities have accelerated the contamination rate of environments (surface, air, soil, sediments, and water) with various toxic pollutants, which are having increasing adverse effects on human health and other living organisms (Danh et al. 2009). Hence, it is necessary to remediate hazardous compounds in our surroundings. Various physical, chemical, and biological approaches, such as electrokinesis, chemical oxidation and leaching, thermal desorption, soil washing, and bio-detoxifying have been widely applied to clean up ecosystems (Liang et al. 2017). Phytoremediation has become a popular strategy compared with all other approaches due to the fact that it is an inexpensive, facile, esthetic, ecofriendly, and widely adaptable method that can be utilized for purifying sites polluted with heavy metals, chlorinated solvents, explosives, pesticides, crude oil, and radionuclides (Macek, Macková, and Káš 2000; Song et al. 2019).

The word phytoremediation comes from the Greek prefix "phyto," which stands for plants, and the Latin suffix "remedium," meaning renovate, cure, or restore (Van Aken 2008). Different species of the Compositae, Brassicaceae, Scrophulariaceae, Euphorbiaceae, Fabaceae, and Lamiaceae families are reported to be appropriate plants for phytoremediation from territorial areas and water (Pandey and Bajpai 2019). *Ludwigia peploides* (an aquatic perennial) and *Limnobium laevigatum*

DOI: 10.1201/9781003397861-12

(a floating aquatic plant) have a high capacity to adapt to extremely punitive environments like heavy metal contaminated water (Fernández San Juan et al. 2018).

Despite having many amenities for treating polluted areas, efficient phytoremediation can be restricted by the physiochemical characteristics of soil and water, the bioavailability of contaminants, weather conditions, and plant species (Srivastav et al. 2018). Considering these drawbacks and limitations, many advanced technologies are needed to be included in phytoremediation. Latterly, remarkable improvements have been made in achieving genetically modified plants or finding more adoptive native plant species for enhancing the phytoremediation efficiency of environmental pollutants; however, these practices are time-consuming and not widely available (Cunningham and Ow 1996). Under these circumstances, nanotechnology in phytoremediation could provide innovative concepts and inspiration for an advanced and rapid remediation strategy. Nano-phytoremediation can effectively accelerate and intensify the decontamination process of various types of pollutants in water and soil (Song et al. 2019). Over the span of years, numerous studies have demonstrated that nanoparticles such as biochar-supported nano-hydroxyapatite, nanocarbon black, tin (IV) oxide NPs, nano-nickel, nano-cobalt, zinc nanopowder, cerium (IV) oxide NPs, nano zero-valent iron (zvFeNPs), nano-silver, nano-Fe_3O_4, and TiO_2 nanopowder can be incorporated with phytoremediation, which is capable of removing heavy metals from soil and water (Zhu et al. 2019). This book chapter therefore focused on providing insight into a broad discussion of different aspects of phytoremediation and the relations of nanomaterials within plants and the environment, along with the latest progresses of nanotechnology in the application of phytoremediation. It also provided the forthcoming predictions and challenges of nano-phytoremediation.

12.2 PHYTOREMEDIATION

Utilization of living plants for removing, degrading, or absorbing hazardous contaminants from air, soil, sediment, and water are known as phytoremediation technologies use living (Reichenauer and Germida 2008). This plant-based approach is an eco-friendly, inexpensive, effective, solar energy-driven green technology for environmental remediation that exploits the capability of plants for distilling toxic components from the environment and detoxifying various compounds. Many vascular plants were found to possess remediation capacity and are capable of accumulating pollutants from spring water, sediments, surface water, and soils. For instance, *Arabis gemmifera, Betula occidentalis, Brassica juncea, Brassica nigra, Cicer arietinum,Corrigiola telephiifolia, Deschampsia cespitosa, Festuca rubra, Helianthus annuus, Helianthus tuberosus, Ipomoea alpine, Marrubium vulgare, Phyllanthus serpentinus, Phytolacca americana, Rumex induratus, Thlaspi caerulescens*, etc. (Kubota and Takenaka 2003; Kucharski et al. 2005; Koptsik 2014; García-Salgado et al. 2012; Peng et al. 2008; Lasat 2002; Zhao et al. 2019; Wang et al. 2012).

Plants adopt different approaches for remediating contamination, such as phytoextration, phytodegradation, phytovolatilization, phytostabilization, rhizodegradation, and rhizofiltration (Figure 12.1). Phytoextraction is a translocation process where pollutants are moved and deposited in aerial parts without any transformation, and afterwards, they are eliminated from the side by harvesting these aerial parts.

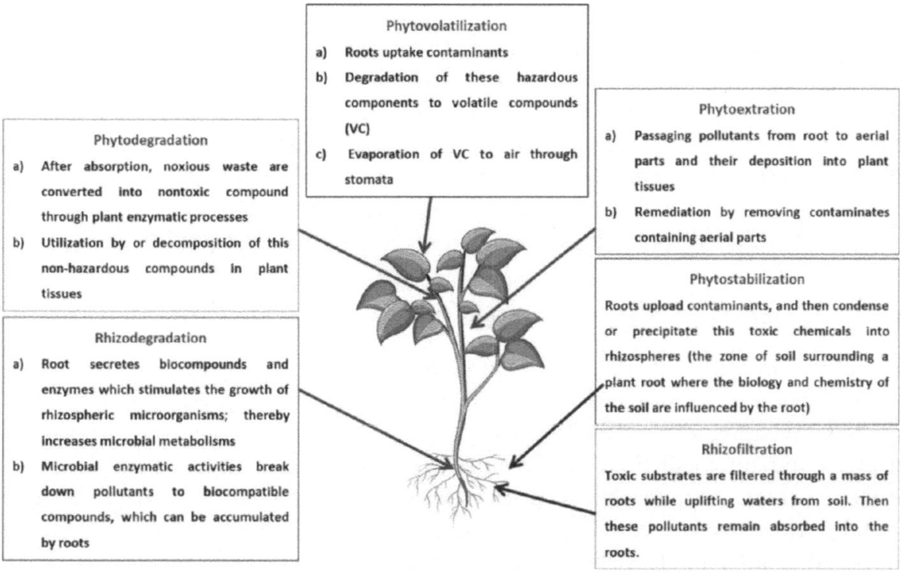

FIGURE 12.1 Schematic representation of strategies plants use for remediating contamination.

This approach is the most effective and prominent system for lessening metallic pollutants from soils. Indian mustard and sunflower are widely used plant species for phytoextraction (McGrath and Zhao 2003). On the other hand, in phytodegradation, plant cells produce multifunctional enzymes, i.e., dehalogenase, peroxidases, and nitroreductase, and complex organic pollutants in plant tissues are disintegrated into simple molecules through the enzymatic catalysis of these enzymes (Lee 2013). Alternatively, plants are capable of altering or degrading certain contaminants into volatile products, which can be released into the air through the diffusion process in open leaf stomata. This gas exchange process of phytoremediation is known as phytovolatilization and is widely used for cleaning up inorganic volatile forms of arsenic and mercury and many organic compounds such as benzene, toluene, ethylbenzene and xylene (BTEX) complexes, trichloroethylene, and chlorinated benzenes (Bankston et al. 2002; Ma and Burken 2003).

Instead of removing pollutants, phytostabilization helps to decrease the dispersion and relocation of pollutants in sediments and soils. Roots of monocots (especially plants of the grass family) uptake heavy meals and condense or precipitate them into rhizosphere zones to avoid their interaction with associated biota (Bolan et al. 2011). Furthermore, rhizodegradation is a plant-microbe-associated remediation process where accumulated organic pollutants are degraded by the microbial metabolisms of symbiotic bacteria and fungi available in the rhizosphere zone of certain vascular plants. Previously published studies suggest that plants of the Fabaceae are capable of remediating heavy metals from contaminated soils since they retain a symbiotic association with arbuscular mycorrhizal fungi (AMF), i.e., glomeromycotan fungi (Yang et al. 2015). Nevertheless, rhizofiltration is an advantageous hydroponic remediation strategy for purifying ground and spring water. While plants pump up contaminated water through the vascular system, root tissues are capable of accumulating and depositing organic and inorganic polluted compounds and subsequently eliminating those toxic compounds from the side by extruding the whole plant (Kristanti et al. 2021). Plants that have fibrous root systems and can grow rapidly, such as *Helianthus annuus*, *Spinacia oleracea, Brassica* spp., *Secale cereale*, and *Nicotiana tabacum* L., are suitable for rhizofiltration remediation (Pandey and Bajpai 2019).

12.3 NANOTECHNOLOGY AND PHYTOREMEDIATION

In recent times, a combination of phytoremediation and nanomaterials has gained huge attention as an auspicious approach for treating contaminated water, soil and air (S.P. McGrath, Zhao, and Lombi 2001). This is due to the fact that the solo use of plants (phytoremediation) for remediating environmental contaminants is a slow process that takes a long time to accomplish, even though it is considered as an eco-friendly, clean-green technology. For example, within a short period of time, the fipronil degradation rate increased by 2.5 folds when biologically synthesized (from dry seeds of *Brassica alba*) silver nanoparticles at a very low concentration (1.16 mg/L) were added to contaminated soil and water (Romeh 2018).

An atom or molecule with dimensions between 1 and 100 nm that possesses altered and unique physicochemical characteristics in parallel with their bulk molecules is known as a nanoparticle (NP) (De, Ghosh, and Rotello 2008). Nanomaterials are broadly categorized according to numerous aspects such as chemical configuration (organic and inorganic); sources (natural NPs from natural deposits, lunar dust, and volcanic dust; incidental, i.e., originate NPs from anthropogenic actions such as exhaustion of diesel engines and combustion of fossil fuels; engineered NPs, fabricated through physical, chemical or biological approaches); dimensional structure (zero, one, two and three dimensions); and usages in research and industrial areas (Saleh and Gupta 2016: 83–133; Ingale 2013). Additionally, based on physiochemical properties NPs can also be categories as metal-based nanomaterials (metal and metal oxide NPs), carbon-based nanomaterials (e.g. carbon nanotubes), dendrimers, quantum dots, and nanocomposites (X. Chen et al. 2015).

Moreover, NPs retain high surface energy and large surface areas, which facilitate their rapid absorption of large amounts of degraded contaminants, and thereby decrease energy consumption during phytoremediation. Various NPs under moderate concentrations are capable of stimulating plant growth and upsurge polluted uptake through plants, especially from heavy metal-contaminated soils, which might intensify phytoremediation efficiency (Zhu et al. 2019). Several studies have also indicated that plants can uptake nano-sized mineral nutrients more efficiently since nanoformulations might increase soil solubility and dispersal capacity of insoluble nutrients, thereby minimizing fixation and absorption by soil and subsequently intensifying the bioavailability of nano-nutrients by plants (Souri et al. 2017; Nair et al. 2010). Besides, nanomaterials can also be used for their repercussions in discharging contaminants from the ecosystem or inhibiting the formation of pollutants (Yan et al. 2020).

A number of organic complex substances, e.g., organochlorines, hydrocarbons, and long chain, are predominantly resistant to microbial and plant degradation. This limitation can be overcome by integrating nanotechnology with phytoremediation, where nano-encapsulated enzymes can be supplied in contaminated soil or water to boot up microbes and plant activity, thereby helping to degrade these complex organic compounds into simpler substrates (Ali, Khan, and Sajad 2013).

Artificially fabricated nanoparticles (Table 12.1) such as zinc oxides, titanium oxides, cerium oxides, manganese oxides, and iron oxides hold a pronounced attraction for metalloid/metal absorption available in contaminated soil, groundwater, and sediment (Li et al. 2019). However, nanosized zero-valent iron (zvFeNPs) are considered as the most effective and integral remediation materials since they retain reducing properties as electron donors and can be utilized for remediating pollutants despoiled by reduction (Han et al. 2019). A number of studies have confirmed that nZV1 can degrade organic pollutants, e.g., chlorpyrifos, molinate, and atrazine, from soil (Chaturvedi and Dave 2018). In addition, granular zero-valent iron nanoparticles (zvFeNP) are capable of reducing chlorinated carbon-based substrates, e.g., dense non-aqueous phase liquids (DNAPLs), from groundwater (Mdlovu et al. 2019). Another study showed that soil treated with zvFeNP (concentration of 500 mg/kg) increased 2, 4, 6-trinitrotoluene (TNT) uptake by roots in e-waste-contaminated soil (Jiamjitrpanich et al. 2013). In a study on the other hand, it was observed that trichloroethylene (TCE) (available in petroleum waste) uptake by eastern cottonwood was raised by 26% and 82% after adding fullerene nanoparticles at concentrations of 2 and 15 mg/L, respectively (Ma and Wang 2010). Additionally, iron oxide NPs support as key scavengers of various extremely noxious heavy metals such as zinc [Zn(II)], lead [Pb(II)], copper Cu(II), chromium [Cr(VI)], cadmium [Cd(II)], and arsenic [As(III)] (Chen and Li 2010; Zhang et al. 2010; Lin et al. 2012; Jiang et al. 2013).

Furthermore, adding an appropriate amount of nanomaterials to the soil also influences plant growth in contaminated surroundings and, thereby, increases the phytoremediation process. For instance, in chromium (Cr)-contaminated soil, not only is the amount of Cr decreased in plants, but also plant development is stimulated by the combination of nanosized zero-valent iron (nZV1) with cardboard mustard (Su et al. 2016). Under high contamination, nanosized zero-valent iron increased the growth of *Panicum maximum* by 53.5%, whereas it was only 30.3% under clean soil (Gao and Zhou 2013). Similarly, titanium dioxide (TiO_2) nano-anatase is able to increase photosynthetic rate, decrease UV-B-induced oxidative stress in spinach chloroplasts, enhance the accumulation of nitrate and its substitutes through spinach root cells after converting soil inorganic nitrogen compounds into organic substances, and hence, change the organic formulation of soil (Lei et al. 2008; Yang et al. 2006).

On the other hand, in contaminated agricultural fields, engineered NPs can be utilized to improve the biomass of crops by minimizing heavy metal bioavailability since they are capable of interacting with metals/metalloids from contaminated soils (Tafazoli et al. 2017). For instance, zero-valent iron nanoparticles (zvFeNP) are capable of removing heavy metals directly by absorbing equal or highly negatively charged standard redox potential metals (Cd and Zn), whereas positively charged

TABLE 12.1
Nanomaterials Used for Nano-Phytoremediation.

Nanomaterials	Plant Used for Phytoremediation	Impacts	References
Salicyclic acid nanoparticle (SNAPs)	*Isatis cappadocica*	Degradation of arsenic-contaminated soil	Souri et al. (2017)
Zero-valent iron nanoparticles (zvFeNPs)	*Alpinia calcarata* *Cymbopogon citratus* *Ocimum sanctum*	Removal of endosulfan from soil	Pillai and Kottekottil (2016)
zvFeNPs	*Eucalyptus dunnii*	Decontamination of chlorinated ethanes and chlorinated methane in soil	Gotelli et al. (2020)
Fullerene nanoparticles	*Populus deltoids W. Bartram ex Marshall*	increased uptake of Trichloroethylene (TCE) from diesel oil contained food waste	Ma and Wang (2010)
zvFeNPs	*Panicum maximum*	Trinitrotoluene (TNT) uptake by root from soil is more influential in presence of NPs	Jiamjitrpanich et al. (2013)
zvFeNPs	*Impatiens balsamina*	Proliferated plant growth by 53.5% under high contamination	Gao and Zhou (2013)
Titanium dioxide (TiO_2) nano-anatase	*Spinacia oleracea*	Reduced the amount of soil inorganic nitrogen based compound by enhancing accumulation of organic nitrate and its substitutes through roots	Yang et al. (2006)
Powdered FeNPs	*Lupinus albus L.*	Declined metal leaching up to 83% and 98% for Cd and Zn, respectively.	Wang et al. (2022)
zvFeNPs	*Lolium perenne* L.	Increased lead (Pb) accumulation in plants at the presence of very low concentration of zvFeNPs (100, 200, and 500 mg/kg) in contaminated soil.	Huang et al. (2018)
Nano hydroxyapatites (NHAP)	*Lolium perenne* L.	Decreasing copper (Cu) and lead (Pb) toxicity of plant by immobilizing Pb and increasing soil pH	Jin et al. (2016)
Biochar-supported hydroxyapatite nanocompounds (nHAP@BC)	–	Increase the immobilization rates of lead in the soil	Yang et al. (2016)
Biosynthesized silver NPs (AgNPs)	–	Decreased the concentration of Fipronil in contaminated soil and water.	Romeh (2018)
Nano-silver	Corn field	Promotes remediation in soil by stimulating phytohormones production	Khan and Bano (2016)
Manganese nanoparticles	*Vigna radiata*	At 50–1,000 mg/L concentration, increased organic compound uptakes by elongating root and shoot	Pradhan et al. (2014)
Single- and multi-walled carbon nanotubes	*Allium cepa,* *Cucumis sativus,* Lycopersicon esculentum	Increased seed germination and biomass production	Lahiani et al. (2013)

standard redox potential metals (Pb and Ni) are adsorbable and reduced (Tafazoli et al. 2017; Naorem et al. 2022). Utilization of zero-valent iron nanoparticles on rice fields spiked with Cadmium (Cd) indicated that a significant amount of Cd accumulation in rice seeds and leaves was reduced in the presence of a very low concentration (0.01%) of zvFeNP (Watanabe et al. 2009). Likewise, another study indicated that iron NPs supplemented soils can decrease the bioavailability of cadmium to plants by immobilizing this metal, reducing Cd toxicity to agronomic plants (Lei et al. 2008). A study by Jin et al. (2016) showed that nanohydroxyapatites (NHAP) are capable of decreasing copper (Cu) and lead (Pb) toxicity of ryegrass (*Lolium perenne* L.) by immobilizing Pb and increasing soil pH, which help this plant to persist and grow in soil contaminated with a high concentration (800 mg/kg) of lead and increase the ryegrass biomass extensively.

12.4 FACTORS AFFECTING NANO-PHYTOREMEDIATION

Phytoremediation with the combination of nanotechnology principally comprises two principals, i.e., (i) how NPs are influencing the morphological, physiological, and molecular structures of plants, e.g., the photosynthetic rate, carotenoid and chlorophyll contents, germination and growth rates, etc.; and (ii) what are the mechanisms involved in nano-phytoremediation to adsorb organic, inorganic, heavy metals, or metal-like pollutants from contaminated ecosystems (Ibrahim et al. 2016).

In the same plant, mechanisms such as nano-phytoaccumulation, nano-phytostabilization, and nano-phytodegradation can take place synchronously, which offers nano-phytoremediation as an effective method of environmental restoration (Jiamjitrpanich et al. 2013). However, nanomaterials used in nano-phytoremidation should not be hazardous to the plants and their microenvironment and rhizosphere microbiota (AM mycorrhizal fungi and N-fixing bacteria) (Jiamjitrpanich et al. 2013). In addition, NPs should be biocompatible with unrivaled binding capacity to pollutants, which may accelerate these phytoextraction processes. During the nano-phytoremediation process, no toxic by-products should be produced throughout the deterioration process of toxic pollutants; otherwise, they might infect the groundwater or ecosystem or enter the food chain through their uptake and accumulation by plant tissues (Zhu et al. 2019; Feng et al. 2017). In addition, nanomaterials at exceedingly high concentrations retain damaging impacts on soil remediation and plant growth (Chen et al. 2017). High concentrations of zero-valent iron nanoparticles (more than 500 mg/kg) impede the total biomass content of ryegrass (*Lolium perenne* L.), albeit they stimulate root growth at very low concentrations (Huang et al. 2018). Likewise in soybean (*Glycine max*), oil amended with high concentration (1.0 g kg^{-1}) of cerium(IV) oxide nanoparticles (CeO_2NPs) aggravated lipid peroxidation, escalated reactive oxygen species (ROS) activity, reduced total chlorophyll contain, increased visible chlorosis, and foliar damages (Priester et al. 2017). Furthermore, the phytoremediation potential of nanomaterials depends not only on plant spices and the amount of nanoparticles but also on the period of the remediation process. For instance, the immobilization capacities of biochar-supported hydroxyapatite nanocompounds (nHAP@BC) were increased when the remediation time was delayed, i.e., the immobilization rates were 44.2%, 49.9%, and 56.8% after 1, 7, and 28 days, respectively (Yang et al. 2016).

On the other hand, different ecological factors, such as the physico-chemical features of contaminated water and soil, the quantity and quality of pollutants, and the nature of the symbiotic association between plant and root rhizosphere, also impact the uptake and deposition of pollutants inside the plant root cytoplasm (Rai et al. 2019). The coexistence of engineered nanoparticles (ENPs) and plant effects on these environmental, chemical and biological conditions accelerates the uptake of contaminants inside the cell cytoplasm of plant roots (Yan et al. 2020).

12.5 CONCLUSION AND FUTURE PERSPECTIVES

There is no doubt that nano-phytoremediation offers a promising and eco-friendly remediation for the environmental cleaning process without harming nature. However, evaluation of nanoparticles in remediation is mostly confined to pot-culture trials or microcosm experiments in the laboratory; hence, extended trials are necessary to observe the large-scale capacity of nano-phytoremediation in contaminated sites. Further experiments are also required to evaluate the substantiated effects of nanomaterials on the properties (pH, mineral balance, temperature, nutritional value, and moisture content) of actual contaminated field soil and water. Besides, more suitable plant species and microbial strains should be identified that retain excellent accumulation capabilities for large amounts of contaminants.

REFERENCES

Ali, Hazrat, Ezzat Khan, and Muhammad Anwar Sajad. 2013. "Phytoremediation of Heavy Metals-Concepts and Applications." *Chemosphere* 91 (7): 869–81.

Bankston, Jamie L., Daniel L. Sola, Andrew T. Komor, and Daryl F. Dwyer. 2002. "Degradation of Trichloroethylene in Wetland Microcosms Containing Broad-Leaved Cattail and Eastern Cottonwood." *Water Research* 36 (6): 1539–46.

Bolan, Nanthi S., Jin Hee Park, Brett Robinson, Ravi Naidu, and Keun Young Huh. 2011. "Phytostabilization: A Green Approach to Contaminant Containment." In *Advances in Agronomy*, vol. 112, 145–204. Amsterdam: Elsevier.

Chaturvedi, Shalini, and Pragnesh N. Dave. 2018. "Water Purification Using Nanotechnology an Emerging Opportunities." *Chemical Methodologies* 3 (1): 1–144.

Chen, Ming, Xiaosheng Qin, and Guangming Zeng. 2017. "Biodiversity Change Behind Wide Applications of Nanomaterials?" *Nano Today* 17 (December): 11–13.

Chen, Xu, Xiaoquan Huang, Chuping Zheng, Yanan Liu, Taoyuan Xu, and Jie Liu. 2015. "Preparation of Different Sized Nano-Silver Loaded on Functionalized Graphene Oxide with Highly Effective Antibacterial Properties." *Journal of Materials Chemistry B* 3 (35): 7020–29.

Chen, Yen-Hua, and Fu-An Li. 2010. "Kinetic Study on Removal of Copper(II) Using Goethite and Hematite Nano-Photocatalysts." *Journal of Colloid and Interface Science* 347 (2): 277–81.

Cunningham, Scott D., and David W. Ow. 1996. "Promises and Prospects of Phytoremediation." *Plant Physiology* 110 (3): 715–19.

Danh, Luu Thai, Paul Truong, Raffaella Mammucari, Tam Tran, and Neil Foster. 2009. "A Choice Plant for Phytoremediation of Heavy Metals and Organic Wastes." *International Journal of Phytoremediation* 11 (8): 664–91.

De, Mrinmoy, Partha S. Ghosh, and Vincent M. Rotello. 2008. "Applications of Nanoparticles in Biology." *Advanced Materials* 20 (22): 4225–41.

Feng, Nai-Xian, Jiao Yu, Hai-Ming Zhao, Yu-Ting Cheng, Ce-Hui Mo, Quan-Ying Cai, Yan-Wen Li, Hui Li, and Ming-Hung Wong. 2017. "Efficient Phytoremediation of Organic Contaminants in Soils Using Plant-Endophyte Partnerships." *Science of the Total Environment* 583 (April): 352–68.

Fernández San Juan, M. Rocío, Carolina B. Albornoz, Karen Larsen, and Roberto Najle. 2018. "Bioaccumulation of Heavy Metals in Limnobium Laevigatum and Ludwigia Peploides: Their Phytoremediation Potential in Water Contaminated with Heavy Metals." *Environmental Earth Sciences* 77 (11): 404.

Gao, Yuan Yuan, and Qi Xing Zhou. 2013. "Application of Nanoscale Zero Valent Iron Combined with Impatiens Balsamina to Remediation of E-Waste Contaminated Soils." *Advanced Materials Research* 790 (September): 73–76.

García-Salgado, Sara, David García-Casillas, Ma. Angeles Quijano-Nieto, and Ma. Milagros Bonilla-Simón. 2012. "Arsenic and Heavy Metal Uptake and Accumulation in Native Plant Species from Soils Polluted by Mining Activities." *Water, Air, & Soil Pollution* 223 (2): 559–72.

Gotelli, M. J., A. Lo Balbo, G. M. Caballero, and C. A. Gotelli. 2020. "Hexachlorocyclohexane Phytoremediation Using Eucalyptus Dunnii of a Contaminated Site in Argentina." *International Journal of Phytoremediation* 22 (11): 1129–36.

Han, Yitong, Subhasis Ghoshal, Gregory V. Lowry, and Jiawei Chen. 2019. "A Comparison of the Effects of Natural Organic Matter on Sulfidated and Nonsulfidated Nanoscale Zerovalent Iron Colloidal Stability, Toxicity, and Reactivity to Trichloroethylene." *Science of the Total Environment* 671 (June): 254–61.

Huang, Danlian, Xiang Qin, Zhiwei Peng, Yunguo Liu, Xiaomin Gong, Guangming Zeng, Chao Huang, et al. 2018. "Nanoscale Zero-Valent Iron Assisted Phytoremediation of Pb in Sediment: Impacts on Metal Accumulation and Antioxidative System of Lolium Perenne." *Ecotoxicology and Environmental Safety* 153 (May): 229–37.

Ibrahim, Rusul Khaleel, Maan Hayyan, Mohammed Abdulhakim AlSaadi, Adeeb Hayyan, and Shaliza Ibrahim. 2016. "Environmental Application of Nanotechnology: Air, Soil, and Water." *Environmental Science and Pollution Research* 23 (14): 13754–88.

Ingale, G. Arun. 2013. "Biogenic Synthesis of Nanoparticles and Potential Applications: An Eco- Friendly Approach." *Journal of Nanomedicine & Nanotechnology* 04 (02).

Jiamjitrpanich, Waraporn, Preeda Parkpian, Chongrak Polprasert, and Rachain Kosanlavit. 2013. "Trinitrotoluene and Its Metabolites in Shoots and Roots of Panicum Maximum in Nano-Phytoremediation." *International Journal of Environmental Science and Development*, 04: 7–10.

Jiang, Wenjun, Miguel Pelaez, Dionysios D. Dionysiou, Mohammad H. Entezari, Dimitra Tsoutsou, and Kevin O'Shea. 2013. "Chromium(VI) Removal by Maghemite Nanoparticles." *Chemical Engineering Journal* 222 (April): 527–33.

Jin, Yu, Wei Liu, Xi-liang Li, Shi-gang Shen, Shu-xuan Liang, Cunqi Liu, and Liyuan Shan. 2016. "Nano-Hydroxyapatite Immobilized Lead and Enhanced Plant Growth of Ryegrass in a Contaminated Soil." *Ecological Engineering* 95 (October): 25–29.

Khan, Naeem, and Asghari Bano. 2016. "Modulation of Phytoremediation and Plant Growth by the Treatment with PGPR, Ag Nanoparticle and Untreated Municipal Wastewater." *International Journal of Phytoremediation* 18 (12): 1258–69.

Koptsik, G. N. 2014. "Problems and Prospects Concerning the Phytoremediation of Heavy Metal Polluted Soils: A Review." *Eurasian Soil Science* 47 (9): 923–39.

Kristanti, Risky Ayu, Wei Jie Ngu, Adhi Yuniarto, and Tony Hadibarata. 2021. "Rhizofiltration for Removal of Inorganic and Organic Pollutants in Groundwater: A Review." *Biointerface Research in Applied Chemistry* 11 (4): 12326–47.

Kubota, Hiroshi, and Chisato Takenaka. 2003. "Field Note: Arabis Gemmifera Is a Hyperaccumulator of Cd and Zn." *International Journal of Phytoremediation* 5 (3): 197–201.

Kucharski, R., A. Sas-Nowosielska, E. Małkowski, J. Japenga, J. M. Kuperberg, M. Pogrzeba, and J. Krzyżak. 2005. "The Use of Indigenous Plant Species and Calcium Phosphate for the Stabilization of Highly Metal-Polluted Sites in Southern Poland." *Plant and Soil* 273 (1–2): 291–305.

Lahiani, Mohamed H., Enkeleda Dervishi, Jihua Chen, Zeid Nima, Alain Gaume, Alexandru S. Biris, and Mariya V. Khodakovskaya. 2013. "Impact of Carbon Nanotube Exposure to Seeds of Valuable Crops." *ACS Applied Materials & Interfaces* 5 (16): 7965–73.

Lasat, Mitch M. 2002. "Phytoextraction of Toxic Metals: A Review of Biological Mechanisms." *Journal of Environmental Quality* 31 (1): 109–20.

Lee, Jae Heung. 2013. "An Overview of Phytoremediation as a Potentially Promising Technology for Environmental Pollution Control." *Biotechnology and Bioprocess Engineering* 18 (3): 431–39.

Lei, Zheng, Su Mingyu, Wu Xiao, Liu Chao, Qu Chunxiang, Chen Liang, Huang Hao, Liu Xiaoqing, and Hong Fashui. 2008. "Antioxidant Stress is Promoted by Nano-Anatase in Spinach Chloroplasts under UV-B Radiation." *Biological Trace Element Research* 121 (1): 69–79.

Li, Bangkai, Xi Chen, Kun Li, Chenghao Zhang, Yang He, Runhong Du, Jianzu Wang, and Li Chen. 2019. "Coupling Membrane and Fe-Pd Bimetallic Nanoparticles for Trichloroethene Removing from Water." *Journal of Industrial and Engineering Chemistry* 78 (October): 198–209.

Liang, Shu-xuan, Yu Jin, Wei Liu, Xiliang Li, Shi-gang Shen, and Ling Ding. 2017. "Feasibility of Pb Phytoextraction Using Nano-Materials Assisted Ryegrass: Results of a One-Year Field-Scale Experiment." *Journal of Environmental Management* 190 (April): 170–75.

Lin, Sen, Diannan Lu, and Zheng Liu. 2012. "Removal of Arsenic Contaminants with Magnetic γ-Fe2O3 Nanoparticles." *Chemical Engineering Journal* 211–212 (November): 46–52.

Ma, Xingmao, and Joel G. Burken. 2003. "TCE Diffusion to the Atmosphere in Phytoremediation Applications." *Environmental Science & Technology* 37 (11): 2534–39.

Ma, Xingmao, and Chen Wang. 2010. "Fullerene Nanoparticles Affect the Fate and Uptake of Trichloroethylene in Phytoremediation Systems." *Environmental Engineering Science* 27 (11): 989–92.

Macek, T., M. Macková, and J. Káš. 2000. "Exploitation of Plants for the Removal of Organics in Environmental Remediation." *Biotechnology Advances* 18 (1): 23–34.

McGrath, S.P., F.J. Zhao, and E. Lombi. 2001. "Plants and Rhizosphere Processes Involved in Phytoremediation of Metal Contaminated Soils." *Plant and Soil* 232 (1/2): 207–14.

McGrath, Steve P., and Fang-Jie Zhao. 2003. "Phytoextraction of Metals and Metalloids from Contaminated Soils." *Current Opinion in Biotechnology* 14 (3): 277–82.

Mdlovu, Ndumiso Vukile, Kuen-Song Lin, Chung-Yu Chen, Fikile Agath Mavuso, Sikhumbuzo Charles Kunene, and Maria Janina Carrera Espinoza. 2019. "In-Situ Reductive Degradation of Chlorinated DNAPLs in Contaminated Groundwater Using Polyethyleneimine-Modified Zero-Valent Iron Nanoparticles." *Chemosphere* 224 (June): 816–26.

Nair, Remya, Saino Hanna Varghese, Baiju G. Nair, T. Maekawa, Y. Yoshida, and D. Sakthi Kumar. 2010. "Nanoparticulate Material Delivery to Plants." *Plant Science* 179 (3): 154–63.

Naorem, Anandkumar, Abhishek Patel, and Shiva Kumar Udayana. 2022. "Exploitation of Soil Amendments to Remediate Heavy Metal Toxicity for Safe Cultivation of Crops." In *Microbial Biotechnology*, edited by Pankaj Chowdhary, Sujata Mani, and Preeti Chaturvedi, 1st ed., 175–88. Hoboken, NJ: Wiley.

Pandey, Vimal Chandra, and Omesh Bajpai. 2019. "Phytoremediation: From Theory toward Practice." In *Phytomanagement of Polluted Sites*, edited by Vimal Chandra Pandey and Kuldeep Bauddh, 1–49. Amsterdam: Elsevier.

Peng, Kejian, Chunling Luo, Wuxin You, Chunlan Lian, Xiangdong Li, and Zhenguo Shen. 2008. "Manganese Uptake and Interactions with Cadmium in the Hyperaccumulator-Phytolacca Americana L." *Journal of Hazardous Materials* 154 (1–3): 674–81.

Pillai, Harikumar P. S., and Jesitha Kottekottil. 2016. "Nano-Phytotechnological Remediation of Endosulfan Using Zero Valent Iron Nanoparticles." *Journal of Environmental Protection* 07 (05): 734–44.

Pradhan, Saheli, Prasun Patra, Shouvik Mitra, Kushal Kumar Dey, Sneha Jain, Samapd Sarkar, Shuvrodeb Roy, Pratip Palit, and Arunava Goswami. 2014. "Manganese Nanoparticles: Impact on Non-Nodulated Plant as a Potent Enhancer in Nitrogen Metabolism and Toxicity Study Both in Vivo and in Vitro." *Journal of Agricultural and Food Chemistry* 62 (35): 8777–85.

Priester, John H., Shelly Cole Moritz, Katherine Espinosa, Yuan Ge, Ying Wang, Roger M. Nisbet, Joshua P. Schimel, A. Susana Goggi, Jorge L. Gardea-Torresdey, and Patricia A. Holden. 2017. "Damage Assessment for Soybean Cultivated in Soil with Either CeO2 or ZnO Manufactured Nanomaterials." *Science of the Total Environment* 579 (February): 1756–68.

Rai, Prabhat Kumar, Sang Soo Lee, Ming Zhang, Yiu Fai Tsang, and Ki-Hyun Kim. 2019. "Heavy Metals in Food Crops: Health Risks, Fate, Mechanisms, and Management." *Environment International* 125 (April): 365–85.

Reichenauer, Thomas G., and James J. Germida. 2008. "Phytoremediation of Organic Contaminants in Soil and Groundwater." *ChemSusChem* 1 (8–9): 708–17.

Romeh, Ahmed Ali Ali. 2018. "Green Silver Nanoparticles for Enhancing the Phytoremediation of Soil and Water Contaminated by Fipronil and Degradation Products." *Water, Air, & Soil Pollution* 229 (5): 147.

Saleh, Tawfik Abdo, and Vinod Kumar Gupta. 2016. *Nanomaterial and Polymer Membranes*. Amsterdam: Elsevier.

Song, Biao, Piao Xu, Ming Chen, Wangwang Tang, Guangming Zeng, Jilai Gong, Peng Zhang, and Shujing Ye. 2019. "Using Nanomaterials to Facilitate the Phytoremediation of Contaminated Soil." *Critical Reviews in Environmental Science and Technology* 49 (9): 791–824.

Souri, Zahra, Naser Karimi, Marzeyh Sarmadi, and Elham Rostami. 2017. "Salicylic Acid Nanoparticles (SANPs) Improve Growth and Phytoremediation Efficiency of Isatis Cappadocica Desv., under as Stress." *IET Nanobiotechnology* 11 (6): 650–55.

Srivastav, Akansha, Krishna Kumar Yadav, Sunita Yadav, Neha Gupta, Jitendra Kumar Singh, Ravi Katiyar, and Vinit Kumar. 2018. "Nano-Phytoremediation of Pollutants from Contaminated Soil Environment: Current Scenario and Future Prospects." In *Phytoremediation*, edited by Abid A. Ansari, Sarvajeet Singh Gill, Ritu Gill, Guy R. Lanza, and Lee Newman, 383–401. Cham: Springer International Publishing.

Su, Huijie, Zhanqiang Fang, Pokeung Eric Tsang, Jianzhang Fang, and Dongye Zhao. 2016. "Stabilisation of Nanoscale Zero-Valent Iron with Biochar for Enhanced Transport and in-Situ Remediation of Hexavalent Chromium in Soil." *Environmental Pollution* 214 (July): 94–100.

Tafazoli, Mahya, Seyyed Mohammad Hojjati, Pourya Biparva, Yahya Kooch, and Norbert Lamersdorf. 2017. "Reduction of Soil Heavy Metal Bioavailability by Nanoparticles and Cellulosic Wastes Improved the Biomass of Tree Seedlings." *Journal of Plant Nutrition and Soil Science* 180 (6): 683–93.

Van Aken, Benoit. 2008. "Transgenic Plants for Phytoremediation: Helping Nature to Clean up Environmental Pollution." *Trends in Biotechnology* 26 (5): 225–27.

Wang, Xinbin Feng, Christopher W.N. Anderson, Ying Xing, and Lihai Shang. 2012. "Remediation of Mercury Contaminated Sites: A Review." *Journal of Hazardous Materials* 221–222 (June): 1–18.

Wang, Ping, Feng Shen, Ronghua Li, Di Guo, Wen Liang, Tao Liu, and Zengqiang Zhang. 2022. "Remediation of Cd and Zn Contaminated Soil by Zero Valent Iron (Fe 0): A Field Trial." *Environmental Technology & Innovation* 28 (November): 102603.

Watanabe, Toshihiro, Yasutoshi Murata, Takashi Nakamura, Yuki Sakai, and Mitsuru Osaki. 2009. "Effect of Zero-Valent Iron Application on Cadmium Uptake in Rice Plants Grown in Cadmium-Contaminated Soils." *Journal of Plant Nutrition* 32 (7): 1164–72.

Yan, An, Yamin Wang, Swee Ngin Tan, Mohamed Lokman Mohd Yusof, Subhadip Ghosh, and Zhong Chen. 2020. "Phytoremediation: A Promising Approach for Revegetation of Heavy Metal-Polluted Land." *Frontiers in Plant Science* 11 (April): 359.

Yang, Fan, Fashui Hong, Wenjuan You, Chao Liu, Fengqing Gao, Cheng Wu, and Ping Yang. 2006. "Influences of Nano-Anatase TiO2 on the Nitrogen Metabolism of Growing Spinach." *Biological Trace Element Research* 110 (2): 179–90.

Yang, Zhanqiang Fang, Liuchun Zheng, Wen Cheng, Pokeung Eric Tsang, Jianzhang Fang, and Dongye Zhao. 2016. "Remediation of Lead Contaminated Soil by Biochar-Supported Nano-Hydroxyapatite." *Ecotoxicology and Environmental Safety* 132 (October): 224–30.

Yang, Yurong, Xiaozhen Han, Yan Liang, Amit Ghosh, Jie Chen, and Ming Tang. 2015. "The Combined Effects of Arbuscular Mycorrhizal Fungi (AMF) and Lead (Pb) Stress on Pb Accumulation, Plant Growth Parameters, Photosynthesis, and Antioxidant Enzymes in Robinia Pseudoacacia L." *PLoS One* . Article e0145726.

Zhang, Xin, Shen Lin, Xiao-Qiao Lu, and Zu-liang Chen. 2010. "Removal of Pb(II) from Water Using Synthesized Kaolin Supported Nanoscale Zero-Valent Iron." *Chemical Engineering Journal* 163 (3): 243–48.

Zhao, Anqi, Lingyun Gao, Buqing Chen, and Liu Feng. 2019. "Phytoremediation Potential of Miscanthus Sinensis for Mercury-Polluted Sites and Its Impacts on Soil Microbial Community." *Environmental Science and Pollution Research* 26 (34): 34818–29.

Zhu, Yi, Fang Xu, Qin Liu, Ming Chen, Xianli Liu, Yanyan Wang, Yan Sun, and Lili Zhang. 2019. "Nanomaterials and Plants: Positive Effects, Toxicity and the Remediation of Metal and Metalloid Pollution in Soil." *Science of the Total Environment* 662 (April): 414–21.

13 Role of Nanophotocatalysts in Water Remediation

Mona Mittal
Galgotias College of Engineering and Technology

Atanu Jana
Dongguk University

CONTENTS

13.1 INTRODUCTION

Water pollution is one of the most challenging environmental issues because it substantially affects all aspects of human life, including food, health, economy, and energy. Water is the most essential substance for the sustenance of life; however, it is currently heavily polluted. Industrial effluents, especially dyes and antibiotics, pesticides, pharmaceuticals, pathogenic bacteria, and heavy metals, contaminate the water continuously and affect the survival of aquatic organisms and humans (Mahlaule-Glory and Hintsho-Mbita 2022). According to estimates, 80% of untreated water is released into various water streams (Cui et al. 2021), which leads to water pollution. Synthetic and natural organic dyes are used to color or change the color of various items, including food, clothing, furniture, cosmetics, bedding, leather, rubber, and paper. Synthetic dyes are disseminated in various industries, such as the textile, printing, tanning, pharmaceutics, food, and cosmetic industries. After consumption, synthetic dyes induce serious conditions in humans such as cancer, kidney failure, brain damage, and skin irritation (Lee et al. 2016). Bacterial strains such as *Enterobacter faecalis* (*E. faecalis*), *Escherichia coli* (*E. coli*), *Pseudomonas*

DOI: 10.1201/9781003397861-13

aeruginosa (*P. aeruginosa*), and *Staphylococcus aureus* (*S. aureus*) have also recently attracted attention as natural contaminants. Industrial effluents and hospital waste, which produce sewage water, are the sources of these natural contaminants. These bacterial strains are captured, gathered, multiplied, and, hence, kept growing into pathogens in wastewater. This renders these bacterial strains highly pathogenic, poisonous, and dangerous, which pose a severe threat to human health and the ecosystem (Ashwini et al. 2021). Hence, it becomes essential to treat polluted water by removing these natural and artificial contaminants from it.

Currently, researchers are attempting to create sustainable approaches that can be employed to prevent inadequate water management and water pollution. Adsorption, reverse osmosis, and chemical coagulation are some common wastewater treatment methods. However, additional treatments are required for all of the above procedures for recycling the material, and hence, the accompanying costs are comparatively high (Saini et al. 2022). The advanced oxidation process, especially heterogeneous photocatalysis of pollutants under the irradiation process, is an important method for the successful treatment of wastewater owing to its cost-effectiveness and ease of operation. Photocatalysis involves the conversion of highly harmful organic contaminants into benign inorganic salts via the formation of extremely reactive species such as superoxide anions, hydroxyl radicals, singlet oxygen, and hydrogen peroxide. The photocatalytic degradation of organic pollutants such as antibiotics and dyes under the illumination of sunlight is also a superior alternative owing to the renewable nature of sunlight (Alharbi et al. 2020).

Researchers are utilizing the emerging science of nanotechnology as a tool for the treatment of polluted water. Semiconductor nanophotocatalysts have been suggested as potential photocatalysts for the treatment of various water pollutants due to their unique physicochemical properties (Lu and Astruc 2018). Catalytic activity, optical absorption, and electrical characteristics of semiconductor nanomaterials can be enhanced by reducing particle size, which results in a high surface-to-volume ratio. A large number of semiconductor nanomaterials such as ZnO, TiO_2, CdS, Fe_2O_3, ZnS, MoO_3, bismuth oxyhalide, perovskite, and WO_3 have been utilized as effective photocatalysts for the degradation of various organic pollutants (Dhariwal and Mittal 2023; Singh et al. 2023; Sharma, Yadav and Mittal 2023). It is worth noting that ZnO and TiO_2 nanoparticles (NPs) are the most extensively utilized photocatalysts owing to their high photochemical stability, photosensitivity, efficiency, insolubility, and non-toxicity (Jabbar and Esmail Ebrahim 2022).

ZnO is a promising n-type semiconductor with a relatively direct 3.37 eV (wide) bandgap, 4.2 eV electron affinity, and 60 meV (high) exciton-binding energy, enabling it to be used in a variety of applications, including solar cells, photocatalysis, gas sensors, biosensors, photodetectors, and nanogenerators. Furthermore, ZnO NPs can be considered an effective alternative photocatalyst to TiO_2 NPs owing to their similar band gap value and lower cost. As a result, ZnO NPs could be one of the best photocatalysts for large-scale wastewater treatment because they have more active sites, a large surface-to-volume ratio, faster reaction rates, and are also more effective in producing hydrogen peroxide (Nahi, Radhakrishnan and Beena 2021; Nazir et al. 2022).

To date, enormous efforts have been made to synthesize ZnO NPs in various forms and proportions using diverse approaches such as wet chemical, hydrothermal, sol–gel, vapor phase, solvothermal, microwave, chemical microemulsion, sonochemical technique, and direct precipitation (Brayner et al. 2010). Various attempts have been made to create hierarchical and directed ZnO nanostructures for photocatalytic applications. These conventional synthesis strategies involve the use of toxic chemicals and consume a lot of time and energy in the completion of the reaction. However, a simple and cost-effective process for producing ZnO NPs with the appropriate shape at a high yield is still needed. In this regard, green synthesis methods have acquired a lot of interest for the large-scale preparation of metal oxide nanomaterials as they utilize natural and organic sources. Plants, bacteria, algae, actinomycetes, and fungi are some of the biological organisms that can help achieve this goal. Green synthesis methods have various advantages over conventional chemical methods because they are simple, sustainable, reliable, eco-friendly, energy-efficient, non-toxic, and cost-effective. It also eliminates the rigidity and difficulties of traditional methods. As a result, there is considerable interest in using green synthesis to make ZnO NPs with narrow particle size distributions for critical applications (Kaviya and Prasad 2015; Patil et al. 2016).

This chapter highlights the recent advancement in the green and eco-friendly synthesis of ZnO-based nanomaterials by utilizing plant extract as reducing and stabilizing agents. Plant extract-based green synthesis technologies are preferred as plant extracts are easily available, simple to handle, cost-effective, non-toxic, and environmentally friendly. Furthermore, the phytochemicals allow for the development of NPs with desired properties. Here, we examine the green synthesis method using the plant parts (leaves, fruits, and stems) that have previously been used. It will enable researchers to gain a fundamental understanding of the process used to synthesize plant extracts. There is also a discussion on their application in the photocatalytic degradation of organic pollutants and antibacterial activity. Finally, the chapter discusses the difficulties and potential prospects associated with synthesizing ZnO-based nanomaterials from plant extracts. This chapter will help scholars in their future endeavors and contribute to the evolution of literature in this field.

13.2 BIOSYNTHESIS OF ZnO-BASED NANOMATERIALS

The plant parts that have both reducing and antioxidant properties are used to synthesize the NPs. Fortunately, all medicinal plants show antioxidant characteristics, and biomolecules such as flavonoids, carboxylic acids, proteins, and phenols in plant extracts have reducing properties. For the synthesis of ZnO NPs, the precursor salt solution of zinc (zinc sulfate, zinc acetate, zinc nitrate, zinc nitrate) and different plant parts such as flower, root, fruit, peel, gum, leaf, rhizome, bark, and seed in powdered forms are required. The powdered form of the plant part is dissolved into distilled water, resulting in the formation of plant extract, and then the same is added to the zinc salt solution with continuous stirring. Furthermore, the mixture is heated to a specific temperature and stirred over a magnetic stirrer for about an hour. The synthesis of NPs is confirmed by observing precipitates of ZnO NPs, which are collected by centrifugation. The obtained precipitates are washed three to four times with distilled water to remove the impurities (Patil et al. 2016; Mahlaule-Glory and

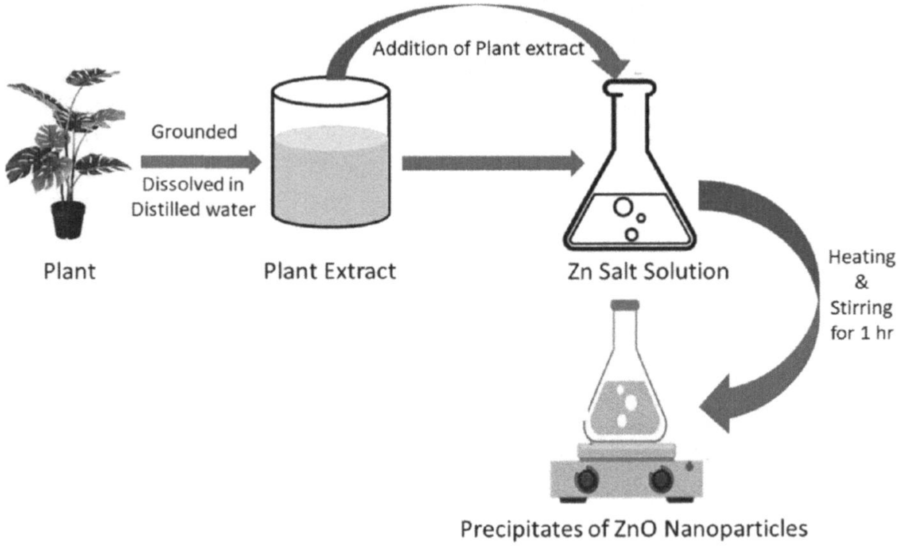

FIGURE 13.1 A schematic representing the green synthesis of ZnO NPs by utilizing plants/plant extracts.

Hintsho-Mbita 2022; Lee et al. 2016). The schematic representing the green synthesis of ZnO NPs utilizing plant extracts or plant materials is shown in Figure 13.1.

According to research, green synthesis using plants or plant extracts appears to occur more quickly than with the help of bacteria and fungi. Plants/plant extracts have attracted interest in green synthesis because of their quick expansion, single-step methodology, cost-effective protocol, and non-pathogenic nature. The various green synthesis methods for the preparation of ZnO NPs with different plant extracts are summarized in Table 13.1.

13.3 CHARACTERIZATION OF ZnO NANOPARTICLES

The as-synthesized ZnO NPs are characterized to determine their structure, morphology, particle size, thermal characteristics, surface chemistry, and functionality. Fourier transform infrared (FT-IR) spectroscopy in the 4000–400 cm^{-1} wavenumber region is used to determine the functionality of NPs surfaces. Absorption and emission features of ZnO NPs are measured using a UV–Vis absorption and fluorescence spectrometer, respectively. The band gap and size of ZnO NPs can be estimated by the position of the excitonic band in the absorption spectrum. Due to quantum confinement, the bandgap of ZnO NPs widens as their size shrinks, which is observed by a shifting of the absorption band toward a lower wavelength. Two emission bands are observed in the emission spectra of ZnO NPs one in the 370–395 nm (UV) region and another in the 530–600 nm (visible) region. Band-edge emission is associated with a sharp emission band in the ultraviolet region, which corresponds to a 3.3 eV band gap of ZnO NPs. Due to oxygen vacancies on its surface, the visible region

TABLE 13.1
Recent Studies on the Photocatalytic Activity of ZnO NPs Toward Organic Dyes under UV/Vis Light Irradiation

NPs	Plant Extract	Size and Shape of NPs	Organic Pollutants	Degradation Efficiency	Reference
DP-ZnO	*Date pulp waste (DPW)*	30 nm, spherical	Methylene blue Eosin yellow	90.5% in 180 minutes 90.6% in 180 minutes	Rambabu et al. (2021)
ZnO/ nanocellulose	*Tinospora Cordifolia leaves*	26 nm, nanoflowers	Enrofloxacin (EF) antibiotic	97% in 120 minutes	Nahi, Radhakrishnan and Beena (2021)
Spongy Nano-ZnO	*Pomegranate seeds molasses*	13 nm, spherical	Flumequine (FL) antibiotic	97.6% in 90 minutes at pH 8.7	Essawy et al. (2020)
ZnO NPs	*Linum Usitatissimum extract*	50 nm, spherical	Methylene blue	83% under UV light after 90 minutes	Alkasir et al. (2020)
ZnO NPs	*Durian waste*	280 nm, spherical	Methylene blue Sulfanilamide	96.70% in 3 hour 84% in 40 minutes	Ravichandran et al. (2020)
ZnO NPs	*Fruit extract of Amomum longiligulare*	50 nm, tetrameric structure	Methylene blue Malachite green	66% in 60 minutes 38.1% by 60 minutes	Liu et al. (2020)
ZnO NPs	*Quince seed Mucilage*	25 nm, spherical	Methylene blue	80% within 2 hour	Tabrizi Hafez Moghaddas, Elahi and Javanbakht (2020)
ZnO NPs	*Phoenix roebelenii palm*	15 nm, spherical	Methylene blue	98% after 105 minutes of UV illumination	Aldeen, Ahmed Mohamed and Maaza (2022)
ZnO NPs	*Codonopsis lanceolata root*	500 nm, flower shape	Methylene blue	90.3% within 40 minutes	Lu et al. (2019)
ZnO NPs	*Rueli tuberosa*	45 nm, rod shape	Methylene blue Malachite green	94% in 2.5 hour 92% in 2.5 hour	Vasantharaj et al. (2019)
ZnO NPs	*Bridelia retusa*	11 nm, wurtzite structure	Rhodamine B	94.74% in 165 nm	Vinayagam et al. (2023)
ZnO NPs	*Gynostemmapenta phylum*	35nm, hexagonal	Malachite green	89% in 180 minutes	Park et al. (2021)
ZnO NPs	*Mimosa tenuiflora*	40 nm, Hexagonal	Methyl orange	100% under UV illumination	Quevedo-Robles, Vilchis-Nestor and Luque (2022)
ZnO NPs	*Solanum trilobatum*	25 nm, spherical	Methylene blue	94.07% under sunlight illumination	Venkatesan et al. (2022)

experiences a large surface-emission peak. As a result of extensive surface emission, ZnO NPs glow green when exposed to UV light. The transmission electron microscopy (TEM) technique is utilized to determine the size and morphology of ZnO NPs. Scanning electron microscopy (SEM), which determines the size, morphology, and nanostructure of micro-sized particles, can also be used to study the nanostructure of ZnO NPs. The crystalline properties of the ZnO NPs are determined using powder x-ray diffraction (PXRD) analysis, while the hydrodynamic particle size and the net charge of the ZnO NPs are investigated by zeta potential analysis. The hexagonal/wurtzite structure of the ZnO NPs has a prominent peak that corresponds to the 101 planes in the XRD pattern. The sample's crystallinity is shown by the PXRD pattern's sharp peaks. Structural defects of NPs are examined by a Raman spectrometer (Shaikhaldein et al. 2021; Talam, Karumuri and Gunnam 2012).

13.4 WASTE WATER TREATMENT BY ZnO NANOPARTICLES

To evaluate the potential of ZnO NPs for industrial effluent treatment, these NPs are applied to the treatment of synthetic wastewater. The biocidal ability/antibacterial activity and photocatalytic activity of NPs are evaluated by killing pathogenic bacteria and decomposing hazardous dye molecules and antibiotics, respectively.

13.4.1 Photocatalytic Activity of ZnO NPs by Photocatalytic Degradation of Organic Pollutants

Organic pollutants, i.e., organic dyes such as methyl orange, rhodamine B, methylene blue, malachite green, and eosin yellow, are widely utilized in the textile and pharmaceutical industries. These dyes undergo anaerobic degradation to produce mutagenic and carcinogenic intermediates despite their great stability (Payra et al. 2019). Prolonged exposure to these hazardous dyes causes a wide range of harmful effects in humans, such as skin and eye irritation, liver, lung, kidney, and retinal damage, and damage to the central nervous system. Furthermore, the presence of these dyes in streams of industrial effluent that combine with domestic and river water sources reduces the penetration of solar light and alters photosynthesis, which damages water ecosystems and contaminates food chains. Therefore, it is very desirable and reassuring for industrial practice if these dye-polluted aqueous streams can be effectively treated using a simple, efficient, and cost-effective technology (Anitha, Senthil Kumar and Sathish Kumar 2016).

The photocatalytic performance of NPs is evaluated by determining their photocatalytic efficiencies in the degradation of different dyes. The test protocol entails the dispersion of a fixed mass of solid NPs in a specific volume of wastewater containing a fixed concentration of organic dye. The solution was kept for one hour under complete darkness while being constantly stirred, to acquire adsorption equilibrium between the dye and NPs. Then, the dye solution containing NPs is subjected to UV or visible light for determining the photocatalytic activity of the NPs toward dye degradation. To ensure the nanoparticle's homogeneous catalytic activity throughout the dye solution, the solution is constantly stirred. At various time intervals of 30, 60, 90, 120, 150, and 180 minutes, the dye-degraded samples are collected and centrifuged to separate the NPs. The supernatant is then subjected to absorbance measurement by a UV–vis spectrophotometer to monitor the extent of dye degradation in the samples (Lam et al. 2013). The photocatalytic dye degradation efficiency, η , is determined as follows:

$$\eta = \left(\frac{A_0 - A_t}{A_0} \right) \times 100 \tag{13.1}$$

where A_0 is the absorbance of the dye solution before starting the experiment and A_t is the absorbance of the dye sample at a given time, t, for photocatalytic degradation. The recent photocatalytic performance of various ZnO NPs prepared by the green synthesis method has been summarized in Table 13.1.

The photocatalytic dye degradation occurs in a quick initial phase followed by a slow secondary phase. The rapid initial phase is facilitated by the abundance of catalytic sites and the wider concentration gradient (Nirmala et al. 2021). Hence, the secondary phase proceeds slowly due to a reduction in the concentration gradient and catalytic sites. The mechanism of photocatalytic dye degradation involves three main steps:

i. the formation and transmission of electron-hole pairs;
ii. the production of peroxide radicals; and
iii. the breakdown of the organic dyes.

The first step entails the formation of electrons and holes in the conduction band (CB) and valance band (VB) of ZnO NPs, respectively, under UV or visible light illumination. After that, electron-hole pairs separate and move toward the surface of NPs. In the subsequent step, surface electrons and oxygen molecules react to form superoxide radicals, which are then combined with surface electrons to form peroxide radicals. Moreover, the surface hole reacts and oxidizes the water molecule to produce $HO^{\cdot}$ (hydroxyl radicals). These radicals are highly active and hence react with dye molecules, causing their degradation into CO_2 and H_2O (Udayabhanu et al. 2017). The reactions for the photocatalytic degradation of organic dyes are shown below:

$$\text{ZnO NPs} + h\nu \rightarrow e^- \ (\text{CB}) + h^+ \ (\text{VB}) \tag{13.2}$$

$$O_2 + e^- \ (\text{CB}) \ \rightarrow \ O_2^{\cdot -} + H^+ \ \rightarrow \ HOO^{\cdot} \tag{13.3}$$

$$HOO^{\cdot} + e^- \ (\text{CB}) \ \rightarrow H_2O_2 \tag{13.4}$$

$$h^+ \ (\text{VB}) + H_2O \rightarrow HO^{\cdot} + H^+ \tag{13.5}$$

$$\text{Organic Dyes} + \left(HOO^{\cdot}, H_2O_2, HO^{\cdot} \right) \rightarrow CO_2 + H_2O \tag{13.6}$$

According to several reports, the photocatalytic effectiveness of semiconductor nanophotocatalysts varies with their physical properties, such as their morphology, surface area, crystallinity, and band gap energy. The adsorption of dyes on the surface of NPs and their photocatalytic degradation can be improved either by increasing the area of the reactive/active sites or by reducing the size of the photocatalyst. Also, the number of surface-active species is crucial to the effectiveness of the degradative process, and these are largely dependent on the proportions of electrons and holes in the CB and VB of NPs, respectively. It is well known that narrowing the energy band gap of the ZnO NPs enhances their photocatalytic performance by increasing the number of electrons and holes produced when the catalyst is illuminated by the light source (Aldeen, Ahmed Mohamed and Maaza 2022).

13.4.2 Antibacterial Activities Towards Pathogenic Bacteria

The presence of various pathogenic bacteria, such as gram-negative and gram-positive bacteria, in water and wastewater streams is not suitable for human health. It leads to various waterborne and water-related diseases in humans. Hence, the removal of pathogenic bacteria is also an important step in water remediation (Bowen et al. 2014).

Antibacterial activity is a procedure that involves the introduction of antibacterial compounds and the monitoring and control of the growth of pathogenic bacteria. The antibacterial activity of NPs is determined by examining their bactericidal capacity, which is done through in vitro diffusion techniques. The test protocol entails the diffusion of different concentrations of NP solution into the bacterial culture. Water and ciprofloxacin are applied as negative and positive controls, respectively, for the interpretation of the antibacterial activity. Then, all these samples are incubated for 24 hours at 37°C, and the zone of inhibition is evaluated (Elumalai and Velmurugan 2015). The recent antibacterial activities of various ZnO NPs synthesized by the environmentally friendly synthesis method have been summarized in Table 13.2.

The ZnO NPs show bactericidal activity because of the formation of a significant number of oxygen-reactive species such as hydrogen peroxide, hydroxyl radicals, and superoxide radicals on their surface. Particularly, the large specific surface area to volume ratio of NPs contributes to the production of more reactive oxygen species (Talam, Karumuri and Gunnam 2012; Rad, Sani and Mohseni 2019). Due to their negative charge, hydroxide ions and superoxide anions could not enter the cell, but they would stay on the cell walls of bacteria and damage their identity. It would lead to the destruction of the bacterial cell wall and cause the intracellular contents to leak out, which would ultimately cause cell death. Hydrogen peroxide, on the other hand, destroys the respiratory enzymes of bacteria by seeping into the cell matrix and, hence, killing them. In certain circumstances, prolonged interaction occurs between bacteria and ZnO NPs via electrostatic interaction, and hence, the abrasive surface characteristics of the ZnO NPs damage the bacterial cell walls and increase the cell membrane's susceptibility toward Zn ions, which results in bacterial toxicity. Another mechanism is the buildup of ZnO NPs on the bacteria's surface, which results in the rupturing of the cell envelope and the discharge of cellular contents (Lee et al. 2016; Sirelkhatim et al. 2015).

TABLE 13.2
Recent Studies on the Antibacterial Activity of ZnONPs Toward Pathogenic Bacteria

NPs	Plant Extract	Size and Shape of NPs	Bacterial Strains	Zone of Inhibition (mm)	Reference
DP-ZnO	*Date pulp waste (DPW)*	30 nm, spherical	*S. aureus*	15.9 at 100 µg/ml	Rambabu et al. (2021)
			S. pyogenes	14.5 at 100 µg/ml	
			P. aeruginosa	18.4 at 100 µg/ml	
			P. mirabilis	18.7 at 100 µg/ml	
ZnO NPs	*Durian waste*	280 nm, spherical	*E. coli*	20 mm at 1 mg/ml	Ravichandran et al. (2020)
			S. aureus	17 mm at 1 mg/ml	
ZnO NPs	*Phoenix roebelenii palm*	15 nm, spherical	*S. aureus*	15 mm at 4 mg/ml	Aldeen, Ahmed Mohamed and Maaza (2022)
			S. pneumoniae	15.1 mm at 4 mg/ml	
			E. coli	15.8 mm at 4 mg/ml	
			S. typhi	21.4 mm at 4 mg/ml	
ZnO NPs	*Rueli tuberosa*	45 nm, rod shape	*E. coli*	12 mm at 1 mg/ml	Vasantharaj et al. (2019)
			S. aureus	10 mm at 1 mg/ml	
ZnO NPs	*Vitis labruska*	20 nm, hexagonal	*E. coli*	39.6 mm	Mostoni et al. (2017)
			S. aureus	30 mm	
ZnO NPs	*Withaniacoagulans*	Nanoflowers	*S. aureus*	85%	Saif et al. (2021)
			P. aeruginosa	94% inhibition	

13.5 CONCLUSION

ZnO nanoparticles prepared by the green synthesis method exhibit enhanced photocatalytic and antibacterial activity. The biological compound present in the plant extract plays a crucial role in their synthesis as a stabilizing and reducing agent. This energy-efficient synthetic method can be scaled up for industrial use owing to the low-temperature calcination and shortened reaction time. Furthermore, the green synthesis method is also more appealing due to the use of inexpensive precursors and the lack of byproducts. These ZnO NPs are suitable for the treatment of industrial and municipal wastewater, especially organic dyes, antibiotics, and pathogenic bacteria-polluted discharge streams. This is corroborated by their encouraging uses in photocatalytic pollutant destruction and antimicrobial activities.

REFERENCES

Aldeen, Thana Shuga, Hamza Elsayed Ahmed Mohamed, and Malik Maaza. 2022. "ZnO Nanoparticles Prepared via a Green Synthesis Approach: Physical Properties, Photocatalytic and Antibacterial Activity." *Journal of Physics and Chemistry of Solids* 160 (January): 110313.

Alharbi, Njud S., Baowei Hu, Tasawar Hayat, Samar Omar Rabah, Ahmed Alsaedi, Li Zhuang, and Xiangke Wang. 2020. "Efficient Elimination of Environmental Pollutants through Sorption-Reduction and Photocatalytic Degradation Using Nanomaterials." *Frontiers of Chemical Science and Engineering* 14 (6): 1124–35.

Alkasir, Maryam, Naser Samadi, Zahra Sabouri, Zahra Mardani, Mehrdad Khatami, and Majid Darroudi. 2020. "Evaluation Cytotoxicity Effects of Biosynthesized Zinc Oxide Nanoparticles Using Aqueous Linum Usitatissimum Extract and Investigation of Their Photocatalytic Activityackn." *Inorganic Chemistry Communications* 119 (September): 108066.

Anitha, T., P. Senthil Kumar, and K. Sathish Kumar. 2016. "Synthesis of Nano-Sized Chitosan Blended Polyvinyl Alcohol for the Removal of Eosin Yellow Dye from Aqueous Solution." *Journal of Water Process Engineering* 13 (October): 127–36.

Ashwini, Jayachandran, Thankamani Ravikumar Aswathy, Anil Babu Rahul, Gautham M. Thara, and Achuthsankar S. Nair. 2021. "Synthesis and Characterization of Zinc Oxide Nanoparticles Using Acacia Caesia Bark Extract and Its Photocatalytic and Antimicrobial Activities." *Catalysts* 11 (12): 1507.

Bowen, Asha C., Steven Y.C. Tong, Mark D. Chatfield, and Jonathan R. Carapetis. 2014. "The Microbiology of Impetigo in Indigenous Children: Associations between Streptococcus Pyogenes, Staphylococcus Aureus, Scabies, and Nasal Carriage." *BMC Infectious Diseases* 14 (1): 727.

Brayner, Roberta, Si Amar Dahoumane, Claude Yéprémian, Chakib Djediat, Michaël Meyer, Alain Couté, and Fernand Fiévet. 2010. "ZnO Nanoparticles: Synthesis, Characterization, and Ecotoxicological Studies." *Langmuir* 26 (9): 6522–28.

Cui, Li, Hong-Zhe Li, Kai Yang, Long-Ji Zhu, Fei Xu, and Yong-Guan Zhu. 2021. "Raman Biosensor and Molecular Tools for Integrated Monitoring of Pathogens and Antimicrobial Resistance in Wastewater." *TrAC Trends in Analytical Chemistry* 143 (October): 116415.

Dhariwal, Sunny, and Mona Mittal. 2023. "Wastewater Treatment with Perovskite-Based Photocatalysts: Environmental Sustainability from a Green Perspective." *Materials Today: Proceedings*. doi.10.1016/j.matpr.2023.03.048.

Elumalai, K., and S. Velmurugan. 2015. "Green Synthesis, Characterization and Antimicrobial Activities of Zinc Oxide Nanoparticles from the Leaf Extract of Azadirachta Indica (L.)." *Applied Surface Science* 345 (August): 329–36.

Essawy, Amr A., Ibrahim Hotan Alsohaimi, Mosaed S. Alhumaimess, Hassan M.A. Hassan, and Mahmoud M. Kamel. 2020. "Green Synthesis of Spongy Nano-ZnO Productive of Hydroxyl Radicals for Unconventional Solar-Driven Photocatalytic Remediation of Antibiotic Enriched Wastewater." *Journal of Environmental Management* 271 (October): 110961.

Jabbar, Zaid H., and Shahlaa Esmail Ebrahim. 2022. "Recent Advances in Nano-Semiconductors Photocatalysis for Degrading Organic Contaminants and Microbial Disinfection in Wastewater: A Comprehensive Review." *Environmental Nanotechnology, Monitoring & Management* 17 (May): 100666.

Kaviya, S., and Edamana Prasad. 2015. "Biogenic Synthesis of ZnO-Ag Nano Custard Apples for Efficient Photocatalytic Degradation of Methylene Blue by Sunlight Irradiation." *RSC Advances* 5 (22): 17179–85.

Lam, Sze-Mun, Jin-Chung Sin, Ahmad Zuhairi Abdullah, and Abdul Rahman Mohamed. 2013. "Green Hydrothermal Synthesis of ZnO Nanotubes for Photocatalytic Degradation of Methylparaben." *Materials Letters* 93 (February): 423–26.

Lee, Kian Mun, Chin Wei Lai, Koh Sing Ngai, and Joon Ching Juan. 2016. "Recent Developments of Zinc Oxide Based Photocatalyst in Water Treatment Technology: A Review." *Water Research* 88 (January): 428–48.

Liu, Ying Chun, JingFeng Li, JongChan Ahn, JianYu Pu, Esrat Jahan Rupa, Yue Huo, and Deok Chun Yang. 2020. "Biosynthesis of Zinc Oxide Nanoparticles by One-Pot Green Synthesis Using Fruit Extract of Amomum Longiligulare and Its Activity as a Photocatalyst." *Optik* 218 (September): 165245.

Lu, Feng, and Didier Astruc. 2018. "Nanomaterials for Removal of Toxic Elements from Water." *Coordination Chemistry Reviews* 356 (February): 147–64.

Lu, Jing, Hashmoonah Ali, Joon Hurh, Yaxi Han, Indra Batjikh, Esrat Jahan Rupa, Gokulanathan Anandapadmanaban, Jin Kyu Park, and Deok-Chun Yang. 2019. "The Assessment of Photocatalytic Activity of Zinc Oxide Nanoparticles from the Roots of Codonopsis Lanceolata Synthesized by One-Pot Green Synthesis Method." *Optik* 184 (May): 82–89.

Mahlaule-Glory, Louisah M., and Nomso C. Hintsho-Mbita. 2022. "Green Derived Zinc Oxide (ZnO) for the Degradation of Dyes from Wastewater and Their Antimicrobial Activity: A Review." *Catalysts* 12 (8): 833.

Mostoni, S., V. Pifferi, L. Falciola, D. Meroni, E. Pargoletti, E. Davoli, and G. Cappelletti. 2017. "Tailored Routes for Home-Made Bi-Doped ZnO Nanoparticles. Photocatalytic Performances towards o-Toluidine, a Toxic Water Pollutant." *Journal of Photochemistry and Photobiology A: Chemistry* 332 (January): 534–45.

Nahi, J., Asha Radhakrishnan, and B. Beena. 2021. "Green Synthesis of Zinc Oxide Incorporated Nanocellulose with Visible Light Photocatalytic Activity and Application for the Removal of Antibiotic Enrofloxacin from Aqueousmedia." *Materials Today: Proceedings* 41: 583–89.

Nazir, Arif, Mohsan Raza, Mazhar Abbas, Shaista Abbas, Abid Ali, Zahid Ali, Umer Younas, Samiah H. Al-Mijalli, and Munawar Iqbal. 2022. "Microwave Assisted Green Synthesis of ZnO Nanoparticles Using Rumex Dentatus Leaf Extract: Photocatalytic and Antibacterial Potential Evaluation." *Zeitschrift Für Physikalische Chemie* 236 (9): 1203–17.

Nirmala, Gnanasundaram, Thanapalan Murugesan, K. Rambabu, K. Sathiyanarayanan, and Pau Loke Show. 2021. "Adsorptive Removal of Phenol Using Banyan Root Activated Carbon." *Chemical Engineering Communications* 208 (6): 831–42.

Park, Jin Kyu, Esrat Jahan Rupa, Mohammad Huzaifa Arif, Jin Feng Li, Gokulanathan Anandapadmanaban, Jong Pyo Kang, Mia Kim, et al. 2021. "Synthesis of Zinc Oxide Nanoparticles from Gynostemma Pentaphyllum Extracts and Assessment of Photocatalytic Properties through Malachite Green Dye Decolorization under UV Illumination-A Green Approach." *Optik* 239 (August): 166249.

Patil, Santosh S., Mukund G. Mali, Mohaseen S. Tamboli, Deepak R. Patil, Milind V. Kulkarni, Hyun Yoon, Hayong Kim, et al. 2016. "Green Approach for Hierarchical Nanostructured Ag-ZnO and Their Photocatalytic Performance under Sunlight." *Catalysis Today* 260 (February): 126–34.

Payra, Soumitra, Swapna Challagulla, Yamini Bobde, Chanchal Chakraborty, Balaram Ghosh, and Sounak Roy. 2019. "Probing the Photo- and Electro-Catalytic Degradation Mechanism of Methylene Blue Dye over ZIF-Derived ZnO." *Journal of Hazardous Materials* 373 (July): 377–88.

Quevedo-Robles, R.V., A.R. Vilchis-Nestor, and P.A. Luque. 2022. "Study of Optical and Morphological Properties of Nanoparticles Semiconductors of Zinc Oxide Synthesized Using Mimosa Tenuiflora Extract for Photodegradation of Methyl Orange." *Optical Materials* 128 (June): 112450.

Rad, Samira Shahriyari, Ali Mohamadi Sani, and Sharareh Mohseni. 2019. "Biosynthesis, Characterization and Antimicrobial Activities of Zinc Oxide Nanoparticles from Leaf Extract of Mentha Pulegium (L.)." *Microbial Pathogenesis* 131 (June): 239–45.

Rambabu, K., G. Bharath, Fawzi Banat, and Pau Loke Show. 2021. "Green Synthesis of Zinc Oxide Nanoparticles Using Phoenix Dactylifera Waste as Bioreductant for Effective Dye Degradation and Antibacterial Performance in Wastewater Treatment." *Journal of Hazardous Materials* 402 (January): 123560.

Ravichandran, Veerasamy, Samuggam Sumitha, Cheah Yi Ning, Ooi Yi Xian, Ung Kiew Yu, Neeraj Paliwal, Syed Adnan Ali Shah, and Minaketan Tripathy. 2020. "Durian Waste Mediated Green Synthesis of Zinc Oxide Nanoparticles and Evaluation of Their Antibacterial, Antioxidant, Cytotoxicity and Photocatalytic Activity." *Green Chemistry Letters and Reviews* 13 (2): 102–16.

Saif, Muhammad Saqib, Ayesha Zafar, Muhammad Waqas, Shahbaz Gul Hassan, Ain ul Haq, Tuba Tariq, Sana Batool, Momina Dilshad, Murtaza Hasan, and Xugang Shu. 2021. "Phyto-Reflexive Zinc Oxide Nano-Flowers Synthesis: An Advanced Photocatalytic Degradation and Infectious Therapy." *Journal of Materials Research and Technology* 13 (July): 2375–91.

Saini, Mona, Sapna Yadav, Nutan Rani, Asifa Mushtaq, Seema Rawat, Kalawati Saini, and Dipak Maity. 2022. "Biosynthesized Zinc Oxide Nanoparticles Using Seed and Bark Extract of Azadirachta Indica for Antibacterial, Photocatalytic and Supercapacitor Applications." *Materials Science and Engineering: B* 282 (August): 115789.

Shaikhaldein, Hassan O., Fahad Al-Qurainy, Salim Khan, Mohammad Nadeem, Mohamed Tarroum, Abdalrhaman M. Salih, Abdel-Rhman Zakaria Gaafar, et al. 2021. "Biosynthesis and Characterization of ZnO Nanoparticles Using Ochradenus Arabicus and Their Effect on Growth and Antioxidant Systems of Maerua Oblongifolia." *Plants* 10 (9): 1808.

Sharma, Tania, Sushma Yadav, and Mona Mittal. 2023. "Recent Advancement in the Sustainable Synthesis of BiOX (X = I, Br, and Cl) Nanomaterials and Their Applications in the Photocatalysis." *Materials Today: Proceedings*, March, S2214785323007484.

Singh, Shreya, Rahul Garg, Atanu Jana, Chinna Bathula, Soniya Naik, and Mona Mittal. 2023. "Current Developments in Nanostructurally Engineered Metal Oxide for Removal of Contaminants in Water." *Ceramics International* 49 (5): 7308–21.

Sirelkhatim, Amna, Shahrom Mahmud, Azman Seeni, Noor Haida Mohamad Kaus, Ling Chuo Ann, Siti Khadijah Mohd Bakhori, Habsah Hasan, and Dasmawati Mohamad. 2015. "Review on Zinc Oxide Nanoparticles: Antibacterial Activity and Toxicity Mechanism." *Nano-Micro Letters* 7 (3): 219–42.

Tabrizi Hafez Moghaddas, Seyyed Mohammad, Behrouz Elahi, and Vahid Javanbakht. 2020. "Biosynthesis of Pure Zinc Oxide Nanoparticles Using Quince Seed Mucilage for Photocatalytic Dye Degradation." *Journal of Alloys and Compounds* 821 (April): 153519.

Talam, Satyanarayana, Srinivasa Rao Karumuri, and Nagarjuna Gunnam. 2012. "Synthesis, Characterization, and Spectroscopic Properties of ZnO Nanoparticles." *ISRN Nanotechnology* 2012 (May): 1–6.

Udayabhanu, G. Nagaraju, H. Nagabhushana, D. Suresh, C. Anupama, G.K. Raghu, and S.C. Sharma. 2017. "Vitis Labruska Skin Extract Assisted Green Synthesis of ZnO Super Structures for Multifunctional Applications." *Ceramics International* 43 (15): 11656–67.

Vasantharaj, Seerangaraj, Selvam Sathiyavimal, Palanisamy Senthilkumar, Felix LewisOscar, and Arivalagan Pugazhendhi. 2019. "Biosynthesis of Iron Oxide Nanoparticles Using Leaf Extract of Ruellia Tuberosa: Antimicrobial Properties and Their Applications in Photocatalytic Degradation." *Journal of Photochemistry and Photobiology B: Biology* 192 (March): 74–82.

Venkatesan, S., S. Suresh, P. Ramu, J. Arumugam, S. Thambidurai, and N. Pugazhenthiran. 2022. "Methylene Blue Dye Degradation Potential of Zinc Oxide Nanoparticles Bioreduced Using Solanum Trilobatum Leaf Extract." *Results in Chemistry* 4 (January): 100637.

Vinayagam, Ramesh, Shraddha Pai, Thivaharan Varadavenkatesan, Arivalagan Pugazhendhi, and Raja Selvaraj. 2023. "Characterization and Photocatalytic Activity of ZnO Nanoflowers Synthesized Using Bridelia Retusa Leaf Extract." *Applied Nanoscience* 13 (1): 493–502.

14 Agri-Food Wastes to Nanomaterials for Environmental Remediation

Shaziya H Siddiqui, Saima H Khan, and Prerna Higgins
Sam Higginbottom University of Agriculture Technology and Sciences

CONTENTS

14.1 INTRODUCTION

The agro-food industry produces a large amount of non-edible waste, mainly vegetable and fruit waste, which needs to be converted into economical products for the protection of the environment; otherwise, it will compile as garbage (Missau et al. 2021). Due to their unique characteristics, such as feasibility, economics, eco-friendliness, and abundance, they make sustainable materials for environmental remediation. With advancements in research, these agricultural wastes can be converted into nanobio-composites by reacting with various metal oxides, polymers, clays, etc. A nanomaterial is a single unit ranging between 1 and 100 nm. Due to their broad

DOI: 10.1201/9781003397861-14

area and strong ion exchange magnitude, nanomaterials come into existence. Due to the presence of various secondary pollutants found in wastewater, it is necessary to develop an adsorbent using greener technology that cannot contribute to further pollution.

Water is a vital source of human life, without which we cannot survive. Due to fast industrialization and abrupt population inflation, water is contaminated day by day by various pollutants discharged through industries, sewage discharge, and household systems. Various purification techniques such as advanced oxidation processes, adsorption, ion exchange, membrane filtration, ozonation, and coagulation are emergent requirements for decontamination (Adegoke et al. 2022). Adsorption is one of the most effective processes for the hospitality of contaminated water because it is economical, feasible, eco-friendly, does not produce much waste, and has improved accessibility to different adsorbents (Siddiqui et al. 2022). Carbon in activated form is one of the most commonly used adsorbents because of its good adsorptive ability. Some environment-safe adsorbents have been prepared using agro-waste materials due to the exorbitant cost and regeneration ability of activated carbon (Isaac and Siddiqui 2022; Foo and Hameed 2012; Ali et al. 2016; Lingamdinne et al. 2016; Siddiqui et al. 2022).

Various agri-food wastes have been converted into nanocomposites using various metal oxides, polymers, and clays to increase their adsorptive power and maintain the eco-friendly behavior of the adsorbents. Various bio-nanocomposites for the removal of contaminants have been reported, including Chrysanthemum-$ZnFe_2O_4$@ZnO nanocomposites (Nguyen et al. 2023), CaAl-LDH/biochars nanocomposites (Missau et al. 2021), CBCS@LDH nanocomposites (Higgins et al. 2022), cellulosic nano-adsorbent (Biswal et al. 2022), sulphonated gum Arabic-powdered mushroom composite (Das et al. 2015), biodegradable nanocellulose (Manchala et al. 2022), and Ca-rich biochar from food waste digestate (Liu et al. 2022). The various toxicants, such as dyes, phenols, heavy metals, and pharmaceutical drugs, are the toxic contaminants discharged into river beds (Xiao et al. 2019). These pollutants can affect the human system by affecting the liver, kidneys, and central nervous system. They are carcinogenic in nature. This book chapter discusses the use of bio-nanocomposites prepared by a green synthetic approach for the removal of these pollutants.

14.2 SYNTHESIS OF BIO-NANOCOMPOSITES

Five grams of kaolinite clay was dissolved in 100 mL of deionized water and agitated at 60°C for 12 hours at 420 rpm. To this solution, 0.1 M L-methionine was added, and the mixture was agitated at 700 rpm for 6 hours. 2.5 g of cellulose was dissolved in a DMAC solution and heated at 80°C for 2 hours. The prepared cellulose solution was slowly added to the solution of L-methionine-modified kaolinite and stirred for 12 hours at 60°C. The resulting (Meth-Kal@Cellulose) bio-nanocomposites were precipitated in 250 mL of acetone, centrifuged, and cleaned with deionized water. Moisture was then removed in a hot air oven at 60°C. The resulting adsorbent was used for further studies (Ahmed and Ansari 2022). The Nerium oleander leaves were washed, dried, and ground. The resulting 10 g of fine powder was warmed in a flask with 150 mL of water for 90 minutes at 70°C under moderate stirring. The colored mixture obtained was allowed to stand at ambient temperature, followed by filtration and washing. The extract obtained was then centrifuged at a force of attraction for 20 minutes. The resulting 100 mL extract was mixed with 100 mL of 1M copper sulfate solution under continuous swirling at 70°C for 2 hours. The extract changed color to a light green. The resulting copper oxide composites were evaporated in a hot air oven at 100°C for 12 hours and used for methylene blue and crystal violet adsorption (Sebeia et al. 2019).

Similar studies were also reported (Sebei et al. 2020). The floral discharge of *Chrysanthemum* spp. was collected and washed with water. The waste flower was then boiled in 800 mL of distilled water and ultrasonicated at 80°C for 1 hour. Fifteen grams of $Zn(NO_3)_2{\cdot}6H_2O$ and 41 g of $Fe(NO_3)_3{\cdot}9H_2O$ were dissolved in 400 mL of the *Chrysanthemum* spp. floral extract. The mixture was magnetized and swirled to dissolve the substrates. To this mixture, 6 g of $Zn(NO_3)_2{\cdot}6H_2O$ solution was added, and the solution was then heated to 70°C at a rate of 240 minutes under continuous stirring. The prepared sample was evaporated at 100°C and calcined at 500°C for 180 minutes to obtain $ZnFe_2O_4$@ZnO nanocomposites, which were then used for dyes (Nguyen et al. 2023).

The black cumin seeds were washed with distilled water and then carbonized at 450°C for 1 hour. The prepared carbonized seeds were dried, ground, and used for the study. The layer double hydroxide was manufactured using the co-precipitation method from compounds such as copper sulfate ($CuSO_4$), sodium carbonate (Na_2CO_3), aluminum nitrate ($Al(NO_3)_3$), and manganese sulfate ($MnSO_4$). Solutions of $MnSO_4$, $Al(NO_3)_3$, Na_2CO_3, and $CuSO_4$ were prepared in 25 mL and mixed to make a volume of 100 mL in a 100 mL volumetric flask. The logarithm of the H^+ ion of the solution (pH) was stabilized using a molar solution of NaOH to 11 and agitated at 500 rpm, 298 K for 5 hours. Four grams of Nigella sativa carbonized were intermixed into the layer double hydroxide solution, followed by agitation for a period of 3 hours. A dark brown precipitate was obtained, which was washed and used for the removal of ciprofloxacin drug (Higgins et al. 2022). The FESEM image of CBC@LDH before and after adsorption of ciprofloxacin was reported (Figure 14.1). An irregular, rough end surface with crystal settlement was observed as the surface of the particles increased, indicating the absorption of particles.

The leftover eggshell was rinsed with deionized water. The drained eggshell waste was crushed and filtered to a 100-mesh size. Ten grams of eggshell powder were dipped in a 50 mL solution of iron sulfate (10 g) and kept on a magnetic stirrer for 1 hour. Then, 20 mL of 5 N NaOH was added dropwise until the precipitate of iron oxide penetrated the coating formed on the area of eggshell powder. The iron oxide-coated eggshell powder bio-adsorbent was filtered, washed, and dried before being used for adsorption studies for the removal of Cu^{2+} ions (Ahmad et al. 2012).

The food waste digestate was collected from the fermentation plant of food waste and then carbonized in an alundum boat in a tubular furnace in a nitrogen atmosphere at a temperature of 400°C–1000°C for different types of samples designated accordingly for 1,000 minutes. This food waste Ca-rich biochar digestate is used for the removal of Arsenic (III) (Liu et al. 2022). The black seed pomace was collected as waste, ground, sieved (35 mesh), and washed with deionized water. A feasible

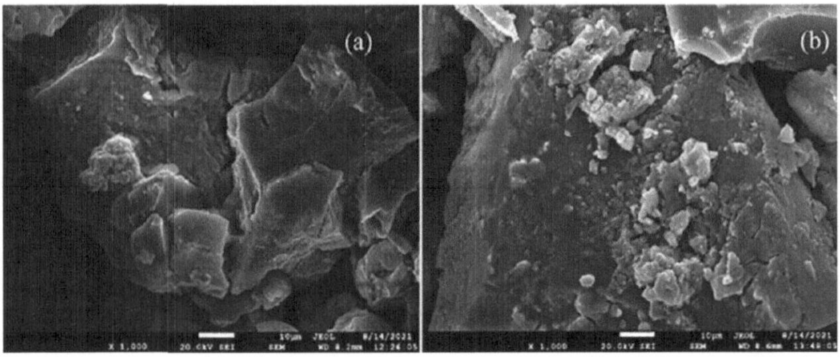

FIGURE 14.1 FESEM image of CBCS@LDH surface. (a) Before adsorption of CPN. (b) After adsorption of CPN (Higgins et al. 2022).

amount of waste was treated with the required concentration of ferric nitrate solution, sonicated for 1 hour at 50°C and dried overnight. The prepared material was carbonized in a tube furnace at 800°C for 2 hours. The iron-impregnated biochar adsorbent was further washed with a standard solution of hydrochloric acid, dried, and labeled (Mustafa and Aziz, 2023). Various other syntheses have been reported, such as Machrouchi et al. (2023), who reported the removal of methyl orange using *Thapsia transtagana* stem activated carbon/zinc oxide (ZnO/ACTTS) composite.

14.3 FACTORS AFFECTING ADSORPTION OF POLLUTANTS

The adsorption process in batch mode is conducted by mixing a considerable amount of the adsorbents in solution under fixed continuous shaking conditions. The absorbance of the aqueous solutions is inculcated by a spectrophotometer, and a curve of calibration between absorbance and concentration deduces the concentration of the unknown solutions. Variations in dosage, concentration effect, temperature, and reaction time were analyzed. The experimentation is carried out to obtain the thermodynamic parameters, isotherm, and kinetic data. Isotherms of different categories are used to find out the uptake yield of the biosorbent. Kinetic data can be analyzed using pseudo-first-order, pseudo-second-order, intra-particle, and Elovich rate equations to find a suitable rate for adsorption. The thermodynamic category investigates the practicality and spontaneity of the process of adsorption.

14.3.1 Solution pH Effect

The solution pH is particularly important for sorption, as the pH of a solution affects the electrostatic charges provided by the ionized molecules of the adsorbate (Higgins et al. 2022). This, in turn, affects the degree of sorption. The effect of pH on sorption can be investigated by preparing surface-assimilative-chemisorptive ions with alterations using 1 molar NaOH or HCl solutions. At higher pH, the positive charge deficit on the solution surface interaction makes the adsorbable surface negatively charged, enhancing the uptake yield. Conversely, at lower pH, the positive charge enhances the liquid-solid interaction, making the adsorbable surface positively charged and reducing the sorption process. Higgins et al. (2022) investigated the pH effect on the adsorbable interaction of ciprofloxacin using a layer of double hydroxide/cumin seed nanocomposite and found that the highest uptake occurred at pH 6, with a decrease in uptake at higher pH.

14.3.2 Dosage Effect

Dosage is a vital characteristic to be considered for systematic adsorption. The increase in adsorptive efficiency with increasing adsorbent quantity can be attributed to the contribution of more adsorption active sites (Shahnaz et al. 2022). However, as the adsorbent dosage increases, the uptake decreases. This is because the adsorbable molecule has an empty amount of active sites at lower concentrations, but with an increase in dosage, the particles agglomerate and the uptake decreases (Silva et al. 2020). The dosage's consequence on adsorbent uptake is due to the interactivity between the adsorbable-adsorptive phase, with different categories of adsorbent required to set the first conglomeration of particles and then centrifuge until the attainment of the equilibrium period. The keen observation of the dosage effect arouses the effectiveness of an adsorbent with the identification of its economical ability. Sonawane and Shrivastava (2009) exhibited the dosage consequences on Malachite Green sorption by maize Cob, where the sorption yield enhanced from 90% to 98.5% for dosage concentrations ranging from 0.5 to 12 g/L.

14.3.3 Effect of Reaction Time

The adsorption rate is a significant factor in adsorption studies, as uptake occurs over time through interaction between the adsorbate and the adsorbent. Typically, experiments involve stirring solutions in a rotary orbital shaker for specified intervals of time to determine the equilibrium time for adsorption (Silva et al. 2020). Table 14.1 shows adsorption studies for various adsorbents used for contaminant removal. The kinetic parameters not only determine the reaction rate but also the adsorptive mechanism, which depends on the physicochemical nature of the sorbent (Ahmad and Ansari 2022). Various kinetic models are used to investigate this, and the feasibility of the order of reaction is exhibited by the correlation coefficient (R^2) value. Different kinetic models are used to observe the contact time for adsorption. The adsorption study was investigated for 5–360 minutes, and uptake was initially fast, becoming slower as the adsorbent reached complete impregnation of the required pores (Haseena et al. 2022). The computed q_e and experimental q_e for both pseudo-first (Lagergren 1898) and second-order models (Mckay and Ho, 1999) indicate that they should be precisely similar, with R^2 being higher toward unity (Ahmad et al. 2012).

The rate-examining procedure for uptake is determined by regressing the kinetic data to the Weber-Morris model, where the value of C determines the ductility of the outermost surface. The model elaborates adsorptive kinetics into three variations: outermost surface layer, internal-particle diffusion, and adsorptive equilibrium stage. Initially, the superficial uptake over the adsorbent begins, and then the penetration of adsorbate molecules diffusively exchanges the outermost surface layer and

TABLE 14.1
Monolayer Adsorption Uptake for the Adsorption of Contaminants

Adsorbent	Contaminant	Monolayer Adsorption Capacity (mg/g)/% Removal	Experimental Condition	Reference
CBCS@LDH	Ciprofloxacin	526.31	pH=6, 120 minutes	Higgins et al. (2022)
Iron oxide-coated egg shell powder	Cu^{2+}	36.17	pH=6,180 minutes, 100ppm	Ahmad et al. (2012)
Met-Kal@Cellulose	Amido black10B	172.14	pH=2,180 minutes, 30°C	Ahmed and Ansari (2022)
Rice Husk carbon-chitosan composite gel(CCRH)	Cu^{2+}	90.90	pH=7, 480 min utes and 25°C	Masih et al. (2018)
$ZnFe_2O_4$@ZnO using*Chrysanthemum* spp. floral waste	Congo red	94.85%	90 minutes, pH=7	Nguyen et al. (2023)
Ca-Al/LDH/Biochar	Crystal Violet	496.55	pH=8, 50°C	Missau et al. (2021)

occupies their positions in the adsorbent sites, ultimately reaching an equilibrium stage. Therefore, the highest value of *C*, with more k_{id} and R^2, represents the process as a rate identification step (Weber and Morris 1963), specifically intra-particle diffusion (Higgins et al. 2022). Additionally, higher A and B values indicate that the area of sorption involves chemical interactions between the adsorbate and the adsorbent (Sebeia et al. 2019). The result of the agglomeration of molecules during sorption was identified by the number of agglomeration molecules. The supernatant obtained after continuous stirring is measured (Sebei et al. 2020).

14.3.4 Isothermal Study on the Change in Concentration

Adsorptive isotherms are vital factors to decipher the interaction between the adsorption molecules and the adsorbent and the degree of accumulation, providing an idea of the adsorption capacity (Siddiqui et al. 2022). The data is regressed into Langmuir (1918), Freundlich, and Temkin isotherm models to observe the feasibility of the interaction of the adsorbent with the adsorbate. The concentration effect on adsorption is seen by varying the concentration of the aqueous solution. The adsorptive capacity generally shows a sharp increase, which can be attributed to the driving force that overcomes the bulk transfer resistance of the adsorbate molecules between the solid and liquid phases, which is the concentration of the solution. The enhancements in the uptake capacity are also applicable to more vacant sites' availability on the adsorbent's surface for easy binding and fast adsorption processes (Missau et al. 2021). Another probable reason for the high uptake capacity is the porous nature of the adsorbent. Due to porosity, diffusion of the molecules in the pores is easily facilitated (Das et al. 2015). A non-linear plot of C_e vs q_e gives the values for q_m, k_L, and R^2. The maximum adsorption capacity q_{max} and Langmuir constant K_L are the monolayer adsorption, and K_L determines the feasibility of the monolayer formation. The value for constant n and k_F gives the value for constant n, which lies in the range 1–10 for feasible adsorption uptake. The Temkin isotherm model assumes a uniform distribution of binding energies and that the heat of adsorption of all molecules in the layer decreases linearly with surface coverage due to adsorbent-adsorbate interaction. The value B1, which is related to the heat of adsorption, when high (above 200 kJ/mol), suggests a chemical adsorption process (Temkin and Pyzhev, 1940).

14.4 ADSORPTION OF POLLUTANTS

14.4.1 Adsorption of Heavy Metals

Heavy metals are those elements possessing a high atom weight, high density, and high toxicity at low concentrations. Various food wastes have been utilized in the research for the remediation of metals of high atomic weight. These food wastes were further modified into nanocomposites using polymers, clay, carbon nanotubes, graphene, metal nanoparticles, biochar, etc. Low-cost adsorbents or food waste are the need of the hour because of their eco-friendly, cost-effective, and abundant availability. These food wastes are highly porous and have good uptake capacity, which, when combined with derived compounds, enchances their uptake of heavy metals due to their increased surface properties (Siddiqui et al. 2022; Lingamandine et al. 2016). Various studies have been reported for the use of food waste-converting nanocomposites, such as Zhong et al. (2014), who synthesized the amphoteric-based wheat-based cellulose composite for copper and chromium removal. It was categorized that the monolayer uptake yield for Cu(II) ions was 241.13 and 90.9 mg/g for Cr(VI) ions. The adsorbent was regressed by the Langmuir isotherm, and Mckay and Ho's second-order kinetics is the best-fit model (Zhong et al. 2014).

Gokila et al. (2018) synthesized the alginate-chitosan composite at the nanoscale for the separation of Cr(VI) ions. The adsorbent shows a monolayer uptake yield of 33.58 mg/g and is regressed by pseudo-second-order kinetic model (Gokila et al. 2018). Ma et al. (2019) synthesized Bacterial cellulose (BC)@Zeolithic imidazolate framework 8(ZIF 8) aerogel composite in the sorption of Pb(II) ions with maximum removal of 81%. Masih et al. (2018) synthesized a chitosan-rice husk carbon composite gel for the sorption of copper ions with a homogenous uptake yield of 90.90 mg/gram. The adsorbent follows Langmuir isograms and is best fitted by pseudo-second-order kinetics. The exothermic behavior and decrease in practicality with a rise in temperature were observed (Masih et al. 2018). Xue et al. (2019) utilized various food wastes and prepared their biochars for the removal of ammoniacal nitrogen. Various food wastes include bones, carbohydrates, stems and leafy vegetables, nut husks, and fruit pericarp. The food wastes were then pyrolyzed at different temperatures: 300°C, 400°C, 500°C, 600°C, and 700°C. The batch equilibrium study confirms that FWBB (food waste bone biochars) shows the highest removal of 92.6% with a maximum adsorption capacity of 7.174 at 25°C (Xue et al. 2019).

14.4.2 Adsorption of Dyes

Dyes are a class of organic compounds that impart color to solutions. Dye molecules are composed of two components: chromophores and auxochromes. Chromophores are responsible for imparting color to fibers, while auxochromes increase the bond between dyes and fibers (Salleh et al. 2011). Dyes are classified as anionic, cationic, and nonionic and are used by textile industries, leather industries, and pulp and paper industries to impart color to materials. However, these industries discharge their waste into river bodies, which can be carcinogenic even at low concentrations. About 1mg/L of dye concentration is unfit for drinking by humans and aquatic organisms (Malik et al. 2007), so remediation is necessary. Food waste and their nanocomposites show better results than synthetic materials because food waste consists of lignocellulosic material consisting of lignin, cellulose, and hemicelluloses. These materials have a high molecular weight and are renewable, requiring a lower production cost (Demirbas, 2009). Therefore, various studies have been done to eradicate dye pollution from wastewater using food waste nanocomposites.

Haseena et al. (2022) synthesized Fe_2O_3 photocatalyzed nanoparticles prepared from leaf extracts of *Leucas aspera* and *Jatropha podagrica* for the removal of MB and RhB dyes, showing above 95% removal. Raza et al. (2022) opted for an ecologically sound method for the preparation of Mn-doped ZnO/*Phoenix dactylifera* bionanocomposite for the sorption of methyl orange. The adsorbent was mixed in a definite proportion and was best fitted by Mckay and Ho's order kinetics and Freundlich's adsorption isogram model (Freundlich, 1906) with a heterogeneous uptake capacity of 96.15 mg/g at pH 5 and 303 K. Sebeia et al. (2020) prepared eco-friendly nanocomposites using Cynomorium coccineum extract/CuO for the detachment of MB dye from the solution. The monolayer uptake yield of methylene blue was 64 mg/gram, and the XRD spectra of prepared nanoparticles depict the crystal size as 14.2 nm. The process is chemisorption in nature. Many researchers have synthesized hemicellulose-based aerogels for the sorption of organic dyes. Sun et al. (2014) prepared inorganic and organic hemicellulose derived from the straw of wheat for MB adsorption with a final uptake yield of 70 mg/g.

Sun et al. (2015) synthesized a hydrogel made up of xylan/pol(acrylic acid)/Fe_3O_4 nanoparticles for the removal of dyes and possessing paramagnetic properties with a final uptake yield of 439 mg/g. The banana peel was converted into activated carbon using a stratified approach with line-up by the hydrothermal method and activation by potassium hydroxide. The prepared adsorbent was helpful in the removal of MB, as studied by Saigl and Ahmad (2021). Abdullah et al. (2018) synthesized colloidal gels from the leftover skin of pitahaya (dragon fruit) using microwave radiation followed by gamma rays. Pectins were removed by treatment with acrylic acid and then polymerized by changing the concentration of pectin/acrylic ratio.

14.4.3 Adsorption of Pharmaceutical Drugs

Pharmaceutical contaminants are the contaminants that are discharged by hospitals, pharmaceutical manufacturing industries, and household sewage discharge. Pharmaceutical compounds are a technical class of contaminant that is hard to remove by traditional water treatment methods due to their good water dissolving properties and low biodegradability (Abed and Faisal, 2023). Pharmaceutical contaminants include antibiotics, analgesics, and anti-inflammatory drugs. Food waste composed of lignocellulosic compounds and polymeric content leads to the production of functional groups such as aldehydes, ketones, phenols, ethanols, and carboxylic acids, which are helpful in the erection of pharmaceutical drugs. When this food waste combines with other compounds, it shows a good uptake and removal yield of drugs (Bankole et al. 2023).

Husein et al. (2019) prepared copper oxide nanoparticles using a green synthetic approach. They added Tilia leaf extract to a 0.01M copper sulfate solution and heated it at 90°C with regular agitation for 30 minutes. The prepared solution was infused without light and then centrifuged. The prepared nanoadsorbents were used for the removal of NSAIDs such as Ibuprofen, Naproxen, and Diclofenac. The highest homogeneous uptake yield was 36 mg/g (Diclofenac) and 33.9 mg/g (Naproxen and Ibuprofen). Silva et al. (2020) prepared a Kalonite/cashew gum bio-nanocomposite for the removal of Doxazosin mesylate with the highest uptake capacity of 31.5 mg/g at pH 3.0 and 25°C. Ahsan et al. (2018) synthesized polymerized coffee waste for the removal of MB, tetracycline, and chromium ions. Coffee waste was treated with sulfuric acid, heated in an oil bath at about 70°C for 3 hours, and then cooled down in ice. The mixture was then dried and used for adsorption studies. The maximum sorption yield of 462 mg/g for tetracycline was obtained. The adsorbent was best regressed by Langmuir's homogeneous model and Mckay and Ho's second-order model. Abed and Faisal (2023) prepared an iron-layered double hydroxide/Calcium/sodium alginate for the removal of tetracycline. The adsorbent shows maximum removal above 95% at 90 minutes and pH 7.

14.5 FUTURE PROSPECTS

This chapter highlights future perspectives in the use of food waste as a sorbent for contaminant discharge, which are:

1. The development of research on the use of agricultural waste to form biomass in carbon derivative form.
2. The highly efficient and modified greener method is the mode of biological adsorption.
3. Future research to provide the use of agricultural land waste as nanocomposites for dealing with engineering-related problems.

To summarize, the materials used for adsorption based on agricultural waste, such as food leftovers, show a tremendous and economical approach for removing contaminants from the wastewater stream. This provides a broad application prospect for the utilization of agricultural waste.

14.6 CONCLUSION

Over the last few years, several studies have been done focusing on the adsorption of various pollutants by agricultural raw waste. Therefore, a lot of analysis and presuppositions exist. This chapter gives a detailed statement about the working, making, and consequences of different pollutants on the process of adsorption and reviews some of the studies of adsorption on agricultural waste. The best sorbents are waste derived from agricultural materials. As economic development happens, regional environmental pollution and ecological devastation become alarming, leading to the survival and development of humankind. A wide range of economical and reusable sorbents are in high demand. Waste derived from plant crops is soft and has large gaps, making it more efficient for binding, which can remove pollutants in a feasible way. The adsorption is affected by pH, temperature, initial concentration, and other factors. Modified agricultural waste can have a better adsorption effect. This chapter demands the engineering of green-modified technology for adsorption in biological mode. The waste of plants as biomass can not only eliminate the environmental burden but also achieve the goal of "treating waste by waste." In the future, environmental approaches will have broad space for development.

REFERENCES

Abdullah, M.F., Azfaralariff, Ahmad, and Lazim, Azwan.Mat. 2018. "Methylene blue removal by using pectin-based hydrogels extracted from dragon fruit peel waste using gamma and microwave radiation polymerization techniques." *Journal of Biomaterial Science Polymer Edition* 29(14): 1745–1763.

Abed, Marwa F., and Faisal Ayed A. H., 2023. "Calcium/iron-layered double hydroxides-sodium-alginate for removal of tetracycline antibiotic from aqueous solution." *Alexandria Engineering Journal* 63: 127–142.

Adegoke, Kayode Adesina, Adegoke Oyeladon Rhoda, Araoye Abimola Oluyomi, Ogunmodede, John, Agboola Oluwatobi Samuel, and Bello Olugbenga Solomon, 2022. "Engineered raw, carbonaceous, and modified biomass-based adsorbents for Rhodamine B dye removal from water and wastewater". *Bioresource Technology Reports* 18: 101082.

Ahmed, Rais, and Ansari Khalid, 2022. "Novel in-situ fabrication of L-methionine functionalized bionanocomposite for adsorption of Amido Black 10B dye." *Process Biochemistry* 119: 48–57.

Ahmad, Rais, Kumar, Rajeev, and Haseeb Shaziya, 2012. "Adsorption of Cu^{2+} from aqueous solution onto iron oxide coated eggshell powder: Evaluation of equilibrium, isotherms, kinetics, and regeneration capacity." *Arabian Journal of Chemistry* 5(3): 353–359.

Ahsan, Md. Ariful, Jabbari Vahid, Islam Md. Tariqul, Kim Hoejin, Hernandes -Viezas Jose Angel, Lin Yirong, Díaz-Moreno Carlos A., Lopez Jorge, Gardea-Torresdey Jorge, and Noveron, Juan C. 2018 "Green synthesis of a highly efficient biosorbent for organic, pharmaceutical, and heavy metal pollutants removal: Engineering surface chemistry of polymeric biomass of spent coffee waste". *Journal of Water Process Engineering* 25: 309–319.

Ali, Rehab M., Hamad Hesham A., Hussein Mohamed M., Malash Gihan F., 2016. "Potential of using green adsorbent of heavy metal removal from aqueous solutions: Adsorption kinetics, isotherm, thermodynamic, mechanism and economic analysis". *Ecological Engineering Journal* 91: 317–332.

Biswal, Achyuta Kumar, Sahoo Madhusudan, Suna Pramod Kumar, Panda Laxmipriya, Lenka Chandrashree, and Mishra Pramila Kumar, 2022. "Exploring the adsorption efficiency of a novel cellulosic material for removal of food dye from water." *Journal of Molecular Liquids* 350: 118577.

Bankole, Deborah T., Oluyuri Abimbola P. and Inyinbor Adejumoke A. 2023. "The removal of pharmaceutical pollutants from aqueous solution using agro waste". *Arabian Journal of Chemistry* 16: 104699.

Das, Devlina, Vimala R., Das Nilanjana. 2015. "Removal of Ag(I) and Zn(II) ions from single and binary solution using sulfonated form of gum arabic-powdered mushroom composite hollow semispheres: Equilibrium, kinetic, thermodynamic and ex-situ studies". *Ecological Engineering Journal* 75: 116–122.

Demirbas, Ayhan 2009. "Agricultural based activated carbons for the removal of dyes from aqueous solutions: A review". *Journal of Hazardous Material* 167: 1–9.

Freundlich, Herbert Max Finley. 1906. "Over the adsorption in solution." *Journal of Physics and Chemistry*, 57(385471): 1100–1107.

Foo, Keng Yan, and Hameed, Bassim H. 2012. "Preparation of activated carbon by microwave heating of langsat (Lansium domesticum) empty fruit bunch waste". *Bioresource Technology* 116: 522–525.

Gokila, S, Gomathi, T, Sudha, P.N, and Anil, Sukumaran, 2018. "Removal of the heavy metal ion chromiuim(VI) using Chitosan and Alginate nanocomposites." *International Journal of Biological Macromolecule* 104: 1459–1468.

Haseena, Sheik., Jaymani N., Shanavas Shajahan., Duraimurugan J., Abu Hajja Mohammad., Suresh Kumar Govindan., Senthil Kumar Arumugam, Prabhuraj T., Maadeswaran P. and Acevedo R. 2022. "Bio-synthesize of photocatalytic Fe_2O_3 nanoparticles using Leucas aspera and Jatropha podagrica leaf extract for an effective removal of textile dye pollutants." *Optik: International Journal for Light and Electron Optics* 249: 168275.

Higgins, Prerna, Siddiqui, Haseeb, Shaziya, and Kumar, Rajeev, 2022. "Design of novel metal hydroxide bio-nanocomposite (CBCS@LDH) for the scavenging of ciprofloxacin (CPN) drug from its aqueous solution: Kinetic, isotherm, thermodynamic and reusability studies." *Current Research in Green and Sustainable Chemistry* 5: 100298.

Husein, Dalal Z., Hassanein, Reda, and Al- Hakkani Mostafa F. 2019. "Green-synthesized copper nano-adsorbent for the removal of pharmaceutical pollutants from real wastewater samples." *Heliyon* 5 : e02339.

Isaac, Runit and Siddiqui Shaziya, 2022. "Sequestration of Ni(II) and Cu(II) using FeSO 4 modified Zea mays husk magnetic biochar: Isotherm, kinetics, thermodynamic studies and RSM." *Journal of Hazardous Material Advances* 8: 100162.

Machrouchi, Aicha, Khiar Habiba, Elhalil Alaâeddine, Sadiq Mhamed, Abdennouri Mohamed, and Barka Noureddine, 2023. "Synthesis, characterization, and photocatalytic degradation of anionic pollutant using a novel ZnO/activated carbon composite." *Watershed Ecology and the Environment* 5: 80–87.

Ma, Xiaoting, Lou Yue, Chen Xiao Bo, Shi Zhan, and Xu Yan, 2019. "Multifunctional flexible composite aerogels constructed through in-situ growth of metal-organic framework nanoparticles on bacterial cellulose." *Chemical Engineering Journal* 356: 227–235.

Malik, R., Ramteke D. S. and Wate S. R., 2007. "Adsorption of malachite green on groundnut shell waste based powdered activated carbon". *Waste Management Journal* 27: 1129–1138.

Masih, Manish, Anthony Prashant, Siddiqui Shaziya Haseeb. 2018. "Removal of Cu (II) ion from aqueous solutions by Rice Husk Carbon-Chitosan Composite gel (CCRH) using response surface methodology." *Environmental Nanotechnology, Monitoring & Management* 10: 189–198.

Manchala, Saikumar, Gandamalla, Ambedkar, Rudrarapu, Aravind, 2022. "Biodegradable Nanocelluolose for removal of hazradous organic pollutants from wastewater". In G.A.M. Ali, A.S.H. Makhlouf (eds) *Handbook of Biodegradable Materials*. Springer: Cham.

Mckay, Gordon and Ho Yuh Shan. 1999. "Pseudo-second order model for sorption processes." *Process Biochemistry* 34(5): 451–465.

Missau, Juliano, Bertuol Daniel Assumpção, and Tanabe Eduardo Hiromitsu, 2021. "Highly efficient adsorbent for removal of crystal violet dye from aqueous solution by CaAl/LDH supported on Biochar." *Applied Clay Science* 214: 106297.

Mustafa, S. Frayad, and Aziz, Hama, Hikmat, Kausar 2023. "Heterogeneous catalytic activation of persulfate for the removal of rhodamine B and diclofenac pollutants from water using iron-impregnated biochar derived from the waste of black seed pomace". *Process Safety and Environmental protection* 170: 436–448.

Nguyen, Ngoan Thi Thao, Nguyen Luan Minh, Nguyen Thuy Thi Thanh, Nguyen Ngoc Hoi, Nguyen Dai Hai, Nguyen Duyen Thi Cam, and Tran Thuan Van, 2023. "Green synthesis of ZnFe2O4@ZnO nanocomposites using chrysanthemum spp. floral waste for photocatalytic dye degradation." *Journal of Environmental Management* 326: 116746.

Lagergren, S.K. 1898. "About the theory of so-called adsorption of soluble substances." *Sven Vetenskapsakad Handingarl* 24: 1–39.

Langmuir, Irving 1918. "The adsorption of gases on plane surfaces of glass, mica and platinum". *Journal of the American Chemical Society* 40(9): 1361–1403.

Liu, Jingxin, Zhang Wenjuan, Mei Meng, Wang Teng, Chen Si and Li Jinping, 2022. A Ca-rich biochars derived from food waste digestate with exceptional adsorption capacity for arsenic(III) removal via a cooperative mechanism. *Separation and Purification Technology* 295: 121359.

Lingamdinne, Lakshmi Prasanna, Yang Jae-Kyu, Chang Yoon-Young, Koduru Janardhan Reddy 2016. "Low-cost magnetized Lonicera japonica flower biomass for the sorption removal of heavy metals." *Hydrometullurgy* 165: 81–89.

Raza, Azam, Shoeb Mohd, Maskoor Fauzia, Rahaman Sabiar, Mobin Mohammad, Jeong Changyoon, Ansari Mohd Yusuf, and Ahmad Absar, 2022. "Phoenix dactylifera mediated green synthesis of Mn doped ZnO nanoparticles and its adsorption performance for methyl orange dye removal: A comparative study." *Materials Chemistry and Physics* 286: 126173. doi.org/10.1016/j.matchemphys.2022.126173.

Saigl, Zainab Mohammad, and Ahmed Arwa Mohammad, 2021. "Separation of rhodamine b dye from aqueous media using natural pomegranate peels." *Indonesian Journal of Chemistry* 21(1): 212–224.

Salleh, Mohamed Amran Mohd, Mahmoud Dalia Khalid, Karim Wan Azlina Wan Abdul, and Idris, Azni 2011. "Cationic and anionic dye adsorption by agricultural solid waste". *Desalination* 280: 1–13.

Siddiqui, Shaziya Haseeb, Uddin Mohammad Kashif, Isaac Runit, and Aldosari Obaid F. 2022. "An effective biomass for the adsorption of methylene blue dye and treatment of river water." *Adsorption Science and Technology*.doi.10.1155/2022/4143138.

Silva, Maura C.C., Santos Maria S.F., Bezarra Roosevelt D.S., Araujo-Junior Edgar A., Osajima Josy A., Santos Maria R.M.C., Fonseca Maria G., and Silva-Filho Edson C., 2020. "Kalonite/cashew gum bio-nanocomposite for doxazosin incorporation and its release." *International Journal of Biological Macromolecules* 161: 927–935.

Sebeia, Nouha, Jabli Mahjoub, and Ghith Adel, 2019. "Biological synthesis of copper nanoparticles, using nerium oleander leaves extract: Characterization and study of their interaction with organic dyes." *Inorganic Chemistry Communications* 105: 36–46.

Sebeia, Nouha, Jabli Mahjoub, Ghith Adel and Saleh Tawfik A. 2020. "Eco-friendly synthesis of cynomorium coccineum extract for controlled production of copper nanoparticles for sorption of methylene blue dye." *Arabian Journal Chemistry* 13(2): 4263–4274.

Shahnaz, Tasrin, Bedadeep Das and Narayanaswamy Selvaraju, 2022. "Investigation of the adsorptive removal of methylene blue using modified nanocellulose." *International Journal of Biological Macromolecule* 200: 162–171.

Sonawane, Gunvant H. and Shrivastava Vinod. 2009. "Kinetics of decolourization of malachite green from aqueous medium by maize cob (Zea maize): A agricultural solid waste". *Desalination* 247: 430–441.

Sun, Xiao- Feng, Ye Qing, Jing Zhanxin and Li Yajing, 2014. "Preparation of hemicellulose-g-poly (methacrylic acid)/carbon nanotube composite hydrogel and adsorption properties." *Polymer Composites* 35: 45–52.

Sun, Xiao-Feng, Liu Baichen, Jing Zhanxin, and Wang Haihong, 2015. "Preparation and adsorption property of xylan/poly(acrylic acid) magnetic nanocomposite hydrogel adsorbent." *Carbohydrate Polymers*, 118: 16–23.

Temkin, Misha J., and Pyzhev V. 1940. "Recent modifications to Langmuir isotherms", *Acta Physico-Chimica Sinica, USSR*, 12: 217–222.

Weber, Walter J. and Morris J. Carrell 1963. "Kinetics of adsorption on carbon from solution." *Journal of the Sanitary Engineering Division* 89(2): 31–60.

Xiao, He, Shahab Asfandyar, Li Jieyue, Xi Beidou, Sun Xiaojie, He Huijun, and Yu Go, 2019. "Distribution, ecological risk assessment and source identification of heavy metals in surface sediments of Huixian karst wetland, China". *Ecotoxicology Environmental Safety* 185, 109700.

Xue S., X. Zhang, H. Ngo, W. Guo, H. Wen, C. Li, Y. Zhang, C. Ma. 2019. "Food waste based biochars for ammonia nitrogen removal from aqueous solutions." Bioresource Technology 292: 121927.

Zhong, Qian-Qian, Yue Qin-Yan, Li Qian, Gao Bao-Yu, and, Xu Xing, 2014. "Removal of Cu(II) and Cr(VI) from wastewater by an amphoteric sorbent based on cellulose-rich biomass." *Carbohydrate Polymers* 111: 788–796.

15 Nanotechnology for Environmental Remediation

Kamel Fahmi Bou-Hamdan and Jana Wissam Al-Ashkar
Beirut Arab University

Ali Samir Hamdan
Norwegian University of Science and Technology

CONTENTS

DOI: 10.1201/9781003397861-15

15.1 INTRODUCTION

Environmental pollution is one of the main challenges the globe is currently facing. It became apparent during the industrial revolution and increased afterward, causing irreversible damage to the environment (Ukaogo, Ewuzie, and Onwuka 2020). Environmental pollution can be classified as soil, water, or air pollution. Nowadays, with the world shifting toward advanced industries, the environment is becoming even more contaminated with various kinds of pollutants. These pollutants include carbon monoxide, chlorofluorocarbons, heavy metals, hydrocarbons, nitrogen oxides, organic compounds, sulfur dioxide, and others. Moreover, the combustion of oil, coal, and gas has a significant impact on pollution. The impact of natural resources can cause water pollution due to oil spills, leakage of fertilizers, and combustion of fossil fuels.

Environmental remediation is the process of reducing pollution by removing contaminants. Remediation repairs brownfield sites, which improves their quality and returns them to their natural condition. However, these sites have been used to deposit hazardous waste, which makes cleaning them challenging (Kalita and Baruah 2020). Environmental remediation is performed over several environmental structures, such as water and soil. For example, in contrast to soil remediation, which involves topsoil, subsoil, and sediment, water remediation involves both groundwater and surface water. Depending on the nature and severity of the contamination, the soil and water cleanup might be done individually or together (Kalita and Baruah 2020). Environmental remediation mainly depends on technologies such as adsorption, absorption, chemical reactions, photocatalysis, and filtration to remove contaminants. Nonetheless, new methods that can be more efficient in environmental remediation are still needed. This encouraged researchers to consider nanotechnology-based materials due to their distinct characteristics (Guerra et al. 2018).

Nanotechnology is one of the recent advancements that has been found useful in different field applications. For instance, nanotechnology and nanomaterials were found to be useful in different oil and gas applications, including enhanced oil recovery and well stimulation operations (Bou-Hamdan 2022; El-Masry et al. 2023). Other uses of nanotechnology are in photography, aerospace, electronics, pharmaceuticals, medicine, and others (You et al. 2013; Xu et al. 2015; Shen et al. 2014; Otsuka, Nagasaki, and Kataoka 2012; Mout et al. 2012). In terms of environmental pollution, nanotechnology proposes a wide range of abilities and technologies that enhance the quality of the environment (Yunus et al. 2012). They can regulate matter at the nanoscale due to their nanosized particles having a high surface-area-to-volume ratio. It has three major abilities that can be used in areas of the environment, including remediation and purification, detection of contaminants, and prevention of pollution (Yunus et al. 2012).

The nanotechnology-related method linked to cleaning the environment is known as nano-remediation. This approach includes the application of some reactive materials to detoxify and transform the contamination into a more environmentally friendly material. Some major benefits of nano-remediation include purifying the air for nontoxic breathing and the water for healthy drinking in agriculture and different industries. Furthermore, nanomaterials, such as nanoscale oxides, can be used on-site to clean up oil spills that could have occurred from leaking underground oil tanks. This application is unlike the remediation method mentioned earlier. In addition to that, nanoscale devices are tools that can be used to examine and sense contaminants present in the environment (Auffan et al. 2019). The benefit of nano-remediation, as compared to other remedial methods, is the ability of nanoparticles to move into polluting areas where others cannot. Therefore, nano-remediation approaches are utilized to detect pollutants and remedy air, soil, sediment, and water. The procedure of nano-remediation includes sensing the pollutants with the nanoparticle, treating them, and finally immobilizing their effect. Nano-remediation techniques can be successfully used to disinfect, desalinate, and remove heavy metal ions and organic pollutants.

This chapter classifies conventional remediation techniques accordingly and discusses the advancements of utilizing nanotechnology in environmental remediation. Moreover, it also covers how nanoparticles may be used to remove numerous environmental pollutants, including dyes, heavy metals, and other organic compounds.

15.2 CONVENTIONAL REMEDIATION

Conventional remediation is the procedure of eliminating contaminants from sites that have been previously polluted by industrial, manufacturing, mining, or even commercial activities (Auffan et al. 2019). Figure 15.1 shows several types of environmental remediation approaches, which will be discussed in this section.

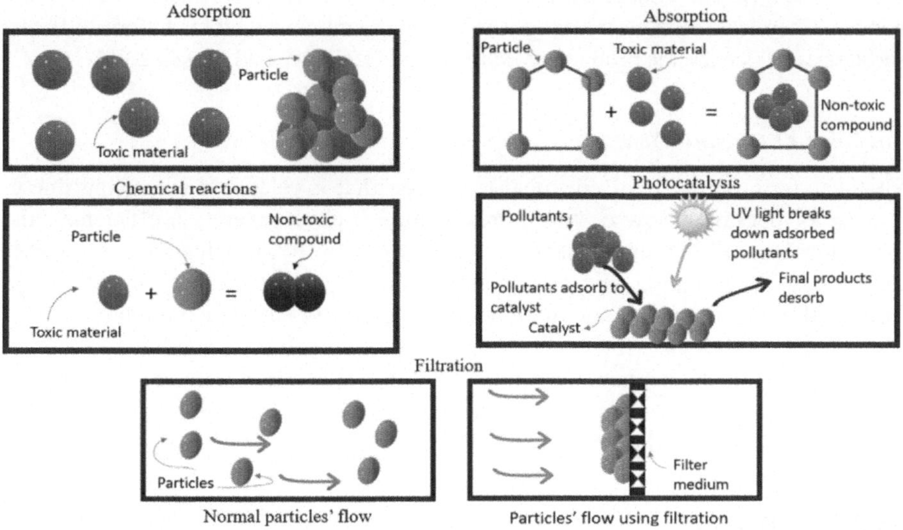

FIGURE 15.1 Conventional remediation techniques

15.2.1 Adsorption

In the area of environmental engineering, adsorption plays the most significant role among other remediation techniques in treating contaminants due to its remarkable ability to manage a range of pollutants as well as its ease of operation. It has become one of the most exceptional treatment techniques, as it is the most commonly used among other treatments, especially for wastewater containing organic as well as inorganic contaminants (Busetty 2019).

This process includes a mass transmission procedure of solutes, known as adsorbates, to solid surfaces, known as adsorbents, in a solution that has both the adsorbates and adsorbents. In this phenomenon, the adsorbates become physically and chemically bound to the surface of the adsorbent (Adekeye et al. 2019).

15.2.2 Absorption

Absorption is a chemical or physical process where the molecules, atoms, and ions of the material being absorbed enter the bulk phase of the substance that is being consumed. It is the method or technique by which a substance gets mixed with or even completely absorbed into another. In this phenomenon, a substance transforms the energy it captures, which will be distributed through the surface by the absorbent. Processes, including liquids or gases that enter the absorbent bulk, are mainly known as absorption (Mhemeed 2018).

15.2.3 Chemical Reactions

A chemical reaction is a method by which one reactant, or, even more, is transformed into one or more different products. These substances can be chemical elements or even compounds. Chemical reactions rearrange the reactant's component atoms to form different substances, which are the products (Michalisková, Haláková, and Prokša 2019).

15.2.4 Photocatalysis

Photocatalysis is a type of chemical technique that is relatively promising due to its ease of operation, low cost, nontoxic nature, high degeneration efficiency, and exceptional stability. Photocatalysis has been extensively investigated as a possible technology for regulating environmental pollution. The contaminants in wastewater, discussed in other sections, are divided into organic and inorganic pollutants.

TiO_2 photocatalysis, which is considered a green catalytic technology, can aid in reducing most of the contaminants. According to studies presented in the literature, the TiO_2 photocatalytic technique is a potential method for degrading more than 3,000 different types of hazardous degradable organic compounds (Yang and Wang 2018).

15.2.5 Filtration

Filtration is the procedure by which solid particles that are present in a liquid or even a gaseous fluid are separated by a filter medium. This allows the passage of the fluid but collects the solid particles. All the particles suspended in the water will be removed in this process through several mechanisms, including flocculation, surface capture, straining, and sedimentation. Filters are characterized by the basic capture technique, which includes particle elimination at the filter media's surface (Lofrano, Libralato, and Brown 2017).

15.3 APPLICATIONS OF NANOTECHNOLOGY IN ENVIRONMENTAL REMEDIATION

15.3.1 Nanotechnology in Soil Remediation

Recently, environmental remediation specialists questioned the reliability of using nanotechnology for remediation purposes. Moreover, with the great interest in agriculture soil remediation, a substantial interest was given to studying the effectiveness of implementing nanomaterials in thc soil rcmediation processes.

Nanomaterials are usually implemented in the earth's upper crust (soil) through direct or indirect methods. Industries usually directly impose nanomaterials while disposing of, using, or recycling their materials in manufacturing units. Indirect release of nanomaterials could take place when producing or disposing of chemical products that are rich in those nanomaterials.

Nanomaterials, also known as engineered nanomaterials, are of great interest in the agriculture industry because plants' natural mechanisms can easily absorb nanomaterials from the soil. In addition to the fact that nanomaterials are effective in soil remediation processes, they can also play a great role in enhancing the plants' roots, depending on the materials that were taken by them. The prior enhancement mentioned is directly subject to the type of materials present and the plant itself. For example, silver nanoparticles can have a positive impact on *Allium cepa* plant health, whereas TiO_2 negatively affects this plant. This is not the case for TiO_2 when absorbed by other plants; from here comes the "Engineered" part of their naming.

Soil remediation techniques should take into consideration the physiochemical properties of nanomaterials to ensure effective remediation without environmental side effects. The physiochemical properties to be considered are: Type, particle size, particle mobility, charge, and materials' surface area. Those properties alter the interactions of nanomaterials with the soil, which, although they have a great effect on soil remediation, pose a serious threat to the beneficial microorganisms that are crucial for the food chain. The toxicity of nanomaterials is highly variable; for example, Ag, Ti, Ce, and Zn were found to be highly toxic for some microbes in the soil, which is not the case for Si, Fe, Au, or Pd, which were found to be less toxic than the previously mentioned materials and even beneficial for other microbes.

The effective remediation of soil using nanomaterials can then be done by combining nanotechnology with soil analysis to avoid any harm or threat to the ecosystem and remediate the soil with the most environmentally friendly technique (Figure 15.2).

In fact, nanomaterials used in soil remediation can be classified into one of two categories: reactive materials or adsorptive materials. The use of adsorptive nanomaterials can help limit any negative side effects on the soil, as they have fewer heavy metals compared with reactive materials and therefore less toxicity toward the soil and its microorganisms. The unique properties of nanoparticles, such as their large surface combined with their small size, make them a reliable asset for soil

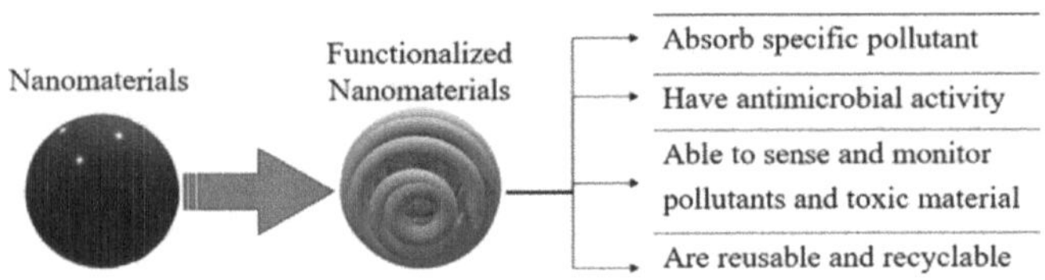

FIGURE 15.2 Advantages of functionalized nanomaterials

remediation, and there have been numerous nanoparticle classes for this concern, which will be detailed below (Lofrano, Libralato, and Brown 2017).

15.3.1.1 Titanium Dioxide Nanoparticles

TiO_2 particles are considered an asset for soil remediation procedures due to their nontoxic nature in phytoremediation processes. TiO_2 has a great ability to decontaminate soil rich in heavy metals, especially cadmium, as it can suppress its effect on plant and microbial growth in the soil (Lofrano, Libralato, and Brown 2017).

After seeing its excellent performance in soil remediation, scientists started investigating the optimal concentration of TiO_2 in remediation procedures. Hu et al. (Wang et al. 2016) conducted a study on coriander plants using different concentrations of TiO_2. It was seen that the increase in TiO_2 concentration from 50 to 200 mg/L didn't show any representative enhancement. On the contrary, it reduced earthworm reproduction, and even at a concentration of 400 mg/L, the process retarded plant growth and caused moisture production in the soil (Wang et al. 2016).

15.3.1.2 Silica Nanoparticles

Si nanoparticles are also considered nontoxic in phytoremediation processes. They are usually used to remove diesel and are mixed with foams to improve their stability and enhance the diesel removal process.

Soil remediation using Si nanoparticles displayed a significant increase in the live cells' reproduction, even in the presence of high concentrations of Cd. These particles help safely reach the required soil remediation standards, especially removing highly toxic heavy metals while unaffected by plant and microbial growth in the soil (Lofrano, Libralato, and Brown 2017).

15.3.1.3 Carbon-Based Nanoparticles

Carbon-based nanoparticles are the most commonly utilized particles for soil remediation due to their high adsorption ability, which helps the decontamination of toxic and heavy pollutants in the soil, in addition to their ability to degrade pesticides and herbicides. Additionally, carbon-based nanoparticles are also known to be effective in the removal of organochlorine pesticides, which are dangerous because of their impact on the soil for an extended time and their hydrophilic property, which puts water bodies at big risk of contamination (Del Prado-Audelo et al. 2021).

The CNT is the most commonly used form of carbon-based nanoparticles, as it has a hexagonal shape, which favors porosity in the material, hence better adsorption ability. Similar to other nanoparticles, carbon-based ones' effectiveness is related to their concentration and exposure time. Referring to the literature, moderate concentrations have no effect on the bioavailability of microbes or the biodegradation of pollutants, whereas high concentrations have an impact on the bacterial population in the soil (Shrestha et al. 2013).

15.3.1.4 Silver Nanoparticles

Ag nanoparticles are unique in the soil remediation process due to their antifungal and antibacterial characteristics, added to their size-dependent chemical and physical properties, which limit the usage of nanoparticles during remediation processes. Green synthesized silver nanoparticles are environment-friendly particles that were considered cost-effective and reliable for decontamination processes. At concentrations of 50 ppm, their impact on the plants was found to be unmeasurable, as it showed an enhancement in the formation of chlorophyll because it changed the pH of the soil from acidic to neutral, increasing the cation-exchange capacity and thus improving the plants' absorption of essential nutrients. On the other side, Ag nanoparticles, at concentrations of 75 ppm and above, started disturbing the biodegradation mechanisms in the soil and increased Ag toxicity (Lofrano, Libralato, and Brown 2017).

15.3.2 Water Treatment

The latest studies done by the UNICEF and WHO show that almost 74 million individuals lack sufficient water supplies and almost 50% of the entire population relies on groundwater. Moreover, inappropriate wastewater remediation as well as effluent release have deteriorated the condition. This is due to the presence of a broad variety of both inorganic and organic compounds, including dyes, pharmaceuticals, heavy metals, and others, that are present on the water's surface. Recently, different water treatment technologies, including membrane filtration, coagulation, biodegradation, and oxidation, have been applied for the remediation of water. Knowing that these methods require some energy consumption and maintenance and occasionally produce secondary by-products by the end of the reaction, nanotechnology (NT) has significantly fixed its origin in the area of the remediation of water. Nanomaterials are known as materials with a size of less than 100 nm, making them highly reactive with a high surface area as well. Nanotechnology (NT) has been used as an effective technique for water treatment. The following chapter will detail various applications of NT as a means for the removal of organic and inorganic materials from water.

Nanomaterials (NMTs) are divided into three groupings based on their dimension, remediation, and chemical composition. NMTs are synthesized by three methods: green methods, top-down, and bottom-up. In the top-down technique, a bulk substance is cut to nano-size with the help of grinding to produce NMT. However, in the bottom-up method, NMTs are processed by the chemical reduction method. Nevertheless, to prevent the usage of chemicals and decrease the consumption of energy, a green synthesis approach was conducted for the preparation of NMTs. In the mentioned technique, pieces of plants having high antioxidant characteristics were used for the synthesis of nanoparticles (Del Prado-Audelo et al. 2021). Table 15.1 highlights different categories of nanomaterials.

TABLE 15.1
A Summary of the Main Nanomaterials Categories

Kind	Characteristics
Size	• 0D: hollow nanoparticles; sized as nanoscale in dimensions (<100nm)
	• 1D: sized as two dimensions in nanoscale (nanofibers)
	• 2D: sized as nanoscale in dimensions (layered nanostructures)
	• 3D: carbon fullerenes; sized as zero nanoscale in dimensions (>100 nm)
Chemical	• Metallic such as iron, zinc, and titanium
Composition	• Combined from hybrid nanofiber, metal–metal, and metal–carbon
	• Carbonaceous, including carbon nanotubes, carbon nanofibers, and graphene
Morphology	• Rounded and flat

15.3.2.1 Heavy Metal Removal

Nanoparticles are regularly used as an adsorbent to remove heavy metals. NMTs, such as magnetite, carbon nanotubes, graphene, and zero-valent, have revealed exclusive properties, such as stability, a high surface-to-volume ratio, inertness, biocompatibility, and reactivity, which make them favorable substances for the adsorption process. These metals are non-biodegradable and accumulate in food chains, which in turn impacts animals and humans as well. In the current stage, iron-based nanomaterials are known as promising materials for extracting heavy metal pollution from an aqueous solution. From these, zero-valent iron is widely applied for chromium, nickel, cadmium, copper, lead, and zinc adsorption. This is due to its reduction property, easy synthesis, and solid nature. Originally, metal ions were physically adsorbed on the surface of zero-valent iron and then precipitated together.

Another phenomenon detected during this process is the reduction mechanism. For the reduction of chromium in the hexavalent state, it was recognized as a toxic contaminant and needed to be reduced to its nontoxic form. The zero-valent particles absorbed these toxics, and then, by the transfer of electrons, these iron atoms got oxidized to ions of iron(III), which then reduced the hexavalent chromium to nontoxic ones. However, zero-valent particles suffer from aggregation because of their nature, which is magnetic, thus lessening the nanoparticles' reactivity. This issue is tackled by restraining nanomaterials on a certain substrate, which can be a metal, polymer, or even a porous material.

Other magnetic nanomaterials, such as Fe_3O_4, are also incorporated into heavy metal removal. Fe_3O_4 nanoparticles were synthesized through the process of green synthesis using a flavonoid polymer compound and used for Pb^{2+}, Cu^{2+}, and Cd^{2+} adsorption from water (Zhang et al. 2021). Moreover, membrane filtration with nanofiber was used for the treatment of metal ions as well. For example, cadmium and chromium-polluted water were removed using nano-fibrous composite membranes altered by polyvinyl alcohol (PVA) and polyacrylonitrile (PAN). The groups of amino acids found at the surface of the PAN-altered membrane end in higher adsorption rates of Cr^{+6} and Cd^{+2} rates. In addition, nanomaterials can be used as catalysts to change the phenomenon of degradation (Del Prado-Audelo et al. 2021).

15.3.2.2 Dye Removal

Dye wastewater is referred to as a biorecalcitrant because of its wide use in different industries as well as its toxicity. Almost 50% of the dye wastewater is discharged into the surface water, which affects the quality of the water and the health of people. Dye is considered to be highly stable and non-biodegradable. According to their usage, dyes are divided into azo, direct, reactive, acid, and original dyes. Azo dyes are known by their –N=N– group presence, which cannot be cracked by any conventional remediation techniques. Azo dyes occupied 70% of the fabric dye wastewater as well. First, the azo bond found in the Congo red dye was degraded by nanoparticles of iron due to the zero-valent iron's structure. Also, the solution's pH ascends with a resilient dropping state that ends up in the dye molecules' transformation. Next, the waste from the initial stage moves into another one, which is made up of an aerobic membrane bioreactor. The incorporation of zero-valent iron improved biodegradability by up to 99% (Liu et al. 2020).

Furthermore, graphene oxide (GO) was synthesized to degrade cationic dyes. The integration of GO into the membrane increases heat resistance and stability. It also overcomes the fouling problem, as the degree of adsorption is affected by the membrane's surface charge, where the dyes with negative charges repel more and hence increase the rate of rejection. At a later stage, the pore size is considered to be very essential in the dyes' extraction. For that, the adsorption process began with dyes' adsorption at the membrane's surface depending on their surface charge and later relying on the membrane's hole size.

15.3.2.3 Organochlorine Compounds (OCCs) Removal

OCCs (organochlorine) compounds are known as those with alkyl and aryl organic chlorides, such as dioxins, polychlorinated biphenyls, and dichlorodiphenyltrichloroethane (DDT). The processes of industrialization, metal degassing, manufacturing, and dry cleaning of herbicides and pesticides overstated the manufacture of chlorinated organic compounds. The existence of carbon–chlorine bonds ensures high structural stability. Moreover, OCCs have a lipophilic nature, having the ability to dissolve in the fatty tissues of living cells. OCC condensation was noticed between 0.07 and 4,493 ng/L in the water at the surface. Knowing that the chlorine atoms are high, it requires intensive energy for the process of dehalogenation. However, earlier research on OCC dehalogenation utilized the adsorption of zero-valent iron atoms. For that reason, dichlorination reactivity, metal leaching, and a lack of stability were noticed in this process (Zhang et al. 2021).

15.3.2.4 Inorganic Anions

The existence of anions such as nitrate, phosphate, nitrate, carbonate, and sulfate highly affected the value of water. High phosphorous and nitrogen amounts in water lead to a reduction of dissolved oxygen. However, agricultural run-off, fertilizers, and sewage waste are the basic reasons behind the pollution of anions. For that, treatment techniques such as reverse osmosis, chemical precipitation, activated sludge process, and electro-dialysis were useful. On the other hand, chemical consumption, effluent neutralization problems, poor selectivity of ions, and the problem of sludge disposal took place during this process. Usually, adsorption by nanomaterials is favored over other treatment techniques because of its simple operation (Velu et al. 2021).

15.3.3 Air Pollution Remediation

Air pollution is the focus nowadays of all nations, especially when the WHO displays the huge effect of the undesired gases and particles on human beings' health and the sustainability of life on earth. Well-known air contaminants such as nitrogen oxides, hydrogen sulfides, and volatile organic compounds pose serious health risks to humans. Therefore, scientists urged developing numerous techniques to limit and manage air pollution, and with the great advantage that nanotechnology offers, especially on the atomic scale, environmental nanotechnology played an important role in improving conventional remediation techniques on all levels and especially in managing the air pollution phenomenon (Luo et al. 2017).

The unique characteristics of nanomaterials make them an effective and reliable tool to mitigate the harmful discharge of gases from both industrial and domestic activities, and they can be considered a sustainable asset to contribute to remediating the environment. Nanotechnology was used in many applications of air pollution remediation, such as removing dangerous contaminants from the air using nanoadsorbents, limiting pollution diffusion using nanofilters and nanostructured membranes, and purifying the atmosphere using nanocatalysts (Lofrano, Libralato, and Brown 2017).

15.3.3.1 Nanoadsorbents

Owing to a large porous surface area with high and efficient activity, nanoadsorbents show a sustainably enhanced removal of gases, bacteria, and other organic compounds from the surrounding environment. The remediation's effectiveness depends directly on the nanomaterials used in the adsorbent. For that reason, each type of contamination is associated with a certain type of nanoadsorbent.

Carbon-based materials have a high affinity for CO_2 as they are chemically stable in the presence of CO_2, and have a unique advantage because of their reversible nature, so they can effectively adsorb CO_2. This is the same for amine groups, which can react with CO_2. Zinc oxide nanostructures, on the other hand, have the required physical and chemical properties, as they are stable at high temperatures and have an extended adsorption time to remove hydrogen sulfide from the air. This is a big concern because SO_2 is very harmful and can be released into the environment during petroleum drilling activities and even in other industrial activities, which puts workers at risk (Lofrano, Libralato, and Brown 2017).

15.3.3.2 Nanofilters and Nanostructured Membranes

Filters and membranes were part of the conventional remediation techniques, and with the advancements in nanotechnology, researchers planned to combine those two, to create complex applications that change the working conditions of air filtering in favor of enhancing the air filtration performance and functional stability of the process.

TiO_2 nanowire-based filters were fabricated using different nanoscales to alter the surface area over volume ratio of the filtration process, thus reaching lower pressure across the filter and hence better separation and filtration performances. This hierarchy structure of the filters allows a higher flow of cleaned air with better exposure of contaminants to nanoadsorbents (Lofrano, Libralato, and Brown 2017).

Such a technique is of great interest when added to nanoadsorbents because, as mentioned previously, nanoadsorbents are an effective tool to remove sulfides, but the release of such contaminants is accompanied by bigger particulate matters that usually impact the performance of nanoadsorbents, which can be mitigated using those nanostructured membranes, increasing the contact area between sulfides and toxic gases with the nanoadsorbents.

Green nanomaterials can also be used to fabricate nanofilters, but antibacterial and temperature considerations should be taken into consideration to avoid diminishing the filtration efficiency (Vohra et al. 2005).

15.3.3.3 Nanocatalysts and Nanosensors

Purification of the environment and surroundings using nanotechnology included the use of nanocatalysts to chemically react with toxic gases and contaminants and degrade them. The success of this remediation process depends on the type of contaminant present in the atmosphere, the chemical ability of the nanomaterial to react, and its porous structure (Lofrano, Libralato, and Brown 2017). Therefore, functionalized nanomaterials were introduced for solving specific environmental problems while limiting the potential risks and uncertainties associated with nanomaterials' usage for environmental remediation (Lofrano, Libralato, and Brown 2017).

Improved catalysis processes using TiO_2 nanomaterials to separate volatile organic compounds proved efficient, and they were shown to be able to remediate the surrounding environment from microbes, viruses, and bacteria. However, considerations about the humidity level should be taken into consideration, as it stops photo-oxidation because H_2O molecules compete with the contaminant for the adsorption sites (Binas et al. 2017). On the other side, the effective remediation process starts with the early detection of certain gases at low concentrations, and here comes the role of nanosensors to detect their presence. CNT, graphene, and other nanoparticles are the base for nanosensor applications to detect toxic gases such as H_2S, NO_2, SO_2, and others (Lofrano, Libralato, and Brown 2017).

15.4 CHALLENGES AND LIMITATIONS

Nanotechnology is similar to any other potential solution to the environment in that it always has its own challenges. With the great improvements that nanotechnology has offered to the remediation sector, researchers have to question certain challenges and work on overcoming them. Other than the general challenges accompanied by nanotechnology, below is a list of the ones specific to environmental remediation:

1. Stability: This is because the environment where nanotechnology is used for environmental remediation is harsh and hazardous. Therefore, it is hard to predict how nanomaterials will react to such surroundings, and the real

question is whether it is possible to limit the decomposition of nanomaterials and reach a certain stability during remediation operations to avoid nanopollution (Sun 2019).

2. Activity and Selectivity: Although nanotechnology offers the advantage of sensing pollutants to selectively remediate them, in many cases, materials that are efficient at certain pollutants are harmful in the presence of other toxic materials (Binas et al. 2017).
3. Scaling: Until now, there has been no standardized methodology to scale the effectiveness, toxicity, safety, and sustainability of the proposed nano-remediation techniques. The subjectivity of researchers is dominant during their experiments. For example, stable nanomaterials in the soil for 5 years can be effective in some places, whereas in others, this is not the case. In addition, the cost of nano-remediation is high, recycling and reusing nanoparticles are still under investigation, and there's no standard to qualify a certain technique as reliable and cost-effective one (Lofrano, Libralato, and Brown 2017).
4. Nanopollution and the fate of nanomaterials: Although nanoparticles showed low levels of intactness in the soil, water, and air, their long-term effect needs further research to know their final fate.. In addition, nanopollution was confirmed to be a risk and thus a challenge to overcome. Metal-based catalysts and nanomaterials, for example, can be used in highly active environments, causing them to break metal particles and allow them to form new intermediates that become secondary contaminants (Sun 2019).

15.5 CONCLUSION

Due to the harm that contaminants and pollutants can cause to human health, environmental remediation services were implemented. The conventional remediation techniques were discussed in this chapter, but they are still not fully effective in the treatment operation. Hence, nanotechnology is being utilized to strengthen conventional techniques and overcome their challenges. Selecting the appropriate nanomaterial to attenuate a specific pollutant requires a detailed analysis that considers the pollutant type to be removed, the efficiency of the used nanomaterials, and the impact of not recovering the used nanomaterials on the environment.

Nanoparticles were found to be efficient in remediating soil contaminants, water impurities, and air pollution. Their improved reactivity and larger surface area gave them an advantage over conventional remediation techniques in soil, water, and air remediation. More studies are still needed to confirm their economic feasibility for use in large field operations.

REFERENCES

Adekeye, Damilola K., Olugbenga Kayode Popoola, Samuel Sunday Asaolu, Adepoju A. Adebawore, Olusola Isaac Aremu, and K. O. Olabode. 2019. "Adsorption and Conventional Technologies for Environmental Remediation and Decontamination of Heavy Metals: An Overview." *International Journal of Research & Review.* 6 (8). www.ijrrjournal.com.

Auffan, Mélanie, Armand Masion, Catherine Mouneyrac, Camille de Garidel-Thoron, Christine Ogilvie Hendren, Alain Thiery, Catherine Santaella, et al. 2019. "Contribution of Mesocosm Testing to a Single-Step and Exposure-Driven Environmental Risk Assessment of Engineered Nanomaterials." *NanoImpact* 13:66–69.

Binas, Vassilios, Danae Venieri, Dimitrios Kotzias, and George Kiriakidis. 2017. "Modified TiO2 Based Photocatalysts for Improved Air and Health Quality." *Journal of Materiomics* 3 (1):3–16.

Bou-Hamdan, Kamel Fahmi 2022. "Applications of Nanomaterials in the Oil and Gas Industry." In *Handbook of Research on Green Synthesis and Applications of Nanomaterials*, edited by R. Garg, R. Garg, and N. Eddy, 173–98. IGI Global: Pennsylvania.

Busetty, Subramanyam. 2019. "Environmental Treatment Technologies: Adsorption." In *Handbook of Environmental Materials Management*, edited by Chaudhery Mustansar Hussain. Springer, Cham. DOI:https://doi.org/10.1007/978-3-319-73645-7_37.

Del Prado-Audelo, M. L. Del, I. García Kerdan, L. Escutia-Guadarrama, J. M. Reyna-González, J. J. Magaña, and G. Leyva-Gómez. 2021. "Nanoremediation: Nanomaterials and Nanotechnologies for Environmental Cleanup." *Frontiers in Environmental Science* 9: 7. https://doi.org/10.3389/fenvs.2021.793765

El-Masry, Jamil Fadi, Kamel Fahmi Bou-Hamdan, Azza Hashim Abbas, and Dmitriy A. Martyushev. 2023. "A Comprehensive Review on Utilizing Nanomaterials in Enhanced Oil Recovery Applications." *Energies* 16 (2):691.

Guerra, Fernanda D., Mohamed F. Attia, Daniel C. Whitehead, and Frank Alexis. 2018. "Nanotechnology for Environmental Remediation: Materials and Applications." *Molecules* 23 (7):1760.

Kalita, Eeshan, and Julie Baruah. 2020. "16- Environmental Remediation." In *Metal Oxides. Colloidal Metal Oxide Nanoparticles,* edited by S. Thomas, A. T. Sunny, and V. Prajitha , 525–76. Elsevier. http://dx.doi.org/10.1016/B978-0-12-813357-6.00014-0

Liu, Xiangxiang, Bingyin Jiang, Xing Yin, Hongyang Ma, and Benjamin S. Hsiao. 2020. "Highly Permeable Nanofibrous Composite Microfiltration Membranes for Removal of Nanoparticles and Heavy Metal Ions." *Separation and Purification Technology* 233:115976.

Lofrano, Giusy, Giovanni Libralato, and Jeanette Brown. 2017. *Nanotechnologies for Environmental Remediation*. Springer International Publishing. http://dx.doi.org/10.1007/978-3-319-53162-5

Luo, Lei, Yangyang Guo, Tingyu Zhu, and Yang Zheng. 2017. "Adsorption Species Distribution and Multicomponent Adsorption Mechanism of SO2, NO, and CO_2 on Commercial Adsorbents." *Energy and Fuels* 31 (10):11026–11033.

Mhemeed, Amal H. 2018. "A General Overview on the Adsorption." *Indian Journal of Natural Sciences* 9 (51): 16127–16131.

Michalisková, Renata, Zuzana Haláková, and Miroslav Prokša. 2019. "Concept of Chemical Reaction in Chemistry Textbooks." *Chemistry Teacher International* 1 (2):20180027.

Mout, Rubul, Daniel F. Moyano, Subinoy Rana, and Vincent M. Rotello. 2012. "Surface Functionalization of Nanoparticles for Nanomedicine." *Chemical Society Reviews* 41 (7):2539–2544.

Otsuka, Hidenori, Yukio Nagasaki, and Kazunori Kataoka. 2012. "PEGylated Nanoparticles for Biological and Pharmaceutical Applications." *Advanced Drug Delivery Reviews* 55 (3):403–419.

Shen, Wenfeng, Xianpeng Zhang, Qijin Huang, Qingsong Xu, and Weijie Song. 2014. "Preparation of Solid Silver Nanoparticles for Inkjet Printed Flexible Electronics with High Conductivity." *Nanoscale* 6 (3):1622–1628.

Shrestha, Babina, Veronica Acosta-Martinez, Stephen B. Cox, Micah J. Green, Shibin Li, and Jaclyn E. Cañas-Carrell. 2013. "An Evaluation of the Impact of Multiwalled Carbon Nanotubes on Soil Microbial Community Structure and Functioning." *Journal of Hazardous Materials* 261:188–197.

Sun, Hongqi. 2019. "Grand Challenges in Environmental Nanotechnology." *Frontiers in Nanotechnology*. doi.10.3389/fnano.2019.00002.

Ukaogo, Prince O., Ugochukwu Ewuzie, and Chibuzo V. Onwuka. 2020. "Environmental Pollution: Causes, Effects, and the Remedies." In *Microorganisms for Sustainable Environment and Health*, 419–429. Elsevier. https://doi.org/10.1016/B978-0-12-819001-2.00021-8

Velu, Manikandan, Balamuralikrishnan Balasubramanian, Palanivel Velmurugan, Hesam Kamyab, Arumugam Veera Ravi, Shreeshivadasan Chelliapan, Chew Tin Lee, and Jayanthi Palaniyappan. 2021. "Fabrication of Nanocomposites Mediated from Aluminium Nanoparticles/Moringa Oleifera Gum Activated Carbon for Effective Photocatalytic Removal of Nitrate and Phosphate in Aqueous Solution." *Journal of Cleaner Production* 281:124533.

Vohra, Amit, D. Yogi Goswami, Deepak Anant Deshpande, and Sandra S. Block. 2005. "Enhanced Photocatalytic Inactivation of Bacterial Spores on Surfaces in Air." *Journal of Industrial Microbiology and Biotechnology* 32 (8):364.

Wang, A-nan, Ying Teng, Xue-feng Hu, Long-hua Wu, Yu- juan Huang, Yong-ming Luo, and Peter Christie. 2016. "Diphenylarsinic Acid Contaminated Soil Remediation by Titanium Dioxide (P25) Photocatalysis: Degradation Pathway, Optimization of Operating Parameters and Effects of Soil Properties." *Science of the Total Environment* 541:348–355.

Xu, Siyang, Ashfaque H. Habib, Andrea D. Pickel, and Michael E. McHenry. 2015. "Magnetic Nanoparticle-Based Solder Composites for Electronic Packaging Applications." *Progress in Materials Science* 67:95–160.

Yang, Xiaogang, and Dunwei Wang. 2018. "Photocatalysis: From Fundamental Principles to Materials and Applications." *ACS Applied Energy Materials* 1 (12):6653–6693.

You, Hongjun, Shengchun Yang, Bingjun Ding, and Hong Yang. 2013. "Synthesis of Colloidal Metal and Metal Alloy Nanoparticles for Electrochemical Energy Applications." *Chemical Society Reviews* 42 (7):2880–2904.

Yunus, Ian Sofian, Harwin, Adi Kurniawan, Dendy Adityawarman, and Antonius Indarto. 2012. "Nanotechnologies in Water and Air Pollution Treatment." *Environmental Technology Reviews* 1 (1):136–148.

Zhang, Jingya, Chao Lei, Wenqian Chen, Qianqian Xie, Qian Guo, and Binbin Huang. 2021. "Electrochemical-Driven Nanoparticulate Catalysis for Highly Efficient Dechlorination of Chlorinated Environmental Pollutant." *Journal of Catalysis* 395:362–374.

16 Ecotoxicological aspects of Nanotechnology

Nnabuk Okon Eddy
University of Nigeria

Anduang Odiongenyi
Akwa Ibom State University

Rajni Garg
Galgotias College of Engineering and Technology

Richard Alexis Ukpe
Federal University Otuoke

Rishav Garg
Galgotias College of Engineering and Technology

Emeka ChimaOgoko and Henrietta Ijeoma Kelle
National Open University of Nigeria
Paul Ocheje Ameh Nigerian Police Academy

Raphael Ogbodo
University of Iowa

CONTENTS

DOI: 10.1201/9781003397861-16

16.1 INTRODUCTION

In modern times, the subject of toxicity is a primary concern in all research, which figuratively translates to the need to consider the environmental consequences of existing and newly designed research, projects and development (Zhang et al. 2022). Toxicity is an index for assessing the harmful tendency of a compound or any substance. Such evaluation may require a short-term approach when the basis is acute toxicity or a long-term approach to evaluate chronic toxicity (Boros and Ostafe 2020). Some of the listed toxicological types are summarized in Figure 16.1. The severity of a toxicant can vary based on factors that will encourage or discourage the operation of the toxic cycle from the point of initiation to termination (Garg et al. 2021).

The major concern about the toxicity of any product is its impact on the ecosystem, which can technically be defined to include the biological and physical components under interaction with the abiotic and biotic factors such that energy and materials are transferred among the cycle (see Figure 16.1). Therefore, our major concern in this review will be the ecotoxicology of nanomaterials. Nanomaterials are one of the fast-growing disciplines in science and technology that is gradually replacing classical materials in various sectors such as electrical, chemical, environmental, etc. (Eddy et al. 2022). The appreciable success of nanomaterials that has sustained the need for continuous usage, improvement and advanced research cannot be separated from their outstanding properties, which can be listed to include thermal and mechanical stability, optical properties, electrical properties, catalytic properties, dimensional and pore properties, etc. (Garg et al. 2021). Identified forms of nanomaterials are broader and cover systems such as metal, metal oxide, nonmetals, nonmetal oxide, composites, advanced or hybrid forms, etc. (Garg et al. 2022).

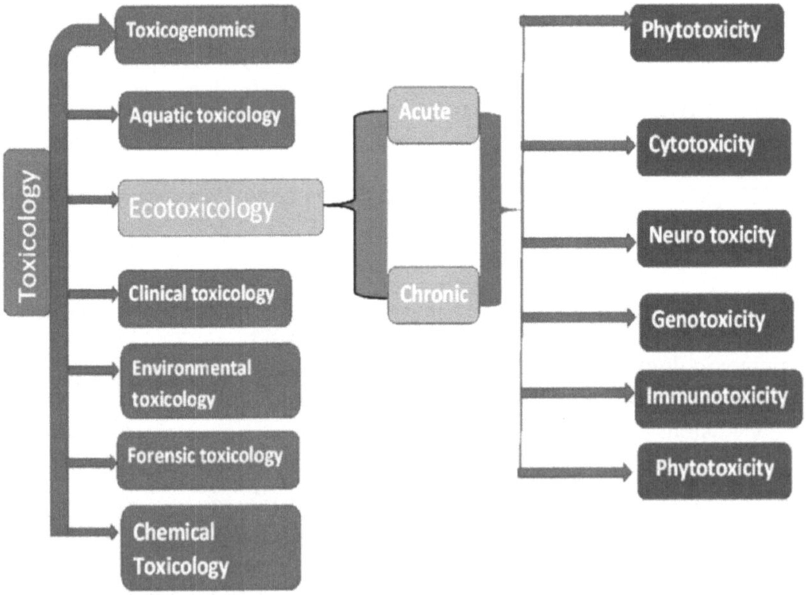

FIGURE 16.1 Levels and impacts of toxicology and ecotoxicology.

Some nanomaterials have been reportedly used directly by humans in application aspects of pharmaceuticals, water treatment, etc. (Eddy and Garg 2022). Literature also abounds in indirect usages such as the purification of water, enhancement of soil fertility, catalysis of consumable products and others (Hammami et al. 2021). Nanoparticles can have the size and other properties that can make environmental scenarios worse compared to classical materials because they can show excellent penetration capacity, retention capacity, reactivity, absorption/adsorption, etc. (Garg et al. 2022). In view of the wider scope of applications recorded for nanomaterials in global society and the ever-expanded and continuous future inflow expected shortly, it becomes necessary to appraise the environmental impacts of current nanomaterials in the environment. Consequently, we have designed this study to assess the ecotoxicity of nanomaterials.

16.2 ECOTOXICITY OF METAL NANOPARTICLES

Metal nanoparticles are commonly formed by noble elements such as Ag, Au, Cu, etc. Silver nanoparticles (Ag-NPs) are useful as disinfectants for the treatment of cancer and as drug carriers, among other functions (Zhang et al. 2021). The toxicity of Ag-NPs can operate through mechanisms that conform to DNA damage, oxidative stress and cytokine induction (Garg et al. 2021). Ferdous and Nemmar (2020) evaluated the health impact of Ag—NPs and highlighted some mammalian organs that target the operation of Ag-NPs toxicity, including the brain, the liver, the vascular system, the reproductive system, the skin and the lung. Chemotherapeutic agent-associated toxicity (i.e., cytotoxicity) has been reported as an active route for Ag-NPs (Mukherjee 2019). Some independent investigations have indicated that the cytotoxicity of Ag-NPs has a strong correlation with the dimension, oxidation state, geometry, zeta potential/surface charge, dosages, degree of agglomeration and nature of the target organism (Dayem et al. 2017). Prasher et al. (2022) also reported that Ag-NPs can be retained in the environment and undergo several changes in chemical and physical properties depending on the influence of biotic and abiotic factors in the ecosystem. Consequently, the presence of organic matter, metal ions and other contaminants can enhance their tendency to progress to their final fate in the environment. Also, Ivask et al. (2014) reported an in vitro investigation that confirmed that the level of toxicity of Ag-NPs to some algae, bacteria, crustaceans and mammals is a function of the size of the nanoparticles.

Some purposeful research considerations have provoked several studies on specific organisms concerning the toxicity of Ag-NPs. For example, Lekamge et al. (2018) investigated the toxicity of Ag-NPs in three species of aquatic animals and observed a wider range of variability in the different species, but a compromised remark indicated that tyrosine-coated Ag-NPs are toxic to *Daphnia carinata, Hydra vulgaris* and *Paratya australiensis.* Based on the work published by Kakakhel et al. (2021) on the toxicity of Ag-NPs to some samples of *C. carpio fish,* 0.09 mg/L of Ag-NPs exerted a toxicity impact on different organs of the fish, especially the liver, the intestinal tract, the gills and the muscles. Confirmed histopathological changes were necrosis and damage to the gills. Judging from the wider application scope of Ag-NPs, Mao et al. (2018) employed a research design for the investigation of the toxicity of Ag-NPs on *Drosophila melanogaster* and reported that Ag-NPs have chronic and acute toxicities on the growth, survival and reactive O_2 induced stress responses of the organism. Delayed development and lethality were manifested in the adult and growing organisms. The organism's tolerance to oxidative stress was also reduced by the lethal dosage of Ag-NPs. Other observed impacts were damage to the DNA, apoptosis and autophagy, among others. Bobyk et al. (2021) remarked that acute and chronic exposure of lung epithelial cells to Ag-NPs inflicted intracellular reactive O_2 accumulation, which was a precursor for the blockage of cell cycle continuity. Alterations in the metabolism of the cell as well as DNA damage were also observed.

Documented evidence has also created a good overview that confirms that gold nanoparticles (Au-NPs) are also toxic. Studies conducted and published by Borase et al. (2019) indicated that Au-NPs exhibited ecotoxicity on *Moinamacrocopa* with an LC_{50} value of about 14 mg/L, causing a decrease in the activity of some enzymes, especially amylase and trypsin. The degree of toxicity was noted to be a concentration-dependent but generally observable increase in the activities of enzymes associated with antioxidation and metal burden. The ecotoxicity of the AU-NPs was mainly aligned with the generation of reactive oxygen species, which arose due to the accumulation of the Au-NPs in the gut of *Moinamacrocopa*. Davarpanah and Guilhermino (2019) also showed that marine algae *Tetraselmis chuii*'s growth rate was severely inhibited by the presence of microplastics. Au-NPs synthesized. Kamaraj et al. (2022) also recorded that various concentrations of AuNP synthesized from the aqueous extract of *G. crassa* displayed high ecotoxicity in *Anopheles stephensi* larvae. The level of ecotoxicity was proportional to the concentration of the Au-NPs. Other reported cases of ecotoxicity are recorded in Table 16.1. It is essential to state that Au-NPs have the tendency to be biomagnified along the food chain, making their toxicity of significant interest (Judy et al. 2011).

Nickel nanoparticles (Ni-NPs) have received significant application in nanomedicine, electrochemical and the magnetic industries. However, cases of reported ecotoxicity of Ni-NPs have been documented for several tested cells and organ systems, including reproductive, cardiovascular and respiratory. Studies have also confirmed signs and symptoms such as damage to the DNA and organs and an increase in the production of reactive oxygen species (Wu and Kong 2020). However, a comprehensive review documented by Meyer et al. (2020) indicated that most reported cases of ecotoxicity from Ni-NPs indicate low levels. Also, most researchers have failed to establish the exact facts on the toxicity of nickel nanoparticles (Kong et al. 2014).

16.3 ECOTOXICOLOGY OF NONMETAL NANOPARTICLES

Some nonmetal nanoparticles have been found to possess unique properties that have given rise to a wider scope of applications. The properties of nanodiamonds (D-NPs) have been found to be related to favourable properties within the nanoframework (Zadeh Mehrizi and Ardestani 2022). However, nanodiamond has been shown to have some levels of ecotoxicology concerning the retardation of the reproductive and developmental progress of *Acheta domesticus* (Karpeta-Kaczmarek et al. 2018). Some impaired antioxidant activities and oxidative stress enzymes in *Acheta*

TABLE 16.1
Ecotoxicology Data for Au-NPs

Toxicity	Target Agent	Reference
No significant toxicity on the activity of the investigative enzymes	Soil microorganism	Maliszewska (2016)
Observed induction of metallothionein, enhanced superoxide dismutase, catalase and glutathione S-transferase activities while oxidative stress and burrowing activities were suppressed	Bivalve *Scrobicularia plana*	Pan et al. (2012)
Significant phytotoxicity was observed	Aquatic macrophytes *Ceratophyllum demersum*	Ostroumov et al. (2014)
Genotoxicity, changes in physicochemical characteristics due to surface functionalization of proteins on VAuNPs and/or and gold bioaccumulation	*Allium cepa* root meristematic cells	Arya et al. (2022)
Reduced aquatic population, impairment of enzyme activity and other ecotoxicological signs	*Macquaria ambigua*	Shuster et al. (2021)
Morphological damage, cytotoxicity on cancer cell	Human hepatocellular cell	Nandhini et al. (2021)
Cytotoxicity on *Folsomia candida* and *Enchytraeuscrypticus*	Soil organisms, *Folsomia candida* and *Enchytraeus crypticus*	Guimarães et al. (2022)
Cytotoxicity observed in the human cell and in the investigated organism	Murine fibroblast cells and *Daphnia magna*	Pem et al. (2019)

domesticus due to exposure to nanodiamonds have been reported. The degree of impairment was observed to be progressively greater as the dosages of the D-NPS increased (Karpeta-Kaczmarek et al. 2016). Increased radical oxygen species (ROS) production and subsequent inducement of oxidative stress and lipid peroxidation in the membrane of enterocytes have also been linked to the ecotoxicity of D-NPs (Domínguez et al. 2018).

Research attention has also been given to the ecotoxicity of carbon nanotubes (CNT). For example, Edgington (2011) enumerated ecotoxicological impacts of CNT on *Daphnia magna* to include oxidative stress inducement, production of ROS, among others. Freixa et al. (2018) concluded that carbon-based nanomaterials are safer at low concentrations but can modify the toxicity of other micropollutants in aquatic systems. Their work also highlighted that the toxicity of carbon-based nanomaterials (including fullerenes, CNT and graphene) is a function of the nature of the organism, concentration and the method used in the preparation. Ziembińska and Łukowiec (2022) have tested the ecotoxicity of some carbon-based nanoparticles (CNT, graphene oxide and reduced graphene oxide) on *Brachionus calyciflorus*, *Artemia franciscana*, *Thamnocephalus platyurus* and *Escherichia coli*. They found that these nanoparticles exhibited toxicity at the three trophic levels, and consumers were more affected than producers. Other impacts were the initiation of ROS and damage to the cell membrane.

Another form of widely applied nonmetal nanoparticles is hydroxyapatite nanoparticles (HA-NPs). Studies conducted by Guimarães et al. (2022) showed that these nanoparticles can affect the reproductive rate of a tested soil organism (*Folsomia candida*) only at higher concentrations and after a prolonged period of exposure. They found that this nanomaterial is relatively less toxic but has the tendency to decrease the size of the animal when capped with some metals, such as titanium. Remya et al. (2017) also confirmed that HA-NPs are non-immunogenic, nontoxic and have a non-significant inflammatory damage inducement. An investigative study conducted by Gao et al. (2018) on the toxicity of HA-NPs using Zebra fish reveals that HA-NPs showed a mild increase in enzyme activity (especially for superoxide dismutase, catalase and peroxidase), but when capped with Cd^{2+} (which is a known environmental contaminants), the ecotoxicity caused damage to the DNA and liver of the fish. This suggests that HA-NPs may be relatively nontoxic, but their tendency to complex with heavy metal ions in the environment can create severe toxicity.

Silica nanoparticles (Si-NPs) can also be considered nonmetal nanoparticles. Consequently, its unique porosity, particle size and others have supported a wider application scope concerning its usage (Garg et al. 2022). Book and Backhaus (2022) linked the ecotoxicity of Si-NPs with the number of OH groups on the surface of the Si-NPs. They observed that algae and bacteria have a protective cell wall that makes them less sensitive to the penetration of Si-NPs. Such a protective wall was also ascribed to the low sensitivity of embryos of fishes, but significant sensitivity to the ecotoxicity of Si-NPs was observed for *Oncorhynchus mykiss* and *Labeorohita* fishes. Wei et al. (2022) investigated the possible impact of Si-NPs on the brain using mice. Intramuscular and intravenous injections of different dosages of Si-NPs were administered at a defined interval, and there was evidence that Si-NPs injected via the intravenous route resulted in induced autophagy in the brain compared to brain apoptosis and autophagy arising from injection through the muscle.

16.4 ECOTOXICITY OF METAL OXIDE AND OTHER NANOPARTICLES

The literature on the ecotoxicological profile of some metal oxide nanoparticles is reviewed in this section. Reported information on other metal-non-metal nanoparticles is also reviewed based on their ecotoxicity records. Bhavya et al. (2016) showed that CaO-NPs showed an ecotoxicological impact on *Cyprinus carpio* (a freshwater fish) at an LC_{50} value of 235 mg/L. Such concentrations do not provide strong support for the placement of CaO-NPs as an eco-toxicant in this regard. The ecotoxicity of CaO-NPs on *Chlorella sp.* concerning antioxidant enzymes, growth and photosynthetic capacity was investigated by Abdel-Kader (2022). The results published by the authors showed no observable ecotoxicity regarding the investigated parameters but rather a significant increase in photosynthesis, growth and antioxidant enzymes when *Chlorella sp.* was exposed to CaO-NPs. Literature reports the beneficial roles of CaO-NPs in remediating the toxicity of barley by the arsenic ion. The arsenic ion was found to impair photosynthesis and the growth rate of barley plants, but these impaired conditions were significantly increased when CaO-NPs were added to the soil (Nazir et al. 2022).

The gerontological impact of Al_2O_3, SiO_2 and TiO_2 on a freshwater fish, *Oreochromis mossambicus*, was investigated by Vidya and Chitra (2017), and they observed signs of acute toxicity after 96 hours of exposure and projected that signs of chronic toxicity are most likely if the exposure period is enhanced. Evaluated lethal concentrations were 40, 120 and 164 mg/L for nanoparticles of Al_2O_3, SiO_2 and TiO_2, respectively. These data portray a significant advancement in the ecotoxicity of the listed metal oxide nanoparticles, especially when compared to the value reported for CaO-NPs. Ahn et al. (2022) showed that AlO-NPs can cause a severe ecotoxicological impact on freshwater microalgae, *Scenedesmus obliquus*. The nanoparticles reduced the chlorophyll and carotenoid content of the algae by 23% and 22%, respectively. The study also reveals that the activity of the genes responsible for ATP synthesis was lowered while that of those that biodegrade xenobiotic chemicals was energized, which explains why the ecotoxicology of sulfacetamide in the same medium was impaired by Al-NPs. The enterotoxin effect of Al-NPs against *Sitophilus oryzae* and the subsequent toxicological effect on albino rats were investigated by Ismail et al. (2021). At 1,000 mg/L of Al-NPs, all the plant species were dead. The nanoparticles also impacted subacute toxicity by increasing the weight of the liver, although they enhanced the concentrations of alanine aminotransferase, aspartate aminotransferase and creatinine.

The ecotoxicity of Ti-NPs on *Chlorella pyrenoidosa* and freshwater Zebra fish was investigated, and the observed data showed acute toxicity symptoms in *C. pyrenoidosa*. Cytotoxicity effects were also observed in both species at reduced particle sizes of the nanomaterials (AL-Ammari et al. 2021). Vijayaraj et al. (2018) conducted a study on the ecotoxicity of Ti-NPs for a series of organisms within the aquatic and terrestrial ecosystems. The summary of the study indicated depletion of bacterial and archaeal amounts with the abundance of nitrifiers. Also observed was the enhancement of the translocation of the nanoparticles from the root to the leaves of *Medicago truncatula*, followed by a reduction in plant height, leaf number and biomass above ground. In the snail, uptake of the nanoparticles by the foot and viscera and excretion in faeces were observed, while genotoxicity was directed to *Xenopus laevis* larvae.

Zinc oxide nanoparticles (ZnO-NPs) are among others that have wider applications in environmental, industrial and other purposes. Some ecotoxicological reports have been documented concerning ZnO-NPs. Wu et al. (2019) investigated the toxicity of ZnO-NPs on different organisms, namely, *Chlamydomonas reinhardtii* (algae), *Escherichia coli* (a bacteria), *Daphnia magna* (a crustacean) and *Danio rerio* (Zebrafish). Their results showed a low penetrating rate and less toxicity that was independent of the size of the ZnO-NPs but a function of the variability in the species. However, while investigating the ecotoxicity of ZnO-NPs on the amphypod *Corophium insidiosum* and copepod *Tigriopus fulvus*, Vimercati et al. (2020) established a mild on the tested organisms. Some published works have indicated that ZnO-NPs can exert cytotoxicity and genotoxicity if ROS is generated (Coll et al. 2016). On this basis, ZnO-NPs are considered to be the most toxic metal oxide nanoparticles, especially when interacting with some environmental variables and having a reduced particle size. Reported cases of neurotoxicity due to exposure of some organisms to ZnO-NPs have also been reported (Schulte et al. 2018). Mohammad et al. (2021) have also shown that ZnO-NPs exhibited a severe ecotoxicological impact on the terrestrial slug *Lehmannia nyctelia*, such that significant histological modifications were recorded.

Ecotoxicological investigations of magnesium oxide nanoparticles (MgO-NPs) have also been reported. However, the adverse impact of MgO-NPs has not been extensively documented. For example, Rempel et al. (2020) deduced from their experimental results that MgO-NPs are not toxic in the terrestrial environment. A similar conclusion on the non-ecotoxicity of MgO-NPs was deduced by Mazaheri et al. (2019) from their experimental results, rather than increasing red and white blood cells as well as haemoglobin and haematocrit. Also observed was the enhancement of alkaline phosphatase, aminotransferase profile and aspartate, although the levels of alanine aminotransferase were significantly unchanged. Rempel et al. (2020) also reported the low ecotoxicological effect of MgO-NPs on some edaphic organisms, namely, *Enchytraeus crypticus* (enchytraeids), *Eisenia andrei* (earthworm) and *Folsomia candida* (springtails). Similar findings of the low toxicity of MgO-Ps have also been reported by Wang et al. (2022) on the oomycete pathogen *Phytophthora infestans* and its host *Solanum tuberosum*. On the contrary, studies conducted by Mangalampalli et al. (2017) using Wister rats to test the ecotoxicity of MgO-NPs indicated that at excessive concentrations, MgO-NPs cause DNA damage and change the biochemistry of the body system. Also observed were induced oxidative stress and significant accumulation of Mg ions in the tissues of the kidney and liver as well as in the urine.

Fe_3O_4 nanoparticles have wider areas of application but are most prominent in magnetic resonance imaging. Drug and gene carriers, in the treatment of hyperthermia-related cancer, as a micronutrient in fertilizers, etc. (Ali et al. 2016). Studies have shown that the toxicity of iron oxide nanoparticles (FeO-NPs) varies with size, with the smaller particles exhibiting a higher ecotoxicity tendency. Guided by this principle, Wu et al. (2022) carried out an ecotoxicological study (using mice) to investigate the effect of different sizes of FeO-NPs and deduced from their results that a severe ecotoxicological profile was recorded for the nanoparticles with smaller particle sizes (2.3–4.2 nm), and the manifestations were oxidative stress, an increase in ROS, heart failure, etc.

16.5 CONCLUSIONS

In this review, the ecotoxicity of several metals, nonmetal and metal oxide nanoparticles has been reviewed. The review reveals the following,

i. The ecotoxicity profile of the investigated nanomaterials showed an increasing degree as the particle size of the nanomaterials decreased.
ii. The ecotoxicity of nanoparticles shows varying degrees of strength for the same nanoparticles synthesized by different methods.
iii. The ecotoxicity of the investigated nanoparticles shows a toxicity range covering cytotoxicity, neurotoxicity, phytotoxicity, genotoxicity and immunotoxicity depending on concentration, particle size and other variables.
iv. The nanoparticles based on Fe, Ca, Mg and a few others have a low ecotoxicological impact.

Given the above and the ever-expanding research and applications of nanoparticles, we recommend that scientists include ecotoxicological assays during research aimed at synthesizing or applying nanoparticles. The need for a review of the ecotoxicology of nanocomposite is also highlighted for further engagement.

FUNDING

This work is fully sponsored by the Tertiary Education Trust Fund of Nigeria through the National Research Fund (NRF), with Prof. NnabukOkon Eddy as the Principal Investigator and others as team members.

REFERENCES

Abdel-Kader, Huwida. 2022. "Short- and Long-Terms Stimulatory Impacts of Calcium Oxide Nanoparticles (CaONPs) on the Growth, Photosynthesis and Antioxidant Enzymes of Chlorella Sp." *Assiut University Journal of Multidisciplinary Scientific Research* 1 (1): 242–62.

Ahn, Hyun Jo, Yongtae Ahn, Mayur B. Kurade, Swapnil M. Patil, Geon Soo Ha, Paul O. Bankole, Moonis A. Khan, et al. 2022. "The Comprehensive Effects of Aluminum Oxide Nanoparticles on the Physiology of Freshwater Microalga Scenedesmus Obliquus and It's Phycoremediation Performance for the Removal of Sulfacetamide." *Environmental Research* 215: 114314.

AL- Ammari, Abdulrahman, Lei Zhang, Jiazhi Yang, Feng Wei, Chuntao Chen, and Dongping Sun. 2021. "Toxicity Assessment of Synthesized Titanium Dioxide Nanoparticles in Fresh Water Algae Chlorella Pyrenoidosa and a Zebrafish Liver Cell Line." *Ecotoxicology and Environmental Safety* 211: 111948.

Ali, Attarad, Hira Zafar, Muhammad Zia, Ihsan ul Haq, Abdul Rehman Phull, Joham Sarfraz Ali, and Altaf Hussain. 2016. "Synthesis, Characterization, Applications, and Challenges of Iron Oxide Nanoparticles." *Nanotechnology, Science and Applications* 9: 49–67.

Arya, Sagar S., James E. Rookes, David M. Cahill, and Sangram K. Lenka. 2022. "Reduced Genotoxicity of Gold Nanoparticles with Protein Corona in Allium Cepa." *Frontiers in Bioengineering and Biotechnology* 10 (April): 1–15.

Bhavya, Channappa , Kambalagere Yogendra, K. M. Mahadevan, Narayanappa Madhusudhana. 2016. "Synthesis of Calcium Oxide Nanoparticles and Its Mortality Study on Fresh Water Fish Cyprinus Carpio." *IOSR Journal of Environmental Science* 10 (January): 55–60.

Bobyk, Laure, Adeline Tarantini, David Beal, Giulia Veronesi, Isabelle Kieffer, Sylvie Motellier, Eugenia Valsami-Jones, et al. 2021. "Toxicity and Chemical Transformation of Silver Nanoparticles in A549 Lung Cells: Dose-Rate-Dependent Genotoxic Impact." *Environmental Science: Nano* 8 (3): 806–21.

Book, Frida, and Thomas Backhaus. 2022. "Aquatic Ecotoxicity of Manufactured Silica Nanoparticles: A Systematic Review and Meta-Analysis." *Science of the Total Environment* 806: 150893.

Borase, Hemant P., Abhijeet B. Muley, Satish V. Patil, and Rekha S. Singhal. 2019. "Nano-Eco Toxicity Study of Gold Nanoparticles on Aquatic Organism Moina Macrocopa: As New Versatile Ecotoxicity Testing Model." *Environmental Toxicology and Pharmacology* 68: 4–12.

Boros, Bianca Vanesa, and Vasile Ostafe. 2020. "Evaluation of Ecotoxicology Assessment Methods of Nanomaterials and Their Effects." *Nanomaterials* 10 (4): 610.

Coll, Claudia, Dominic Notter, Fadri Gottschalk, Tianyin Sun, Claudia Som, and Bernd Nowack. 2016. "Probabilistic Environmental Risk Assessment of Five Nanomaterials (Nano-TiO2, Nano-Ag, Nano-ZnO, CNT, and Fullerenes)." *Nanotoxicology* 10 (4): 436–44.

Davarpanah, Elham, and Lúcia Guilhermino. 2019. "Are Gold Nanoparticles and Microplastics Mixtures More Toxic to the Marine Microalgae Tetraselmis Chuii than the Substances Individually?" *Ecotoxicology and Environmental Safety* 181 (June): 60–68.

Dayem, Ahmed Abdal, Mohammed Kawser Hossain, Soo Bin Lee, Kyeongseok Kim, Subbroto Kumar Saha, Gwang Mo Yang, Hye Yeon Choi, and Ssang Goo Cho. 2017. "The Role of Reactive Oxygen Species (ROS) in the Biological Activities of Metallic Nanoparticles." *International Journal of Molecular Sciences* 18 (1): 1–21.

Domínguez, Gustavo A., Marco D. Torelli, Joseph T. Buchman, Christy L. Haynes, Robert J. Hamers, and Rebecca D. Klaper. 2018. "Size Dependent Oxidative Stress Response of the Gut of Daphnia Magna to Functionalized Nanodiamond Particles." *Environmental Research* 167 (March): 267–75.

Eddy, Nnabuk Okon, and Rajni Garg. 2022. "CaO Nanoparticles: Synthesis and Applications in Water Remediation." In *Handbook of Research on Green Synthesis and Applications of Nanomaterials*, edited by Rajni Garg, Rishav Garg and nnabuk Okon Eddy, 247–68. IGI Global.

Eddy, Nnabuk Okon, Richard Alexis Ukpe, Paul Ameh, Rapheal Ogbodo, Rajni Garg, and Rishav Garg. 2022. "Theoretical and Experimental Studies on Photocatalytic Removal of Methylene Blue (MetB) from Aqueous Solution Using Oyster Shell Synthesized CaO Nanoparticles (CaONP-O)." *Environmental Science and Pollution Research.* doi.10.1007/s11356-022-22747-w.

Ferdous, Zannatul, and Abderrahim Nemmar. 2020. Health Impact of Silver Nanoparticles: A Review of the Biodistribution and Toxicity Following Various Routes of Exposure. *International Journal of Molecular Sciences* 21 (7): 2375.

Freixa, Anna, Vicenç Acuña, Josep Sanchís, Marinella Farré, Damià Barceló, and Sergi Sabater. 2018. "Ecotoxicological Effects of Carbon Based Nanomaterials in Aquatic Organisms." *Science of the Total Environment* 619–620: 328–37.

Gao, Minling, Yujuan Yang, Mengting Lv, Wenhua Song, and Zhengguo Song. 2018. "Oxidative Stress and DNA Damage in Zebrafish Liver Due to Hydroxyapatite Nanoparticles-Loaded Cadmium." *Chemosphere* 202: 498–505.

Garg, Rajni, Rishav Garg, and Nnabuk Okon Eddy. 2022. *Handbook of Research on Green Synthesis and Applications of Nanomaterials*, edited by Rajni Garg, Rishav Garg, and Nnabuk Okon Eddy. Advances in Chemical and Materials Engineering. IGI Global.

Garg, Rishav, Rajni Garg, Nnabuk Okon Eddy, Abdulaziz Ibrahim Almohana, Sattam Fahad Almojil, Mohammad Amir Khan, and Seung Ho Hong. 2022. "Biosynthesized Silica-Based Zinc Oxide Nanocomposites for the Sequestration of Heavy Metal Ions from Aqueous Solutions." *Journal of King Saud University: Science* 34 (4): 101996.

Garg, Rishav Rajni, Priya Rani, Rishav Rajni Garg, and Nnabuk Okon Eddy. 2021. "Study on Potential Applications and Toxicity Analysis of Green Synthesized Nanoparticles." *Turkish Journal of Chemistry* 45 (6): 1690–1706.

Guimarães, Bruno, Susana I.L. Gomes, Janeck J. Scott-Fordsmand, and Mónica J.B. Amorim. 2022. "Impacts of Longer-Term Exposure to AuNPs on Two Soil Ecotoxicological Model Species." *Toxics* 10 (4): 153.

Hammami, Inès, Nadiyah M. Alabdallah, Amjad Al jomaa, and Madiha kamoun. 2021. "Gold Nanoparticles: Synthesis Properties and Applications." *Journal of King Saud University - Science* 33 (7): 101560.

Ismail, Tamer, Mohamed A. Salama, and Mostafa El-Ebiary. 2021. "Entomotoxic Effects of Synthesized Aluminum Oxide Nanoparticles against Sitophilus Oryzae and Their Toxicological Effects on Albino Rats." *Toxicology and Industrial Health* 37 (10): 594–602.

Ivask, Angela, Imbi Kurvet, Kaja Kasemets, Irina Blinova, Villem Aruoja, Sandra Suppi, Heiki Vija, et al. 2014. "Size-Dependent Toxicity of Silver Nanoparticles to Bacteria, Yeast, Algae, *Crustaceans and Mammalian Cells in Vitro." PLoS One* 9 (7): e102108.

Judy, Jonathan D., Jason M. Unrine, and Paul M. Bertsch. 2011. "Evidence for Biomagnification of Gold Nanoparticles within a Terrestrial Food Chain." *Environmental Science and Technology* 45 (2): 776–781.

Kakakhel, Mian Adnan, Fasi Wu, Wasim Sajjad, Qi Zhang, Ikram Khan, Kalim Ullah, and Wanfu Wang. 2021. "Long-Term Exposure to High-Concentration Silver Nanoparticles Induced Toxicity, Fatality, Bioaccumulation, and Histological Alteration in Fish (Cyprinus Carpio)." *Environmental Sciences Europe* 33 (1). doi10.1186/s12302-021-00453-7.

Kamaraj, Chinnaperumal, Sengodan Karthi, Appadurai Daniel Reegan, Govindasamy Balasubramani, Govindaraju Ramkumar, Kandaswamy Kalaivani, A. Abduz Zahir, et al. 2022. "Green Synthesis of Gold Nanoparticles Using Gracilaria Crassa Leaf Extract and Their Ecotoxicological Potential: Issues to Be Considered." *Environmental Research* 213: 113711.

Karpeta-Kaczmarek, Julia, Marta Dziewięcka, Maria Augustyniak, Magdalena Rost-Roszkowska, and Mirosława Pawlyta. 2016. "Oxidative Stress and Genotoxic Effects of Diamond Nanoparticles." *Environmental Research* 148: 264–72.

Karpeta-Kaczmarek, Julia, Andrzej Kędziorski, Maria A. Augustyniak-Jabłokow, Marta Dziewięcka, and Maria Augustyniak. 2018. "Chronic Toxicity of Nanodiamonds Can Disturb Development and Reproduction of Acheta Domesticus L." *Environmental Research* 166 (February): 602–9.

Edgington, Aaron. 2011. "The Toxicity of Carbon Nanotubes to Daphnia magna." All Dissertations. 864

Kong, Lu, Meng Tang, Ting Zhang, Dayong Wang, Ke Hu, Weiqi Lu, Chao Wei, Geyu Liang, and Yuepu Pu. 2014. "Nickel Nanoparticles Exposure and Reproductive Toxicity in Healthy Adult Rats." *International Journal of Molecular Sciences* 15 (11): 21253–69.

Lekamge, Sam, Ana F. Miranda, Amanda Abraham, Vivian Li, Ravi Shukla, Vipul Bansal, and Dayanthi Nugegoda. 2018. "The Toxicity of Silver Nanoparticles (AgNPs) to Three Freshwater Invertebrates with Different Life Strategies: Hydra Vulgaris, Daphnia Carinata, and Paratya Australiensis." *Frontiers in Environmental Science* 6 (December): 1–13.

Maliszewska, Irena. 2016. "Effects of the Biogenic Gold Nanoparticles on Microbial Community Structure and Activities." *Annals of Microbiology* 66 (2): 785–94.

Mangalampalli, Bhanuramya, Naresh Dumala, and Paramjit Grover. 2017. "Acute Oral Toxicity Study of Magnesium Oxide Nanoparticles and Microparticles in Female Albino Wistar Rats." *Regulatory Toxicology and Pharmacology* 90: 170–84.

Mao, Bin-Hsu, Zi-Yu Chen, Ying-Jang Wang, and Shian-Jang Yan. 2018. "Silver Nanoparticles Have Lethal and Sublethal Adverse Effects on Development and Longevity by Inducing ROS-Mediated Stress Responses." *Scientific Reports* 8 (1): 1–16.

Mazaheri, Nafiseh, Nooshin Naghsh, Akbar Karimi, and Hossein Salavati. 2019. "In Vivo Toxicity Investigation of Magnesium Oxide Nanoparticles in Rat for Environmental and Biomedical Applications." *Iranian Journal of Biotechnology* 17 (1): 1–9.

Meyer, Joseph S., Tara Lyons-Darden, Emily R. Garman, Elizabeth T. Middleton, and Christian E. Schlekat. 2020. "Toxicity of Nanoparticulate Nickel to Aquatic Organisms: Review and Recommendations for Improvement of Toxicity Tests." *Environmental Toxicology and Chemistry* 39 (10): 1861–83.

Mohammad, Wafaa A., Safaa M. Ali, Nasser Farhan, and Shimaa Mohamed Said. 2021. "The Toxic Effect of Zinc Oxide Nanoparticles on the Terrestrial Slug Lehmannia Nyctelia (Gastropoda-Limacidae)." *The Journal of Basic and Applied Zoology* 82 (1): 1–9.

Mukherjee, Pulok K. 2019. "Quality Evaluation of Herbal Medicines: Challenges and Opportunities." In *Quality Control and Evaluation of Herbal Drugs*, 53–77. Elsevier.

Nandhini, Joseph T., Devaraj Ezhilarasan, and Shanmugam Rajeshkumar. 2021. "An Ecofriendly Synthesized Gold Nanoparticles Induces Cytotoxicity via Apoptosis in HepG2 Cells." *Environmental Toxicology* 36 (1): 24–32.

Nazir, Muhammad Mudassir, Qi Li, Muhammad Noman, Zaid Ulhassan, Shafaqat Ali, Temoor Ahmed, Fanrong Zeng, and Guoping Zhang. 2022. "Calcium Oxide Nanoparticles Have the Role of Alleviating Arsenic Toxicity of Barley." *Frontiers in Plant Science* 13 (March): 843795.

Ostroumov, S. A., V. A. Poklonov, S. V. Kotelevtsev, and S. N. Orlov. 2014. "Toxicity of Gold Nanoparticles for Plants in Experimental Aquatic System." *Moscow University Biological Sciences Bulletin* 69 (3): 108–12.

Pan, Jin-Fen, Pierre-Emmanuel Buffet, Laurence Poirier, Claude Amiard-Triquet, Douglas Gilliland, Yolaine Joubert, Paul Pilet, et al. 2012. "Size Dependent Bioaccumulation and Ecotoxicity of Gold Nanoparticles in an Endobenthic Invertebrate: The Tellinid Clam Scrobicularia Plana." *Environmental Pollution* 168: 37–43.

Pem, Barbara, Igor M. Pongrac, Lea Ulm, Ivan Pavičić, Valerije Vrček, Darija Domazet Jurašin, Marija Ljubojević, Adela Krivohlavek, and Ivana Vinković Vrček. 2019. "Toxicity and Safety Study of Silver and Gold Nanoparticles Functionalized with Cysteine and Glutathione." *Beilstein Journal of Nanotechnology* 10 (September): 1802–17.

Prasher, Parteek, Mousmee Sharma, Harish Mudila, Amit Verma, and Pankaj Bhatt. 2022. "Silver Nanoparticles in Natural Ecosystems: Fate, Transport, and Toxicity." In *Green Synthesis of Silver Nanomaterials*, edited by Kamel A. Abd-Elsalam, 649–68. Elsevier.

Rempel, Suzana, André Júnior Ogliari, Elizeu Bonfim, Glaucea Warmeling Duarte, Humberto Gracher Riella, Luciano Luiz Silva, Josiane Maria Muneron de Mello, Carolina Riviera Duarte Maluche Baretta, and Márcio Antônio Fiori. 2020. "Toxicity Effects of Magnesium Oxide Nanoparticles: A Brief Report." *Revista Materia* 25 (4): 1–9.

Remya, N. S., S. Syama, A. Sabareeswaran, and P. V. Mohanan. 2017. "Investigation of Chronic Toxicity of Hydroxyapatite Nanoparticles Administered Orally for One Year in Wistar Rats." *Materials Science and Engineering C* 76: 518–27.

Schulte, Paul A., Veruscka Leso, M. Niang, and Ivo Iavicoli. 2018. "Biological Monitoring of Workers Exposed to Engineered Nanomaterials." *Toxicology Letters* 298 (December): 112–24.

Shuster, Jeremiah, Maria A.D. Rea, Bhanu Nidumolu, and Anupama Kumar. 2021. " Toxicity Assessment of Gold Ions and Gold Nanoparticles to Golden Perch Larvae (Macquaria Ambigua)." *Mineralogical Magazine* 85 (1): 94–104.

Vidya, P. V., and K. C. Chitra. 2017. "Assessment of Acute Toxicity (LC 50-96 h) of Aluminium Oxide, Silicon Dioxide and Titanium Dioxide Nanoparticles on the Freshwater Fish, Oreochromis." *International Journal of Fisheries and Aquatic Studies* 5 (1): 327–32.

Vijayaraj, Vinita, Clarisse Liné, Stéphanie Cadarsi, Clément Salvagnac, David Baqué, Arnaud Elger, Maialen Barret, Florence Mouchet, and Camille Larue. 2018. "Transfer and Ecotoxicity of Titanium Dioxide Nanoparticles in Terrestrial and Aquatic Ecosystems: A Microcosm Study." *Environmental Science and Technology* 52 (21): 12757–64.

Vimercati, Luigi, Domenica Cavone, Antonio Caputi, Luigi De Maria, Michele Tria, Ermelinda Prato, and Giovanni Maria Ferri. 2020. "Nanoparticles: An Experimental Study of Zinc Nanoparticles Toxicity on Marine Crustaceans. General Overview on the Health Implications in Humans." *Frontiers in Public Health* 8 (May): 1–19.

Wang, Ze Le, Xi Zhang, Guang-Jin Fan, Yi Que, Feng Xue, and Ying-Hong Liu. 2022. "Toxicity Effects and Mechanisms of MgO Nanoparticles on the Oomycete Pathogen Phytophthora Infestans and Its Host Solanum Tuberosum." *Toxics* 10 (10): 553.

Wei, Wei, Ziyi Yan, Xuting Liu, Zongming Qin, Xiaoqi Tao, Xiaokang Zhu, Erqun Song, et al. 2022. "Brain Accumulation and Toxicity Profiles of Silica Nanoparticles: The Influence of Size and Exposure Route." *Environmental Science and Technology* 56 (12): 8319–25.

Wu, Fan, Bryan J. Harper, and Stacey L. Harper. 2019. "Comparative Dissolution, Uptake, and Toxicity of Zinc Oxide Particles in Individual Aquatic Species and Mixed Populations." *Environmental Toxicology and Chemistry* 38 (3): 591–602.

Wu, Lin, Wen Wen, Xiaofeng Wang, Danhua Huang, Jin Cao, Xueyong Qi, and Song Shen. 2022. "Ultrasmall Iron Oxide Nanoparticles Cause Significant Toxicity by Specifically Inducing Acute Oxidative Stress to Multiple Organs." *Particle and Fibre Toxicology* 19 (1): 1–14.

Wu, Yongya, and Lu Kong. 2020. "Advance on Toxicity of Metal Nickel Nanoparticles." *Environmental Geochemistry and Health* 42 (7): 2277–86.

Zadeh Mehrizi, Tahereh, and Mehdi Shafiee Ardestani. 2022. "Application of Non-Metal Nanoparticles, as a Novel Approach, for Improving the Stability of Blood Products: 2011-2021." *Progress in Biomaterials* 11 (2): 137–61.

Zhang, Fan, Zhuang Wang, Willie J.G.M. Peijnenburg, and Martina G. Vijver. 2022. "Review and Prospects on the Ecotoxicity of Mixtures of Nanoparticles and Hybrid Nanomaterials." *Environmental Science and Technology* 56 (22): 15238–50.

Zhang, Jie, Sijin Liu, Jinxiang Han, Zhe Wang, and Shuping Zhang. 2021. "On the Developmental Toxicity of Silver Nanoparticles." *Materials and Design* 203: 109611.

Index

EXTENDED REALITY FOR HEALTHCARE SYSTEMS

Extended Reality for HEALTHCARE SYSTEMS

Recent Advances in Contemporary Research

Edited by

SAMIYA KHAN

Postdoctoral Research Fellow, Digital Innovation, School of Mathematics and Computer Science, University of Wolverhampton, Wolverhampton, United Kingdom

MANSAF ALAM

Associate Professor, Department of Computer Science, Jamia Millia Islamia, Okhla, New Delhi, India

SHOAIB AMIN BANDAY

Assistant Professor, Islamic University of Science and Technology, Awantipora, Jammu and Kashmir, India

MOHAMMED SHAUKAT USTA

Advisor, Ministry of Health and Family Welfare, Government of India, India

ELSEVIER

ACADEMIC PRESS

An imprint of Elsevier

Academic Press is an imprint of Elsevier
125 London Wall, London EC2Y 5AS, United Kingdom
525 B Street, Suite 1650, San Diego, CA 92101, United States
50 Hampshire Street, 5th Floor, Cambridge, MA 02139, United States
The Boulevard, Langford Lane, Kidlington, Oxford OX5 1GB, United Kingdom

Notices

Knowledge and best practice in this field are constantly changing. As new research and experience broaden our understanding, changes in research methods, professional practices, or medical treatment may become necessary.

Practitioners and researchers must always rely on their own experience and knowledge in evaluating and using any information, methods, compounds, or experiments described herein. In using such information or methods they should be mindful of their own safety and the safety of others, including parties for whom they have a professional responsibility.

To the fullest extent of the law, neither the Publisher nor the authors, contributors, or editors, assume any liability for any injury and/or damage to persons or property as a matter of products liability, negligence or otherwise, or from any use or operation of any methods, products, instructions, or ideas contained in the material herein.

ISBN: 978-0-323-98381-5

For information on all Academic Press publications visit our website at https://www.elsevier.com/books-and-journals

Publisher: Mara E. Conner
Acquisitions Editor: Yura R. Sonnini
Editorial Project Manager: Zsereena Rose Mampusti
Production Project Manager: Prasanna Kalyanaraman
Cover Designer: Christian J. Bilbow

Typeset by TNQ Technologies

Contents

Contributors

Shamshad Alam
Department of Electronics and Communication Engineering, Dr. B R Ambedkar NIT Jalandhar, Punjab, India

Anshita
Institute of Informatics and Communication, University of Delhi, Delhi, India

Ashima Arya
Department of Information Technology, Samalkha, Haryana, India

Shoaib Amin Banday
Department of Electronics and Communication, Islamic University of Science and Technology, Awantipora, Jammu and Kashmir, India

Neha Bhatia
Department of Information Technology, Samalkha, Haryana, India

Agrima Bhatt
School of Biology, MIT World Peace University, Pune, Maharashtra, India

Manish Biyani
Department of Biotechnology, Biyani Girls College, Jaipur, Rajasthan, India; Department of Bioscience and Biotechnology, Japan Advanced Institute of Science and Technology, Ishikawa, Japan

Tabasum Rasool Dar
Interdisciplinary Centre for Water Research (ICWaR), Indian Institute of Science, Bangalore, Karnataka, India

Gargi Dubey
Department of Biotechnology, Biyani Girls College, Jaipur, Rajasthan, India

Umar Farooq
Department of Electronics and Communication, Islamic University of Science and Technology, Awantipora, Jammu and Kashmir, India

S Roohan Farooq Lala
Department of Materials Engineering, IISC Banglore, Karnataka, India

Hema Garg
School of Interdisciplinary Research, Indian Institute of Technology Delhi, New Delhi, India

Rashi Garg
Department of Genomic Center, Bioinformatics, All India Institute of Medical Sciences, New Delhi, India

Usharani Hareesh Govindarajan
Business School, University of Shanghai for Science and Technology, Shanghai, China

Mohd Javaid
Department of Mechanical Engineering Jamia Millia Islamia, New Delhi, India

Nirat Kandwani
School of Humanities, Mohanlal Sukhadia University, Udaipur, Rajasthan, India

Hushmat Amin Kar
Department of Information Technology, National Institute of Technology, Srinagar, Jammu and Kashmir, India

Ab Rouf Khan
VIT Bhopal University, Ashta, Madhya Pradesh, India

Samiya Khan
Faculty of Science & Engineering, University of Wolverhampton, Wolverhampton, United Kingdom

Ibrahim Haleem Khan
College of Engineering, Northeastern University, Boston, Massachusetts, United States

Shahbaz Khan
Institute of Business Management, GLA University, Mathura, Uttar Pradesh, India

Deepika Koundal
School of Computer Science, University of Petroleum & Energy Studies, Dehradun, Uttarakhand, India

Tarun Kumar Kumawat
Department of Biotechnology, Biyani Girls College, Jaipur, Rajasthan, India; Department of Botany, University of Rajasthan, Jaipur, Rajasthan, India

Sapna Juneja
Department of Computer Science, KIET Group of Institutions, Ghaziabad, Uttar Pradesh, India

Mitu Sehgal
Department of Information Technology, Samalkha, Haryana, India

Tawseef Ayoub Shaikh
Department of Computer Science & Engineering, Baba Ghulam Shah Badshah University, Rajouri, Jammu & Kashmir, India

Vishnu Sharma
Department of Biotechnology, Biyani Girls College, Jaipur, Rajasthan, India

Shabir Sofi
ITE Department, National Institute of Technology, Srinagar, Jammu & Kashmir, India

Viraj Uttamrao Somkuwar
Department of Textile and Fibre Engineering, Indian Institute of Technology Delhi, New Delhi, India

Neetu Sood
Department of Electronics and Communication Engineering, Dr. B R Ambedkar NIT Jalandhar, Punjab, India

Rajiv Suman
Department of Industrial & Production Engineering, G.B. Pant University of Agriculture & Technology, Uttarakhand, India

Sonali Vyas
University of Petroleum and Energy Studies, Dehradun, Uttarakhand, India

Amarah Zahra
Department of Electronics and Communication Engineering, Dr. B R Ambedkar NIT Jalandhar, Punjab, India

Dali Zhang
Sino US Global Logistics Institute, Antai College of Economics and Management, Shanghai Jiao Tong University, Shanghai, China

Preface

Transformative healthcare technologies have really found a footing for themselves in the last few years. This change can be attributed to the fact that COVID-19 pandemic has forced healthcare facilities to rethink and accelerate their journeys of digital transformation. Most efforts in this regard have been focused on understanding and mitigating the impact of COVID-19 on the manner in which healthcare services are conventionally provided.

One of the most crucial needs of the hour is to deploy and enable an infrastructure that can allow remote access to varied healthcare services. AR and VR can play an instrumental role in backing this use case. Having said that, AR and VR was already having a significant impact on the healthcare market even before the pandemic came around. This technology is expected to demonstrate remarkable growth with the AR and VR healthcare market expected to grow over USD $11 billion by 2024.

In addition to the healthcare use case, AR/VR has found significant attention with the sales for headsets rising by over 350% during lockdowns and COVID-19 restrictions. This technology has served as the escape route and anxiety management solution for millions around the world. With the expanding research in this field, the market for XR in not just healthcare but also related avenues is only expected to grow in the near future with many academic institutions and SMEs (Small- and Medium-sized Enterprises) investing in innovation, research, and development in this field. Preliminary trials provide evidence to the efficacy of XR-backed approaches, also manifesting potential cost savings. It is noteworthy that both application-level innovation and development of an evaluation framework for XR approaches are in their nascent stages.

For most healthcare organizations, one of the biggest drivers for technology adoption is cost savings and the fact that any expenditure must be valued for money. This aspect of XR adoption has been investigated by (this report) and it was found that XR-backed services provide a more cost-effective way to providing the same services with significant improvement of achieved outcomes. There are no two ways about that fact that delivering therapies in a remote fashion incurs much lesser costs with costs reduced by 50% and more. In addition, remote therapies are lesser traumatic for the patient considering that there are no wait times.

Another important aspect of healthcare that can substantially benefit from XR is education and training. The cost of training is reduced with estimates that the posttraining performance is improved by more than 200%. While costs are a critical aspect of any service provider, performance and effectiveness cannot be compromised, particularly in healthcare, which makes evaluation of XR applications quintessential for adoption.

One of the most profound reasons for limited research and development in the XR sector is the need for cross-domain and interdisciplinary collaborations. There is no way technologist can create XR applications for application-specific domains without the support of their respective domain experts. This also provides reasoning for the lack of a marketplace for distribution of XR solutions in the healthcare sector. It is extremely hard for products to reach the stage of clinical trials, which is the most crucial evidence to drive adoption. As a result, it is extremely hard for these products to find their route to market.

This book provides an elaborate introduction to extended reality and its multidimensional applications in healthcare. It also attempts to explore the potential value of this technological intervention for future research, potential impacts, and benefits of XR adoption in healthcare. Finally, it also looks at use cases and case studies around the existing applications of XR in healthcare.

Samiya Khan
Mansaf Alam
Shoaib Amin Banday
Mohammed Shaukat Usta

Acknowledgments

We would like to take this opportunity to thank everyone involved in the making of this book. Our sincerest thanks to our respective families, friends, and colleagues for their unconditional support and encouragement. We extend our gratefulness to all the contributing authors who took out time from their busy schedules to provide their valuable inputs to this book.

We would also like to thank the Elsevier team for their support and expert help on the different aspects of the publication process.

Samiya Khan
Mansaf Alam
Shoaib Amin Banday
Mohammed Shaukat Usta

CHAPTER ONE

Extended reality: bringing the 3Rs together

Samiya Khan
Faculty of Science & Engineering, University of Wolverhampton, Wolverhampton, United Kingdom

1.1 Introduction

Extended reality (XR) is an umbrella term that jointly includes augmented (AR), virtual (VR), and mixed realities (MR). The past few decades have witnessed remarkable advancements in research and technology adoption for these fields. The multifaceted benefits of using XR technologies ranging from developing user-friendly applications to streamlining business processes to improve productivity and efficiency have led to extensive industry investments and research efforts in this domain (Du et al., 2018; Miettinen and Paavola, 2014; Stanton et al., 2020; Radianti et al., 2020; Wang et al., 2018).

XR technology has lately garnered immense attention with sale of XR devices expected to rise to USD $9.1 billion by 2021 from USD $1.5 billion in 2017 (Flavián et al., 2019). The fact that this technology enables a unique user experience makes it not just useful, but it also makes this technology extremely fascinating. Conventionally, this technology has most extensively been used in the gaming industry. However, recent advances have broadened the realm to education, businesses, and healthcare sectors, in addition to many others (Arnaldi et al., 2018). The growing market value and use-case-based applications in diverse domains make XR a technology for the future.

The XR technology in the most basic way can support use cases for remote collaboration, experience, and intervention. Therefore, any use case that requires human intervention can greatly benefit from this technology with the added benefit that the individual need not be physically present onsite. This inevitably reduces costs and saves time. In addition to mediating the constraints of space and location, the use of these technologies has additional advantages such as safety and resource optimization.

Existing literature on XR research (Radianti et al., 2020; Zahabi and Abdul Razak, 2020; Seymour et al., 2018; Yan et al., 2011; Nikolić et al., 2019) focuses on one of the component technologies (AR, VR, or MR) or the use of XR for specific use cases. This chapter aims to provide the theoretical and methodological overview of the three component technologies, also providing insights on their evolution and applications. This chapter provides a literature review on XR and the three constituent technologies elaborating on the foundations and applications of these technologies.

Extended Reality for Healthcare Systems, Volume 1
ISBN 978-0-323-98381-5, https://doi.org/10.1016/B978-0-323-98381-5.00014-3

The rest of the chapter is organized in the following manner: Section 1.2 provides history of XR, while Section 1.3 introduces the three main Rs including AR, VR, and MR. Section 1.4 gives a theoretical background on definitions of XR and its component technologies, and Section 1.5 elaborates on the new Rs introduced in research. Finally, Section 1.6 discusses the different technologies and platforms associated with XR, and Section 1.7 concludes with insights on scope for future research.

1.2 History

Interestingly, the history and evolution of XR has its roots in art. The first immersive system was developed by Giovani Fontana, an Italian engineer in the 15th century. The system was called CAVE, and it used light from lanterns to make wall projections of images (Arnaldi et al., 2018). Many engineers and artists contributed to the concept of illusionary world in the years to follow. The modern concept of VR is known to have been defined by Antonin, a theater artist, in 1958 whose work majorly focused on characters, images and objects for theatrical purposes.

One of the most significant contributors to popularity and development of VR concept is science fiction. The era of the 80 and 90s witnessed many groundbreaking works in this domain; some of which include True Names and Snow Crash (Arnaldi et al., 2018). Historically, science has evolved through science fiction and what was imaginary and impossible in one age become the reality of another. Recent works of fiction in this domain such as Minority Report have open doors for commercial adoption of XR (Dick, 2002).

On the technology front, the first machine created to play videos with immersive experience was Sensorama. This machine was developed by Heilig and used stereographic audio and images to enable the experience (Heilig, 1962) and is known to have been inspired by Antonin's concept of a virtual theater. It is for these reasons that this machine is considered one of the first VR systems developed. The limitation of this machine was its size, which was overcome by a device called "The Ultimate Display," which was developed in 1965 by Ivan Sutherland and Robert Sproull (Sutherland, 1965). The merger of portability to VR gave birth to the HMD concept, which was used by many companies and scientists to develop advanced VR devices.

The next milestone in the evolution of XR was the launch of "The Virtual Environment Workstation" in 1984 by NASA (McGreevy, 1991). This machine improved its predecessor by incorporating functionality such as tracking of gestures made using fingers and position of head. In addition, the graphics of this machine were the most powerful in its time. In comprehension of the limitations of HMD concept, Milgram and Kishino (Milgram, 1994) proposed the "Reality to Virtual" concept in 1994.

The widespread adoption of XR is known to have occurred after 2010, more so after the 2013 launch of the first popular VR device called Oculus Rift (Arnaldi et al., 2018).

Many other devices such as Valve Index and HTC Vive were launched later (Huang et al., 2019). HoloLens, an MR device, was launched in the commercial space in 2015 (Arnaldi et al., 2018). Moreover, the availability of AR development platforms such as ARKit and ARCore has facilitated development of AR-backed mobile applications.

1.3 The 3Rs

XR is a combination of 3Rs, namely, AR, VR, and MR.

1.3.1 Virtual reality

VR allows users to interact with a digitally created world in a completely immersed mode. There are three main characteristics of VR that includes interactivity, real-time experience, and a full 360-degree view of the space. The 360-degree view of the virtually created world is typically enabled by the device (Speicher et al., 2019). To ensure that users have a pseudonatural experience, the virtually created objects and entities must have a real-time behavior to fulfill the real-time criteria (Arnaldi et al., 2018). Another significant aspect of providing pseudonatural experience is interactivity. Users must be able to position themselves and have an active interaction with the virtual world (Arnaldi et al., 2018). An image of an individual experiencing VR using a headset can be seen in Fig. 1.1A.

VR has found numerous applications in fields such as manufacturing, construction, education, healthcare, and designing. The core of VR benefits lies in the fact that it allows high communication efficiency. With that said, the use of VR is also known to have an impact on humans with conditions such as headaches and stomach awareness reported (Weiβker et al., 2018). Additionally, the cost of devices required for VR is still high in comparison with conventional entertainment equipment even though there has been gradual and considerable reduction in costs over time (Avila and Bailey, 2014).

1.3.2 Augmented reality

AR is a technology that extends real environment in such a manner that knowledge and perception of the real world are reinforced (Arnaldi et al., 2018). One of the simplest examples of AR in action is applications that allow a user to open the camera and view additional information about an object in the field of view. The three main characteristics of AR include interactivity, combined operation, and real-time responsiveness.

AR does not create a digital or virtual world. Instead, it simply appends a layer on top of the real world to provide additional information to the user (Ro et al., 2018). Therefore, AR enabled combination of real and virtual world in such a manner that one supplements the other to provide enhanced user experience. An image of a girl using AR app on the handheld device can be seen in Fig. 1.1B. For seamless user experience, AR

(A)
(B)
(C)

Figure 1.1 XR user experience. (A) Virtual Reality; (B) Augmented Reality; (C) Mixed Reality.

needs to work in real time and allows interaction between user and system to the level that cooperating elements to support interaction are provided.

Uses of AR are found in specialized domains such as driving assistance, gaming, professional gesture assistance, and tourism, to name a few (Kimura et al., 2017; Mgbemena et al., 2016). Similar to VR, one of the biggest strengths of AR is that it enhances communication efficiency by providing visual information and enhanced user experience. In comparison with VR, the health-related concerns with VR are not there in AR, but then AR experience is not as immersive and fascinating as VR.

1.3.3 Mixed reality

MR can be best described as a concept that lies on the intersection of VR and AR (Speicher et al., 2019). While AR presents a combination of the real and virtual world, VR completely replaces the real world with a digitally created world. However, MR creates a digital or virtual world under and working along with the real world. Like VR, a head-mounted display is required for MR experience. The limitation of AR that it cannot simulate the virtual world without the assistance of practical support from the environment is mitigated by MR. Thus, MR is more inclined toward VR than AR.

Therefore, in a way, MR inherits the advantages of AR and VR solving their disadvantages to effect. However, MR has its own disadvantages and limitations, which include high cost of equipment (Kun et al., 2017) and limited capabilities of the available MR devices. Any form of information consumption in this digital age can be impacted greatly by MR. One of the most popular products in this domain is HoloLens (Flavián et al., 2019). An image of an individual experiencing MR using HoloLens can be seen in Fig. 1.1C. MR also happens to be one of the best candidate technologies for the healthcare sector in view of the multiple available scenarios that require merging of real and virtual world to communicate and comprehend in the most effective manner.

1.3.4 Extended reality

The concept of XR was given by Charles Wyckoff in 1961 (Wyckoff, 1961, 1962). The introduction of this concept was a result of the patent filed for XR film that could enable observance of phenomena beyond the capabilities of normal human vision. The modern trademark for this concept is registered in the name of Sony, who extensively use this term to refer to mobile AR. Several definitions of XR have come into existence, which are described in the following:

- Type 1 XR

This type of XR defines X mathematically as a variable that can take any number of the real number line, which is an axis for one of the following:

- o Type 1a XR

This type of reality augments human sensory capabilities with the help of wearable devices and thus is seen as an "extrapolation" (Wyckoff, 1961, 1962). Therefore, this form of XR is a superset of MR.

o Type 1b XR

This type of reality blends human sensory capabilities with virtual elements being seen as an "interpolation" (Mann et al., 2018). Therefore, this form of XR is similar to MR.

- Type 2 XR

In this type of XR, the real and virtual aspects of reality are provisioned by sensors/ actuators and an online platform, respectively. Therefore, this form of XR is a subset of MR.

1.4 Concepts and definitions

As mentioned previously, XR is a term that typically refers to the technological foundation for AR, VR as well as MR (Fast-Berglund et al., 2018). This makes XR a rather loosely defined term with some references also using the term cross reality to refer to the same concept (Sherman and Craig, 2019). One of the first milestones in the technological chronology of XR includes conceptualization of VR in the 1960s (Aukstakalnis, 2017; Berg and Vance, 2017; Whyte and Nikolić, 2018). However, it was not until the 1990s when multiple solutions in the VR space came about (see, e.g. Moore, 1995, 1998).

The categorization of VR is classically performed using the virtuality continuum (Milgram and Kishino's, 1994). The virtuality continuum states that there are two environments namely the virtual and real environments, and MR lies between these two environments. Therefore, when the observer or participant is completely immersed in a synthetic, digitally created environment, the type of reality is referred to as VR.

The concept of AR was introduced in the 1990s (Aukstakalnis, 2017). When the real environment is "augmented" or expanded with objects that are virtual and digitally produced, the type of reality is called AR. Another definition of AR refers to it as augmentation of virtual, digitally produced environment with elements from the real environment (Milgram and Kishino, 1994).

In conclusion, the key difference between AR and VR is the fact that AR supplements reality, whereas VR totally replaces it. MR is a term that cumulatively refers to both forms of augmentation: virtuality and reality (Fast-Berglund et al., 2018). There is literature-level contradiction on whether AR and VR are sibling technologies (Wang et al., 2018) or AR is a subset technology of VR (Sherman and Craig, 2019). Key differences between the three constituent technologies and their place in the XR tech domain are provided in Table 1.1.

Two of the main features of XR include immersion and interactivity (Radianti et al., 2020; Sherman and Craig, 2019). The definition of these two features differs on the basis

Table 1.1 Differences between 3Rs.

Source	Features	Platform	Device	Prospect
Augmented reality	1. Real world augmented by virtual/digitally created content. 2. Needs supporting sound graphics and sensory input for realistic experience.	Digitally created content is augmented onto real world and is not a part of it. Commonly used with camera to provide additional information such as recipes based on identified ingredients.	A headset can improve the experience, but AR is commonly used in mobile apps and a technology accessed on handheld devices.	Extensively used in training, assistive applications, and gaming.
Virtual reality	Provides totally immersive experience, and cameras/headsets are used to completely block the real world.	Extensively used in gaming, medical training, military, and other applications that require the individual to get completely absorbed into a virtual world.	Typically uses a headset, and individuals can interact with other users in the same experience as a game and digitally created characters.	Virtual reality has been around for a long time. However, its multidimensional use cases such as flight and scenario simulations and treatment of specific health issues are some of the novel applications that have been developed recently.
Mixed reality	Brings together virtual and real world by integrating objects and scenarios to real world.	The sense of realism is better in MR and is commonly used for gaming and training purposes.	The use of headset provides better sense of realism. However, the use of headset may hamper interaction with other individuals in the same setting. Therefore, shared level interaction between individuals is better suited in some cases.	This is the most advanced form of new age experiences. However, devices and applications for MR are expensive, which limits the reach of this technology in the market.

of XR component and the devices being used. For instance, if we consider VR, then VR is actually a medium and the content being mediated is the VR world (Sherman and Craig, 2019). Thus, in this case, the medium, VR, will have two characteristics, namely, immersion and interactivity. In addition to medium and content, the space will also constitute of content creators and participants. For the other two kinds of realities (Rs), the definitions can be extended from the VR context because all the Rs have an aspect of VR in them whether the whole reality is virtualized or augmented.

The sense of being within an environment is referred to as immersion. This can be accomplished via physical or mental means (Sherman and Craig, 2019, p.10). There have been many perspectives on how presence is related to immersion. According to Slater and Wilbur (1997), presence is immersion-related state of consciousness, while Nykänen et al. (2020) considered presence as a state that is consequential to immersion. These aspects make immersion a significant facet of XR. Since VR is a medium of entering and interacting with digitally created world/content, the physically approached immersion excludes the sense of any interaction with the real (Wang et al., 2018). Thus, VR is usually associated with and referred to as immersive technology (Whyte and Nikolić, 2018) where immersion can also be understood as sense of presence.

The extent of immersion may vary from nonimmersive to completely immersive (Jamei et al., 2017). The most widely used classification criterion for deciding if a technology is immersive or not is based on the VR hardware used for the same. For instance, if VR experience involves the use of headsets, it is immersive. On the other hand, desktop-based VR is classified as nonimmersive. In fact, existing literature also question if monitor or 3D glasses-based experience can even be called VR because it does not involve immersion (Whyte and Nikolić, 2018). In contrast to VR, AR, and its relatedness to immersion is confined to usability (Aukstakalnis, 2017). From a generic point of view, high-level usability is considered to have been achieved if a technology can be accessed in an effortless manner without any scope for unreasonable breaks.

Interactivity or interaction is the second core feature of XR technologies and typically refers to the interaction between the medium and the user or among users through the medium (Pan and Hamilton, 2018). Therefore, from a user's point of view, he or she must be able to change perspectives and position, on choice (Berg and Vance, 2017, p. 3; Whyte and Nikolić, 2018, pp. 20–21). Recent advances in XR research have enabled real-time manipulation of the environment (Radianti et al., 2020, p. 3). To make the experience comprehensible, the virtual environment may also consist of temperature, vibration, lightning, temperature and wind (Berg and Vance, 2017, p. 2).

1.5 Other realities: beyond 3Rs and XR

The taxonomy of realities is not limited to AR, VR, MR, and XR. Recently, many new concepts have been proposed in this domain. Some of these concepts include meditated reality, which blends and modifies reality and all reality or multimediated reality that is the multiscale multidimensional, multimodal, and multisensory form of

mediated reality (Mann et al., 2018). All reality extends the interactive multimedia-based reality that targets the five senses of human with the use of not just wearable devices but also integrative applications that work in conjunction with the concepts of smart cities. Although research is underway in these domains, they have been included in this chapter for comprehensibility.

1.5.1 Mediated reality

Technologies act as an intentional or unintentional intermediate between the environment and humans. Therefore, they modify or mediate our reality in a deliberate or accidental manner. On the basis of this definition, two types of mediated reality exist, namely, unintentionally mediated reality and deliberately mediated reality. One of the earliest examples of deliberate mediation in reality is the invention of upside-down vision glasses (Mann, 2016). Although first works in this domain were psychological experiments, there are many practical applications of deliberately mediated reality which include filtering out irrelevant visual information such as advertisements (Davies, 2012). There have been some works that have investigated the use of this technology for other applications such as surveillance and wearable computing (Tan, 2013).

In AR applications, there is usually an attempt made to keep the reality unaltered. However, in case of applications such as video see-through implementation, digitally generated, virtual objects are overlaid onto an unintentionally mediated form of reality. This is typically a result of placing this technology between the real world and our own selves. Therefore, the mediated reality continuum extends the conventional concept by stating that a continuum lies in not just the degree to which the reality is virtualized but also the degree to which it is altered (Tan, 2013).

1.5.2 Multisensory reality

The existing continuum such as Milgram's (1994) and Mann's (1994) did not take into account the effects of sensory attenuation technologies such as sleep masks and sunglasses on realities. Most experiments in this domain use darkness for sensory attenuation. In this scenario, virtualized aspects of reality do not originate from computer-based simulations. Instead, the use of light, a quantifiable sensory measure, drives the display mechanism. This type of reality is also referred to as phenomenal reality (Mann et al., 2018).

1.5.3 Multimediated reality

Multimediated reality considers the interactions between the environment and human through the use of technology and how the same augments and overlays onto human mind and body. This concept also drifts toward the idea of humanistic intelligence (Minsky et al., 2013), according to which technology must act as an intermediate layer between environment and humans, and intelligence should evolve with the help of a computational feedback loop, which includes not just the technology but also humans. In other words, multiple scales will exist in multisensory reality ranging from

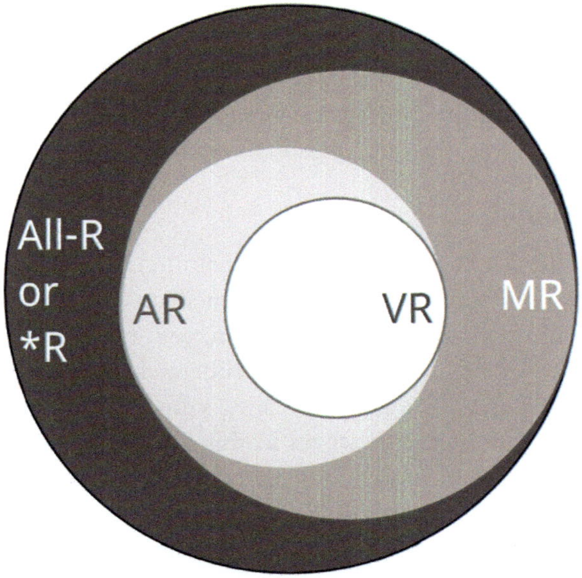

Figure 1.2 Venn diagram for AR, VR, MR, and *R.

environmental scales such as smart rooms and wearable devices. This form of reality is also referred to as ⋆R, All-R, or ZR (Mann et al., 2018). The Venn diagram showing the hierarchical coverage of different technologies is shown in Fig. 1.2.

1.6 Technologies and platforms

1.6.1 Haptic and biofeedback technologies

Some of the devices used for XR have sensors that can acquire physiological user data. These data can be used in different ways to support applications. For instance, these data can be sent to the healthcare provider for monitoring and further intervention. Moreover, these data may also be fed back to the user in real time so that functions such as brain activity and breathing can be consciously controlled by the user. This mechanism is also referred to as biofeedback.

The availability of biosensing devices and communication technologies on XR devices makes it possible to gather valuable data, which can then be used to back a plethora of applications such as consultation, diagnosis, and treatment of physical and psychological conditions. The use of these technologies is also expected to take us closer to personalized treatment and precision medicine by backing technological use cases such as validation of approaches.

Having said that, some state-of-the-art XR devices support biosensing that goes beyond collection of basic biometric data such as heart rate. In fact, some of these

modern devices are also capable of collecting data associated with emotions, expression, and respiration. These data can be analyzed and is specifically important in assessing the impact of VR experiences on individual users. Furthermore, it is now possible to interface VR headsets with devices such as breath belts, eye tracking devices, heart rate monitors, and galvanic skin response devices. These devices aid further analysis and contextualization of available data. For instance, the use of eye-tracking devices enables analysis of parameters such as user focus, attention, and presence during a VR experience.

1.6.2 Development platforms

Several platforms are available for development of XR applications. Some of the commonly used and popular options include Unity[1] and Unreal.[2] While most XR applications are built on these development platforms, several startups have come up with their own offerings. These solutions allow developers to create and distribute XR applications in a simplified manner.

To make the XR technology accessible to users and to further simplify this process line, WebXR technology was introduced. The use of this technology allows users to access XR applications via a Web browser instead of downloading them onto the device before use. The quick and easy access provided by this technology has led to its increasing popularity among users and developers alike. Having said that, the dependence of the application on Internet can lead to potential accessibility issues, particularly with unreliable and restricted access in setups such as healthcare facilities. One of the examples of this technology in use is Mozilla Hubs.

1.7 Conclusion

XR collectively refers to the three established realties, namely, MR, AR and VR. However, the application-level classification of a use case under one of the three categories is not simplistic. Several definitions and concepts have been proposed in existing literature with regard to the 3Rs and XR. This chapter comprehensively covers these aspects, outlining the differences between these technologies, their history, and how the development of one has impacted advancement of others. In addition, this chapter also explores the extensions of these technologies in contemporary research in the form of multimediated and multisensory forms of reality, covering the taxonomy and terminologies associated with the same. Finally, for the sake of comprehensibility, this chapter also includes the different technologies and platforms available for XR development.

1 https://unity.com/.
2 https://www.unrealengine.com/en-US/.

Future research in this domain shall focus on development of a continuum for ⋆R and conceptual refinement of existing taxonomy.

References

Arnaldi, B., Guitton, P., Moreau, G. (Eds.), 2018. Virtual Reality and Augmented Reality: Myths and Realities. John Wiley & Sons.

Aukstakalnis, S., 2017. Practical Augmented Reality: A Guide to the Technologies, Applications, and Human Factors for AR and VR. Addison-Wesley Professional.

Avila, L., Bailey, M., 2014. Virtual reality for the masses. IEEE Annals of the History of Computing 34 (05), 103–104.

Berg, L., Vance, J., 2017. Industry use of virtual reality in product design and manufacturing: a survey. Virtual Reality 21 (1), 1–17. https://doi.org/10.1007/s10055-016-0293-9.

Davies, C., September 12, 2012. Quantigraphic Camera Promises HDR Eyesight from Father of AR.

Dick, P.K., 2002. Tiptree J. Minority Report. Gollancz, London.

Du, J., Shi, Y., Zou, Z., Zhao, D., 2018. CoVR: cloud-based multiuser virtual reality headset system for project communication of remote users. Journal of Construction Engineering and Management 144 (2). https://doi.org/10.1061/(ASCE)CO.1943-7862.0001426.

Fast-Berglund, Å., Gong, L., Li, D., 2018. Testing and validating Extended Reality (xR) technologies in manufacturing. Procedia Manufacturing 25, 31–38. https://doi.org/10.1016/j.promfg.2018.06.054.

Flavián, C., Ibáñez-Sánchez, S., Orús, C., 2019. The impact of virtual, augmented and mixed reality technologies on the customer experience. Journal of Business Research 100, 547–560.

Heilig, M.L., 1962. Sensorama Simulator: U.S. Patent 3,050,870, pp. 8–28.

Huang, Y., Shakya, S., Odeleye, T., 2019. Comparing the functionality between virtual reality and mixed reality for architecture and construction uses. Journal of Civil Engineering and Architecture 13, 409–414.

Jamei, E., Mortimer, M., Seyedmahmoudian, M., Horan, B., Stojcevski, A., 2017. Investigating the role of virtual reality in planning for sustainable smart cities. Sustainability 9 (11). https://doi.org/10.3390/su9112006.

Kimura, R., Matsunaga, N., Okajima, H., et al., 2017. Driving Assistance System for Welfare Vehicle Using Virtual Platoon Control with Augmented Reality, 2017 56th Annual Conference of the Society of Instrument and Control Engineers of Japan (SICE). IEEE, pp. 980–985.

Kun, A.L., Meulen, H., Janssen, C.P., 2017. Calling while Driving: An Initial Experiment with HoloLens.

Mann, S., Havens, J.C., Iorio, J., Yuan, Y., Furness, T., 2018. May. All reality: values, taxonomy, and continuum, for virtual, augmented, eXtended/MiXed (X), Mediated (X, Y), and multimediated reality/intelligence. In: Presented at the AWE 2018.

Mann, S., June 2016. Surveillance (Oversight), Sousveillance (Undersight), and Metaveillance (Seeing Sight Itself), pp. 1408–1417.

Mann, S., 1994. Mediated Reality. TR 260. M.I.T. M.L. vismod, Cambridge, Massachusetts. http://wearcam.org/mr.htm.

McGreevy, M.W., 1991. The Virtual Environment Display System.

Mgbemena, C.E., et al., 2016. Gesture detection towards real-time ergonomic analysis for intelligent automation assistance. In: Advances in Ergonomics of Manufacturing: Managing the Enterprise of the Future. Springer, Cham, pp. 217–228.

Miettinen, R., Paavola, S., 2014. Beyond the BIM utopia: approaches to the development and implementation of building information modeling. Automation in Construction 43, 84–91. https://doi.org/10.1016/j.autcon.2014.03.009.

Milgram, P., 1994. Augmented Reality: A Class of Displays on the Reality-Virtuality Continuum.

Milgram, P., Kishino, F., 1994. A taxonomy of mixed reality visual displays. IEICE - Transactions on Info and Systems 77 (12), 1321–1329.

Minsky, M., Kurzweil, R., Mann, S., 2013. The society of intelligent veillance. In: IEEE ISTAS.

Moore, G.E., 1998. Cramming more components onto integrated circuits. Proceedings of the IEEE 86 (1), 82–85. https://doi.org/10.1109/JPROC.1998.658762.

Moore, G.E., 1995. Lithography and the future of Moore's Law. In: Paper presented at the Integrated Circuit Metrology, Inspection, and Process Control vol. IX, pp. 2–17. https://doi.org/10.1117/12.209151, 2439.

Nikolić, D., Maftei, L., Whyte, J., 2019. Becoming familiar: how infrastructure engineers begin to use collaborative virtual reality in their interdisciplinary practice. Journal of Information Technology in Construction (24), 489–508. https://doi.org/10.36680/j.itcon.2019.026.

Nykänen, M., Puro, V., Tiikkaja, M., Kannisto, H., Lantto, E., Simpura, F., et al., 2020. Implementing and evaluating novel safety training methods for construction sector workers: results of a randomized controlled trial. Journal of Safety Research. http://doi:10.1016/j.jsr.2020.09.015.

Pan, X., Hamilton, A., 2018. Why and how to use virtual reality to study human social interaction: the challenges of exploring a new research landscape. British Journal of Psychology 109 (3), 395–417.

Radianti, J., Majchrzak, T., Fromm, J., Wohlgenannt, I., 2020. A systematic review of immersive virtual reality applications for higher education: design elements, lessons learned, and research agenda. Computers and Education 147, 103778. https://doi.org/10.1016/j.compedu.2019.103778.

Ro, Y.K., Brem, A., Rauschnabel, P.A., 2018. Augmented Reality Smart Glasses: Definition, Concepts and Impact on Firm Value creation. Augmented Reality and Virtual Reality. Springer, Cham, pp. 169–181.

Seymour, M., Riemer, K., Kay, J., 2018. Actors, avatars and agents: potentials and implications of natural face technology for the creation of realistic visual presence. Journal of the Association for Information Systems 19 (10), 4. Retrieved from. https://aisel.aisnet.org/jais/vol19/iss10/4.

Sherman, W.R., Craig, A.B., 2019. Understanding Virtual Reality: Interface, Application, and Design, second ed. Morgan Kaufmann, San Francisco, CA.

Slater, M., Wilbur, S., 1997. A framework for immersive virtual environments (FIVE): speculations on the role of presence in virtual environments. Presence: Teleoperators and Virtual Environments 6 (6), 603–616.

Speicher, M., Hall, B.D., Nebeling, M., 2019. What is mixed reality?. In: Proceedings of the 2019 CHI Conference on Human Factors in Computing Systems, pp. 1–15.

Stanton, N.A., Plant, K.L., Roberts, A.P., Allison, C.K., Howell, M., 2020. Seeing through the mist: an evaluation of an iteratively designed head-up display, using a simulated degraded visual environment, to facilitate rotary-wing pilot situation awareness and workload. Cognition, Technology and Work 22 (3), 549–563. https://doi.org/10.1007/s10111-019-00591-2.

Sutherland, I.E., 1965. The Ultimate Display. Multimedia: From Wagner to Virtual Reality, vol. 1.

Tan, L., 2013. Mediated reality in bearable prosthesis: a tool for surveillance and intervention. In: Proceedings of the 1st Fascinate Conference. Falmouth University, Cornwall, UK, pp. 1–17.

Wang, P., Wu, P., Wang, J., Chi, H., Wang, X., 2018. A critical review of the use of virtual reality in construction engineering education and training. International Journal of Environmental Research and Public Health 15 (6). https://doi.org/10.3390/ijerph15061204.

Weißker, T., Kunert, A., Fröhlich, B., et al., 2018. Spatial updating and simulator sickness during steering and jumping in immersive virtual environments. In: 2018 IEEE Conference on Virtual Reality and 3D User Interfaces (VR). IEEE, pp. 97–104.

Wyckoff, C.W., 1962. An Experimental Extended Response Film. SPIE Newslett, pp. 16–20.

Wyckoff, C.W., 1961. An Experimental Extended Response Film. Technical Report NO. B-321. Edgerton, Germeshausen & Grier, Inc., Boston, Massachusetts.

Whyte, J., Nikolić, D., 2018. Virtual Reality and the Built Environment. Routledge, London. https://doi.org/10.1201/9781315618500.

Yan, W., Culp, C., Graf, R., 2011. Integrating BIM and gaming for real-time interactive architectural visualization. Automation in Construction 20 (4), 446–458. https://doi.org/10.1016/j.autcon.2010.11.013.

Zahabi, M., Abdul Razak, A., 2020. Adaptive virtual reality-based training: a systematic literature review and framework. Virtual Reality: The Journal of the Virtual Reality Society. https://doi:10.1007/s10055-020-00434-w.

CHAPTER TWO

Clinical applications of extended reality

Samiya Khan
Faculty of Science & Engineering, University of Wolverhampton, Wolverhampton, United Kingdom

2.1 Introduction

Recent past has witnessed steep growth in the extended reality (XR) market as the world moves out of the COVID-19 pandemic. Moreover, this growth has been recorded not just at the enterprise level but also in the public domain. There has been worldwide spending on XR, and according to a forecast, it is expected to increase from $12 billion (2020) to $72.8 billion (2024).[1] In contrast to other sectors and businesses, the COVID-19 pandemic has proven to be a catalyst for augmented reality/virtual reality (AR/VR) adoption because of the remote working model of the entire world for almost 2 years. Statistics suggest that spendings in XR sector will increase by 70% in public safety, 74% in education, and 75% in healthcare in the span of next 5 years.[2]

A significant aspect of the XR market and one of the most profound reasons for its rapid adoption is the dropping prices of XR equipment. It is in fact much more affordable now, and every household can potentially own an XR headset. The growing uptake of the XR devices is evident from the fact that 2022 witnessed 14.94 million devices being shipped to users, which is a 55% increase from2021.[3] These numbers are expected to witness a sevenfold rise in the year range of 2020–25. Within the United Kingdom, it was reported in January 2021 that 1 year of lockdown had resulted in XR devices sales to shoot up by 350%.[4] From Europe's perspective, a study performed in 2020 estimated the XR market share to be $50.55 billion.

The growing adoption of the XR technology has also resulted in acceleration of research within this field from technology to haptics and headsets. The integration of immersive experiences with technologies such as artificial intelligence and 5G has improved the quality, reliability, and efficacy of this technology. The wide range of applicability is also one of the primary reasons for accelerated research and adoption. Primary adoption has demonstrated that XR has the most impactful use cases in public

1 https://www.idc.com/getdoc.jsp?containerId=prUS47012020.
2 https://www.xrhealthuk.org/the-growing-value-of-xr-in-healthcare.
3 https://www.idc.com/promo/arvr.
4 https://www.thetimes.co.uk/article/virtual-reality-pandemic-leads-to-rise-in-headset-sales-to-escape-lockdown-jhhn8wghn.

Extended Reality for Healthcare Systems, Volume 1
ISBN 978-0-323-98381-5, https://doi.org/10.1016/B978-0-323-98381-5.00013-1

services. Moreover, the contribution of XR in minimizing the multifaceted impact of COVID-19 pandemic has attested that this technology is capable of mitigating grave challenges that plague the global society in the modern era.

If we consider the healthcare market, XR has remarkable benefits and demonstrated impact in the mental health sector. Mental illness is an unfortunate and economically draining condition, which costs the UK economy as much as £70—100 billion annually.[5] To add to the number of people with known mental issues, the COVID-19 pandemic is expected to result in approximately 10 million people requiring additional mental health support.[6] The NHS England is already under immense pressure, and providing this additional support will add to the challenges that healthcare sector is currently facing.

One of the key benefits of XR for healthcare is that it empowers people to engage in well-being and health-related activities in an active manner. XR has been used for several assistive therapeutic interventions for conditions such a posttraumatic stress disorder, phobia, anxiety, and eating disorders, to name a few. Digital apps are widely being used by the NHS to provide patients with support for varied mental health issues, which is not just a cost-effective alternative to traditional treatment, but it is also highly effective. The realm of these solutions is increasingly evolving with the inclusion of biosensors that allow real-time monitoring of patient conditions and understanding the way they react to specific scenarios or situations that cause them stress.

Other assistive uses of XR include skill development for individuals with learning disabilities and difficulties such as autistic children. XR is actively used for improving social skills in autistic people. Moreover, this technology is also being used for diagnostic purposes. For instance, eye movement tracking is used for detecting visual impairments and conditions such as dementia and Alzheimer's disease. The ability of XR to facilitate patient engagement for self-management and skill development allows it to be used for rehabilitation purposes, particularly for pain management, stroke rehabilitation, and occupational therapy, making it a useful tool for preventive care.

Research in the field of patient-centered healthcare and precision medicine has revolutionized with the exploration of XR use cases in healthcare. XR healthcare takes precision medicine to an entirely new level by providing opportunities for data-driven healthcare, which encourages both participation and personalization. As a result, clinicians are now actively prescribing XR headsets to patients so that they can initiate and continue their remote healthcare routine even outside the conventional healthcare setting. While XR-based therapeutic interventions have demonstrated benefits to patient's conditions and well-being, it also allows clinicians to focus on the cases that are complex and inevitably require a face-to-face interaction.

5 https://www.mentalhealth.org.uk/statistics/mental-health-statistics-global-and-nationwide-costs.

6 https://www.centreformentalhealth.org.uk/news/10-million-people-england-may-need-support-their-mental-health-result-pandemic-says-centre-mental-health.

Furthermore, XR is being actively used for training and medical education, which allows medical professionals to learn and practice in their own time, at their own pace, without putting patient life at risk. XR has its roots in the gaming and entertainment industry; it involves creative development and storytelling to engage audiences. This facet of XR makes it an ideal tool for therapeutic intervention because one of the biggest challenges in healthcare is patient interaction and response. Research in XR for healthcare spans across three decades and focuses on evaluation of XR usage for varied applications. Over the years, the affordability and accessibility aspects of XR have evolved at a rapid pace. The clinical areas that are being actively supported by XR are illustrated in Fig. 2.1.

This chapter discusses the core clinical sectors that are supported by XR with case studies. The rest of the chapter is organized in the following manner. Sections 2.2–2.5 elaborate on the use of XR for clinical mental health, mental well-being, pain management, and physiotherapy and rehabilitation, respectively. Finally, Section 2.6 concludes the chapter providing insights into the future and potential of XR for development of novel applications in healthcare.

2.2 Clinical mental health

The use of XR for diagnosis and management of mental health issues has garnered immense research attention from the start. The first clinical trial that attested the use of VR for mental health treatment was performed in 1995 (Opdyke et al., 1995). Early

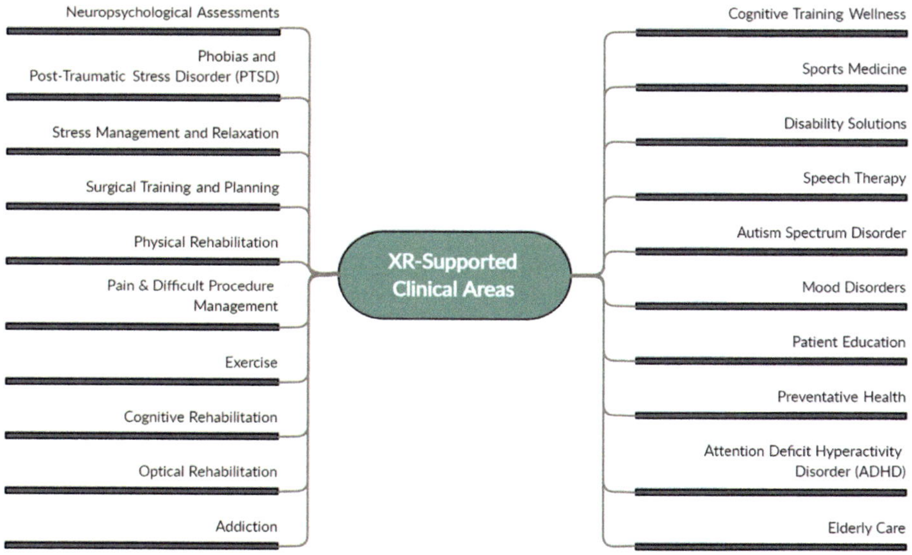

Figure 2.1 Clinical areas supported by XR. *XR*, extended reality.

research in VR application for mental health treatment was centered on recreation of scenarios specific to different phobias using simulations. Since these scenarios are typically difficult to simulate in real life, VR could solve the purpose by presenting patients with stimuli in a controlled manner, to which they would respond, and the responses were correspondingly captured for analysis (Rizzo and Koenig, 2017).

Several analysis and efficacy studies have been performed to evaluate VR for management of conditions such as posttraumatic stress syndrome (Eshuis et al., 2020), depression (Falconer et al., 2016), eating disorders (Ferrer-Garcia et al., 2012), anxiety (Opris et al., 2012), autism (Parsons and Cobb., 2011), and psychosis (Valmaggia, 2017). Moreover, VR has also been used and evaluated for behavioral assessment (Morina et al., 2015). These studies have evidenced the application of VR and its effectiveness for different components of therapeutic intervention including cognitive restructuring and exposure. To start with, VR can safely be used as a supplemental tool to complement existing therapies. With technological and research advancements, the application of VR has expanded within the domains of psychiatry, psychology, and clinical mental health. Therefore, recent research in this domain has focused on developing innovative ways to incorporate VR in conventional therapeutic settings.

While the use of VR is widely established, the potential of AR and mixed reality (MR) awaits investigation and exploration. The primary reason for this research gap stems from the fact that the computational capabilities required for rendering of high-quality graphics demanded by these technologies have only become available at scale, recently. For instance, the latest smartphones can support AR functions (Baus and Bouchard, 2014). Moreover, with the launch of advanced MR headsets and haptic suits, it will now be possible to explore the potential of this technology for therapeutic interventions. Some of the potential applications of these technologies lie in facilitating independent living and autonomy in individuals with neurological disorders (Aruanno et al., 2018) and developing social skills in autistic children (Crowell et al., 2020).

2.2.1 Early detection of Alzheimer's disease and dementia

Howett et al. (2019) reported the development and trial of a VR-based navigation test for individuals who are at risk of acquiring dementia. One of the early signs of Alzheimer's disease includes mild cognitive impairments. However, such impairments may also be caused due to other factors such as anxiety, which makes it difficult to assess. Existing methods used for early detection of dementia include pen-and-paper-based methods for testing cognitive impairments, which lacks accuracy. An alternative method involves extraction and analysis of spinal fluid. Although this method promises high accuracy, it is very painful and costly.

In contrast to these existing methods, VR presents a cost-effective, noninvasive, and accurate assessment method in which the patient needs to wear the VR headset and just walk in the simulated environment. The system analyzes patient's movements to assess his

or her navigation abilities. The navigation task is designed in such a manner that successful completion is achieved by only those people who have a fully functional entorhinal cortex, which is one of the first regions to be damaged by Alzheimer's disease.

There are several benefits of using this tool. Firstly, it is cost-effective and accurate. Secondly, the ecological validity of this method is higher because of its ability to test the patient's behavior across different simulated versions of real-life scenarios. Therefore, it allows accessible and timely assessment, which in turn allows well-timed interventions for an early-stage Alzheimer's disease. In the recent years, VR-based interventions have found their place in the model of care for patients suffering from mental health or cognitive impairment issues.

The mass adoption of these solutions is attributed to the lower prices and heightened awareness among people. It is typically observed that people who have had VR experiences are more optimistic about the use of VR-based interventions. However, a larger fraction of people believe that VR cannot replace face-to-face therapy with the viewpoint that VR can be useful alongside conventional methods. In view of this generic mindset, blended therapeutic interventions are being designed to prove the efficacy of VR-based intervention and reduce patient resistance regarding its usage.

2.2.2 Phobias and posttraumatic stress disorder

As mentioned previously, most of the early VR-based interventions were centered on exposing the patient to a simulated real-life scenario that causes anxiety or stress. Therefore, such applications had a direct impact in treatment of social anxiety disorder, PTSD, and anxiety. In contrast to real-life scenarios, VR-based exposure to stimuli can be completely controlled by the therapist. For instance, if a patient is phobic to spiders, then he or she may be exposed to small spiders. Then, the size of the spiders may be gradually increased. However, the size of the spider cannot be increased beyond a limit. Within the VR space, the size of the spider can be increased to as big as a house.[7] Moreover, there are options such as starting out with a cartoon version of the spider and making it more realistic with increased exposure.

For individuals dealing with PTSD, their anxiety is typically linked to a stressful situation. Such a scenario cannot be practically replicated in the real world. However, VR does not just allow "real" replication of the scenario, but it can also allow viewing of the scenario from different perspectives. It has been observed that people respond to these scenarios in much the same way as they would do if the scenario were replicated in the real world (Martens et al., 2019). Another critical finding, from a therapeutic point of view, is that people typically resist real-world exposure to the scenario that causes them stress. However, they are more likely to agree for a virtual experience (García-Palacios et al., 2004).

[7] https://www.bbc.co.uk/news/technology-27186022.

2.2.3 Depression and psychosis

Recent years have witnessed extensive use of VR for psychological treatment. This usage spans across distinct aspects of therapy including diagnosis and treatment (Ferrer-Garcia and Gutiérrez-Maldonado, 2012; Sorkin et al., 2006). The target conditions include depression, psychosis, and anorexia nervosa. In real-world scenarios, VR is typically used along with traditional therapy in a blended format. Craig et al. (2018) proposed a novel treatment for managing auditory hallucinations, particularly for people do not respond to antipsychotic medication. In this treatment method, a virtual avatar of the entity that people hear is created. The therapist takes this form via the computer and initiates a dialogue in such a way that the individual can overcome his or her fear and challenge this avatar. Leff et al. (2014) have shown that such a solution can completely eliminate auditory hallucinations for some people.

Falconer et al. (2016) proposed a treatment methodology for depression in which the individual embodies self as a virtual child and then as an adult. This method is known to facilitate self-compassion and reduction in the feeling of self-hate, thereby reducing depression symptoms. Research is underway to develop self-management solutions, and the clinical efficacy of these solutions needs to be evaluated and evidenced before this solution can make a way into clinical settings (Freeman et al., 2019). There are several obvious benefits of such a therapy such as allowance for self-management and therapist-supervised remote therapy. Therefore, VR-based therapy is expected to make interventions cost-effective and accessible to a larger population.

In view of the tight adherence of XR-based solutions to traditional treatments and the need for a blended setup for convenient adoption by patients, the research community is focusing on development of treatment packages instead of standalone solutions. While these packages are yet to leave the confines of research labs, efforts are underway to provide toolkits to healthcare facilities for smoother implementation.[8] It is a widely accepted fact that most of the barriers associated with VR implementation and adoption can be managed by increasing people's engagement and eliminating the stigma around virtual therapy. VR-based therapy shall allow people convenient, affordable, and quicker access to services that they require.

One of the pathbreaking works in this domain has been performed by OxfordVR. Their current offerings include three services, namely, gameChange, Social Engagement,[9] and Fear of Heights.[10] The first of these solutions, gameChange,[11] is being trialed by NHS England. The solution primarily targets well-being of psychosis patients who find everyday tasks such as shopping and taking a bus, challenging. As a result, they

8 https://www.england.nhs.uk/digitaltechnology/connecteddigitalsystems/exemplars/gde-blueprints/.
9 https://www.oxfordvr.co/social_engagement/.
10 https://www.oxfordvr.co/how-we-can-help/.
11 https://gamechangevr.com/.

typically end up in social seclusion, affecting their physical and mental health. This solution allows them to overcome anxiety associated with everyday tasks, providing them a cost-effective alternative to traditional therapy.

2.3 Mental well-being

In the United Kingdom, significant investments are being made for well-being. NHS England's 5-year forward plan[12] emphasizes on the need to harness technology and provide easier access to digital services. Moreover, the NHS Long Term Plan[13] recommends the use of arts and creative practice to support ongoing and long-term emotional/physical wellbeing. Mental well-being remains one of the most successful applications of XR, globally. There is documental evidence to prove that immersive technologies can be used to promote engagement in creative activities and exercising, which in turn reduce anxiety and lower the chances of onset of physical and mental conditions.

Immersive experiences ensure a high degree of engagement. This aspect of immersive technologies empowers patients and encourages them to self-manage, which is a long-term solution to such conditions. Immersive experiences work for well-being in much the same way as reading books or watching movies reduces stress. Moreover, these experiences also widen the horizon and allow an individual to understand the world with compassion and empathy. These aspects of immersive experiences work on individuals in multifaceted ways. While it may be relaxing and mood-enhancing, it also increases self-knowledge and motivates the individual to perform activities such as exercise.

There has been an increased need for well-being solutions during the pandemic, and XR-based solutions have proven their metal in these tough times. Organizations such as the NHS have used XR-based well-being solutions for their staff to prevent burnout.[14] The demonstrated success of these solutions has resulted in setting up of well-being centers and prescriptions of XR headsets being made to patients and staff for mental and physical well-being.

2.3.1 Relaxation and mindfulness

To breakdown the benefits of XR, one of the most impacted subsectors is meditation and mindfulness. The exposure of individuals to environments such as beaches and hills has a

12 https://www.england.nhs.uk/wp-content/uploads/2016/02/Mental-Health-Taskforce-FYFV-final.pdf.
13 https://apps.who.int/iris/bitstream/handle/10665/329834/9789289054553-eng.pdf.
14 https://www.med-technews.com/news/nhs-workers-use-xr-tech-for-training-during-covid-19-pandemi/.

relaxing effect on them. The list of environments currently supported by XR is endless, and depending on what an individual finds most relaxing, he or she can select the environment for a relaxation session. XR has become a popular relaxation tool for use at homes and workplaces (Sonney et al., 2021). A use case of this functionality is waiting rooms to reduce preintervention anxiety in patients and, generally, for people to relax before any anxious or stressful situation.

One of the applications that is popular in this domain is Explore Deep,[15] which exposes the individual to a breathing-controlled VR experience for meditative purposes. This VR application uses a combination of biofeedback tools, slow diaphragmatic breathing, and meditation for visualization of internal states. This enables the individual to gain better awareness of his or her inner, physical, and emotional state. For controlling the breathing movement, a breath belt and syncing of the tai chi hand movements with breathing are used.

Meditative breathing is a long-term management techniques for anxiety, and Explore Deep can play an instrumental role in teaching this technique as well. It has been established that the tool can be used to relieve short-term bouts as well as long-term anxiety issues by creating a positive feedback loop that rewards positive behavior with a relaxed body state. This application is being trialed for various use cases such as supporting students with special needs and patient rehabilitation, in addition to others. Some of the known benefits of Explore Deep include relaxation that lasts up to 2 h after a short session and improvements in well-being markers (Weerdmeester et al., 2021; Bossenbroek et al., 2020).

2.3.2 Creative activities for well-being

There are several applications available in VR that allow individuals to indulge in sculpting, art, 3D drawing, and building. These free-form creation tools can potentially be used for art and play therapy (Hacmun et al., 2018). There has been considerable innovation in this domain with companies such as Hatsumi,[16] providing solutions that allow the individual to communicate emotion and pain through 3D drawing and art. Hatsumi is also actively involved in exhibiting these works anonymously to create awareness about the connections between emotional and physical pain, and the presence of invisible conditions that need attention. Other solutions in this domain include Virtual Sandtray,[17] which is the virtual adaptation of the traditional sandtray therapeutic intervention.

15 www.exploredeep.com.
16 https://www.hatsumivr.com/.
17 https://www.virtualsandtray.org/.

Other activities such as group singing in a virtual environment are also known to have enhanced patient outcomes (Daffern et al., 2019). In line with this, a desktop-based nature karaoke was developed in the form of a 360-degree video by Limina Immersive in 2020, calling it Lost in Song.[18] The application allows the individual to access different songs and settings such as beautiful British Isles. User statistics indicate that 81% of the users experienced a boost in mood after the experience.

2.3.3 Alleviating loneliness

Loneliness and isolation can be extremely detrimental to mental and emotional well-being, which ultimately also begins to affect the individual physically. This fact has been realized beyond measure during the COVID-19 pandemic when people across the world were forced to live isolated lives. VR-based social platforms facilitate interaction between friends in the virtual setting and have demonstrated a positive impact (Riva et al., 2020). Although virtual, the sense of connectedness ensured by virtual interactions provides a heightened sense of "being there." VR-based multiplayer games have been extended to develop social settings such as dance games and table tennis.

VR-based solutions are increasingly being used in settings such as care homes to treat apathy and manage conditions such as dementia. One of the biggest problems that care home residents encounter is loneliness and lack of intellectual stimuli. As a result, there is a faster decline in their cognitive abilities, which does not just affect their quality of life, but it also increases the burden on caregiver. A study that used 360-degree videos of places was conducted with the elderly population. It allowed people to travel to their place of choice and interest, forming a virtual form of the reminisce therapy. The study reported positive results because of the application's ability to provide entertaining and relaxing experiences to people who are typically too frail to travel. In another study (Bourdin et al., 2017), VR-based experiences were used to reduce fear of death in people by giving them a higher sense of purpose, self-acceptance, awareness of well-being, and an opportunity to reflect.

2.3.4 Promoting fitness

An individual's well-being is highly dependent on his or her activity levels. A healthy lifestyle with an active routine forms the basis of a healthy body. However, for a majority of the people, the most difficult aspect of exercising is maintaining adherence to routines.[19] Many VR-based applications are commercially available today that provide a virtual fitness coach, motivational music, and inspiring background scenery to keep

18 https://www.lostinsong.com/.
19 https://www.wired.com/story/virtual-reality-fitness-supernatural/.

individuals interested and engaged. One such application is Supernatural VR.[20] It is anticipated that the combined effects of VR and exercise will have a more serious impact on fitness as compared with exercise alone. In a study, it was reported that people involved in VR-based exercise enjoy their routine more and feel lesser tired (Plante et al., 2003). Lotan et al. (2009) suggested that such an exercise regime can be beneficial for people with disabilities as well.

2.4 Pain management

Opioids are typically harmful, and administering them for pain is not advisable. Therefore, there has been an evolving need for alternatives to conventional methods for pain management. The experience offered by immersive technologies engages and distracts the patient. As a result, the patient's perception of pain is considerably reduced. Thus, VR is being considered for management of pain during therapeutic processes.

Pain can be classified into two categories, namely, acute pain and chronic pain. Acute pain is usually the body's response to injury or disease manifested in the form of a sympathetic nervous system activation or muscular spasm. On the other hand, chronic pain is a consequence of a disease and usually lasts longer because of its association with psychological state (Grichnik and Ferrante, 1991). The conventional method of pain management is prescribing opioids. Recently, nonpharmacological alternative in the form of cognitive behavioral therapy (CBT) has also come into existence.

The continued use of opioids can have grave impact on health and bodily functions. The impact of opioid usage can be assessed from the fact that 68,500 Americans died of opioid overuse in 2018.[21] The current state of the affairs presents VR as a viable alternative for acute and chronic pain management. For instance, VR-based interventions can be used for acute pain arising from childbirth, colonoscopy, and cancer pain, to name a few. In addition, it may also be used for chronic pain of the neck or back. However, it is important to note that VR-based pain management solutions are still in their infancy, and development of targeted solutions in this domain can be potential future research work.

Chronic pain is typically a result of prolonged injury or disease. When a patient suffers from chronic pain over prolonged periods of time, it begins to have psychological implications manifesting in the form of depression and anxiety. One of the popular solutions in this domain is Virtual Meditative Walk (Gromala et al., 2015), which is designed for helping patients in improving their interoceptive awareness and modulating pain. In contrast to chronic pain, acute pain is usually the body's response to a disease or injury.

20 https://www.getsupernatural.com/.
21 https://www.reuters.com/article/us-usa-drugs-overdoses-idUSKCN1UC2HZ.

Within this domain, a solution called SnowWorld was developed. This solution introduced the concept of VR analgesia (Hoffman et al., 2000) and is based on the fact that VR experiences can be so attention-capturing that the individual may be completely drawn away from the real world so much so that he or she can tolerate very painful procedures with ease.

Some of the potential uses of VR-based pain management solutions include management of prevaccination anxiety or procedural pain which are performed with local anesthesia such as dental procedures and redressing and procedures performed on patients with burn injuries. Some research studies have proven the efficacy of this proposition. For instance, a study reported lower anxiety and pain for patients during toe-removal procedure with the help of VR.[22] The benefits of a VR-based pain management solution can be summarized as reduced opioid usage, no side effects, and lowered value medical bills.

2.5 Physiotherapy and rehabilitation

Physiotherapy and occupational therapy are collectively referred to as physical therapy. They typically include task-focused active-use therapy and strength training. Typically, this intervention is provided to the patient over a period of 4–8 weeks and involves a therapist who manages and monitors patient's progress. There are some specific types of therapies involving patients with upper limb motor impairment. Such patients have to undergo physiotherapy repetitively to improve movement in the affected area (Ho et al., 2018). Reduction in functioning of the upper limb is seen as an indicator of reduced well-being in patients after they have had a stroke, which is the reason why innovations in provisioning this type of therapeutic interventions are high-priority area in stroke research (Hayward et al., 2019). Moreover, recent research in physiotherapy has also focused on development of nonpharmacological interventions (Pillai Riddell et al., 2015).

VR-based solutions gamify traditional rehabilitative practices and physiotherapy, making them engaging, enjoyable, and more effective. In fact, rehabilitation was the first clinical area in which clinicians prescribed VR to patients in the same way as medication is prescribed, marking the birth of the "virtual pharmacy" approach. This approach has provided more visibility and perceived efficacy to VR-based interventions. As part of this approach, the patient can perform exercises remotely under the supervision of a clinician or medical professional, saving both costs and time. In a VR-based therapy, the patient needs to wear a headset, and limb movements are tracked using sensors.

Rehabilitation for upper limb movement is required by a diverse population and typically includes patients suffering from orthopedic trauma, musculoskeletal conditions,

[22] https://immersive.tsdft.uk/2020/07/podiatry-anxiety-and-pain-distraction-using-vr/.

and neurological disorders. This population additionally includes adults dealing with poststroke conditions and children with cerebral palsy (Chen et al., 2014). In several trials, the effectiveness of VR in reducing costs, lowering procedural pain, and enhancing engagement levels has been demonstrated (Garrett et al., 2014; Sharan et al., 2012; Sun, 2012). According to Phelan et al. (2019), the use of VR in the physiotherapeutic process provides opportunities for movement recovery. VR-based therapeutic interventions are based on traditional therapy. However, virtual simulations are more engaging and enjoyable, which has a positive impact on movement recovery (Gerber et al., 2016). One of the VR trials in this domain was performed by Kiper et al. (2018), and the results of the study demonstrated that the reinforcement feedback in VR-based treatment is effective for adult patients who are dealing with poststroke conditions.

There are several benefits of early intervention in musculoskeletal conditions, which include lowering the risk of the problem transforming into a chronic condition and reducing the time required by the patient to recover (Neilson et al., 2019). One of the most significant deterrents in complying with treatment and achieving the required range of motion is pain associated with existing intervention methodology. The rise in pain also reduces the patient's confidence in the care team, reducing adherence, which is critical for achieving long-term results (Threapleton et al., 2016). Another benefit of using a VR-based treatment stems from the fact that VR allows remote treatment. Therefore, patients can adhere to the schedule of repeated sessions and gain maximum benefit from the intervention. On the other hand, fewer clinic appointments reduce the pressure on healthcare staff.

Research validates the benefits of VR therapy for poststroke patients (Piron et al., 2001), children with cerebral palsy (Sandlund et al., 2009), and upper limb rehabilitation (Laver et al., 2017; Deutsch et al., 2008; Saposnik and Levin, 2011; Lohse et al., 2014). Nonimmersive VR has also been used for motor rehabilitation in children and has proven to be effective (Sharan et al., 2012; Jannink et al., 2008; Sun, 2012). However, Henderson et al. (2007) reported that immersive VR is more effective for poststroke patients than the nonimmersive version of VR.

One of the companies that work in this domain is Immersive Rehab,[23] which focuses on providing therapeutic services to patients with neurological conditions. These conditions may have resulted due to sclerosis, stroke or spinal injury. The company has performed clinical trials across the United Kingdom to prove the efficacy of the solution and is in the final stages of the same. Their aim is to increase access to services and improve patient outcomes for their target group.

Most of the current solutions are centered on increasing engagement and as a result reduce pain experienced during interventions. Exergames is one such game that has been designed for children (Finkelstein et al., 2010). The future of research in this domain is

[23] https://immersiverehab.com/.

looking at development of solutions that can help therapists and clinicians in providing better interventions, improving the diversity of covered scenarios and interactivity. Therefore, such solutions that can be independently used at home solutions will allow patients to maintain exercise schedule and achieve better outcomes (Jolly et al., 2007). As self-care and management is the only long-term solution to these conditions, VR provides a viable solution for rehabilitation and physical therapy. Additionally, VR-based solutions also reduce travel costs, home visit costs, and pressure on healthcare staff (Threapleton et al., 2016; Salawu et al., 2020).

2.6 Conclusion

XR, particularly VR, has been used and trialed for varied clinical use cases. However, most widespread use of VR for clinical mental health lies in therapeutic interventions for social anxiety, phobias, and posttraumatic stress because VR provides therapists the ability to fully control treatment. VR-based interventions can be performed in a completely remote setup and at scale, which makes it a gamechanger for patients who are nonresponsive, clinically, or behaviorally. Immersive technologies offer a plethora of experiences and activities, keeping their audiences engaged. This aspect of their design can be used to good effect for preventative care by engaging people in activities that encourage them to maintain their emotional well-being and perform exercise. As mentioned previously, immersive technologies are already being used for mental health, particularly for people with a sense of loneliness and isolation, and are known to have improved wellness outcomes. However, there is scope for development of novel pathways by active contribution and collaboration between academic researchers and healthcare professionals.

Increased pressure on healthcare staff was one of the most obvious aftereffects of the COVID-19 pandemic. The healthcare staff was subject to immense pressure. Although VR-based applications allowed remote treatment, it also brought forth another challenge, patient monitoring. To mitigate this challenge, Rescape's DR.VR Frontline[24] was developed that enabled a clinician to have complete control of a patient's VR experience through an external device such as a tablet. This solution has proved effective in real-world settings and has been extensively used for pain management and anxiety reduction. The well-being sector is rather wide and ranges in terms of contexts and sections of the society. Therefore, while it can be used to alleviate stress at workplaces, it can be effectively used for anxiety management at home. The COVID-19 pandemic has contributed to a paradigm shift in the way people used to work, socialize, and access services. VR-based applications make their remote interactions as realistic as possible, with easier access and at lower costs. Mental health and well-being, as a sector, presents

[24] https://www.rescape.health/.

several challenges and opportunities to the VR space by playing a significant role in preventative care and imparting skills for emotional regulation and fitness. Other clinical areas such as pain management, physiotherapy, and rehabilitation have also been explored by the VR space. However, collaborative research methods that promote development of high-quality, clinically acceptable solutions are the need of the hour.

References

Aruanno, B., Garzotto, F., Torelli, E., Vona, F., 2018, October. HoloLearn: wearable mixed reality for people with neurodevelopmental disorders (NDD). In: Proceedings of the 20th International ACM SIGACCESS Conference on Computers and Accessibility, pp. 40–51.

Baus, O., Bouchard, S., 2014. Moving from virtual reality exposure-based therapy to augmented reality exposure-based therapy: a review. Frontiers in Human Neuroscience 8, 112.

Bourdin, P., Barberia, I., Oliva, R., Slater, M., 2017. A virtual out-of-body experience reduces fear of death. PLoS One 12 (1). journals.plos.org/plosone/article?id=10.1371/journal.pone.0169343e01.

Bossenbroek, R., Wols, A., Weerdmeester, J., Lichtwarck-Aschoff, A., van Rooij, M.J.W., Granic, I., 2020. Efficacy of a virtual Reality biofeedback game (DEEP) to reduce anxiety and disruptive classroom behavior: single-case study. JMIR Mental Health. https://doi.org/10.2196/16066.

Chen, Y.P., Lee, S.Y., Howard, A.M., 2014. Effect of virtual reality on upper extremity function in children with cerebral palsy: a meta-analysis. Pediatric Physical Therapy 26 (3), 289–300. https://doi.org/10.1097/PEP.0000000000000046. Fall.

Craig, T.K.J., Rus-Calafell, M., Ward, T., Leff, J.P., Huckvale, M., Howarth, E., Emsley, R., Garety, P.A., 2018. AVATAR therapy for auditory verbal hallucinations in people with psychosis: a single-blind, randomised controlled trial. The Lancet Psychiatry 5 (1), 31–40. https://doi.org/10.1016/S2215-0366(17)30427-3.

Crowell, C., Sayis, B., Benitez, J.P., Pares, N., 2020. Mixed reality, full-body interactive experience to encourage social initiation for autism: comparison with a control nondigital intervention. Cyberpsychology, Behavior, and Social Networking 23 (1), 5–9.

Daffern, H., Camlin, D.A., Egermann, H., Gully, A.J., Kearney, G., Neale, C., Rees-Jones, J., 2019. Exploring the potential of virtual reality technology to investigate the health and well being benefits of group singing. International Journal of Performance Arts and Digital Media 15 (1), 1–22. https://doi.org/10.1080/14794713.2018.1558807.

Deutsch, J.E., Borbely, M., Filler, J., Huhn, K., Guarrera-Bowlby, P., 2008. Use of a low-cost, commercially available gaming console (Wii) for rehabilitation of an adolescent with cerebral palsy. Physical Therapy 88 (10), 1196–1207 pubmed.ncbi.nlm.nih.gov/18689607.

Eshuis, L.V., Gelderen, M. J. Van, Zuiden, M. Van, Nijdam, M.J., Vermetten, E., 2020. Efficacy of immersive PTSD treatments: a systematic review of virtual and augmented reality exposure therapy and a meta-analysis of virtual reality exposure therapy. Journal of Psychiatric Research (November) https://doi.org/10.1016/j.jpsychires.2020.11.030.

Falconer, C.J., Rovira, A., King, J.A., Gilbert, P., Antley, A., Fearon, P., et al., 2016. Embodying self-compassion within virtual reality and its effects on patients with depression. BJPsych Open 2 (1), 74–80. https://doi.org/10.1192/bjpo.bp.115.002147.

Ferrer-García, M., Gutiérrez-Maldonado, J., 2012. The use of virtual reality in the study, assessment, and treatment of body image in eating disorders and nonclinical samples: a review of the literature. Body Image 9 (1), 1–11. https://doi.org/10.1016/j.bodyim.2011.10.001.

Finkelstein, S.L., Nickel, A., Barnes, T., Suma, E.A., 2010, March. Astrojumper: designing a virtual reality exergame to motivate children with autism to exercise. In: 2010 IEEE Virtual Reality Conference (VR). IEEE, pp. 267–268.

Freeman, D., Yu, L.-M., Kabir, T., Martin, J., Craven, M., Leal, J., et al., 2019. Automated virtual reality (VR) cognitive therapy for patients with psychosis: study protocol for a single-blind parallel group randomised controlled trial (gameChange). BMJ Open 9 (8).

García-Palacios, A., Hoffman, H.G., Kwong See, S., Tsai, A., Botella, C., 2004. Redefining therapeutic success with virtual reality exposure therapy. CyberPsychology and Behavior 4 (3) pubmed.ncbi.nlm.nih.gov/11710258/.

Garrett, B., Taverner, T., Masinde, W., Gromala, D., Shaw, C., Negraeff, M., 2014. A rapid evidence assessment of immersive virtual reality as an adjunct therapy in acute pain management in clinical practice. The Clinical Journal of Pain 30, 1089–1098 pubmed.ncbi.nlm.nih.gov/24535053.

Gerber, C.N., Kunz, B., van Hedel, H.J., 2016. Preparing a neuropediatric upper limb exergame rehabilitation system for home-use: a feasibility study. Journal of NeuroEngineering and Rehabilitation 13 (33), 1–12. https://doi.org/10.1186/s12984-016-0141-x.

Grichnik, K.P., Ferrante, F.M., 1991. The difference between acute and chronic pain. Mount Sinai Journal of Medicine 58 (3), 217–220. New York.

Gromala, D., Tong, X., Choo, A., Karamnejad, M., Shaw, C.D., April 2015. The virtual meditative walk: virtual reality therapy for chronic pain management. In: Proceedings of the 33rd Annual ACM Conference on Human Factors in Computing Systems, pp. 521–524.

Hacmun, I., Regev, D., Salomon, R., 2018. The principles of art therapy in virtual reality. Frontiers in Psychology 9, 2082. https://doi.org/10.3389/fpsyg.2018.02082.

Hayward, K.S., Kramer, S.F., Thijs, V., et al., 2019. A systematic review protocol of timing, efficacy and cost effectiveness of upper limb therapy for motor recovery post-stroke. Systematic Reviews 8, 187. https://doi.org/10.1186/s13643-019-1093-6.

Henderson, A., Korner-Bitensky, N., Levin, M., 2007. Virtual reality in stroke rehabilitation: a systematic review of its effectiveness for upper limb motor recovery. Topics in Stroke Rehabilitation 14 (2), 52–61.

Ho, E.S., Campos, A.A., Klar, K., Davidge, K., 2018. Evaluation of pain in pediatric upper extremity conditions. Journal of Hand Therapy 31, 206–214. https://doi.org/10.1016/j.jht.2018.02.004 pubmed.ncbi.nlm.nih.gov/17517575.

Hoffman, H.G., Doctor, J.N., Patterson, D.R., Carrougher, G.J., Furness, T.A., March 2000. 3rd. Virtual reality as an adjunctive pain control during burn wound care in adolescent patients. Pain 85 (1–2), 305–309. https://doi.org/10.1016/s0304-3959(99)00275-4. PMID: 10692634. pubmed.ncbi.nlm.nih.gov/10692634/.

Howett, D., Castegnaro, A., Krzywicka, K., Hagman, J., Marchment, D., Henson, R., et al., 2019. Differentiation of mild cognitive impairment using an entorhinal cortex-based test of virtual reality navigation. Brain 142 (6), 1751–1766.

Jannink, M.J., Van der Wilden, G.J., Navis, D.W., Visser, G., Gussinklo, J., Ijzerman, M., 2008. A low-cost video game applied for training of upper extremity function in children with cerebral palsy: a pilot study. CyberPsychology and Behavior 11 (1), 27–32 pubmed.ncbi.nlm.nih.gov/18275309/.

Jolly, K., Taylor, R., Lip, G., Greenfield, S., Raftery, J., 2007. The Birmingham Rehabilitation Uptake Maximisation Study (BRUM). Home-based compared with hospital-based cardiac rehabilitation in a multi-ethnic population: cost-effectiveness and patient adherence. Health Technology Assessment 11 (35) pubmed.ncbi.nlm.nih.gov/17767899.

Kiper, P., Szczudlik, A., Agostini, M., Opara, J., Nowobilski, R., Ventura, L., Tonin, P., Turolla, A., 2018. Virtual reality for upper limb rehabilitation in subacute and chronic stroke: a randomized controlled trial. Archives of Physical Medicine and Rehabilitation 99 (5), 834–842 pubmed.ncbi.nlm.nih.gov/29453980.

Laver, K.E., Lange, B., George, S., Deutsch, J.E., Saposnik, G., Crotty, M., 2017. Virtual reality for stroke rehabilitation. Cochrane Database of Systematic Reviews (11) pubmed.ncbi.nlm.nih.gov/29156493.

Leff, J., Williams, G., Huckvale, M., Arbuthnot, M., Leff, A.P., 2014. Avatar therapy for persecutory auditory hallucinations: what is it and how does it work? Psychosis 6 (2), 166–176. https://doi.org/10.1080/17522439.2013.773457.

Lohse, K.R., Hilderman, C.G., Cheung, K.L., Tatla, S., Van der Loos, H.M., 2014. Virtual reality therapy for adults post-stroke: a systematic review and meta-analysis exploring virtual environments and commercial games in therapy. PLoS One 9 (3), e93318. March 28. www.ncbi.nlm.nih.gov/pmc/articles/PMC3969329.

Lotan, M., Yalon-Chamovitz, S., Weiss, P.L.T., 2009. Improving physical fitness of individuals with intellectual and developmental disability through a Virtual Reality Intervention Program. Research in Developmental Disabilities 30 (2), 229–239.

Martens, M.A.G., Antley, A., Freeman, D., Slater, M., Harrison, P.J., Tunbridge, E.M., 2019. It feels real: physiological responses to a stressful virtual reality environment and its impact on working memory. Journal of Psychopharmacology 33 (10), 1264–1273.

Morina, N., Ijntema, H., Meyerbröker, K., Emmelkamp, P.M.G., 2015. Can virtual reality exposure therapy gains be generalized to real-life? A meta-analysis of studies applying behavioral assessments. Behaviour Research and Therapy 74, 18–24. https://doi.org/10.1016/j.brat.2015.08.010.

Neilson, A.R., Jones, G.T., Macfarlane, G.J., Walker-Bone, K., Burton, K., Heine, P.J., McCabe, C.S., McConnachie, A., Palmer, K.T., Coggon, D., McNamee, P., 2019. Cost-utility of maintained physical activity and physiotherapy in the management of distal arm pain: an economic evaluation of data from a randomized controlled trial. Family Practice 36 (2), 179–186. https://doi.org/10.1093/fampra/cmy047. March 20, PMID: 29878103; PMCID: PMC6425461.

Opdyke, D., Williford, J.S., North, M., 1995. Effectiveness of computer-generated (virtual reality) graded exposure in the treatment of acrophobia. Am J psychiatry 1 (152), 626–628.

Opris, D., Pintea, S., Garcia-Palacios, A., Botella, C., Szamosközi, S., David, D., 2012. Virtual reality exposure therapy in anxiety disorders: a quantitative meta-analysis. Depression and Anxiety 29 (2), 85–93. https://doi.org/10.1002/da.20910.

Parsons, S., Cobb, S., 2011. State-of-the-art of virtual reality technologies for children on the autism spectrum. European Journal of Special Needs Education 26 (3), 355–366. https://doi.org/10.1080/08856257.2011.593831.

Phelan, I., Furness, P.J., Fehily, O., Thompson, A.R., Babiker, N.T., Lamb, M.A., Lindley, S.A., 2019. A mixed-methods investigation into the acceptability, usability, and perceived effectiveness of active and passive virtual reality scenarios in managing pain under experimental conditions. Journal of Burn Care and Research 40 (1), 85–90. https://doi.org/10.1093/jbcr/iry052.

Pillai Riddell, R.R., Racine, N.M., Gennis, H.G., Turcotte, K., Uman, L.S., Horton, R.E., Ahola Kohut, S., Hillgrove Stuart, J., Stevens, B., Lisi, D.M., 2015. Non-pharmacological management of infant and young child procedural pain. Cochrane Database of Systematic Reviews (12), CD006275. https://doi.org/10.1002/14651858.CD006275.pub3.

Piron, L., Cenni, F., Tonin, P., Dam, M., 2001. Virtual reality as an assessment tool for arm motor deficits after brain lesions. Studies in Health Technology and Informatics 81, 386–392.

Plante, T.G., Aldridge, A., Bogden, R., Hanelin, C., 2003. Might virtual reality promote the mood benefits of exercise? Computers in Human Behavior 19, 4 495–509. www.sciencedirect.com/science/article/abs/pii/S0747563202000742.

Riva, G., Mantovani, F., Wiederhold, B.K., 2020. Positive Technology and COVID-19, Cyberpsychology, Behavior, and Social Networking, pp. 581–587. https://doi.org/10.1089/cyber.2020.29194.gri.

Rizzo, A., Koenig, S.T., 2017. Is clinical virtual reality ready for primetime? Neuropsychology 31 (8), 877.

Salawu, A., Green, A., Crooks, M.G., Brixey, N., Ross, D.H., Sivan, M., 2020. A proposal for multidisciplinary tele-rehabilitation in the assessment and rehabilitation of COVID-19 survivors. International Journal of Environmental Research and Public Health 17 (13), 4890. https://doi.org/10.3390/ijerph17134890.

Sandlund, M., McDonough, S., Häger-Ross, C.H., 2009. Interactive computer play in rehabilitation of children with sensorimotor disorders: a systematic review. Developmental Medicine and Child Neurology 51 (3), 173–179 pubmed.ncbi.nlm.nih.gov/19191834.

Saposnik, G., Levin, M., 2011. Virtual reality in stroke rehabilitation: a meta-analysis and implications for clinicians. Stroke 42 (5), 1380–1386. www.ahajournals.org/doi/full/10.1161/STROKEAHA.110.605451.

Sharan, D., Ajeesh, P.S., Rameshkumar, R., Mathankumar, M., Paulina, R.J., Manjula, M., 2012. Virtual reality based therapy for post operative rehabilitation of children with cerebral palsy. Work 4, 3612–3615. https://doi.org/10.3233/WOR-2012-0667-3612.

Sonney, J., Björling, E.A., Rodriguez, S., Carr, N., 2021. Pivoting to "No contact": a protocol for conducting a virtual reality relaxation home study for teens amidst the COVID-19 pandemic. Journal of Paediatric Healthcare. https://doi.org/10.1016/j.pedhc.2021.01.002.

Sorkin, A., Weinshall, D., Modai, I., Peled, A., 2006. Improving the accuracy of the diagnosis of schizophrenia by means of virtual reality. American Journal of Psychiatry 163 (3), 512–520. https://doi.org/10.1176/appi.ajp.163.3.512.

Sun, H., 2012. Exergaming impact on physical activity and interest in elementary school children. Research Quarterly for Exercise & Sport 83, 212–220. https://doi.org/10.5641/027013612800745248.

Threapleton, K., Drummond, A., Standen, P., 2016. Virtual rehabilitation: what are the practical barriers for home-based research? Digital health 2. https://doi.org/10.1177/2055207616641302, 2055207616641302.

Valmaggia, L., 2017. The use of virtual reality in psychosis research and treatment. World Psychiatry 16 (3), 246–247. https://doi.org/10.1002/wps.20443.

Weerdmeester, J., van Rooij, M.M., Maciejewski, D.F., Engels, R.C., Granic, I., 2021. A Randomized Controlled Trial Assessing the Efficacy of a Virtual Reality Biofeedback Video Game: Anxiety Outcomes and Appraisal Processes.

CHAPTER THREE

The role of innovative telehealth system in revolutionizing healthcare

Vishnu Sharma[1], Tarun Kumar Kumawat[1,2] Rashi Garg[3], Agrima Bhatt[4], Nirat Kandwani[5], Gargi Dubey[1] and Manish Biyani[1,6]

[1]Department of Biotechnology, Biyani Girls College, Jaipur, Rajasthan, India
[2]Department of Botany, University of Rajasthan, Jaipur, Rajasthan, India
[3]Department of Genomic Center, Bioinformatics, All India Institute of Medical Sciences, New Delhi, India
[4]School of Biology, MIT World Peace University, Pune, Maharashtra, India
[5]School of Humanities, Mohanlal Sukhadia University, Udaipur, Rajasthan, India
[6]Department of Bioscience and Biotechnology, Japan Advanced Institute of Science and Technology, Ishikawa, Japan

3.1 Introduction

The word "healthcare system" refers to systems, people, and infrastructure responsible for providing health treatment to individuals. The conventional healthcare system has a wide range of features, defined as the practice of the diagnosis, control, and manipulation for patients which promote the prevention of infection and diseases (Baars and Hamre, 2017). The healthcare system has grown into one of the most competitive shareholders in the world during the past decades. New types of equipment, therapeutic standards are being invented in the continued evolution of the worldwide burden and pattern of diseases (Talib et al., 2015). For example, in developed countries, primary health care (PHC) is mostly provided at primary levels with "general practitioners — family physicians (GPs)," but developing nations not have such PHCs (Mariolis et al., 2008). The conventional health system consists mainly of four components at the constitutional level: patients; healthcare providers; hospitals and nursing homes; and financial and political authorities (Sun et al., 2021). Distortion of these components affects the health system directly. In pandemic times, the influx of funds enhanced political recognition and the availability of trained workers responsible for the healthcare industry's upheaval (Moon et al., 2010; Lavallee et al., 2021).

As the population grows and ages, the health system becomes bewildering assortment of highly dispersed industries, where health experts cooperate, but interdisciplinary team does not integrate (Greiner and Knebel, 2003). The obstacles of delivering proper prognosis and management facilities owing to lack of health insurance, expenses of transport, failure in the provision of health services, diagnosis costs, etc. become challenging-full throughout the world (Mansberger et al., 2013; Strasser et al., 2016).

Extended Reality for Healthcare Systems, Volume 1
ISBN 978-0-323-98381-5, https://doi.org/10.1016/B978-0-323-98381-5.00015-5

During this period, rural healthcare systems are widely affected by availability of critical hospitals' infrastructures, lack of qualified physicians, deficiencies of transport facilities, convenience of insured economical flow for medical expenses, as compared with metropolitan healthcare (Panagariya, 2014; Weeks, 2018; Kumar and Kumar, 2018; Malatzky et al., 2020; Leider et al., 2020; Sirili and Simba, 2020; Weiss et al., 2020).

Because of such drawbacks, to fight in the modern pandemic period, the healthcare system is switching to a technology-based communication healthcare system, i.e., "telemedicine." Telemedicine is a phrase meaning 'remote healing.' It offers reliable and rapid health services and enhances therapeutic and health records. Throughout this regard, all professionals use ICT (information and communication technologies) to share medical data used in the diagnosis, treatment, and prevention of diseases (Shirzadfar, 2017; Gogia, 2020). All popular features such as follow-up visits, management of chronic diseases, prescription, professional consultation, etc., which can be rendered remotely, are managed through video teleconferencing (Craig and Patterson, 2005; Ryu, 2010; Mahar et al., 2018; Gupta et al., 2019). It provides huge promises for quicker, better, less costly, and more convenient care of goods and services.

The earliest history of telehealth was chronicled by Alexander Graham Bell through the use of the telephone to get assistance from his colleague Mr Watson after spitting acid on his pants in 1876 (Mullick et al., 2020). Other model of the telehealth was recorded during the American Civil War, in which the telegraph was used to convey losses and order medical supplies (Gogia, 2020). At the time of World War I, radio communication was invented to relay medical information.

Like this, the invention of long-distance ECG transmission in 1905 by the Dutch physician, Willem Einthoven revolutionized in the origins of telemedicine system. Hugo Gernsback, the "Father of Science Fiction," termed this gadget Teledactyl (Tele – distance; Dactyl – finger) during his idea of remote health visits in 1925 (Scott and Mars, 2015). In 1964, the establishment of a closed video system connecting Omaha's Nebraska Psychiatric Institute and Norfolk's state mental hospital provided a new direction toward interactive consultations between specialists and general practitioners, as well as education and training (Whitten et al., 2010; Mullick et al., 2020).

In India, telehealth was introduced by developing the first telecardiology for the teletransmission ECGs at Gajara Raja Medical College, Gwalior, India, in 1975. The device allowed wireless transmission of electrocardiographic data from the patient's home and the central station of the ICU (Gogia, 2020). The technology was also proved to track patients in distant regions by peacekeepers (Serhani et al., 2020). For the past decades, Jiva Ayurveda is providing the first Ayurveda telemedicine facility in India (Singh et al., 2018). The Indian Space Research Organization initiated an endeavor to expand the mobile telemedicine in India's Kumbh festival in compliance with the Ministry of Health and Family Welfare (MOHFW), India, AIIMS Bhopal, NRC and SGPGI, Lucknow (Gogia, 2020).

Today, when the COVID-19 outbreak is increasing with unprecedented challenges across the world, the world's best healthcare systems are in failure. In this increased view of the outbreak, telemedicine has taken a lead in providing treatment for healthcare providers and patients with social distance quickly and continuing management of chronic diseases (Shamasunder et al., 2020; del Rio and Malani, 2020). Telemedicine has shown that it can assist psychologically to patients and caregivers without getting exposed to the virus. The workload of the diagnosis and treatment at tertiary hospital centers has also been lowered by way of telemedicine (Agarwal et al., 2020; Schulz et al., 2020).

In seven components, this chapter addresses the approaches and associated attributes of telemedicine. The first and second sections cover the telemedicine concept, its roots, and growth as well as the different types of telemedicine frameworks. Section 3 discusses the advantages of health records over conventional healthcare paradigms. Section 4 discusses the emerging role of telehealth in the continuing pandemic period, outlining the advantages of remote health care. Artificial intelligence in telehealth is presented in Section 5. Future prospects for telemedicine studies are discussed in Section 6. Section 7 finishes with conclusions.

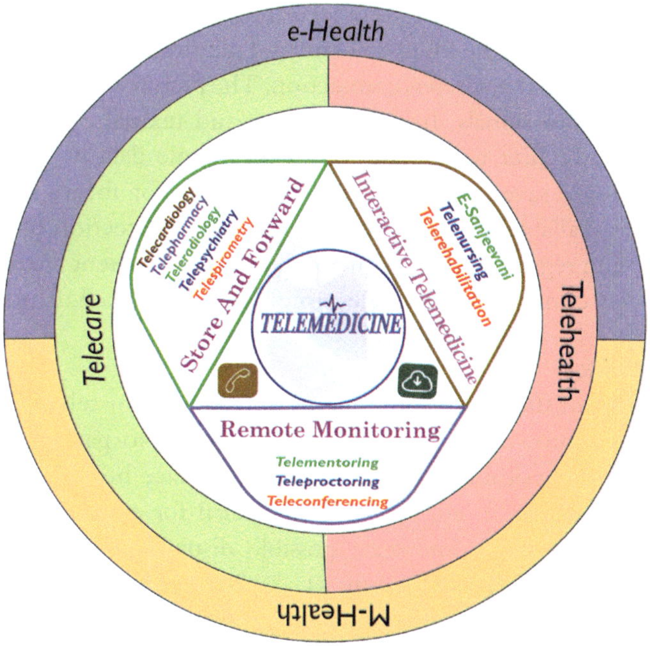

Figure 3.1 Framework conceptualizing the relationships among eHealth, telehealth, telecare, and telemedicine.

3.2 Types of telemedicine

Telemedicine may be divided into three major categories based on its versatility and effectiveness (Fig. 3.1):

3.2.1 Store-and-Forward

Telemedicine exchanges medical data, such as biomedical images and biosignals, for offline assessment with medical experts and physicians (Bhowmik et al., 2013). Patients communicate relevant health records such as blood pressure, weight, pulse rate, oxygen saturation, and diabetes level from their home without an appointment via telephonic conversations (Whitten et al., 2010). During the installation of a remote telemedicine interface, the store-and-forward method is appropriately utilized between a primary care practitioner and a qualified doctor (Smith, 2018). Frequent uses in pathology, radiology, dermatology, and many other specialist medical disciplines are accessible with such type of telemedicine. Email and WhatsApp are also most widely utilized models of store-and-forward interaction (Craig and Petterson, 2005; Gogia, 2020).

3.2.1.1 Telecardiology

By leveraging the strength of telecom, this type of store-and-forward telemedicine provides remote care and prognosis for cardiovascular disease. A 12-lead ECG mobile device is used to record ECGs. After the ECG is obtained, a picture of the tracing as an audio wave is communicated through the wireless connection. The picture on monitor of another side will be reconverted into signals. Informal reports and textual descriptions are sent to patients simultaneously after the assessment of the whole data by professionals. Every individual's records are saved on an electronic database for future reference. See-like single-lead ECG is available for better interpretation and detection of cardiac problems at home. Information regarding the health condition of the patient is stored on the watch and can be remotely controlled to the telecardiology center (Backman et al., 2010).

3.2.1.2 Telepharmacy

Telepharmacy is one of the useful components of telehealth or telemedicine, which is a developing trend for providing pharmaceutical treatment to patients in remote places where they may lack physical access to pharmacists or face healthcare shortages (Rangasamy et al., 2011; Kilova et al., 2021). It is critical for expanding patient access to pharmaceutical therapy and minimizing possible dispensing mistakes in community pharmacies (Poudel et al., 2016; Ibrahim et al., 2020a). Due to the minimal risk of transmission, remote pharmacist interventions garnered significantly greater attention during the coronavirus disease 2019 (COVID-19) outbreak (Ameri et al., 2020; Muflih et al., 2021; Mohamed Ibrahim et al., 2021).

3.2.1.3 *Teleradiology*

In this kind of telemedicine, instead of carrying around the film or sending radiological images to a radiologist for interpretation, digital radiographs and other images, such as computed tomography (CT), magnetic resonance imaging (MRI), ultrasound, and nuclear medicine scans, are sent via a network link from an image review station to the interpretation center (Rangasamy et al., 2011; Krupinski, 2014). All radiography equipment is connected to the DICOM (Digital Imaging and Communications in Medicine) network. Viewing, manipulation, measurements, three-dimensional reconstructions, reporting and storage, and transmission of results are all performed using software such as EFilm (Kumar and Krupinski, 2008; Jankharia and Burute, 2009).

3.2.1.4 *Telepsychiatry*

Telepsychiatry, application of telehealth in the discipline of psychiatry, is defined as the use of digital communication and information technology to treat clinical mental health at a distance (Saeed et al., 2012; Malhotra et al., 2013; Naskar et al., 2017; Perera et al., 2020). Telepsychiatry is known colloquially as telemental health or E-mental health. This concept encompasses various communication modes such as telephony contact, telegraph, Internet mail, and video conferencing in real time (Conn et al., 2015). In the current COVID-19 outbreak, telepsychiatry is stayed in the new world of social distance to encouraging patients and to enhance access to mental healthcare (Ibrahim et al., 2020b; Ramadas et al., 2020; Guaiana et al., 2021).

3.2.1.5 *Telespirometry*

Telespirometry is a newly created technology that utilizes a portable spirometer to communicate the patient's lung ventilatory values by telephone to a distant monitoring center. Therefore, telespirometry can detect early indications of allergies (Bruderman and Abboud, 1997; Fois et al., 2014; Kumar et al., 2020). The continuing epidemic has generated several concerns about the management of COPD patients. In this concern, telespirometry provides an acceptable alternative to improve the lung's breathing capacity (Crimi et al., 2020; Ferenczi et al., 2020; Morais-Almeida et al., 2021; Rutkowski, 2021).

3.2.2 Remote monitoring

Remote monitoring, sometimes referred to as self-monitoring, allows medical practitioners to observe a patient through virtual technology. It is mainly used to treat serious diseases such as cardiovascular disease, blood sugar, and bronchitis, as well as high blood pressure and obesity (El-Rashidy et al., 2021). Numerous scientists are currently developing and testing sensors for health monitoring. Noninvasive or percutaneous wireless "smart" devices are used here that examine the individual in real time (Volterrani and Sposato, 2019). A "smart band-aid," digital drugs, and clothing embedded with

nanostructures are all examples of these sensors. Telemedicine and remote monitoring have accelerated during the ongoing coronavirus pandemic (COVID-19). The modern era's telemedicine and telemonitoring offerings contribute to the development of a ubiquitous, efficient, and sustainable system. Pulse oximetry monitoring is becoming more critical in the management of respiratory diseases, particularly in the present epidemics of respiratory viral infections. Due to the relationship between SpO_2 and lung injury, it became a critical component of clinical evaluation for differentiating those who require close monitoring and hospitalization. During home quarantine, remote clinical pulse oximetry enables patients to objectively report their SpO_2 level and pulse rate and their symptoms (Whittenet al., 2010; Desai and Diamond, 2021).

3.2.3 Real-time interactive telemedicine

Interactive telemedicine, which is part of the telemedicine approach, delivers services to patients through real-time contact distinguished by distance, such as live chat, digital communications, and in-home visits. Face-to-face consultations rely on demographic data such as the patient's disease history, a physical examination, a mental evaluation, and ophthalmic and dental tests (Flodgren et al., 2015; Segato and Masella, 2017; Sasikala et al., 2018; Samonte et al., 2019). Both clinical experts and patients unanimously agreed to use telemedicine for healthcare during the current epidemic (Costich et al., 2021). E-Sanjeevani is a recently launched telemedicine initiative under Ayushman Bharat Pradhan Mantri Jan Arogya Yojana. It was started by Health and Wellness Centers at the Centre for Development of Advanced Computing (C-DAC) in Mohali. Throughout the COVID-19 outbreak, the E-Sanjeevani OPD initiative allowed teleconsultation between patients and clinicians receiving vital healthcare services while remaining at home.

3.2.3.1 Telenursing

Telenursing is remote nursing services for diagnosing and monitor chronic illnesses and symptoms (Schlachta-Fairchild et al., 2010). It is attracting attention as the result of low expenditures and strong accessibility of health to patients, especially in rural areas (Franek, 2012). Telenursing provides its members absolutely limitless opportunities to enhance the overseas nursing profession (Chang et al., 2021; Salles et al., 2017). National Institute of Mental Health and Neurosciences, Karnataka, and Postgraduate Institute of Medical Education and Research, Chandigarh, are educating nursing staff and conducting patient case talks through telenursing to bring telemedicine assistance within a manner of medication monitoring, tracking, data collecting, and pain management. Telenursing's implementation during the COVID-19 outbreak has contributed to a reduction in the total of PPE required for the safeguarding of healthcare professionals (Rawat, 2018). During the recent COVID-19 pandemic, telenursing allowed ICTs to

optimized geometric collect patient data from any distant area, administer medical care via telemedicine, and offer remote support (Barbosa et al., 2016; Li et al., 2017).

3.2.3.2 Telerehabilitation

Telerehabilitation is the delivery of clinical rehabilitation services through digital communications. The video conference and webcams frequently used to communicate symptoms and clinical progress are great resources for this objective (Brennan et al., 2009; Schmeler et al., 2015; Novaes, 2020; Knepley et al., 2021). It has instilled a sense of personal autonomy and empowerment in the patient and empowers them to stay aware of their disease's treatment during the COVID-19 outbreak (Bettger and Resnik, 2020; Albahrouh and Buabbas, 2021). Telerehabilitation is used in physical therapy, psychology, speech pathology, etc. There are individual or group therapies also available such as motor training activities, speech therapy, virtual reality, robotic treatment, goal setting, and group exercise (Theodoros and Russell, 2008; Peretti et al., 2017; Knepley et al., 2021).

3.2.3.3 mHealth

Telecommunication infrastructure is playing a crucial role in enabling people and organizations to stay connected and functioning in this unprecedented crisis period of COVID-19 (Williams et al., 2020; Khan, 2021). Digital technologies, which include smartphones, tablets, laptops, and other mobile devices, as well as hundreds of software programs linked to telemedicine, are in use to provide healthcare facilities in the form of mHealth in the present time (Albabtain et al., 2014; Aslani et al., 2020; Almalki and Giannicchi, 2021). The mHealth method is taken up by users to monitor their health, to encourage good health, to increase awareness, to provide assistance in the clinic, and to help patients make decisions while dealing with the present epidemic (Asadzadeh and Kalankesh, 2021). Numerous health-related apps are now available that can track anything from a diabetic's blood sugar level to an individual's daily water consumption (Kondylakis et al., 2020). mHealth technology has proven to aid in smoking cessation, weight loss, physical activity, diabetic control, STD research and treatment, and hypertension diagnosis and control (Fiordelli et al., 2013, Gogia, 2020).

3.3 Telehealth benefits over traditional healthcare

Scientific advancements and epidemics have resulted in major changes to the healthcare system. During an epidemic, the authorities advised residents to remain at home to prevent the virus from spreading. As a result, healthcare professionals are being pushed to reconsider their approach to providing treatment and clinic services to children and families. Telehealthcare, according to research, has the potential to bridge the gap between current needs and future advancements (Langarizadeh et al., 2017; Maryam Alvandi 2017).

Compared with the conventional healthcare system, remote healthcare services substantially save healthcare service expenditures, minimize needless nonemergency hospital visits, and remove transportation costs for routine checks (Heinzelmann 2005). Additionally, it decreases the amount of time the patient must wait for an appointment owing to the high volume of people already on the appointment list (Willmottand Arrowsmith, 2013). Wireless connectivity of the telehealthcare system enables patients to have one-on-one consultations with their physicians. Occasionally, a patient's condition is serious, and it could be lifesaving (Bhowmik et al., 2013; Maryam Alvandi, 2017). The artificial intelligence of telehealthcare systems helps to create approaches that combine the pool of available healthcare professionals with necessary skills and knowledge (Kuziemsky et al., 2019; Stewart et al., 2018). Clinicians can compile computerized data of most patients and transmit it to several other physicians using machine intelligence methods (Rajkomar et al., 2018; Davenport and Kalakota, 2019).

The existing ingestible sensors in telehealth enable the accurate intervention of various kinds of organ failure and illness that are not traceable using other approaches (Beardslee et al., 2020). Robotic surgery and treatment in telehealth systems provide previously unattainable prospects for automated objective skill evaluation and a fast management tool. These methods provide up new avenues for surgical skill development and customized training (Fard et al., 2018; Murphy, 2012; Reddy et al., 2019).The traditional healthcare system takes lots of time and makes extensive use of laboratory equipment, while telehealthcare conducts onsite testing with minimal equipment and provides point-of-care care (Fig. 3.2) (Nichols, 2020).

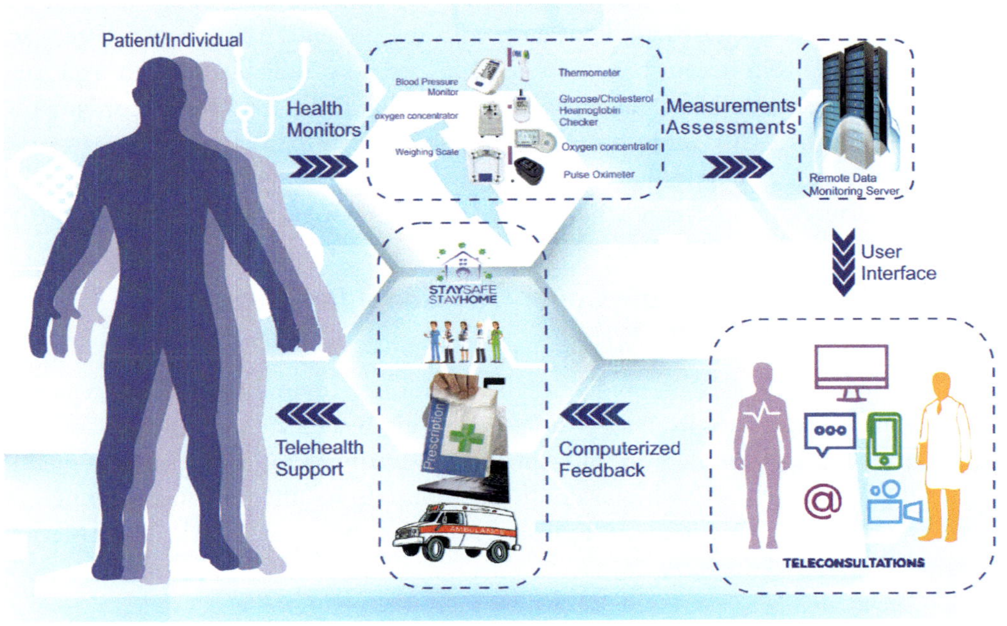

Figure 3.2 Application of the telemedicine technology.

3.4 Role of telehealth in pandemic period

Telemedicine has grown in prominence in recent years as a result of the increasing deployment and development of digital technology. The occurrence of epidemics or infectious diseases has necessitated use of progressively innovative strategies, which has resulted in a higher use of electronic health records at various time of illness, as was the case with the severe acute respiratory syndrome outbreak in 2003, MERS-CoV in 2013, and, most recently, SARS-CoV-2 health crisis (Enfield et al., 2015; Galiero et al., 2020; Vidal-Alaball et al., 2020). Throughout the COVID-19 epidemic, enforced social isolation and a lack of viable therapies have rendered telemedicine the most secure method of communication between patients, infected and uninfected, and doctors (Heymann, 2020; Del Rio and Malani, 2020; Wosik et al. 2020). Telehealth has enabled societies to practice "social distance" or "stay-at-home" policy along with "medical distancing" during this pandemic by separating healthcare personnel from patients on a discretionary basis (Monaghesh and Hajizadeh, 2020; Kutlu et al., 2020; Sharma et al., 2020). The current pandemic has compelled the hospital to restructure the ICU division by including telehealth equipment that is essential for a patient's care (Fig. 3.2) (Wosik et al., 2020; Dale et al., 2021; da Silva and Barbosa, 2021).

Telemedicine's reach will continue to grow, as providers seek worldwide expansion. This will not only assist physician shortages and global physician maldistribution but may also offer patients with uncommon illnesses with more avenues for highly specialized treatment (Kichloo et al., 2020). Telemedicine's critical role in postpandemic recovery is to emerge sustainable systems for securing patient data and assisting patients throughout their recovery (Wosik et al., 2020). The telehealthcare system benefits several patients who want to preserve their privacy and get laboratory results in full security (Reeves et al., 2020). To promise telemedicine's growth in the postpandemic world, efforts toward expanding access and educating consumers must be taken (Kichloo et al., 2020).

3.5 Heterogeneous systems for telehealth

Telehealth has recently begun via the use of artificial intelligence in many areas of the profession including distribution of computerized healthcare and personal consultations and provided people experience. Artificial intelligence potentially helps to address the problem and provide the algorithms for care providers with the necessary clinical abilities to the urgent demand for such competencies (Kuziemsky et al., 2019). Artificial intelligence also assists in addressing the need for seamless communications and connections across various elements of healthcare delivery (Davenport and Kalakota, 2019; Latif et al., 2017). This shift from a purely human-based system to one that is intelligent and automated will provide physicians with bionic assistance by substantially supplementing or enhancing their expertise. Even the most inexperienced nurse or

physician will be able to operate at the same level as the most accomplished specialists (Khan et al., 2020).

There are mainly two kinds of artificial intelligence devices. The first type includes machine learning (ML) techniques for evaluating data sets, such as that found in image analysis and genomics. Such algorithms are used in medical applications to cluster people's characteristics or to predict the probability of disease consequences (Darcy et al., 2016). The second category comprises language processing technologies that retrieve information from unstructured sources such as medical documentation journals to complement and improve organized patient records (Jiang et al., 2017). Artificial intelligence technologies are being utilized in the healthcare sector through robots (Kalaiselvan et al., 2021).

The burgeoning telehealthcare, with a growing variety of wireless communication, necessitates the establishment of a robust network system capable of efficiently linking patients, healthcare providers, hospital instruments, and others for the purpose of data exchange. The sensors will collect data such as the electrocardiogram, heart rate, temperature, and blood glucose through next-generation wristbands and other remote access and primary carer (Dananjayan and Raj, 2021). The next generation of communication networks (5G or future generations) will reshape the medical system by smartly increasing the quality of care, managing resource distribution among developed and developing countries, alleviating the financial cost of healthcare (Li, 2019; Ahad et al., 2019). The development of the next generation of wireless connectivity, fifth-generation (5G) mobile networks, or future Internet connectivity, is able to transform telemedicine and healthcare in broad ranges (Dananjayan and Raj, 2021).

This new era of data transmission will bring enhanced connection, data storage, and a diverse set of interconnected apps and devices. The development of the wireless cloud computing will be enabled by extensive computing power combined with virtual system architecture. Connecting billions of sensors through advanced digital networks will enable advances in healthcare (Magsi et al., 2018). Self-determination medicine will develop in the 5G era as computers increase their capacity to analyze massive amounts of data. Rather than the already prevalent practitioner medical assessment and treatment model, the self-determination biological approach will considerably increase the autonomy of the patient by allowing for patient engagement throughout the therapy (Li, 2019).

3.6 Directions for future research

Telemedicine is an attractive way of managing episodic, acute, and preventive treatment, which improves diagnostic evidence. In today's developing world, telemedicine securely transports healthcare from the hospital or clinic to the patient's house and makes them overcome their sickness. We can anticipate that continuous advancement in

wireless telecommunications, monitoring devices, and nanostructures is paving the path for future acceptance of telehealth medicare. This approach is sustaining the cost-effective, time-effective systems to provide patients with focused access to targeted treatment. Prospective research is required to improve routinely efficiency levels following present and prospective epidemics.

Similarly, after the experience to COVID-19, there is a demand of the developing world or regions with inadequate infrastructure to build structured telemedicine to fulfill the deficiency of the healthcare system. It should be upgraded and should connect providers at primary health centers, referral hospitals, and tertiary facilities. Although telemedicine is now utilizing frequently in a few specialty sectors of the medical system, additionally it could be beneficial for it to become the standard in other fields as well. For example, cancer sufferers seek frequent prescriptions over the Internet to ensure that they are doing all necessary to combat their condition. Likewise, future research for telemedicine can be a milestone for fighting against severe disease in remote areas.

3.7 Conclusions

Telemedicine helps to the redesign of health services by providing society with virtual settings for care and well-being, which benefits both healthcare professionals and patients. It facilitates information sharing and collaboration between health specialists and laymen, with the motto "anyone, anytime, anywhere healthcare at a reasonable cost." It has required an extraordinary push to provide mHealth in distant regions during the current epidemic era. This momentum may be sustained via the use of a wide range of high-speed satellite and terrestrial wireless communications connections, resource centralization and synchronization, and government support for concerns such as inclusive national e-health policy development and legal/ethical considerations.

3.8 Conflicts of interest

The authors declare no conflict of interest.

References

Agarwal, N., Jain, P., Pathak, R., Gupta, R., 2020. Telemedicine in India: a tool for transforming health care in the era of COVID-19 pandemic. Journal of Education and Health Promotion 9, 190.

Ahad, A., Tahir, M., Yau, K.-L.A., 2019. 5G-Based smart healthcare network: architecture, taxonomy, challenges and future research directions. IEEE Access 7, 100747–100762.

Albabtain, A.F., Almulhim, D.A., Yunus, F., Househ, M.S., 2014. The role of mobile health in the developing world: a review of current knowledge and future trends. Journal of Selected Areas in Health Informatics 4 (2), 10–15.

Albahrouh, S.I., Buabbas, A.J., 2021. Physiotherapists' perceptions of and willingness to use telerehabilitation in Kuwait during the COVID-19 pandemic. BMC Medical Informatics and Decision Making 21, 122.

Almalki, M., Giannicchi, A., 2021. Health apps for combating COVID-19: descriptive review and taxonomy. JMIR mHealth and uHealth 9, e24322.

Alvandi, M., 2017. Telemedicine and its role in revolutionizing healthcare delivery. The American Journal of Accountable Care 5 (1), e1–e5.

Ameri, A., Salmanizadeh, F., Bahaadinbeigy, K., 2020. Tele-pharmacy: a new opportunity for consultation during the COVID-19 pandemic. Health Policy and Technology 9, 281–282.

Asadzadeh, A., Kalankesh, L.R., 2021. A scope of mobile health solutions in COVID-19 pandemics. Informatics in Medicine Unlocked 23, 100558.

Aslani, N., Lazem, M., Mahdavi, S., Garavand, A., 2020. A review of mobile health applications in epidemic and pandemic outbreaks: lessons learned for COVID-19. Archives of Clinical Infectious Diseases 15 (4), e103649.

Baars, E.W., Hamre, H.J., 2017. Whole medical systems versus the system of conventional biomedicine: a critical, narrative review of similarities, differences, and factors that promote the integration process. Evidence-based Complementary and Alternative Medicine 1–13, 2017.

Backman, W., Bendel, D., Rakhit, R., 2010. The telecardiology revolution: improving the management of cardiac disease in primary care. Journal of the Royal Society of Medicine 103, 442–446.

Barbosa, I. de A., Silva, K.C., da, C.D., da, S., da, V.A., Silva, M.J.P. da, 2016. O processo de comunicaçãonaTelenfermagem: revisãointegrativa. Revista Brasileira de Enfermagem 69, 765–772.

Beardslee, L.A., Banis, G.E., Chu, S., Liu, S., Chapin, A.A., Stine, J.M., Pasricha, P.J., Ghodssi, R., 2020. Ingestible sensors and sensing systems for minimally invasive diagnosis and monitoring: the next frontier in minimally invasive screening. ACS Sensors 5, 891–910.

Bettger, J.P., Resnik, L.J., 2020. Telerehabilitation in the age of COVID-19: an opportunity for learning health system research. Phys Ther. Physical Therapy & Rehabilitation Journal 100 (11), 1913–1916.

Bhowmik, D., Singh, R.K., Pradesh, A., Advance, M., 2013. Telemedicine- an innovating healthcare system in India. The Pharma Innovation - Journal 2 (4), 1–5.

Brennan, D.M., Mawson, S., Brownsell, S., 2009. Telerehabilitation: enabling the remote delivery of healthcare, rehabilitation, and self management. In: Studies in Health Technology and Informatics.

Bruderman, I., Abboud, S., 1997. Telespirometry: novel system for home monitoring of asthmatic patients. Telemedicine Journal 3, 127–133.

Chang, M.-Y., Kuo, F.-L., Lin, T.-R., Li, C.-C., Lee, T.-Y., 2021. The intention and influence factors of nurses' participation in telenursing. Informatics 8, 35.

Conn, D., Gajaria, A., Madan, R., 2015. Telepsychiatry: effectiveness and feasibility. Smart Homecare Technology and TeleHealth 59–67.

Costich, M., Robbins-Milne, L., Bracho-Sanchez, E., Lane, M., Friedman, S., 2021. Design and implementation of an interactive, competency-based pilot pediatric telemedicine curriculum. Medical Education Online 26, 1911019.

Craig, J., Patterson, V., 2005. Introduction to the practice of telemedicine. Journal of Telemedicine and Telecare 11, 3–9.

Crimi, C., Impellizzeri, P., Campisi, R., Nolasco, S., Spanevello, A., Crimi, N., 2020. Practical considerations for spirometry during the COVID-19 outbreak: literature review and insights. Pulmonology 27 (5), 438–447.

da Silva, F.C.T., Barbosa, C.P., 2021. The impact of the COVID-19 pandemic in an intensive care unit (ICU): Psychiatric symptoms in healthcare professionals. Progress in Neuro-Psychopharmacology and Biological Psychiatry 110, 110299.

Dale, C.R., Starcher, R.W., Chang, S.C., Robicsek, A., Parsons, G., Goldman, J.D., Vovan, A., Hotchkin, D., Gluckman, T.J., 2021. Surge effects and survival to hospital discharge in critical care patients with COVID-19 during the early pandemic: a cohort study. Critical Care 25, 70.

Dananjayan, S., Raj, G.M., 2021. 5G in healthcare: how fast will be the transformation? Irish Journal of Medical Science 190 (1971), 497–501.

Darcy, A.M., Louie, A.K., Roberts, L.W., 2016. Machine learning and the profession of medicine. JAMA 315 (6), 551–552.

Davenport, T., Kalakota, R., 2019. Digital Technology the potential for artificial intelligence in healthcare. Future Healthcare Journal 6 (2), 94–98.

del Rio, C., Malani, P.N., 2020. COVID-19—new insights on a rapidly changing epidemic. JAMA 323 (14), 1339–1340.

Desai, N.R., Diamond, E.J., 2021. Emerging role of remote patient monitoring in pulmonary care. Chest 159, 477–478.

El-Rashidy, N., El-Sappagh, S., Islam, S.M.R., El-Bakry, H., M., Abdelrazek, S., 2021. Mobile health in remote patient monitoring for chronic diseases: principles, trends, and challenges. Diagnostics 11, 607.

Enfield, K., Mehring, B., Carpenter, R., Rheuban, K., Cattell-Gordon, D., Gunnell, D., Corey, T., Burke, V., Rose, R., Sifri, C.D., 2015. Application of a telemedicine platform, isolation communication management system, for the care of dangerous infectious disease: a case series. Open Forum Infectious Diseases 2 (1), 228.

Fard, M.J., Ameri, S., Darin Ellis, R., Chinnam, R.B., Pandya, A.K., Klein, M.D., 2018. Automated robot-assisted surgical skill evaluation: predictive analytics approach. International Journal of Medical Robotics and Computer Assisted Surgery 14, e1850.

Ferenczi, G., Gyoergy), Mahmood, A., Bergmann, R., 2020. Telemedicine pre and post covid-19: lessons for commercialisation based on previous use cases. Journal of the International Society for Telemedicine and eHealth 8 (1–10), e8.

Fiordelli, M., Diviani, N., Schulz, P.J., 2013. Mapping mHealth research: a decade of evolution. Journal of Medical Internet Research 15, e95.

Flodgren, G., Rachas, A., Farmer, A.J., Inzitari, M., Shepperd, S., 2015. Interactive telemedicine: effects on professional practice and health care outcomes. Cochrane Database of Systematic Reviews 9, CD002098.

Fois, A.G., Manca, S., Ostera, S., Zeminian, D., Glisenti, F., Meloni, G.M., Pirina, P., 2014. Telespirometry performed by general practitioners for early diagnosis of obstructive airway disease: our experience. European Respiratory Journal 44.

Franek, J., 2012. Home telehealth for patients with chronic obstructive pulmonary disease (COPD): an evidence-based analysis. Ontario Health Technology Assessment Series 12 (11), 1–58.

Galiero, R., Pafundi, P.C., Nevola, R., Rinaldi, L., Acierno, C., Caturano, A., Salvatore, T., Adinolfi, L.E., Costagliola, C., Sasso, F.C., 2020. The importance of telemedicine during COVID-19 pandemic: a focus on diabetic retinopathy. Journal of Diabetes Research 1–8, 2020.

Gogia, S., 2020. Fundamentals of Telemedicine and Telehealth, Fundamentals of Telemedicine and Telehealth. Elsevier.

Greiner, A.C., Knebel, E., 2003. Executive Summary. Health Professions Education: A Bridge to Quality.

Guaiana, G., Mastrangelo, J., Hendrikx, S., Barbui, C., 2021. A systematic review of the use of Telepsychiatry in depression. Community Mental Health Journal 57, 93–100.

Gupta, A., Dogar, M.E., Zhai, E.S., Singla, P., Shahid, T., Yildirim, H.N., Singh, S., 2019. Innovative telemedicine approaches in different countries: opportunity for adoption, leveraging, and scaling-up. Telehealth and Medicine Today 5, 1–13.

Heinzelmann, P.J., Lugn, N.E., Kvedar, J.C., 2005. Telemedicine in the future. Journal of Telemedicine and Telecare 11, 384–390.

Heymann, D.L., 2020. Data sharing and outbreaks: best practice exemplified. The Lancet 395, 469–470.

Ibrahim, F.A., Pahuja, E., Dinakaran, D., Manjunatha, N., Kumar, C.N., Math, S.B., 2020a. The future of Telepsychiatry in India. Indian Journal of Psychological Medicine 42, 112S–117S.

Ibrahim, O.M., Ibrahim, R.M., Z Al Meslamani, A., Al Mazrouei, N., 2020b. Role of telepharmacy in pharmacist counselling to coronavirus disease 2019 patients and medication dispensing errors. Journal of Telemedicine and Telecare 1–10, 1357633X2096434.

Jankharia, B., Burute, N., 2009. Teleradiology: the Indian perspective. Indian Journal of Radiology and Imaging 19, 16.

Jiang, F., Jiang, Y., Zhi, H., Dong, Y., Li, H., Ma, S., Wang, Y., Dong, Q., Shen, H., Wang, Y., 2017. Artificial intelligence in healthcare: past, present and future. Stroke and Vascular Neurology 2, 230–243.

Kalaiselvan, V., Sharma, A., Gupta, S.K., 2021. "Feasibility test and application of AI in healthcare"—with special emphasis in clinical, pharmacovigilance, and regulatory practices. Health Technology 11, 1–15.

Khan, M.K., 2021. Importance of telecommunications in the times of COVID-19. Telecommunication Systems 76, 1–2.

Khan, Z.H., Siddique, A., Lee, C.W., 2020. Robotics utilization for healthcare digitization in global COVID-19 management. International Journal of Environmental Research and Public Health 17 (11), 3819.

Kichloo, A., Albosta, M., Dettloff, K., Wani, F., El-Amir, Z., Singh, J., Aljadah, M., Chakinala, R.C., Kanugula, A.K., Solanki, S., Chugh, S., 2020. Telemedicine, the current COVID-19 pandemic and the future: a narrative review and perspectives moving forward in the USA. Family Medicine and Community Health 8, e000530.

Kilova, K., Peikova, L., Mateva, N., 2021. Telepharmacy and opportunities for its application in Bulgaria. Pharmacia 68, 333–337.

Knepley, K.D., Mao, J.Z., Wieczorek, P., Okoye, F.O., Jain, A.P., Harel, N.Y., 2021. Impact of telerehabilitation for stroke-related deficits. Telemedicine and e-Health 27, 239–246.

Kondylakis, H., Katehakis, D.G., Kouroubali, A., Logothetidis, F., Triantafyllidis, A., Kalamaras, I., Votis, K., Tzovaras, D., 2020. COVID-19 mobile apps: a systematic review of the literature. Journal of Medical Internet Research 22, e23170.

Krupinski, E., 2014. Teleradiology: current perspectives. Reports in Medical Imaging 7, 5–14.

Kumar, S., Krupinski, E.A., 2008. Teleradiology, Teleradiology. Springer Berlin Heidelberg, Berlin, Heidelberg.

Kumar, P., Kumar, R., 2018. Rural health scenario – role of family medicine: academy of family physicians of India position paper. Journal of Family Medicine and Primary Care 7 (6), 1157.

Kumar, M., Agarwal, A., Singh, M., Singh, M., 2020. Feasibility and accuracy of Tele-spirometry: a community based approach. World Allergy Organization Journal 13 (8), 100151.

Kutlu, O., Gunes, R., Coerdt, K., Metin, A., Khachemoune, A., 2020. The effect of the "stay-at-home" policy on requests for dermatology outpatient clinic visits after the COVID-19 outbreak. Dermatologic Therapy 33, e13581.

Kuziemsky, C., Maeder, A.J., John, O., Gogia, S.B., Basu, A., Meher, S., Ito, M., 2019. Role of artificial intelligence within the telehealth domain. Yearbook of Medical Informatics 28, 035–040.

Langarizadeh, M., Tabatabaei, M., Tavakol, K., Naghipour, M., Moghbeli, F., 2017. Telemental health care, an effective alternative to conventional mental care: a systematic review. Acta Informatica Medica 25 (4), 240–246.

Latif, S., Qadir, J., Farooq, S., Imran, M., 2017. How 5G wireless (and concomitant technologies) will revolutionize healthcare? Future Internet 9, 93.

Lavallee, J., Byrne-Davis, L., Weinman, J., Hart, J., 2021. Effective working in health care. In: Reference Module in Neuroscience and Biobehavioral Psychology. Elsevier.

Leider, J.P., Meit, M., McCullough, J., Mac Resnick, B., Dekker, D., Alfonso, Y.N., Bishai, D., 2020. The state of rural public health: enduring needs in a new decade. American Journal of Public Health 110, 1283–1290.

Li, D., 2019. 5G and intelligence medicine—how the next generation of wireless technology will reconstruct healthcare? Precision Clinical Medicine 2, 205–208.

Li, Z., Moran, P., Dong, Q., Shaw, R.J., Hauser, K., 2017. Development of a tele-nursing mobile manipulator for remote care-giving in quarantine areas. In: IEEE International Conference on Robotics and Automation (ICRA). IEEE, pp. 3581–3586, 2017.

Magsi, H., Sodhro, A.H., Chachar, F.A., Abro, S.A.K., Sodhro, G.H., Pirbhulal, S., 2018. Evolution of 5G in Internet of medical things. In: 2018 International Conference on Computing, Mathematics and Engineering Technologies (ICoMET). IEEE, pp. 1–7.

Mahar, J.H., Rosencrance, G.J., Rasmussen, P.A., 2018. Telemedicine: past, present, and future. Cleveland Clinic Journal of Medicine 85, 938–942.

Malatzky, D.C., Gillespie, J., Couch, D.D.L., Cosgrave, D.C., 2020. Why place matters: a rurally-orientated analysis of COVID-19's differential impacts. Social Sciences & Humanities Open 2, 100063.

Malhotra, S., Shah, R., Chakrabarti, S., 2013. Telepsychiatry: promise, potential, and challenges. Indian Journal of Psychiatry 55, 3–11.

Mansberger, S.L., Gleitsmann, K., Gardiner, S., Sheppler, C., Demirel, S., Wooten, K., Becker, T.M., 2013. Comparing the effectiveness of telemedicine and traditional surveillance in providing diabetic retinopathy screening examinations: a randomized controlled trial. Telemedicine and e-Health 19, 942–948.

Mariolis, A., Mihas, C., Alevizos, A., Mariolis-Sapsakos, T., Marayiannis, K., Papathanasiou, M., Gizlis, V., Karanasios, D., Merkouris, B., 2008. Comparison of primary health care services between urban and rural settings after the introduction of the first urban health centre in Vyronas, Greece. BMC Health Services Research 8, 124.

Mohamed Ibrahim, O., Ibrahim, R.M., Abdel-Qader, D.H., Al Meslamani, A.Z., Al Mazrouei, N., 2021. Evaluation of telepharmacy services in light of COVID-19. Telemedicine and e-Health 27, 649–656.

Monaghesh, E., Hajizadeh, A., 2020. The role of telehealth during COVID-19 outbreak: a systematic review based on current evidence. BMC Public Health 20, 1193.

Moon, S., Szlezák, N.A., Michaud, C.M., Jamison, D.T., Keusch, G.T., Clark, W.C., Bloom, B.R., 2010. The global health system: lessons for a stronger institutional framework. PLoS Medicine 7, e1000193.

Morais-Almeida, M., Barbosa, M.T., Sousa, C.S., Almeida, I., Pimenta, L., Aguiar, R., 2021. Spirometry outside the hospital. Archivos de Bronconeumología 57 (3), 236.

Muflih, S.M., Al-Azzam, S., Abuhammad, S., Jaradat, S.K., Karasneh, R., Shawaqfeh, M.S., 2021. Pharmacists' experience, competence and perception of telepharmacy technology in response to COVID-19. International Journal of Clinical Practice 75, e14209.

Mullick, A., Rayhan, N., Koly, K., Nahar, D., Hossain, I., 2020. Telemedicine and telehealth: a virtual care platform with prospects & importance during COVID-19 outbreak. European Journal of pharmaceutical and medical research 7, 39–45.

Murphy, K.P., 2012. Machine Learning - A Probabilistic Perspective - Table-Of-Contents. The MIT Press.

Naskar, S., Victor, R., Das, H., Nath, K., 2017. Telepsychiatry in India - where do we stand? A comparative review between global and Indian Telepsychiatry programs. Indian Journal of Psychological Medicine 39, 223–242.

Nichols, J.H., Alter, D., Chen, Y., Isbell, T.S., Jacobs, E., Moore, N., Shajani-Yi, Z., 2020. AACC guidance document on management of point-of-care testing. The Journal of Applied Laboratory Medicine 5, 762–787.

Novaes, M. de A., 2020. Chapter 10 - Telecare within different specialties. In: Fundamentals of Telemedicine and Telehealth.

Panagariya, D.A., 2014. The challenges and innovative solutions to rural health dilemmas. Annals of Neurosciences 21 (4), 125–127.

Perera, S., Gambheera, H., Williams, S., 2020. "Telepsychiatry" in the time of COVID-19: overcoming the challenges. Indian Journal of Psychiatry 62 (9), 391–394.

Peretti, A., Amenta, F., Tayebati, S.K., Nittari, G., Mahdi, S.S., 2017. Telerehabilitation: review of the state-of-the-art and areas of application. JMIR Rehabilitation and Assistive Technologies 4, e7.

Poudel, A., Nissen, L.M., 2016. Integrated Pharmacy Research and Practice DovepressTelepharmacy: a pharmacist's perspective on the clinical benefits and challenges. Integrated Pharmacy Research and Practice 5, 83–84.

Rajkomar, A., Oren, E., Chen, K., Dai, A.M., Hajaj, N., Hardt, M., Liu, P.J., Liu, X., Marcus, J., Sun, M., Sundberg, P., Yee, H., Zhang, K., Zhang, Y., Flores, G., Duggan, G.E., Irvine, J., Le, Q., Litsch, K., Mossin, A., Tansuwan, J., Wang, D., Wexler, J., Wilson, J., Ludwig, D., Volchenboum, S.L., Chou, K., Pearson, M., Madabushi, S., Shah, N.H., Butte, A.J., Howell, M.D., Cui, C., Corrado, G.S., Dean, J., 2018. Scalable and accurate deep learning with electronic health records. Npj Digital Medicine 1, 18.

Ramadas, S., Kuttichira, P., 2020. Celebrity suicide—a cause for media celebration? Analysis of three newspaper reports. Indian Journal of Psychological Medicine 42, 498–500.

Rangasamy, M., Balasubramaniam, A., Krishnarajan, D., Kante, N.R., Kumar, N.S., 2011. Role of Telemedicine in Health Care System: A Review. Pharmaceutical Research.

Rawat, G., 2018. Tele nursing. International Journal of Current Research 10 (03), 66185–66187.

Reddy, S., Fox, J., Purohit, M.P., 2019. Artificial intelligence-enabled healthcare delivery. Journal of the Royal Society of Medicine 112, 22–28.

Reeves, J.J., Hollandsworth, H.M., Torriani, F.J., Taplitz, R., Abeles, S., Tai-Seale, M., Millen, M., Clay, B.J., Longhurst, C.A., 2020. Rapid response to COVID-19: health informatics support for outbreak management in an academic health system. Journal of the American Medical Informatics Association 27, 853–859.

Rutkowski, S., 2021. Management challenges in chronic obstructive pulmonary disease in the COVID-19 pandemic: telehealth and virtual reality. Journal of Clinical Medicine 10 (6), 1261.

Ryu, S., 2010. History of telemedicine: evolution, context, and transformation. Healthcare Informatics Research 16 (1), 65.

Saeed, S.A., Bloch, R.M., Diamond, J.M., 2012. Telepsychiatry: overcoming barriers to implementation. Current Psychiatry 11 (12), 28–31.

Salles, N., Lafargue, A., Cressot, V., Glenisson, L., Barateau, M., Thiel, E., Doutre, M.-S., Bourdin, N., Prigent, A.-M., Libert, K., Dupuy, M., Caubet, C., Raynal, F., 2017. Global geriatric evaluation is feasible during interactive telemedicine in nursing homes. European Research in Telemedicine/La RechercheEuropéenneenTélémédecine 6, 59–65.

Samonte, M.J.C., Aguilar, V., Marino, A.K., Tolentino, R., 2019. InTelect: interactive telemedicine communication technologies mobile app. IOP Conference Series: Materials Science and Engineering 482, 012022.

Sasikala, S., Indhira, K., Chandrasekaran, V.M., 2018. Performance prediction of interactive telemedicine. Informatics in Medicine Unlocked 11, 87–94.

Schlachta-Fairchild, L., Varghese, S.B., Deickman, A., Castelli, D., 2010. Telehealth and telenursing are live: APN policy and practice implications. The Journal for Nurse Practitioners 6, 98–106.

Schmeler, M.R., Schein, R.M., McCue, M., Betz, K., 2015. Telerehabilitation and clinical applications: research, opportunities, and challenges. International Journal of Telerehabilitation 12–24.

Schulz, T., Long, K., Kanhutu, K., Bayrak, I., Johnson, D., Fazio, T., 2020. Telehealth during the coronavirus disease 2019 pandemic: rapid expansion of telehealth outpatient use during a pandemic is possible if the programme is previously established. Journal of Telemedicine and Telecare 0 (0), 1–7.

Scott, R., Mars, M., 2015. Telehealth in the developing world: current status and future prospects. Smart Homecare Technology and TeleHealth 3, 25–37.

Segato, F., Masella, C., 2017. Telemedicine services: how to make them last over time. Health Policy and Technology 6, 268–278.

Serhani, M.A., El Kassabi, H., T., Ismail, H., NujumNavaz, A., 2020. ECG monitoring systems: review, architecture, processes, and key challenges. Sensors 20, 1796.

Shamasunder, S., Holmes, S.M., Goronga, T., Carrasco, H., Katz, E., Frankfurter, R., Keshavjee, S., 2020. COVID-19 reveals weak health systems by design: why we must re-make global health in this historic moment. Global Public Health 15, 1083–1089.

Sharma, A., Fölster-Holst, R., Kassir, M., Szepietowski, J., Jafferany, M., Lotti, T., Goldust, M., 2020. The effect of quarantine and isolation for COVID-19 in general population and dermatologic treatments. Dermatologic Therapy 33, e13398.

Shirzadfar, H., 2017. The evolution and transformation of telemedicine. International Journal of Biosensors & Bioelectronics 3 (4), 303–306.

Singh, H., Bhargava, S., Ganeshan, S., Kaur, R., Sethi, T., Sharma, M., Chauhan, M., Chauhan, N., Chauhan, R., Chauhan, P., Brahmachari, S.K., 2018. Big data analysis of traditional knowledge-based Ayurveda medicine. Progress in Preventive Medicine 3, e0020.

Sirili, N., Simba, D., 2020. Understanding the rural–rural migration of health workers in two selected districts of Tanzania. Advances in Public Health 1–9, 2020.

Smith, Y., 2018. Types of Telemedicine. News Medical [WWW Document].

Stewart, J., Sprivulis, P., Dwivedi, G., 2018. Artificial intelligence and machine learning in emergency medicine. Emergency Medicine Australasia 30, 870–874.

Strasser, R., Kam, S.M., Regalado, S.M., 2016. Rural health care access and policy in developing countries. Annual Review of Public Health 37, 395–412.

Sun, S., Xie, Z., Yu, K., Jiang, B., Zheng, S., Pan, X., 2021. COVID-19 and healthcare system in China: challenges and progression for a sustainable future. Globalization and Health 17 (1), 14.

Talib, F., Azam, M., Rahman, Z., 2015. Service quality in healthcare establishments: a literature review. International Journal of Behavioural and Healthcare Research 5 (1/2), 1.

Theodoros, D., Russell, T., 2008. Telerehabilitation: current perspectives. Studies in Health Technology and Informatics 131, 191–209.

Vidal-Alaball, J., Acosta-Roja, R., Pastor Hernández, N., Sanchez Luque, U., Morrison, D., Narejos Pérez, S., Perez-Llano, J., Salvador Vèrges, A., LópezSeguí, F., 2020. Telemedicine in the face of the COVID-19 pandemic. Atención Primaria 52, 418–422.

Volterrani, M., Sposato, B., 2019. Remote monitoring and telemedicine. European Heart Journal - Supplements 21, M54–M56.

Weeks, E., 2018. Medicalization of rural poverty: challenges for access. Journal of Law Medicine & Ethics 46, 651–657.

Weiss, D.J., Nelson, A., Vargas-Ruiz, C.A., Gligorić, K., Bavadekar, S., Gabrilovich, E., Bertozzi-Villa, A., Rozier, J., Gibson, H.S., Shekel, T., Kamath, C., Lieber, A., Schulman, K., Shao, Y., Qarkaxhija, V., Nandi, A.K., Keddie, S.H., Rumisha, S., Amratia, P., Arambepola, R., Chestnutt, E.G., Millar, J.J., Symons, T.L., Cameron, E., Battle, K.E., Bhatt, S., Gething, P.W., 2020. Global maps of travel time to healthcare facilities. Nature Medicine 26, 1835–1838.

Whitten, P., Holtz, B., LaPlante, C., 2010. Telemedicine. Applied Clinical Informatics 1 (2), 132–141.

Williams, S.Y., Adeyemi, S.O., Eyitayo, J.O., Odeyemi, O.E., Dada, O.E., Adesina, M.A., Akintayo, A.D., 2020. Mobile health technology (mhealth) in combating COVID-19 pandemic: use, challenges and recommendations. European Journal of Medical and Educational Technologies 13, em2018.

Willmott, C., Arrowsmith, J.E., 2013. Point of care testing. Core Topics in Cardiothoracic Critical Care 117–122.

Wosik, J., Fudim, M., Cameron, B., Gellad, Z.F., Cho, A., Phinney, D., Curtis, S., Roman, M., Poon, E.G., Ferranti, J., Katz, J.N., Tcheng, J., 2020. Telehealth transformation: COVID-19 and the rise of virtual care. Journal of the American Medical Informatics Association 27, 957–962.

CHAPTER FOUR

Next-generation technologically empowered telehealth systems

Amarah Zahra[1], Neetu Sood[1], S Roohan Farooq Lala[2] and Shamshad Alam[1]
[1]Department of Electronics and Communication Engineering, Dr. B R Ambedkar NIT Jalandhar, Punjab, India
[2]Department of Materials Engineering, IISC Banglore, Karnataka, India

4.1 Introduction

A health system, often known as a healthcare system, is a collection of people, organizations, and facilities that provide medical services to meet the needs of certain populations. Healthcare is both a business and a system for providing individuals with the healthcare they require. Healthcare refers to efforts made by qualified and licensed experts to preserve and enhance physiological, psychological, or behavioral well-being[1] Healthcare system is defined as a process of delivering services for the early prevention of diseases and their treatments. It also deals with the upgradation of physical and mental well-being. In the decades ahead, healthcare will undergo drastic transformations. Healthcare refers to the provision of services to individuals in specific contexts, such as homes, educational institutions, workplaces, public areas, communities, hospitals, and clinics, to contribute to their health. A health system consists of all entities, individuals, and actions with the primary purpose of managing symptoms. Efforts to improve health determinants as well as more direct health-improving behaviors are included.[2] The study conducted by IBIS World reveals the tremendous growth of the revenue of telehealth industry by a whopping 35% from the years 2014–19. The market size of the telehealth services is expected to grow from 45 billion USD in 2019 to 175 billion USD in 2026 which is almost 400%. The capability of growth of this industry is even greater in future. It has several benefits; the patients may be able to see their doctors sooner than normal, and they may have access to a larger variety of specialists that might not be always available in their own hometown. Telehealth can be used as a comprehensive clinical vocabulary to identify the patient's core complaints using natural language processing. With the advent of telehealth, patients will have a digital computational model of themselves enabling doctors to predict, diagnose, and treat illness. Patients can also have an access to use telemedicine. If in future we face some pandemic which also requires

1 https://isilanguagesolutions.com/2019/10/15/health-care-vs-healthcare.
2 https://pallipedia.org/health-care-system.

Extended Reality for Healthcare Systems, Volume 1
ISBN 978-0-323-98381-5, https://doi.org/10.1016/B978-0-323-98381-5.00001-5

social distancing, telehealth can be a savior in such scenarios. It is a real-time monitoring, and it allows Medicare patients to have physical and virtual access to care by phone or video conference at no additional cost. The world now is heading toward a completely networked society wherein many of the operations and human functions would be automated by intelligent machines and necessary algorithms. Many individuals believe that having access to healthcare is a basic human right. Poor healthcare can lead to a worse quality of life and a shorter life span than countries with a robust and efficient healthcare system. Health conditions are often better in countries with affordable and accessible healthcare services. The degree of healthcare quality in any nation is determined by a number of factors. The care process (disease management measures, nutrition services, clinical services, and patient involvement and expectations), availability, sustainable growth, equity, mortality rate, and pharmaceutical results are among them. These parameters were used by the Commonwealth Fund to rate 11 nations based on their healthcare quality. The United Kingdom, Australia, and the Netherlands are the top-ranked countries.

On a range of zero to 100, the Healthcare Access and Quality (HAQ) Index evaluates healthcare outcome ratings, with 100 being the best. Scores range from 90 to 96.1 for countries with the greatest healthcare systems in the world. With a score of 96.1, the Netherlands is in first place. Healthcare in Switzerland is universal, and all residents are obliged to have medical insurance. Basic medical insurance coverage in Switzerland includes 80%—90% of healthcare expenditures, including outpatient care, emergency care, prescriptions, maternity medication, vaccines, postoperative rehabilitation, and more. Switzerland's healthcare system includes private, subsidized private, and public systems to offer its residents a large network of trained physicians, state-of-the-art medical centers and hospitals, and no waiting lists.[3] Fig. 4.1 represents best healthcare in 2021, according to the 2021 edition of the CEOWORLD magazine Health Care Index, countries such as the United Kingdom, Netherlands, Italy, Spain, Japan, Qatar, Canada, New Zealand, South Korea, Taiwan, Belgium, Denmark, and Austria have the best healthcare system in the world.[4]Table 4.1 represents countries with the best healthcare systems, 2021.

4.2 Telehealth

Telehealth encompasses a wide range of digital and telecommunication services and technological solutions that are used to deliver care and treatment over long distances. The American Telemedicine Association defines telehealth as nonclinical healthcare delivered over the Internet. Telemedicine and telehealth encompass

3 https://worldpopulationreview.com/country-rankings/best-healthcare-in-the-world.
4 https://ceoworld.biz/2021/04/27/revealed-countries-with-the-best-health-care-systems-2021/.

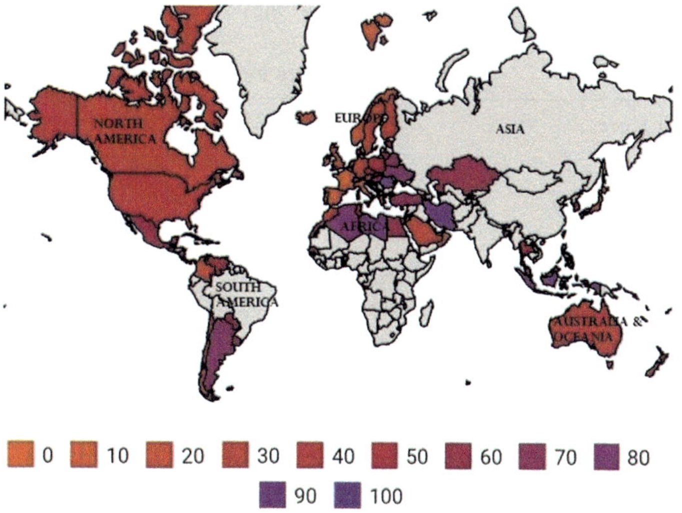

Figure 4.1 Best healthcare in the world 2021

teleconferencing, still image transfer, and e-health, which includes patient portals, remote monitoring of vitals, continuing patient practice, and nurse call centers. The use of electronic communication in transmitting the medical information from one location to another to monitor and enhance patient's health has been defined as telehealth. "Telehealth" has been widely used interchangeably as telemedicine.[5] Telehealth technology, techniques, and services are quickly becoming a crucial part of the healthcare system. The usage of telemedicine by doctors and the benefits to patients are depicted in Fig. 4.2. As doctors and patients largely are using telehealth technology as a potential new generation technology, the evidence base for these new techniques needs to be improved to the satisfaction of all. "Through their professional organizations and institutions, physicians should promote continuing refinement of technology and the establishment of clinical standards for telehealth and telemedicine," according to the American Medical Association (AMA) Council on Ethical and Judicial Affairs (Tuckson et al., 2017). Telehealth has the ability to meet the needs of rural communities that are underserved. This study was designed as a proof-of-concept to see if voice therapy can be provided effectively via the Internet (Mashima et al., 2003). The three interconnected developments play a vital role in shaping the telehealth now. The first step is the improvement in access to healthcare in a way that it offers convenience and finally leads toward cost reduction. The second development is the extrapolation of the telehealth facilities which surpasses beyond the acute illnesses to include episodic and chronic

5 About telemedicine. Washington, DC: American Telemedicine Association (http://www.america ntelemed.org/main/about/about-telemedicine/telemedicine-faqs).

Table 4.1 Countries with the best healthcare systems, 2021

Rank	Country	Health care index (overall)	Medicine availability
1	South Korea	78.72	82.3
2	Taiwan	77.7	78.99
3	Denmark	74.11	92.06
4	Austria	71.32	88.23
5	Japan	70.73	74.18
6	Australia	67.99	67.51
7	France	65.38	55.1
8	Spain	64.66	83.82
9	Belgium	64.63	64.78
10	United Kingdom	61.73	90.25
11	Netherlands	60.16	67.29
12	Finland	59.6	59.65
13	Thailand	59.52	98.74
14	Czech Republic	57.96	71.33
15	Norway	56.71	75.73
16	New Zealand	54.86	73.86
17	Germany	52.3	60.94
18	Switzerland	52.25	69.68
19	India	52.1	97.84
20	United Arab Emirates	51.99	55.08
21	Israel	50.15	74.19
22	Portugal	49.58	83.54
23	Canada	48.64	52.91
24	Singapore	48.54	81.98
25	Ecuador	48.27	66.11
26	Greece	48.13	98.21
27	Argentina	47.15	75.19
28	Sweden	46.24	62.6
29	Mexico	45.84	66.54
30	United States	45.62	76.28
31	Lithuania	45.3	85.77
32	Estonia	45.3	64.48
33	Qatar	45.29	60.69
34	Malaysia	45.1	52.11
35	Colombia	44.72	74.03
36	Hong Kong	44.55	64.34
37	Italy	44.43	98.43
38	Philippines	43.06	72.19
39	Uruguay	42.99	55.79
40	Sri Lanka	42.92	53.28
41	Iceland	42.5	47.94
42	Croatia	42.31	91.25

(*Continued*)

Table 4.1 Countries with the best healthcare systems, 2021—cont'd

Rank	Country	Health care index (overall)	Medicine availability
43	Jordan	41.99	74.42
44	Chile	41.97	69.12
45	Lebanon	41.63	65.42
46	China	41.4	65.36
47	Slovenia	39.85	87.57
48	Latvia	39.65	65.85
49	Hungary	39.37	74.7
50	Costa Rica	39.03	65.81
51	Poland	39.02	47.83
52	Indonesia	38.95	62.74
53	South Africa	38.65	94.29
54	Slovakia	38.48	67.42
55	Saudi Arabia	38.43	50.02
56	Panama	38.13	97.21
57	Belarus	37.94	64.1
58	Russia	37.76	57.85
59	Tunisia	37.71	63.39
60	Macedonia	37.16	73.01
61	Nepal	37.08	73.65
62	Peru	36.74	80.07
63	Brazil	36.31	52.97
64	Puerto Rico	36.26	59.31
65	Turkey	35.96	92.72
66	Vietnam	35.85	43.32
67	Bulgaria	35.64	72.25
68	Algeria	35.61	71.68
69	Romania	35.32	64.15
70	Kenya	35.16	51.23
71	Kuwait	35.09	89.2
72	Dominican Republic	34.97	62.85
73	Nigeria	34.78	82.83
74	Albania	34.78	63.47
75	Bosnia and Herzegovina	34.63	59.3
76	Cyprus	34.61	58.37
77	Serbia	34.37	66.16
78	Iran	34.28	59.64
79	Georgia	33.84	64.43
80	Ireland	33.65	58.24
81	Ukraine	33.38	53.55
82	Kazakhstan	33.22	52.56
83	Morocco	33.01	63.2
84	Egypt	32.94	62.68
85	Bangladesh	32.89	68.18

(*Continued*)

Table 4.1 Countries with the best healthcare systems, 2021—cont'd

Rank	Country	Health care index (overall)	Medicine availability
86	Azerbaijan	32.88	59.35
87	Iraq	32.55	57.45
88	Pakistan	32.52	58.14
89	Venezuela	32.42	53.7

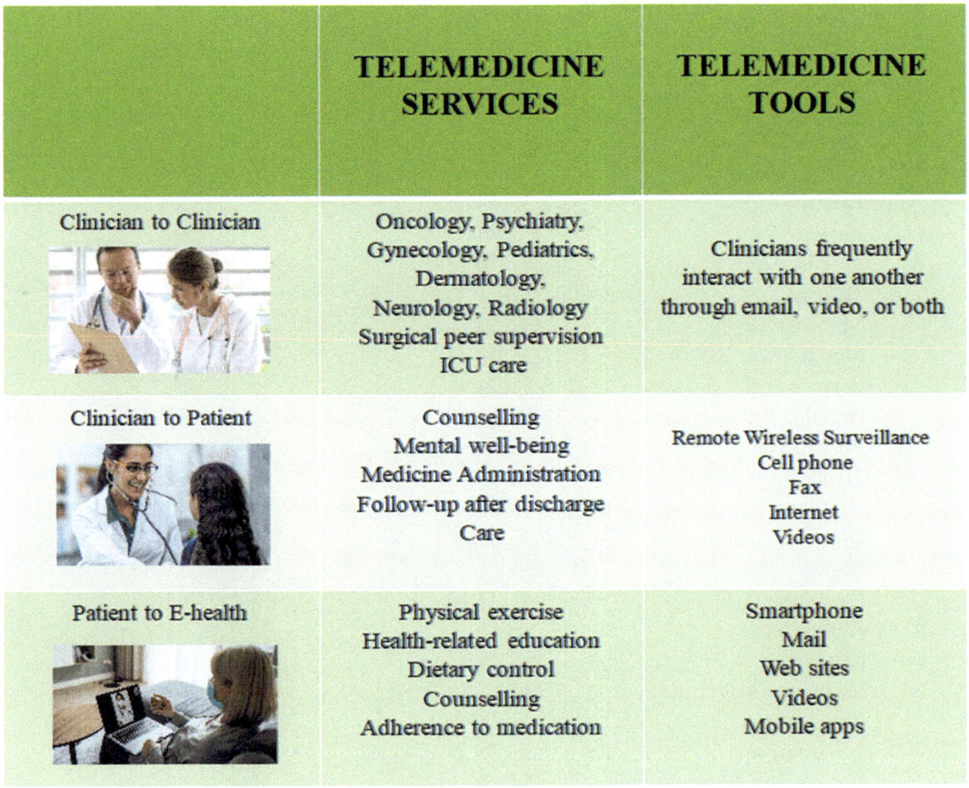

Figure 4.2 Use of telemedicine and patient benefits.

disorders. The third and the most important development is the spread of telemedicine from healthcare institutes to the people's homes and mobile devices (Dorsey and Topol, 2016). Among the first and unrelenting uses of telehealth were the programs offering treatment to people in military, prisoners in jails, and people from rural areas (Brown, 2013 May). An average of 20 days is required to get a 20-minute appointment with a doctor, which includes a waste of 2 h due to travel and waiting time in the clinic (Physician appointment, 2014; Ray et al., 2015). Increased accessibility, cost-

effectiveness, greater educational possibilities, improved quality of life, higher standards of healthcare, a better standard of living, and enhanced social support are all advantages of telehealth (Jennett et al., 2003) (Table 4.1 and 4.2).

4.2.1 Telemedicine

Telemedicine is medicine that is provided via the Internet. Telemedicine is a powerful tool that allows doctors and patients to communicate with one another. Telemedicine is the practice of providing medical treatment via the use of technology. Telemedicine is now available to anybody who has access to the Internet and a smartphone. Technology that most people use on a daily basis may be utilized to conduct extremely sophisticated telemedicine consultations. Lou Gehrig's disease, also known as amyotrophic lateral sclerosis (ALS), affects individuals ranging in age from their 20s to their 80s, and as a result, they are patients who, over time, may experience difficulties in walking, speaking, and breathing, making it difficult for them to come to the clinic. As a result, telemedicine is a lifesaver for people suffering from ALS. Telemedicine finds its application beyond diseases such as ALS as well. The 2019 COVID pandemic required social distancing even between the doctor and the patient and that is where telehealth is a savior allowing for online interaction between the patient and healthcare provider keeping patient care a priority while also not compromising the health of the healthcare worker. Telemedicine is a subset of telehealth that focuses entirely on the delivery of healthcare and education services through the Internet. Telemedicine is telehealth, but not all telehealth is telemedicine. It goes all the way back to the usage of ship to shore radio for delivering health assistance to sea captains in its widely recognized form, in which a doctor—patient contact incorporates telecommunication (Wootton, 2001). Techniques in telehealth have been in progress for over 61 years. In 1959, Wittson and colleagues established a microwave link for telepsychiatry consultations between the Nebraska Psychiatric Institute in Omaha and the state mental hospital 112 miles away, making them the first to use IATV for medical reasons (Wittson et al., 1961a,b; Wittson et al., 1961a,b). Jutra established teleradiology in Montreal, Quebec, in the same year, by sending telefluoroscopic exams via coaxial cable (utra, 1959). Improved telemedicine services are also being used as a justification for expanding service regions, eliminating access fees, and getting other concessions from government authorities. The telemedicine industry has also attracted manufacturers and fabricators of video conferencing, computers, imaging, medical, and multimedia technology (Perednia and Allen, 1995).

Long-distance carriers, local telephone companies, cellular providers, mobile telephone systems, cable operators, and the worldwide Internet computer network all benefitted greatly from the move to digital communication technology in the late 1980s and early 1990s. When transformed to digital forms, information transmission services that were formerly supplied by distinct businesses, such as phone calls, personal letters,

picture and document exchange, and broadcast television, become comparable. As a result, most telecommunications firms' specialist industries have been combined into a single sector in which all providers offer the same commodity—digital bandwidth. Selling in this new digital industry was a cumbersome job. In a commodity market, there are only two ways to increase sales: (1) increase the overall market by increasing bandwidth demand or (2) gain market share through marketing and product differentiation. Telemedicine has the ability to accomplish both. A two-way interactive video image consumes a lot of bandwidth. As a consequence, leading telecommunication carriers have begun to provide and widely advertise their goods and services for telehealth applications (Perednia and Allen, 1995). Digital search engines such as MEDLARS, PUBMED, and others have established the groundwork for a cultural shift that allows any health practitioner to access up-to-date "case-oriented" information through the Internet in seconds (Hjelm, 2005). Comparison of telehealth and telemedicine is shown in Fig. 4.3.

4.2.2 No substitute to telemedicine

In distant locations, such as the Antarctic, and on ships or planes, where providing medical treatment to a patient in a timely manner is difficult, telemedicine has a significant role to play. Telemedicine, on the other hand, allows access to services that would not otherwise be available in countries with unstable or poor economies, where

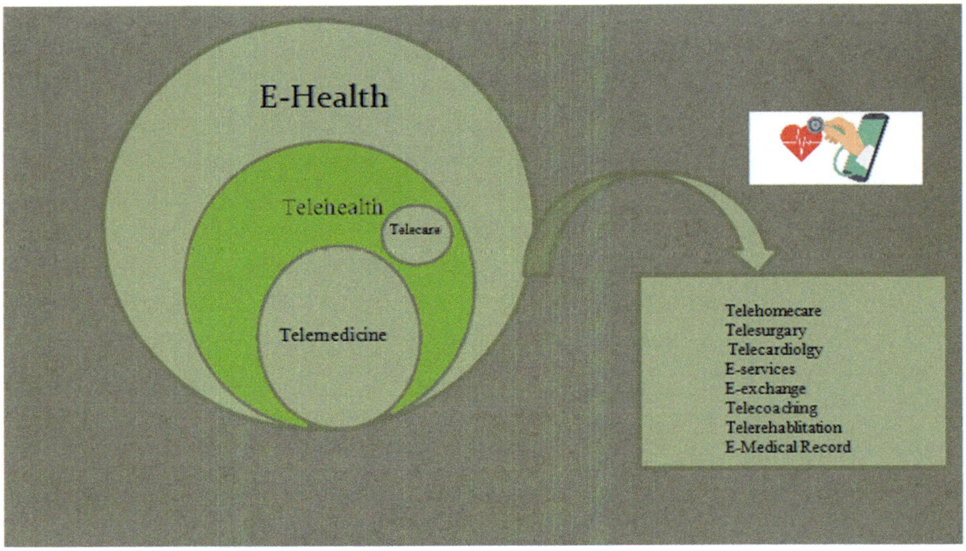

Figure 4.3 Telehealth versus telemedicine.

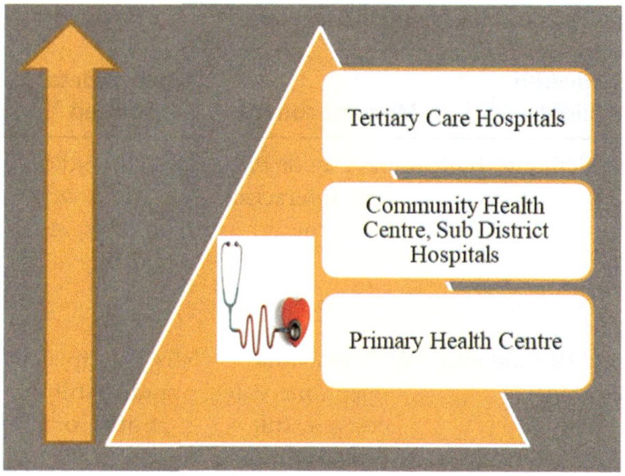

Figure 4.4 Levels of healthcare in India.

healthcare is typically not a priority. Fig. 4.4 represents levels of healthcare in India. Telemedicine is the only option and is superior to existing traditional services (Wootton et al., 2017) Table 4.2 represents the gamut of telehealth.

4.3 Telehealth at present

Citizens all around the world are using technology to better understand their fitness, diagnose sickness, and manage chronic conditions. Technologies such as power wearables, sensors, and robots can help with insomnia, depression, anxiety, as well as improve exercise routines. Insomnia and many other sleep disorders are rising in our society. Getting enough sleep is as vital as getting sufficient water and food during the day. Not getting enough sleep can cause stress. According to the Centers for Disease Control (CDC) and Prevention, over 70 million Americans have chronic sleep difficulties. According to the CDC, those who have chronic sleep difficulties are more likely to report additional health problems including heart disease, stroke, asthma, and cancer. Using a self-supervised technique, several sensors are employed for sleep recognition (Zhao et al., 2020). According to a World Health Organization (WHO) report, almost one-third of the world's population suffers from sleep difficulties, with a global sleep disorder rate of 27%, posing a major threat to people's health and well-being (Crivello et al., 2019). Now, with the help of telehealth and technology, a bed with sensors can now help patients to detect motion and artificial intelligence (AI) can be used to analyze sleep. Data analysis can assist doctors in identifying persistent patterns of restlessness and, as a consequence, finding a cure. Sleep position is closely connected to sleep problems, according to a study published in IEEE Explore. Sleep tracking is also done using smart

Table 4.2 The Gamut of medical telehealth interaction.

Objective	Minimum bandwidth required	Mode of contact	Types of data transmitted	Usage
Investigation and remedial consultation	Moderate to high	One-way or two-way interaction, real time	Speech, audio, motion video pictures, data, and files	Telepsychiatry and psychological health uses, remote surgery, and interactive examination
Investigation and remedial consultation	Low to moderate	With real-time telephone voice contact, still pictures or video clips can be used	Message, speech, audio, still photos, or brief videos	Dermatology, cardiology, otolaryngology, and orthopedics
Investigation and remedial consultation	Low	Still pictures or video clips with text content; "store-and-forward," with data captured and transferred for examination at a later time	Audio, static video pictures or brief video snippets, and text	Dermatology, cardiology, otolaryngology, and orthopedics
Healthcare education	Moderate to high	One-way or two-way real-time or delayed video	Voice, audio, time-lapse pictures, text, and files	Training and education via the Internet
Case management or paper work	Low to high	Digital text, picture, or other data transmission	Text, pictures, documents, and other data	Medical record maintenance, public medical information platforms

watches. Some devices can also be worn as headbands that track and identify our brainwaves' resting times (Sleep Technology). Diabetic patients can also benefit from telehealth. Meal logs, medicine, dosage, and blood sugar readings may now be uploaded to a nurse who replies electronically via a smartphone or other device without any delay or very little delay. Patients can now watch videos on diet control and can now download an app on their phone for the same. They can also use an app to predict how much insulin is needed based on diet and exercise, and online ordering is available for medicines. Patients can now access their lab tests, make an appointment, seek medication

refills, and email their doctors using an online health portal. They can also obtain a remote retinal image screening at their doctor's office instead of making an appointment with a specialist, and they can also get reminders for flu slots, foot exams, and other preventive care by email, message, or Smartphone (Healthy Lifestyle Consumer health). The DIABTel is a telemedicine system tool that uses telemonitoring and telecare services to supplement medical treatment and intensive management of diabetic patients. The system consists of a patient unit (PU) that patients use in their daily activities and a clinical operating system that healthcare professionals and nursing staff use in hospitals. Both applications provide capabilities for gathering, managing, viewing, and interpreting data, as well as exchanging data and messages. Its major objective is to expand the capabilities of standard clinical processes by offering an integrated solution that manages and improves various elements of everyday diabetes treatment for both doctors and patients. All diabetic patient populations (type I, type II, gestational diabetes, and others) can benefit from the system, even those with severe long-term chronic problems and impairments (Gómez et al., 2002). Web-based "consultations" with a physician or clinical pharmacist are another type of digital appointment. These services offered are no less than in a walk-in clinic. Virtual appointments with doctors and nurses are available at various clinics. When a physical visit is not possible or necessary, these appointments play a vital role for patients in receiving the treatment from our usual doctor. Several big corporations as part of their healthcare operations provide access to virtual doctors' offices. This has now made it possible that our primary healthcare clinic has an online patient grievance portal. These websites are an alternative to the conventional emails, which are considered unsafe for communicating sensitive clinical records. We are directed through a set of questionnaires when we login onto a Web-based platform. A physician or staff nurse can provide prescriptions, advise on home care techniques, or refer us to a specialist. A nursing contact center, meanwhile, is staffed with nurses who give guidance for home care based on our questionnaire. But a nurse call center does not diagnose or prescribe medicine to patients. Framework of remote health monitoring is shown in Fig. 4.5.

4.3.1 Telemedicine for heart diseases

Heart failure (HF) is a severe and challenging condition that causes poor quality of life, a wide variety of diseases, and a high rate of hospitalization (McMurray and Stewart, 2002; Lang and Mancini, 2007; Chen et al., 2011). Frequent and recurrent hospitalizations can put a significant financial strain on society (Stewart, 2005). Better self-care can significantly minimize hospitalizations. Home telemonitoring has the potential to improve patients' health. A telemonitoring-assisted self-care model was established, in which a patient uses a mobile phone app to routinely submit their most critical health indicators to the nurse. Fig. 4.6 depicts Login page for uploading different parameters and components used in the telemedicine aided self-care Model. A home-care kit was supplied to

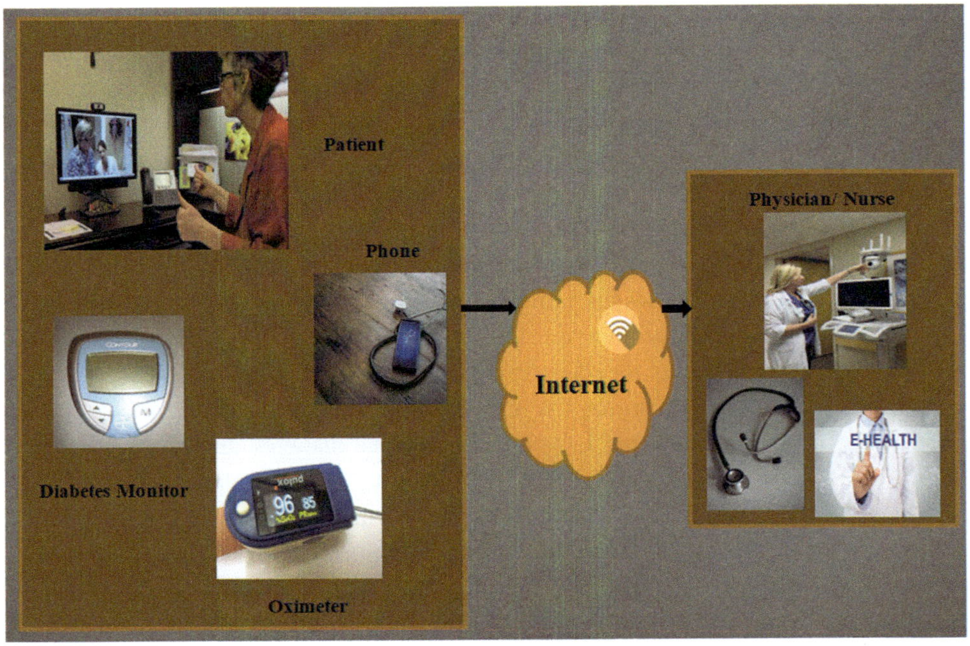

Figure 4.5 Framework for remote health monitoring.

Figure 4.6 Login page for uploading different parameters and components used in the telemedicine-aided self-care model.

the patients, which included a weight scale, a blood pressure meter, a cell phone, and self-care instructions. Patients were instructed to perform and submit measures as well as a symptom evaluation once a week. Diastolic and systolic blood pressure, pulse, body weight, and a symptom evaluation were among the measurements taken at home and uploaded. The patient's dizziness, dyspnea, palpitation, weakness, and edema were all assessed as symptoms. Patients were also asked to judge their general health, i.e., whether their condition had gotten worse, better, or stayed the same. The uploading of measurements and self-assessment of symptoms were made possible via a preinstalled software program on the mobile phone. Because the majority of the patients were seniors, special attention was paid to the user interface's simplicity and convenience of use throughout the creation of the mobile app. The patient received an automated machine-based response on whether the reported value was within their own limits set by the physician (Vuorinen et al., 2014).

4.3.2 Telehealth for asthma

Asthma is a respiratory illness that causes disturbed and difficult breathing, as well as a chronic cough, wheezing, and heaviness in the chest. Asthma is not a significant condition for some people. However, in certain cases, it is a life-threatening condition that might obstruct daily duties, making them impossible to complete (What are the causes). Asthma is a common condition that has a significant impact on people's quality of life as well as the financial impact it has on society. However, the exact etiology of asthma and its pathogenesis are unknown (Martin et al., 2001). Patients with asthma have noticed that their symptoms worsen on days when there is a lot of pollution in the air. In their apartment, they are now using an electronic nose to supervise dirt particles, smog, and other irritants to which they may be exposed (Aisha Manages Her Asthma). A device with an array of electronic chemical sensing and a pattern recognition algorithm is known as an electronic nose. Sensors in electronic noses detect a combination of gases, commonly known as a breath print. The gases in the air might react with the chemical sensors' surfaces, producing a change in conductivity. The variations are subsequently translated into electrical impulses, which result in a unique response pattern for the gas combination. This technique can compare gases into the atmosphere to breath prints that have been previously learned. The nose will be able to distinguish between air pollution and pollen since their response patterns differ. Volatile organic compound (VOC) molecules and concentrations exist in human breath. The combination of VOCs gives crucial information about the physiological and pathological processes involved in respiratory illnesses. Asthmatic individuals have certain VOCs in their breath. An electronic nose can identify whether or not someone has asthma or another respiratory disease. Fig. 4.7 shows the electronic nose. The nose may compare response patterns with various human breaths once it knows the breath print. Through breath print

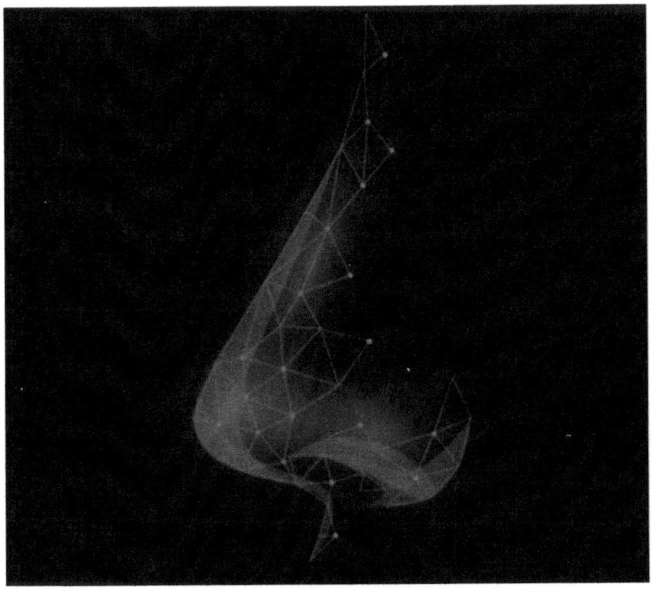

Figure 4.7 Electronic nose.

monitoring, electronic noses can also identify various diseases and chronic ailments (Can Electronic Noses Help Asthma Symptoms?). E-nose has the advantages of being inexpensive and simple to use, but it is not designed for evaluating breath odor (Novel Breath Analysis System Based on Electronic Olfaction, 2010). The electronic nose can aid in the early and quick detection of changes in a patient's condition (Byun et al., 2014).

4.3.3 Teletherapy for anxiety

There is a significant increase in stress and mental health issues. COVID-19, a respiratory illness, has wreaked havoc on people all over the world. It has caused severe damage to mental healthcare, in addition to the obvious physical symptoms in infected instances (Rehman et al., 2021). Anxiety is a state of mind indicated by tense sensations, anxious thoughts, and hormonal issues such as elevation of blood pressure, high rate of heartbeat, etc. The defining characteristics of anxiety are characterized by recurrent intrusive thoughts or concerns, which become stuck in the mind and can cause distress. The affected people due to their over concerned nature might avoid certain situations. Physical symptoms such as shivering, nervousness, disorientation, or rapid heart palpitations may also be present (Anonymous, 2006). As per a Kaiser Family Foundation study, over half of American people believe that the epidemic is affecting their mental health (Panchal et al., 2020). People are worried about being sick, losing loved ones, and dealing with financial loss and loneliness. COVID-19 has been related to neurological

issues, as well as anxiousness, frustration, and insomnia (Tedros Adhanom Ghebreyesus, 2020). Anxiety is being treated with technology methods such as teletherapy and e-health (mobile health) application. Teletherapy may also make it easier for patients to keep their therapy visits (COVID-19, 2020; IEEE transmitter). Wearable gadgets might be used to keep track of a person's mental health (Yang & et al., Apr. 2019). New technologies (smartphone, watches, machines, and glasses) can be used to gather information on ambulatory activity, interaction patterns, electrodermal activity (EDA), body movements, face expression, heart rate variability (HRV), location, interpersonal interactions, speaking style, and technological usage, all of which are relevant to a variety of psychological diseases (Abdullah & Choudhury, Jan. 2018). Because telehealth researchers can track anxiety, the researchers used data from the smart device's sensors to examine a variety of activities, including time spent alone, hours spent in bed, changes in sleep patterns, interaction with family members and friends, and hours spent on social networking sites. With the advancement in technology, now the evaluation of depression can be done using aforementioned parameters at home only (Ueafuea et al., 2021). Telehealth system allows patients to get treatment without needing to see a doctor in person.

4.3.4 Telehealth for loneliness

Loneliness is defined as a state of isolation or being solitary, but it is essentially a mental condition. People who are lonely feel hopeless, abandoned, and unloved. Lonely people by the nature want human interaction, but due to their present mental state, it is much harder for them to develop relationships. It is a common human feeling that is both complex and personal to each person. This potentially detrimental mental state might differ greatly from person to person in its prevention and management (The Health Consequences of Loneliness). As most personal stories demonstrate, loneliness is a terrible experience (Perlman and Peplau, 1981). Engineers and scientists, scholars, and healthcare professionals are utilizing the benefits of employing "robot buddies" as shown in Fig. 4.8 and virtual reality for loneliness to assist individuals who are suffering from

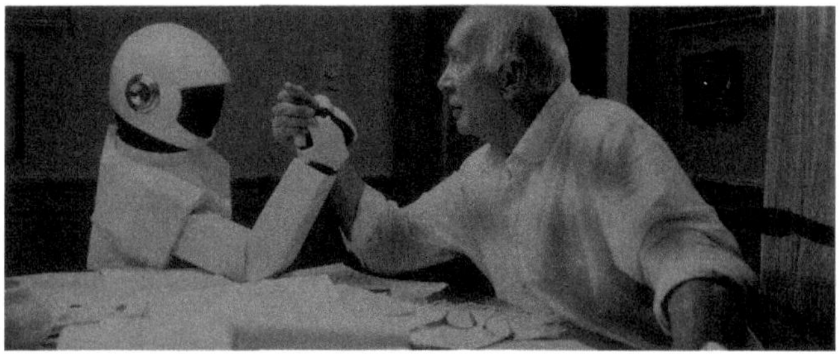

Figure 4.8 A robotic friend.

these sorts of symptoms. When human engagement is inadequate, social robots, or artificial buddies, have given an alternative for social interaction. These health management robots generally employ AI and machine learning to strengthen their communications with people (Robot Friends). Streaming video may be mentally taxing since it amplifies some of the social stresses that occur in real-world conversations. Also our sense of screen boundaries is a problem with virtual meetings. When we are in person, we are engaged in a setting, and humans have learned and developed to anticipate this sensation of immersion.

4.4 Smart instruments for telehealth

Smart watches and other wearable health gadgets are becoming increasingly popular. These gadgets, which can monitor our health information, can allow us to understand our own health and wellness. Now, even at home, one can keep track of all activities related to health. This has been made possible only because of telehealth. Wearables like smartwatch as shown in Fig. 4.9 employ sensors integrated into their

Figure 4.9 Smart watch.

processors—software algorithms functioning on the software layer—to assist us and track our activities such as breathing and heart rate variability. We can monitor our breathing and heart rates using a variety of approaches, including photoplethysmography for blood pressure monitoring, accelerometer sensing for movement, and so on. Edge computing, IoT, and cloud computing are three of the most common technologies that are used with wearables. We can now monitor our exercise on our own. Tracking vital signs and alterations to them as one exercises, as well as offering real-time feedback, may help one stay stress free. Exercises may be recommended with more confidence for those with chronic or other illnesses on the basis of the feedback system. Activity monitoring aids in raising public awareness of the need to maintain a healthy balance of workouts, sleep, and food habits. Wearables, on the other hand, are not just for healthy individuals; they are also for those with illnesses such as heart disease and obesity. Athletes may get fractures during practicing, which may be tracked using exercise diaries and other self-tracking approaches using wearables. Self-tracking gadgets allow health-conscious individuals to check their personal levels of fitness through applications and messages that can be viewed on portable wearables, which require even to not take the device out of their pockets. The essential impact of these wearables is that they provide a comprehensive perspective of one's health in one location, saving time, and money. Despite the fact that the data acquired by these wearables are highly sensitive, data security and privacy rules remain. However, with correct agreement, an individual's data may be exchanged with their physicians, which is crucial information for a patient's condition. Technology is progressing. Apart from the current wearable technology such as smart watches, intelligent badges, smart glasses, intelligent bracelets, and smart head-mounted gadgets, it is believed that exoskeleton suits for heavy weight lifting and 3D-printed prosthetics for creating human body parts will be used in the manufacturing and healthcare sectors to endorse the same (5 Reasons to Use a Wearable Health Tracker).

4.5 The future of telehealth

Telehealth opens up new avenues for the advancement of the healthcare system. Telehealth allows individuals to receive healthcare at a much cheaper cost. Consumers may access health information, goods, and services remotely because of digital technology. Telehealth also includes things such as reading health blogs. People with disabilities and the elderly can perform their therapies remotely with the aid of telehealth. Better treatment, better services, and better outcomes are all benefits of telehealth. According to a report, the cost of telehealth is $40–$50 each visit, but the cost of an in-person appointment is $176. According to another study, the overall cost of remote hospital treatment is 32% cheaper than regular hospital care, and telehealth patients are expected to grow exponentially in the upcoming decades. Telehealth will make a bold and positive contribution to our society, and we will see the advantages of telehealth in

the coming years. Transdisciplinary networking is easier with a smartphone, and smartphone apps can help with telemedicine (Mars and Scott, 2017). Human activities, the economy, and innovation will all play a role in the future of telemedicine. Individual, organizational, and social changes are all influenced by technological behaviors (Heinzelmann et al., 2005). Since the 19th century, telephony systems have been utilized to deliver healthcare knowledge for treatment. The use of the telephone to prevent needless medical visits was mentioned in *The Lancet* in 1878, and a tele-stethoscope had previously been described in 1910 (Bashshur et al., 2009). NASA employed remote monitoring devices to evaluate astronauts' physiological functioning in the mid-20th century. With the Papago Indians in the southwestern United States, the Satellite Technology Integrated to Rural Papago Advanced Health Care initiative advanced this subject (Freiburger et al., 2007). Telehealth is being used more frequently than it has ever been since the emergence of COVID-19. With ultrareliable and high-mobility communication, 5G is poised to revolutionize the healthcare business. It is ushering in a new era of individualized healthcare and patient care. 5G and 6G can transform the way healthcare is delivered when AI is combined with fifth- and sixth-generation wireless technologies, such as augmented reality, virtual reality, and edge intelligence architecture. Capgemini created a 5G-enabled real-time tele-ultrasound system to achieve the goal of an effective point-of-care system. There are two key aspects of the developed solution. First, a medical cooperation system that links a doctor, a primary care provider, and a patient for real-time monitoring, anomaly detection, remote visualization, and consultation. The second is AI-based physician assistance and triage, which uses a customizable decision engine to provide faster diagnosis, suggestions, and automated reporting. To satisfy real-time performance demands, microservices architecture with distributed edge—cloud architecture, data aggregation, and 5G multiaccess edge computing is the answer. A primary care provider can use this solution to upload a patient's history data and scans before setting up a live ultrasound feed and audio/video conference with a faraway doctor. In real time, the doctor examines the patient's condition using high-resolution 3D/4D ultrasound recordings. The specialist will be able to direct the healthcare provider in probe positioning via live streaming over 5G with the help of detecting sensitivity. During the meeting, the physician can record, take measurements, make notes, or capture pictures for later study. The AI-powered medical aid system automatically detects important artifacts from an ultrasound image and generates a preliminary diagnosis report and recommendations. In Web-based offline mode, the doctor can evaluate it and write a final report. The solution may be a savior for patients at rural medical clinics who require professional assistance, as well as a rescue for those who require emergency care. An AI based doctor assistance system is shown in Fig. 4.10.

The healthcare sector has invested in healthcare monitoring, telemedicine, robots, and data gathering and analysis to enhance patient access and results, as well as speed the

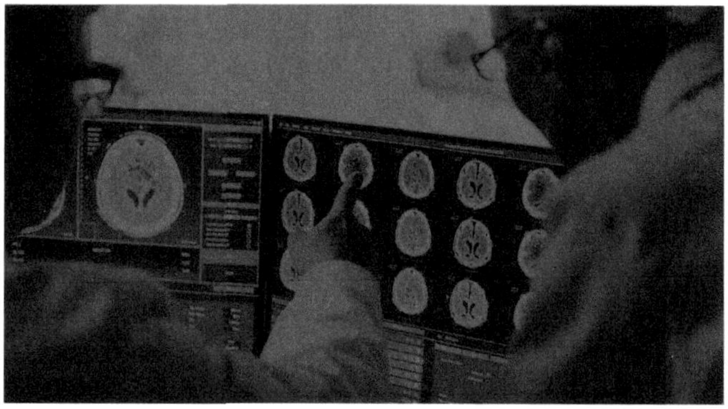

Figure 4.10 AI-based doctor assistance system. AI, artificial intelligence.

collecting of huge volumes of medical data for future study. With the ongoing implementation of ultrafast 5G networks in developing countries, which enable the near-instantaneous transfer of massive quantities of data, and the release of new, more efficient digital and IoT devices, the mobile sector is doing its part. The infrastructure required for comprehensive 5G coverage is being built at a cost of billions of dollars by telecom companies. This will shorten the effective distance between patients and their healthcare professionals, improve access to treatment, and change the quality of care provided. Extremely high network dependability and 5G coverage will be in great demand. For healthcare mission-critical tasks, access to a 5G network for less than 90% of the time is inadequate. The promise of 5G in terms of speed, capacity, and connection will serve as the foundation for rethinking how the healthcare sector addresses well-being. The healthcare professional and the patient will be able to spot patterns earlier and diagnose concerns more precisely than ever before. As the healthcare sector moves toward more customized treatment, 5G can aid in the incorporation of more data into the decision-making process, resulting in better outcomes. Today, high-speed streaming of almost any sort of medical picture is used, allowing clinicians to see these scans almost immediately after they are acquired. This will broaden the talent pool by introducing highly specialized and responsive health expertise to areas where it would otherwise be absent (5G and the future of healthcare (Reader Forum)). AI and 5G are becoming more common across the country. According to a recent Accenture survey, 82% of public health executives believe 5G would transform their business by providing new methods to deliver products and services (Accenture,; The Ascent of 5G Holds Bold Promise for Healthcare).

4.5.1 Telehealth and 6G

The sixth-generation technology is referred to as 6G technology. It is planned that 5G technology will be integrated for worldwide coverage. Satellites are used in sixth-

generation (6G) digital wireless communication networks to provide worldwide coverage. All cellular providers will be connected to a single core, which might be a mix of noncore and AI (Arockia Panimalar et al., 2017). We are evolving toward a fully digital society in which intelligent technology and processes will automate many civic and social activities. 6G would provide the foundation for complicated processes and interfaces to run smoothly with minimal human intervention. Machine learning, augmented reality, increased system capacity, faster data rates, ultrareliable low latency networks, huge machine type connectivity, and enhanced data security and user experience are some of the main technologies anticipated to be used on 6G capable networks. Wearable electronic gadgets, space and undersea communication, surgical implants, and robots are some of the other technologies involved. The 6G system-driven wireless connection is expected to be up to 1000 times more efficient than 5G. Smart wearable headsets, integrated gadgets, and body implants are all part of the wireless brain—computer interface (BCI) applications. The human mind will be able to interact with external devices using BCI technology, which will analyze and translate brain signals. 6G will make it easier to transfer data from the five human senses to interact with the environment intelligently and distantly. By removing time and location barriers through telemedicine and enhancing healthcare system operations and workflow, 6G will benefit the health system (Zhang et al., 2018). 6G will enable e-healthcare services while also meeting stringent quality of service (QoS) criteria such as 99.999% communication reliability, ultralow latency (less than 1 ms), and mobility robustness (Shahraki et al., 2021). To address difficult network optimization challenges, AI will be an important element of the 6G wireless infrastructure (Khan et al., 2020). 6G will enable remote operations utilizing robots and AI in the healthcare industry. Machines' intelligent learning abilities would be aided by AI. AI will make data transfer easier and more efficient. Metamaterials, intelligent networks, self-sustaining wireless networks, and built-in machine learning will all help it succeed. The advancement of nanotechnology will be aided by THz bandwidth, allowing nanodevices to function inside the human body using smart and remote instructions. Around 2030, 6G is projected to enable real-world products and services. Doctors will wear augmented reality eyeglasses with a camera, and biosensors will be attached to patients. The data from all of the biosignals will be shown locally on the augmented reality glasses, and the data will be captured and transferred over 6G to the hospital. The doctor will next keep an eye on the sensor data and prescribe the appropriate medicine. All of this requires extremely high levels of security and reliability, which are only feasible with sixth-generation technology. We have a data limit with present wireless technology, and we wish to accomplish more, which is only feasible because of 6G.

4.5.2 Future of surgery

Surgeons have always been early adopters of new technology and innovators. Surgery is both an art and an innovation in and of itself. Robot-assisted surgery will be possible in

the future. The technology of 3D printing is rapidly evolving. Many more people will be able to profit from telerobotic surgery in the future. We may achieve even greater heights with the help of modern technology. Around 70% of the world's population, or 5 billion people, do not have access to safe surgical treatment. People cannot even get routine medical treatments when they need them, according to the WHO's Lancet Commission. According to research, there are just 10 competent surgeons in the United States, or 1 for every 600,000 people. According to another study, the United States will require an additional 100,000 surgeons by 2030 merely to keep up with the demand for basic surgical operations. Digitalization has the ability to do much more than enable us to purchase online, interact with others through social networking sites, and keep updated. The heritage of surgery is littered with advances in how science and engineering have assisted surgeons in overcoming the most difficult challenges. Robotic surgery, like contemporary automated machinery, is characterized by ultraprecision, or the capacity to carry out treatments on the finest scales with a level of precision that exceeds that of human hands. In robotic surgery, the physician does not have to be at the patient's bedside to provide treatment; instead, the doctor can control a robot through a laptop while gazing at a monitor. Remote surgery is the term for this type of procedure. A smartphone, an iPad, or desktop, as well as Internet access and augmented reality collaboration tools are required for robot surgery. A professional surgeon will be able to virtually transport himself or herself into any medical setting easily by using his mobile phone or monitor, and he or she will be able to clearly and virtually interact during surgery from beginning to finish, directing, and mentoring a local physician through the process point by point, using this augmented reality collaboration software. Two persons can interact digitally using augmented reality. Remote surgery saves times, enhances accessibility, and reduces costs.

4.5.3 Future of electronic nose

Although breath detection by electronic nose is hard and difficult, the clinical applications are critical to its advancement. Breath analysis provides information on a patient's metabolism and microbiota, and it is a minimally invasive technique with low costs. It will play an important clinical role in the future.

4.6 Growth of telehealth

Telehealth usage is now 38 times greater than it was before the COVID-19 pandemic. Fig. 4.11 shows the growth in telehealth from January 2020 to February 2021. Following an early increase in April 2020, when telehealth accounted for more than 32% of all office and outpatient visits, utilization levels have mostly normalized, ranging from 13% to 17% across all specialties. Since April 2020, telehealth has

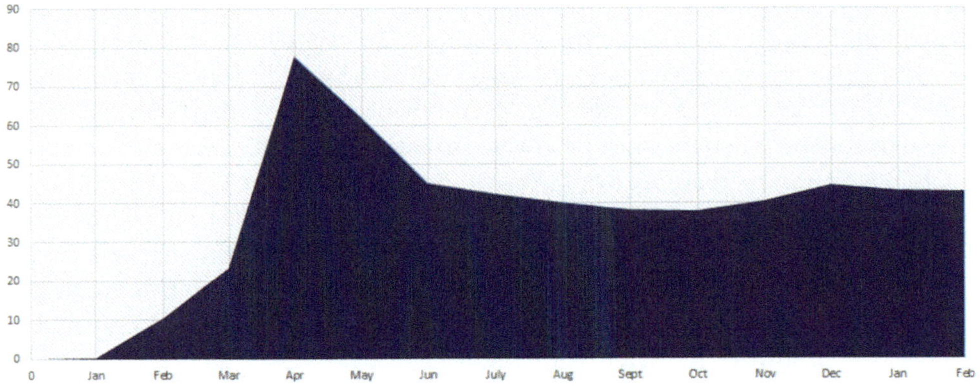

Figure 4.11 Growth in telehealth from January 2020 to February 2021.

accounted for up to 17% of all outpatient/office visit claims involving assessment and managing (E&M) services. Since June 2020, this usage has been pretty constant (Telehealth). By 2030, the use of telehealth will have skyrocketed. In 2050, the term "telehealth" will be obsolete. In 2025, it will no longer exist. It will simply be referred to as "health." It won't be called "virtual care," but just "care" (Telehealth, 2050).

4.7 Conclusion

If we want to enhance our healthcare, we should give telehealth a try.

It will throw forward immense opportunity for healthcare professionals as well as the patient. Wireless technologies such as 5G and 6G will provide us higher data rates and low latency, allowing us to do things in telehealth that are currently unachievable.

References

5 Reasons to Use a Wearable Health Tracker - IEEE Transmitter, IEEE transmitter (https://transmitter.ieee.org/5-reasons-to-use-a-wearable-health-tracker).

5G and the future of healthcare (Reader Forum), Tasman Technologies LLC Scott Schelle, RCR Wireless News (https://www.rcrwireless.com/20200910/opinion/readerforum/5g-and-the-future-of-healthcare-reader-forum).

Abdullah, S., Choudhury, T., Jan. 2018. Sensing technologies for monitoring serious mental illnesses. IEEE MultimediaMag 25 (1), 61–75.

About Telemedicine. Washington, DC: American Telemedicine Association (http://www.americantelemed.org/main/about/about-telemedicine/telemedicine-faqs).

Accenture.com (https://www.accenture.com/_acnmedia/PDF-102/Accenture-Digital-Health-Tech-Vision-2019.pdf#zoom=50).

Aisha Manages Her Asthma - IEEE Transmitter, IEEE Transmitter (https://transmitter.ieee.org/health-wellness-2021/technology-for-asthma/).

Arockia Panimalar, S., Monica, J., Amala, S., Chinmay, V., 2017. 6G Technology. International Research Journal of Engineering and Technology (IRJET).

Bashshur, R.L., Shannon, G.W., Krupinski, E.A., Grigsby, J., Kvedar, J.C., Weinstein RS, R.S., et al., 2009. National telemedicine initiatives: essential to healthcare reform. Telemedicine Journal and e-Health 15 (6), 600–610 [Medline].

Brown, E.M., May 2013. The ontario telemedicine network: a case report. Telemedicine Journal and e-Health: The Official Journal of the American Telemedicine Association 19 (5), 373–376. https://doi.org/10.1089/tmj.2012.0299. PMID: 23301768.

Byun, H.G., Yu, J.B., Huh, J.S., Lim, J.O., 2014. Exhaled breath analysis system based on electronic nose techniques applicable to lung diseases. Hanyang Medical Reviews 34 (3), 125–129.

Can Electronic Noses Help Asthma Symptoms? - IEEE Transmitter, IEEE Transmitter (https://transmitter.ieee.org/can-electronic-noses-help-asthma-symptoms/).

https://ceoworld.biz/2021/04/27/revealed-countries-with-the-best-health-care-systems-2021/.

Chen, J., Normand, S.L., Wang, Y., Krumholz, H.M., October 19, 2011. National and regional trends in heart failure hospitalization and mortality rates for Medicare beneficiaries, 1998-2008. JAMA 306 (15), 1669–1678.

COVID-19 Is Straining Mental Health-Could Technology Be the Answer? IEEE Pulse, 2020. Summer Allen.

Crivello, A., Barsocchi, P., Girolami, M., Palumbo, F., 2019. The meaning of sleep quality: a survey of available technologies. IEEE Access 7, 167 374–167 390.

Dorsey, E.R., Topol, E.J., 2016. State of telehealth. New England Journal of Medicine 375 (2), 154–161.

Freiburger, G., Holcomb, M., Piper, D., 2007. The STARPAHC collection: part of an archive of the history of telemedicine. Journal of Telemedicine and Telecare 13 (5), 221–223.

Gómez, E.J., Hernando, M.E., García, A., Del Pozo, F., Cermeño, J., Corcoy, R., Brugués, E., De Leiva, A., 2002. Telemedicine as a tool for intensive management of diabetes: the DIABTel experience. Computer Methods and Programs in Biomedicine 69, 163–177. https://doi.org/10.1016/s0169-2607(02)00039-1.

Healthy Lifestyle Consumer Health, Telehealth: Technology Meets Health Care(https://www. mayoclinic.org/healthy-lifestyle/consumer-health/in-depth/telehealth/art-0044878)

Heinzelmann, P.J., Lugn, N.E., Kvedar, J.C., 2005. Telemedicine in the future. Journal of Telemedicine and Telecare 11 (8), 384–390.

Hjelm, N.M., 2005. Benefits and drawbacks of telemedicine. Journal of Telemedicine and Telecare 11 (2), 60–70.

IEEE transmitter (https://transmitter.ieee.org/health-wellness-2021/technology-for-anxiety/.

https://isilanguagesolutions.com/2019/10/15/health-care-vs-healthcare.

Jennett, P.A., Hall, L.A., Hailey, D., Ohinmaa, A., Anderson, C., Thomas, R., Young, B., Lorenzetti, D., Scott, R.E., 2003. The socio-economic impact of telehealth: a systematic review. Journal of Telemedicine and Telecare 9 (6), 311–320.

Khan, L.U., Yaqoob, I., Imran, M., Han, Z., Hong, C.S., 2020. 6G wireless systems: a vision, architectural elements, and future directions. IEEE Access 8, 147029–147044.

Lang, C.C., M Mancini, D., June 2007. Non-cardiac comorbidities in chronic heart failure. Heart 93 (6), 665–671.

Mars, M., Scott, R.E., 2017. Being spontaneous: the future of telehealth implementation? Telemedicine and e-Health 23 (9), 766–772.

Martin, R.J., Kraft, M., Chu, H.W., Berns, E.A., Cassell, G.H., 2001. A link between chronic asthma and chronic infection. The Journal of Allergy and Clinical Immunology 107 (4), 595–601.

Mashima, P.A., Birkmire-Peters, D.P., Syms, M.J., Holtel, M.R., Burgess, L.P., Peters, L.J., 2003. Telehealth 12 (4), 432–439. https://doi.org/10.1044/1058-0360(2003/089).

McMurray, J.J.V., Stewart, S., 2002. The burden of heart failure. European Heart Journal - Supplements 4 (Suppl. D), 307–333.

A novel breath analysis system based on electronic olfaction. IEEE Transactions on Biomedical Engineering, Dongmin Guo 57, 2010.

https://pallipedia.org/health-care-system.

N. Panchal et al., "The implications of COVID-19 for mental health and substance use," Kaiser Family Foundation. Accessed: Apr. 21, 2020. [Online]. Available (https://www.kff.org/coronavirus-covid-19/issue-brief/the-implications-of-covid-19-for-mental-health-and-substance-use/).

Perednia, D.A., Allen, A., 1995. 1 Telemedicine technology and clinical applications. JAMA 273 (6), 483–488.

Perlman, D., Peplau, L.A., 1981. Toward a social psychology of loneliness. Personal Relationships 3, 31–56.

Physician appointment, 2014. Wait Times and Medicaid and Medicare Acceptance Rates. Merritt Hawkins, Irving, TX. http://www.merritthawkins.com/uploadedFiles/MerrittHawkings/Surveys/mha2014waitsurvPDF.pdf.

Psychosocial measures for clinical trials in spinal cord injury: quality of life, depression, and anxiety. In: Topics in Spinal Cord Injury Rehabilitation, 2006. Scott Richards.

Ray, K.N., Chari, A.V., Engberg, J., Bertolet, M., Mehrotra, A., 2015. Disparities in time spent seeking medical care in the United States. JAMA Internal Medicine 175 (12), 1983–1986. https://doi.org/10.1001/jamainternmed.2015.4468.

Rehman, U., Shahnawaz, M.G., Khan, N.H., Kharshiing, K.D., Khursheed, M., Gupta, K., Kashyap, D., Uniyal, R., 2021. Depression, anxiety and stress among Indians in times of Covid-19 lockdown. Community Mental Health Journal 57 (1), 42–48.

Robot Friends" and VR for Loneliness, IEEE Transmitter, (https://transmitter.ieee.org/robot-friends-and-vr-for-loneliness-during-the-pandemic/).

Shahraki, A., Abbasi, M., Piran, M., Chen, M., Cui, S., 2021. A Comprehensive Survey on 6g Networks: Applications, Core Services, Enabling Technologies, and Future Challenges arXiv preprint arXiv:2101.12475.

Sleep Technology: How it helps us to get better Zzz's, IEEE transmitter (https://transmitter.ieee.org/sleep-technology-how-it-helps-us-get-better-zzzs/).

Stewart, S., March 16, 2005. Financial aspects of heart failure programs of care. European Journal of Heart Failure 7 (3), 423–428.

Tedros Adhanom Ghebreyesus, 2020. Addressing mental health needs: an integral part of COVID-19 response. World Psychiatry 19 (2), 129–130. https://doi.org/10.1002/wps. 20768, 2020.

Telehealth: A quarter-trillion-dollar post Covid-19 reality? (www.mckinsey.com/industries/healthcare-systems-and-services/our-insights/telehealth-a-quarter-trillion-dollar-post-covid-19-reality).

Telehealth 2050: The Future Design of Virtual Care Technology, Healthcare IT News (https://www.healthcareitnews.com/news/telehealth-2050-future-design-virtual-care-echnology).

The Ascent of 5G Holds Bold Promise for Healthcare, BMC Blogs, (https://www.bmc.com/blogs/5g-in-healthcare/).

The Health Consequences of Loneliness, Verywell Mind, (www.verywellmind.com/loneliness-causes-effects-and-treatments-2795749).

Tuckson, R.V., Edmunds, M., Hodgkins, M.L., 2017. Telehealth. New England Journal of Medicine 377 (16), 1585–1592.

Ueafuea, K., et al., 2021. Potential applications of mobile and wearable devices for psychological support during the COVID-19 pandemic: a review. In: IEEE Sensors Journal, vol 21, pp. 7162–7178. https://doi.org/10.1109/JSEN.2020.3046259, 6.

Utra, A., 1959. Teleroentgen diagnosis by means of videotape recording. American Journal of Roentgenology 82, 1099–1102.

Vuorinen, A., Leppänen, J., Kaijanranta, H., Kulju, M., Heliö, T., Van Gils, M., Lähteenmäki, J., 2014. Use of home telemonitoring to support multidisciplinary care of heart failure patients in Finland: randomized controlled trial. Journal of Medical Internet Research 16 (12), e282. https://doi.org/10.2196/jmir.3651.

What are the causes of asthma and the misconceptions associated with it?, Tutorialspoint.com, (https://www.tutorialspoint.com/what-are-the-causes-of-asthma-and-the-misconceptions-associated-with-it).

Wittson, C.L., Affleck, D.C., Johnson, V., 1961a. The Use of Two-Way Television in Group Therapy. Nebraska Psychiatric Institute, University of Nebraska College of Medicine, Omaha.

Wittson, C.L., Affleck, D.C., Johnson, V., 1961b. Two-way television group therapy. Ment Hosp. November 12, 22–23.

Wootton, R., 2001. Telemedicine. BMJ 323 (7312), 557–560.

Wootton, R., Craig, J., Patterson, V., 2017. Introduction to Telemedicine. CRC Press.

https://worldpopulationreview.com/country-rankings/best-healthcare-in-the-world.

Yang, S., et al., Apr. 2019. IoT structured long-term wearable social sensing for mental wellbeing. IEEE Internet of Things Journal 6 (2), 3652–3662.

Zhang, Q., Liu, J., Zhao, G., 2018. Towards 5G Enabled Tactile Robotic Telesurgery arXiv preprint arXiv:1803.03586.

Zhao, A., Dong, J., Zhou, H., 2020. Self-supervised learning from multi-sensor data for sleep recognition. IEEE Access 8, 93907–93921.

CHAPTER FIVE

Extended reality for patient recovery and wellness

Usharani Hareesh Govindarajan[1], Dali Zhang[2] and Anshita[3]
[1]Business School, University of Shanghai for Science and Technology, Shanghai, China
[2]Sino US Global Logistics Institute, Antai College of Economics and Management, Shanghai Jiao Tong University, Shanghai, China
[3]Institute of Informatics and Communication, University of Delhi, Delhi, India

5.1 Introduction

The healthcare sector, in the past decades, has observed significant technological advancements. E-health has empowered medical stakeholders such as doctors, nurses, and patients with increased efficiency and improved access to healthcare systems and healthcare data. Technological advancements in healthcare systems have impacted the medical infrastructure holistically. Data, information, and technology are improving healthcare sectors such as surgical training and planning, patient—doctor communication, and patient recovery and wellness. Patient recovery implies the process of regaining health or fitness after an illness or injury (Lea, 2014). Patient recovery and wellness can be roughly categorized into two types, namely, neurological recovery and physiological recovery, and Fig. 5.1 shows the outline of such a classification. The popular treatment for neurological classification stems from a form of phobia and anxiety treatment and conditions triggered by a terrifying event such as posttraumatic stress disorder (PTSD). Stress management is systematically able to reduce stress levels, especially in chronic cases, to improve everyday functioning. Patient wellness based on a functional classification involves physiological elements such as improving damaged muscles that encapsulate motor rehabilitation and pain management—restoring functional activity after a stroke to a stabilized condition.

Conventional recovery and wellness methods involve counseling, medications, and physiotherapies, usually without the intervention of technology. There has been consistent growth in the study and development of applications of immersive extended reality (XR) in healthcare systems. Trivial applications of immersive technologies such as surgical training and surgical practice are typical and have been widely researched. Postoperative pain management is one such application where pain treatment is facilitated by virtual reality (VR) technology (Cacau et al., 2013). Current techniques for managing acute and chronic pain, such as opioids and physical therapy, are often incomplete or ineffective. This factor of prevalent techniques has made further research

Extended Reality for Healthcare Systems, Volume 1
ISBN 978-0-323-98381-5, https://doi.org/10.1016/B978-0-323-98381-5.00007-6

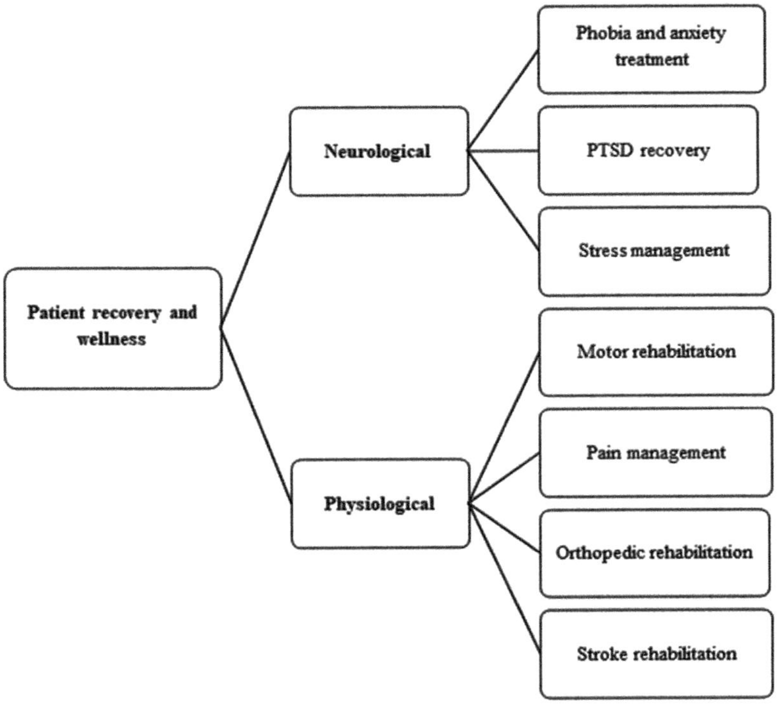

Figure 5.1 Outline of patient recovery and wellness.

and development of newer techniques more significant. VR trials demonstrate the potential to redefine the approach to treating acute and chronic pain in the clinical setting. Patient immersion in interactive VR provides a distraction from painful stimuli and can decrease an individual's perception of the pain (Pourmand et al., 2018).

Immersive XR, an umbrella term used for technologies such as VR, augmented reality (AR), mixed reality (MR), and 3D imaging, has applications such as PTSD treatment and poststroke rehabilitation and psychotherapy. The diagnosis of specific phobias is based on a thorough clinical interview and diagnostic guidelines. The goal of treatment is to improve life quality not limited by phobias. The most popular treatment for phobias worldwide is psychotherapy called cognitive behavioral therapy (CBT). CBT is an enhanced version of exposure therapy that involves gradual and repeated exposure-specific phobia combined with other techniques and/or medication. Medication for phobias includes beta-blockers and sedatives. Beta-blockers obstruct the effects of adrenaline caused by anxiety. Sedatives help reducing irritability or excitement during exposure (Govindarajan et al., 2019). A review of metaanalytic studies examining the efficacy of CBT conducted (Hofmann et al., 2012) demonstrated that this treatment had been used for a wide range of psychological problems. Similarly, virtual reality exposure

therapy (VRET) is used as psychotherapy for phobias and PTSD. In VRET, a patient is immersed in a programmed computer-generated virtual environment that helps the person directly confront feared situations or locations that may not be safe to encounter in real life. There is strong evidence that shows that VRET is useful for treating several different phobias and social anxiety (Ding and Dai, 2018).

The current developments of immersive XR in healthcare are observed based on an extensive search of academic publications from 2015 to 2020 using the query "extended reality healthcare." The number of results obtained was 952, inferring an increasing research trend in immersive XR in healthcare. Research and studies are being conducted to develop virtual clinics and rehabilitation centers to enhance the wellness infrastructure in society further. Therefore, this chapter focuses on systematically exploring the available academic publications and global patent grants to understand the current trends of immersive XR in patient recovery and wellness. The chapter is organized into four sections. Section 8.2 provides a background of immersive XR and understanding of patient recovery and wellness. Section 8.3 describes the recent advancements made by immersive XR in patient recovery and wellness in the context of trivial and nontrivial emerging aspects and all holds analysis of recent global patent grants. Section 8.4 presents discussions and key findings, and Section 8.5 presents the conclusion and discusses the future scope of the study.

5.2 Background

Immersive XR provides a host of systems that combine the semantics of virtual and reality to deliver robust applications that include broad concepts enabling users in the physical world to connect and immerse in the cyber world. XR, VR, AR, and MR are some immersive technologies. VR is now providing innovative ways for specialists from numerous domains to interact and collaborate with the acceleration of productivity. VR hosts technologies that mimic interactive 3D environments. Virtual interactions can be initiated by wearing head-mounted displays (HMDs), VR-enabled helmets, or goggles.

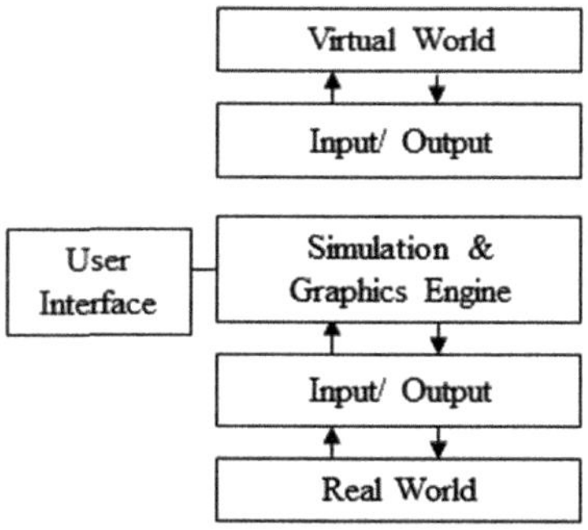

Figure 5.2 Immersive technology block diagram.

AR/MR or substitutional reality (SR) are applications where the virtual world and the real world are blended in immersive settings. Users can interact with real-world virtual content and make distinctions. Fig. 5.2 presents a generic immersive technology block diagram.

The user interface providing the immersion acts as a bridge for interaction between the user and the ecosystem. The components used are proportional to the level of immersion and the type of immersive technology expected from the application. There are three levels of achievable immersion, viz., nonimmersive, semiimmersive, and fully immersive technology. The nonimmersive level represents 3D effects experienced on a desktop computer. The semiimmersive level model represents elements from the real world to construct VR applications, for instance, cognitive training and surgical practice. Fully immersive technology is the goal to achieve the highest level of VR. This level achieves an immersion into the virtual world where the human brain is directly connected to a database and the viewer's current position and orientation in the virtual world evolving from innovations in their respective domain. Prior research on immersive technologies for industrial applications presents AR/MR technologies to lead the way for future innovation across domains (Govindarajan et al., 2018). Some of the key contributions in immersive XR with the key market players, ecosystems, and development platforms are given in Table 5.1.

Immersive XR and its applications are additionally being explored in the healthcare sector. In psychotherapy, VRET is now rapidly incorporated in clinical treatments for addictions and phobias. One application for autism spectrum disorder (ADS) affects children's ability to interpret, communicate, and involve repetitive behaviors. Autistic children have different and specific phobias such as object associations and specific places. The exposure-related treatments for such phobias are expensive and inconvenient and require constant therapist supervision that decreases availability. VRET studies show promising results to alleviate phobias (Sime, 2019). Apart from psychotherapy, there are

Table 5.1 Immersive extended reality market players.

Key player	Popular VR ecosystems	Significant development platforms
Google, Inc	Microsoft HoloLens	Oculus Medium
PTC, Inc	Magic Leap One	Unity
Microsoft Corporation	Oculus Quest/Quest 2	CRYENGINE
Seiko Epson	HTC Vive Cosmos	Unreal Engine 4
Lenovo	Sony PlayStation VR	Amazon Sumerian
Magic Leap, Inc	Samsung Gear VR	GoogleVR
Blippar	Valve Index	Apple ARKit
Maxst		Spark AR Studio
Vuzix		Google ARCore
		Vuforia Engine

continuous studies and testing done to develop applications of immersive XR to complement or substitute standard therapy for physical rehabilitation (Gordo, 2018). Such continuous studies and developments are facilitating the integration of immersive XR with patient recovery and wellness. The background presents a concise discussion of how immersive technologies will play an active role in future therapeutic healthcare. The broad study of the current knowledge frontier and its adjacencies is critical and is presented in the following section.

5.3 Recent advancements in immersive extended reality

The mapping of current advancements of immersive XR and its subsequent technologies in patient recovery and wellness prerequisites a data search across academic and patent databases. The first part of this section focuses on academic manuscripts and theoretical research. The next part aims to gain information about practical innovations and developments using the patent database. The final discussion is on the emerging standards related to the domain and the government and policy in the therapeutic immersive XR.

Table 5.2 Academic data: queries and data size.

Source	Keywords	Year limit	Resulting papers
PubMed	("patient recovery" OR "wellness") AND ("extended reality" OR "virtual reality" OR "augmented reality" OR "mixed reality")	2015-01-01 to 2020-12-31	28
Scopus	Title-Abstract-Keyword (("extended reality" OR "virtual reality" OR "augmented reality" OR "mixed reality") AND ("patient recovery" OR "recovery" OR "wellness")) AND PUBYEAR > 2014 AND PUBYEAR < 2021	2015-01-01 to 2020-12-31	660
Web of Science	"extended reality" OR "virtual reality" OR "augmented reality" OR "mixed reality" (Abstract) and patient recovery OR wellness (Abstract)	2015-01-01 to 2020-12-31	149
IEEE Explore	(("All Metadata":immersive technology OR "All Metadata":virtual reality OR "All Metadata":augmented reality OR "All Metadata":mixed reality) AND ("All Metadata":patient recovery OR "All Metadata":wellness))	2015-01-01 to 2020-12-31	113

5.3.1 Current study in patient recovery and wellness

Understanding the current research trend involving the usage of immersive XR in patient recovery and wellness is achieved by thoroughly mapping the academic publications across 5 years (2015–20). Table 5.2 displays the academic databases used to analyze and map the academic manuscripts relevant to immersive XR and patient recovery and wellness. The extracted data present a considerable study in the implementation of immersive XR in recovery and wellness, especially in motor recovery in stroke patients, in therapies of mental or cognitive diseases, in postoperative pain reduction and management, etc. Additionally, through a keyword-based analysis, the academic documents are analyzed for the most relevant keywords based on a weighting methodology that assigns the weight to the most relevant words in the titles of all the academic documents in the Web of Science and IEEE Explore databases. The scan presents around 92 terms that occur at least five times, and for each of the 92 terms, a relevance score will be calculated. The most relevant terms are selected based on the score and are presented in Fig. 5.3.

Stroke is a global healthcare problem that is common, serious, and disabling (Hoermann et al., 2017). Stroke rehabilitation is an integral part of patient care (Langhorne et al., 2011), which usually focuses on the motor function rehabilitation of patients as most effects of the stroke are physical (Physical Effects of Stroke, 2021), making the study and development of motor rehabilitation therapies vital. Immersive XR has been widely applied to motor rehabilitation and exemplified by the computer

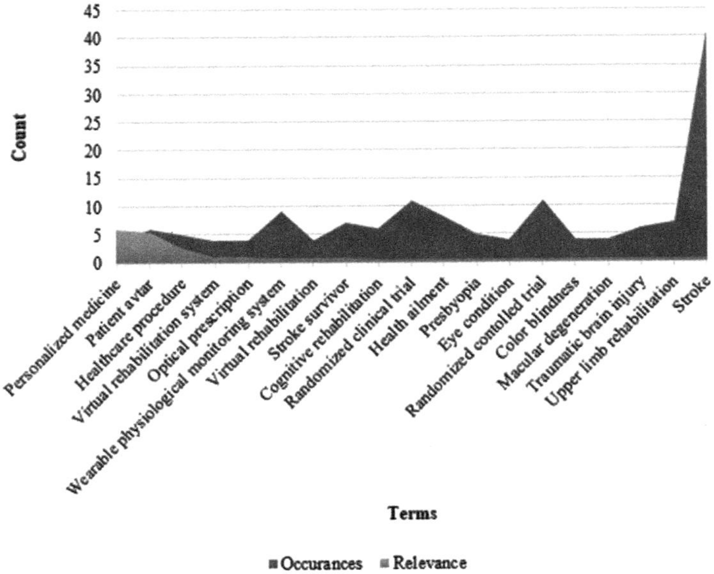

Figure 5.3 Key phrases and weights across current studies.

mirror therapy for early stroke upper limb rehabilitation. The mirror therapy uses a combination of the BeST protocol with augmented reality technology (ART) as a therapeutic intervention for in patients in their first month poststroke. In mirror therapy (MT), recovery from motor impairments after stroke is stimulated by observing the unimpaired hand in a mirror placed perpendicular between the two hands. This creates a mirror visual illusion for the patient that the impaired hand is healthy and moving again. In this way, the observation of the mirrored hand stimulates neurological support and helps in the improvement of various impairments (Hoermann et al., 2017). ART includes all the capabilities of traditional MT and provides a wider range of computer-mediated visual illusions and exercise possibilities (Regenbrecht et al., 2011). Also, due to ART's high immersive nature of ART, it shows better results than the traditional mirror therapy (Regenbrecht et al., 2011).

Another example is leap motion—based VR therapy for functional motor recovery of upper limbs in subacute stroke patients in a similar therapy. The leap motion controller developed by Ultraleap (2021) facilitates capturing and tracking of the fine movements of the hand and fingers while controlling a virtual environment that requires hand—arm coordination as part of the practicing of virtual tasks (Iosa et al., 2015). This technique showed encouraging results in motor function rehabilitation (Wang et al., 2017). Another study of proactive motor functional recovery (Mekbib et al., 2020) stated that custom-build virtual environments are also effective in motor rehabilitation. Many game-based VR systems have also been developed to focus on the same. Research is being conducted frequently to develop VR-facilitated therapies for upper limb rehabilitation and other parts of the body. Lateropulsion rehabilitation using VR is one such example. Lateropulsion is a physical effect of stroke that disturbs the center of gravity of the patient body. The posture of the patient changes, and the body remains inclined to one side. In a study conducted on 121 stroke patients, a VR game was used to observe the effects to VR on the rehabilitation of the center of the body. This study made use of a VR-based game that throws obstacles such as barrels toward the patient. It works on the principle of reflex actions of the human body, and the game aims to throw an obstacle of any kind at high speed. Speedy throw in the virtual environment simulates patient reflexes of avoiding the obstacle. Thus, the patient bends away from the obstacle and creates a movement toward the side of the body that suffers from lateropulsion. This helps shift the center of gravity of the patient's body and gradually improves lateropulsion (Tsatsis et al., 2017). These and several more examples show the use of immersive XR in motor function rehabilitation. A VR-based exercise program (VREp) that provides auditory, visual, and proprioceptive feedback in a VR environment offers individualized exercise training programs. It promotes motor learning in the paralyzed upper extremities of patients. VREp has been developed in a fashion that enables them to adjust the difficulty of the exercise to the individual's level of adaptation (Lee, 2015). Immersive XR can also be used for orthopedic rehabilitation. Motor rehabilitation, especially

orthopedic rehabilitation, requires monitoring since physicians need to comprehend the current state of a patient and his or her progress (Negrillo et al., 2020). Immersive XR can prove to be very effective in rehabilitation if requires feedback is guaranteed. Immersive XR systems also decrease recovery time and cost, which gives them profound stability in patient recovery and wellness.

Therapeutic immersive XR targets neurological rehabilitation for mental disorders too and several mental issues such as PTSD. Several disciplines, including horticultural therapy (HT), ecological psychology, environmental psychology, and medical geography, have supported the significance of views of natural scenery and landscapes, seeking to discover how and why they may alter people's moods, reduce stress, and improve cognitive functioning (Uwajeh et al., 2019). Clinical applications of VR to treat mental health conditions such as PTSD suggest that VR has significant potential to address and alleviate such issues as well as a full range of mental health issues. Results obtained from such studies show that the clinical application of VR offers the potential to create environments that allow for the control of complex, immersive stimulus presentations allowing interaction, behavioral tracking, user response, and performance recording, together with therapeutic intervention from the clinician (Georgieva and Georgiev, 2020).

Several kinds of research have been done to understand the efficacy of immersive XR and its parts in pain management. Immersive XR provides distraction with high immersion and interaction; thus, it can be used as an analgesic. Two very common applications of immersive XR in pain management are virtual environment (VE) and virtual reality training (VRT). In a systematic study with isokinetic training (IKT) and conventional training exercise, VRT showcased a higher improvement in pain and sports performances (Nambi et al., 2020). The following section focuses on reviewing the current innovation scenario through an exhaustive review of the patenting trends through a 5-year database timeline analysis.

5.3.2 Innovations and investments in global patent grants

Innovations in immersive XR for patient recovery and wellness have been documented theoretically and practically. Section 8.3.1 explored the theoretical developments and studies exploring the academic manuscripts from the prestigious databases including Scopus, Web of Science, PubMed, and IEEE Xplore. The section extends to the commercial development trends using global patent grants. The global patent data have been extracted using the patent search and analysis software PatSeer. The database provides wide coverage of more than 136 million records extracted from more than 108 authorities such as the United States Patent and Trademark Office (USPTO), the World Intellectual Property Organization (WIPO), and the European Patent Office (EPO). Table 5.3 tabulates the two different sets of search queries that have been used to extract

Table 5.3 Patent data: queries and data size.

Source	Keywords	Year limit	Resulting papers
PatSeer	Title Abstract Claims (extended reality OR virtual reality OR augmented reality OR mixed reality) AND Title Abstract Claim (patient recovery OR wellness OR healthcare)	2015-01-01 to 2020-12-31	327
	Title Abstract Claims (extended reality OR virtual reality OR augmented reality OR mixed reality) AND Title Abstract Claims (patient recovery OR wellness)	2015-01-01 to 2020-12-31	39

patent grants, where the first query carries the larger data set that holds patents common to immersive XR and healthcare. The second query is more specific and comprises a comparably smaller data set. The section extends to the patent data analysis to extract the knowledge of critical commercial trends from the databases for a global perspective.

The Cooperative Patent Classification (CPC) is an extension of the International Patent Classification (IPC) and is jointly managed by the EPO and the USPTO and is logically divided into nine sections, A-H and Y. The sections are subdivided into classes, subclasses, groups, and sub-groups. There are approximately 250,000 classification entries (European Patent Office, 2017). The IPC scheme is an international invention index standard, and classification characters are allotted to patent documents based on technology coverage. The dynamics of the IPC assignment across the patent office ensure that each patent is assigned to at least one class. There is no upper limit on the number of IPC classes that a patent may belong to. CPC enhances patent searching and analysis as it is a jointly developed classification and is more detailed. An assignee is an entity that has the current ownership rights to a patent grant. First is the analysis of 327 patent grants retrieved by searching the PatSeer patent database for patents titled or containing the keywords "extended reality, virtual reality, augmented reality, mixed reality, patient recovery, wellness, or healthcare" in the title, abstract, and claims section. These data are extracted for the years 2015—2020.

Table 5.4 shows the top 10 assignees that own patents relevant to immersive XR in the healthcare sector. Magic Leap Inc. with a patent count of 85 dominates the industry. Second, in lead is the Samsung Group with the ownership of 20 patents. Third, in

Table 5.4 Patent assignees versus CPC codes.

Top assignees		Top CPC codes	
Assignee	**Patent count**	**CPC (4 characters)**	**Patent count**
Magic Leap Inc.	85	G16H	193
Samsung Group	20	A61B	181
Blast Motion Inc.	11	G06T	155
Health Care Originals Inc.	9	G06K	125
Nant Holdings IP, LLC	9	G06F	105
ChemImage Corp.	8	A61M	102
General Electric Co.	8	A61N	97
IBM Corp	8	G02B	97
Becton Dickinson & Co.	6	A61H	87
Cirtec Medical Corp.	5	A61F	85

standing is assignee Blast Motion Inc. with the patent count 11. The fourth position is shared by Health Care Originals Inc. and Nant Holding IP LLC, and the patent count is 9. Sharing patent count 8 and fifth position from the top are ChemImage Corp., General Electric Co., and IBM Corp.

The leading assignee Magic Leap Inc. is an American company that works actively in the field of AR that manufactures wearable devices supporting AR. The industries of involvement of this top assignee are health, defense and public sector, manufacturing, oil and gas, automotive, transport, architecture, engineering, and construction. Additionally, the table also showcases the analysis of the highest-ranking Cooperative Patent Classification (CPC). The analysis of the patent data set centralized to map the developments of immersive XR in healthcare depicts that maximum patents published in the years 2015–2020 belong to CPC code G16H. CPC code G16H belongs to the Health Informatics, i.e., this subsection of CPC holds the patents that claim innovations in the domains of Information and Communication Technology (ICT) specially adapted for the handling or processing of medical or healthcare data.

Table 5.5 shows the top five CPC (main) codes, their ranks concerning the number of patents, and patent counts along with the corresponding CPC groups. CPC code G16H40 with 152 patents under it deals with information and communication technology (ICT) specially adapted for the management or administration of healthcare resources or facilities. Similarly, G16H50 handles ICT specially adapted for medical diagnosis medical simulation, or medical data mining, and has 60 patents under it, while G16H20 contains patents that deal with ICT specially adapted for therapies or health-improving plans and withholds 52 patents. The second highest patent category is A61B with 181 patents. CPC A61B deals with the diagnosis, surgery, and identification. The top three CPC groups within are A61B are A61B5, A61B2562, and A61B3. A61B5 deals with patents related to detecting measuring or recording for diagnostic purposes

Table 5.5 Top five CPC (main) codes with their subtechnologies and patent count.

Ranking	CPC (main)	CPC (group)	Patent count (CPC group)	Patent count (CPC main)
1.	G16H: Health Informatics	G16H40	152	193
		G16H50	60	
		G16H20	52	
2.	A61B: Diagnosis, Surgery, Identification	A61B5	163	181
		A61B2562	108	
		A61B3	91	
3.	G06T: Image Data Processing or Generation	G06T19	137	155
		G06T11	100	
		G06T3	89	
4.	G06K: Recognition of Data; Presentation of Data; Record Carriers; Handling Records Carriers	G06K9	120	125
		G06K2209	6	
5.	G06F: Electric Digital Data Processing	G06F3	66	105
		G06F16	21	
		G06F19	20	

and identification of persons. A61B2562 consists of patents related to details of sensors, constructional details of sensor housings or probes, and accessories for sensors. A61B3 withholds patents regarding apparatus for testing the eyes and instruments for examining the eyes. Among other CPC groups, G06T19 (Manipulating 3D models or images for computer graphics), G06K9 (Methods or arrangements for reading or recognizing printed or written characters or for recognizing patterns e.g., fingerprints), and G06T11 (2D image generation) have the majority patents with the patent count as 137, 120, and 100 respectively.

5.3.3 Emerging technical standards

The integration of healthcare with technology requires understanding both the healthcare sector and technology ensuring that the technology is fulfilling required medical needs. The process of integration of technology into healthcare gives rise to many concerns such as the privacy and security of medical data, the proper implementation of the technology, the regulation and check on the adverse effects of the technology used, etc. Standards are published documents that establish specifications and procedures designed to maximize the reliability of the materials, products, methods, and/or services that people use every day. Several standardization bodies create standards for immersive XR. The Institute of Electrical and Electronics Engineers Standards Association (IEEE SA) is a leading consensus-building association that develops global standards for many industries, including healthcare (IEEE Standards Association, 2021).

A wide range of issues including protocols to attain maximum product functionality and compatibility, interoperability and consumer safety, and public health are covered under the standards.

Through a catalog of 67 standards in the domain of healthcare systems, the recent and emerging standards are classified as the emerging standards in healthcare systems compiled through a review in Fig. 5.4. The emerging standards in E-health include the common framework of location services for healthcare standards. The sensors and devices categories discuss the medical devices that find portable use or point of care medical devices and lay down the standards for their susceptibility, radiation levels, and detailing in a similar context. The intelligent and autonomous systems include the recent inclusion of artificial intelligence-based technologies and their impact on humans. The standards in the category guide the development of the product, identification of areas for improvement, risk management, performance assessment, and the identification of intended and unintended users, uses, and impacts on human well-being of artificial intelligence and intelligent systems.

The standard detailing in the auspice of the healthcare sector includes in the E-health domain as the IEEE 1847-2019 describes the recommended practices for location services for healthcare. The recent standards related to healthcare devices include the IEEE/ANSI C63.24-2021 that recommends the test methods for assuring the radiofrequency (RF) immunity of electronic devices and systems that might experience susceptibility from general use transceivers or the RF ambient. The data standards secure the patient data and information, and IEEE 1752.1-2021 is a draft that standardizes mHealth data, which has observed increased adoption. It proposes an improvement in the ease and alignment accuracy of aggregating data across multiple mobile health sources. The impact of artificial intelligence or autonomous and intelligent systems (A/IS) on humans is measured by IEEE 7010-2020 standard. The discussions in the chapter are concerned under the standards IEEE 3333.1.1-2015, IEEE 3333.1.2-2015, and IEEE 3333.1.2-

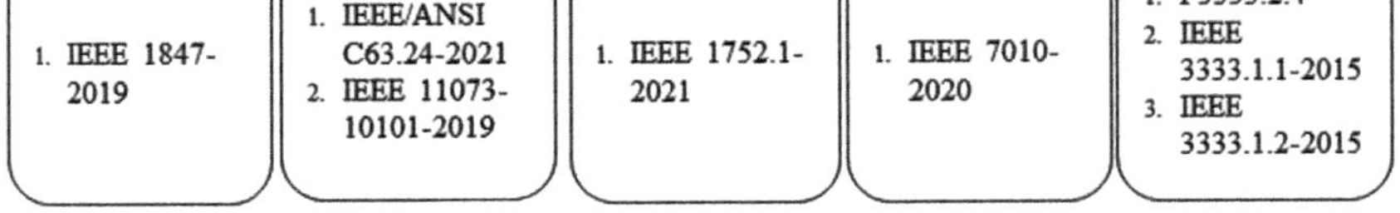

Figure 5.4 Emerging standards in the healthcare systems.

2017 that assess the quality and techniques used to develop the 3D display device and signal-processing engine industries. The characteristics of human perception, display mechanism, and the viewing environment are continuing in the further development of the standard.

5.4 Discussions

The knowledge maps jointly presented in Fig. 5.5 map the current research and development of immersive XR in patient recovery and wellness. Additionally, the extracted knowledge synergistically bridges and summarizes the characteristics of immersive XR technology and its crucial importance due to the data utilization approach in healthcare domains, including, psychotherapy, patient recovery, and wellness. The depicted network visualization map of research articles is based on title and abstract data. The current major adoption areas include poststroke motor recovery and neurorehabilitation.

VOSviewer is utilized to extract the key terms from the corpus of academic documents and extract information. VOSviewer is a software tool for creating maps based on network data such as coauthorship, cooccurrence, citation, bibliographic coupling, or cocitation links and for visualizing and exploring these maps (Van Eck and Waltman, 2013). The objective has been to extract keyword patterns and assess the connections between terms by creating a network visualization map. The map is automatically created where the back-end implementation involves two major stages as term identification and term selection. For the term identification stage, the software uses the Apache OpenNLP library to preprocess text data. In the term selection stage, users can customize the word list. When one chooses to exclude a multiword term, occurrences of this term may be counted as occurrences of a shorter term. Patients, virtual reality, system, health system, and treatment are the most weighted issues, representing the whole study route toward Extended Reality for Patient Recovery and Wellness. The patent node clusters with keywords such as stroke, anxiety, treatment, and so on represent all healthcare-related problems in the network map, and their connectivity with nodes such as VR, AR, and wearable devices represent a clear relationship between health therapy and immersive technologies. The VR and device nodes are crucial in terms of their link with subjects such as AR, wearable devices, and patient experience, which address the present researcher's direction and techniques.

The validation attained in the network visualization graph generated through the corpus is through a three-quadrant structure consisting of patient, system, and applications. The visualization graph connects the patient node to key nontrivial applications such as traumatic brain injury (TBI), where Maggio et al. (2019) explore VR and its role in creating a positive, motivating, and enjoyable learning experience for the TBI patients. Additionally, patents such as the HMD perform regular monitoring, specific

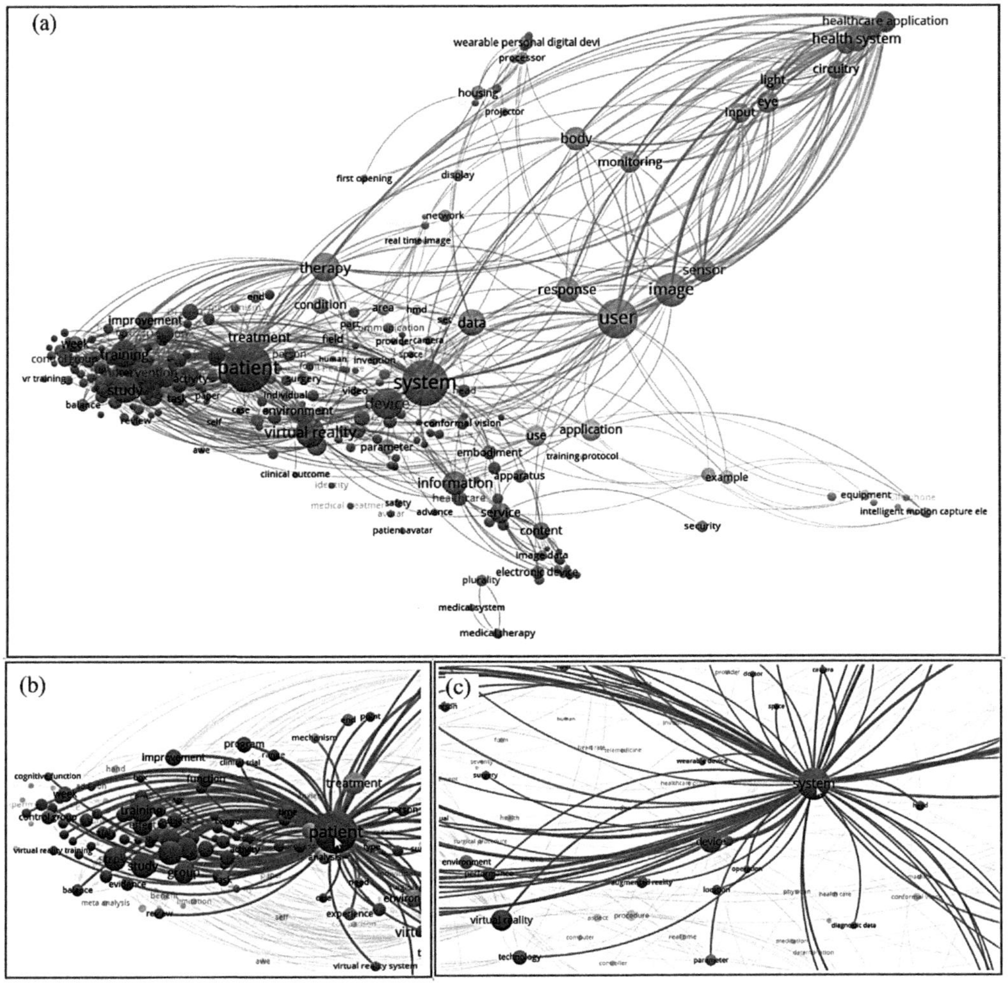

Figure 5.5 (a) The network visualization plot of academic and patent grants data. (b) Keyword cluster connected with "patients" and related medical domain phrases. (c) Keyword cluster associated with the phrase "virtual reality" and related XR technological terms.

examination, surgical intervention, or other care procedures on the eyes automatically or in conjunction with a device of an eye care professional across a computer network to care for the user's eyes in real time or on an ongoing basis. Such examples represent the relationship as a key finding to the chapter. Newer patents such as (Diaz and Haverman 2020) disclose a medical system that allows a healthcare professional to view medical

records as they are associated with a patient avatar. The system additionally can create a three-dimensional view of an avatar representing a patient and attach patient records to specific locations of the avatar as representation of the data within each record and access timeline of treatment to a particular position on the avatar corresponding to medical treatments. A similar wearable monitoring system is disclosed by the patent owner (Dwarika et al., 2018) that acts as a physiological monitoring system to collect and/or detect various parameters, such as cough, wheeze, heart rate, skin temperature, activity, respiration rate, skin impedance, electrocardiogram data, blood pressure, galvanic skin response, and the like. The systematic discussion points to additional development of standards and government policies for the incorporation of more nontrivial applications under the healthcare systems with further development of the XR technologies.

5.5 Conclusion and future scope

The healthcare sector has seen rapid change because of the use of information and communication technologies, which allows for faster access to patient records, more accurate diagnosis, and the opening of new areas of technology-curated treatments such as XR or immersive XR. The characteristics of XR, such as the integration of real and virtual environments and the facilitation of user interaction using previous data, observed accelerated adoption and exploration in its application due to the COVID-19 pandemic. The advantages of XR technologies in healthcare allow greater control over the presentation of stimuli in three dimensions, the creation of complex scenarios allowing audio, haptic, olfactory, and motion to be experienced simultaneously. This enables and offers tremendous scope in the study of situations that can be impractical or dangerous in real life. The application scenario of XR currently includes trivial aspects such as surgical training and surgical practice. The patient treatment recovery along with wellness are emerging nontrivial applications. This chapter explored the nontrivial emerging aspects and the related standards associated with such technologies. The presented discussion in the chapter lays down an exploration path for similar nontrivial applications.

The government policies and emerging standards are an important aspect in the coming future for the evolution of these nontrivial applications. The addition of a human element restricts the development with training, clinical trials, and observance of protocols and additional adoption barrier. The nontrivial applications evolve synergistically with inclusive government policies, which is an extended exploratory scope of the chapter. The XR technology discussed in this chapter offers a wide range of incorporation benefits indicating a gateway for continued future developments. The exploration of government policies and the technical specifications form a basis of the future scope of this chapter for investigation in healthcare-related applications.

Acknowledgments

The authors of this chapter acknowledge the efforts of our team's research intern Mr Gagan Narang for his support in data collection, interpretation, and information organization and research assistant Mr Manish Kumar for cross-verification of the presented results.

References

Cacau, L.D.A.P., Oliveira, G.U., Maynard, L.G., Araújo Filho, A.A.D., Silva Junior, W.M.D., Cerqueria Neto, M.L., Santana-Filho, V.J., 2013. The use of the virtual reality as intervention tool in the postoperative of cardiac surgery. Brazilian Journal of Cardiovascular Surgery 28, 281–289.

Ding, X., Dai, L.I., 2018. Virtual reality-based exposure therapy for anxiety disorders: A meta-analysis for randomized controlled trials. Chinese Mental Health Journal 32 (3), 191–199.

Diaz, O.C., Haverman, Y., 2020. U.S. Patent Application No. 16/084,187.

Dwarika, J., Samjitsingh, S., 2018. U.S. Patent Application No. 15/431, 531.

European Patent Office, May 17, 2017. EPO - Cooperative Patent Classification. (CPC). Copyright © 2007 European Patent Office. All Rights Reserved. https://www.epo.org/searching-for-patents/helpful-resources/first-time-here/classification/cpc.html.

Georgieva, I., Georgiev, G.V., 2020. Reconstructing personal stories in virtual reality as a mechanism to recover the self. International Journal of Environmental Research and Public Health 17 (1), 26.

Gordo, A., 2018. Virtual Reality for Locomotion Rehabilitation.

Govindarajan, U.H., Trappey, A.J., Trappey, C.V., 2018. Immersive Technology for Human-Centric Cyberphysical Systems in Complex Manufacturing Processes: A Comprehensive Overview of the Global Patent Profile Using Collective Intelligence. Complexity, 2018.

Govindarajan, U.H., Trappey, A.J.C., Trappey, C., 2019. 360° technology as a gateway for immersive psychotherapy applications: an intelligent patent mining analysis. In: 15th International Conference on Data Science, vol 19. ICDATA.

Hoermann, S., Ferreira dos Santos, L., Morkisch, N., Jettkowski, K., Sillis, M., Devan, H., Cutfield, N.J., 2017. Computerised mirror therapy with augmented reflection technology for early stroke rehabilitation: clinical feasibility and integration as an adjunct therapy. Disability & Rehabilitation 39 (15), 1503–1514.

Hofmann, S.G., Asnaani, A., Vonk, I.J., Sawyer, A.T., Fang, A., 2012. The efficacy of cognitive behavioral therapy: a review of meta-analyses. Cognitive Therapy and Research 36 (5), 427–440.

IEEE SA - What Are Standards? (n.d.). IEEE Standards Association. Retrieved July 25, 2021, from https://standards.ieee.org/develop/develop-standards/overview.html.

Iosa, M., Morone, G., Fusco, A., Castagnoli, M., Fusco, F.R., Pratesi, L., Paolucci, S., 2015. Leap motion controlled videogame-based therapy for rehabilitation of elderly patients with subacute stroke: a feasibility pilot study. Topics in Stroke Rehabilitation 22 (4), 306–316.

Langhorne, P., Bernhardt, J., Kwakkel, G., 2011. Stroke rehabilitation. The Lancet 377 (9778), 1693–1702.

Lea, D., Bull, V., Webb, S.S., Duncan, R. (Eds.), 2014. Oxford Learner's Dictionary of Academic English. Oxford University Press.

Lee, K.H., 2015. Effects of a virtual reality-based exercise program on functional recovery in stroke patients: part 1. Journal of Physical Therapy Science 27 (6), 1637–1640.

Maggio, M.G., De Luca, R., Molonia, F., Porcari, B., Destro, M., Casella, C., Calabro, R.S., 2019. Cognitive rehabilitation in patients with traumatic brain injury: a narrative review on the emerging use of virtual reality. Journal of Clinical Neuroscience 61, 1–4.

Mekbib, D.B., Zhao, Z., Wang, J., Xu, B., Zhang, L., Cheng, R., Xu, D., 2020. Proactive motor functional recovery following immersive virtual reality–based limb mirroring therapy in patients with subacute stroke. Neurotherapeutics 17 (4), 1919–1930.

Nambi, G., Abdelbasset, W.K., Elsayed, S.H., Alrawaili, S.M., Abodonya, A.M., Saleh, A.K., Elnegamy, T.E., 2020. Comparative effects of isokinetic training and virtual reality training on sports performances in university football players with chronic low back pain-randomized controlled study.

Evidence-based Complementary and Alternative Medicine 2020, 10. https://doi.org/10.1155/2020/2981273, 2981273 2020.

Negrillo-Cárdenas, J., Jiménez-Pérez, J.R., Feito, F.R., 2020. The role of virtual and augmented reality in orthopedic trauma surgery: from diagnosis to rehabilitation. Computer Methods and Programs in Biomedicine 191, 105407.

Physical effects of stroke. (n.d.). Stroke Association. Retrieved July 25, 2021, from https://www.stroke.org.uk/effects-of-stroke/physical-effects-of-stroke.

Pourmand, A., Davis, S., Marchak, A., Whiteside, T., Sikka, N., 2018. Virtual reality as a clinical tool for pain management. Current Pain and Headache Reports 22 (8), 1–6.

Regenbrecht, H.T., Franz, E.A., McGregor, G., Dixon, B.G., Hoermann, S., 2011. Beyond the looking glass: fooling the brain with the augmented mirror box. Presence 20 (6), 559–576.

Regenbrecht, H., McGregor, G., Ott, C., Hoermann, S., Schubert, T., Hale, L., Franz, E., October 2011. Out of reach?—a novel AR interface approach for motor rehabilitation. In: 10th IEEE International Symposium on Mixed and Augmented Reality. IEEE, pp. 219–228, 2011.

Sime, D.W., 2019. Virtual reality therapeutic environments in autism Spectrum disorder (ASD) and alzheimer's: treatment, diagnosis, and refinement. Virtual and Augmented Reality in Mental Health Treatment 51–59.

Tsatsis, C.G., Rice, K.E., Protopopova, V., Ramos, D., Jadav, J., Coppola, J.F., Putrino, D., May 2017. Lateropulsion rehabilitation using virtual reality for stroke patients. In: IEEE Long Island Systems, Applications and Technology Conference (LISAT). IEEE, pp. 1–6, 2017.

Ultraleap. (n.d.). Ultraleap. Retrieved July 25, 2021, from https://www.ultraleap.com/product/leap-motion-controller/.

Uwajeh, P.C., Iyendo, T.O., Polay, M., 2019. Therapeutic gardens as a design approach for optimising the healing environment of patients with Alzheimer's disease and other dementias: a narrative review. Explore 15 (5), 352–362.

Van Eck, N.J., Waltman, L., 2013. VOSviewer manual. Leiden. Univeristeit Leiden 1 (1), 1–53.

Wang, Z.R., Wang, P., Xing, L., Mei, L.P., Zhao, J., Zhang, T., 2017. Leap Motion-based virtual reality training for improving motor functional recovery of upper limbs and neural reorganization in subacute stroke patients. Neural regeneration research 12 (11), 1823.

Further reading

Garcia-Garcia, E., Baeza, P.S.H., Cuesta-Gomez, A., 2019. Effectiveness of the virtual reality in the rehabilitation of the upper limb in the spinal cord injury. A systematic review. Revista de Neurologia 69 (4), 135–144.

CHAPTER SIX

Role of virtual reality for healthcare education

Mohd Javaid[1], Ibrahim Haleem Khan[2], Rajiv Suman[3] and Shahbaz Khan[4]
[1]Department of Mechanical Engineering Jamia Millia Islamia, New Delhi, India
[2]College of Engineering, Northeastern University, Boston, Massachusetts, United States
[3]Department of Industrial & Production Engineering, G.B. Pant University of Agriculture & Technology, Uttarakhand, India
[4]Institute of Business Management, GLA University, Mathura, Uttar Pradesh, India

6.1 Introduction

The development of advanced technologies would be benefitting ongoing medical studies. Information technology (IT) is gradually permeating all aspects of life to enhance and expand their capabilities. IT technologies can be used to optimize time-consuming activities, assisting consumers in identifying gaps in their existing operating model while simultaneously discovering new, more straightforward, and more creative approaches to solving problems. The Med-Data in healthcare is the future to go as it teaches how to deal with medical issues (surgical, medicine, injury, etc.) beforehand with most preventive measures. Healthcare is one of the top industries implementing IT solutions due to its difficulty and coping with classified details. Among these technologies, virtual reality (VR) is getting attention from medical practitioners and educators, as it creates an innovative network to map and adapt various activities for medical applications. There is an improvement in the efficacy of using VR in treatment. This technology assists physicians, nurses, and students develop mastery by real-life scenarios, sharing experience, preparing to solve challenges, and saving lives. VR is driving learning experience creativity by delivering unique ways to learn-by-doing for improved retention and proficiency, sharing available resources internationally, and saving money on the high cost of preparation and time. It is a new way to interact with knowledge using VR solutions, offering an innovative learning opportunity in the healthcare field (Falah et al., 2014a,b; Pottle, 2019; O'Sullivan et al., 2021).

VR has found its way into various areas, serving as realistic solutions for both the user and industry. This technology is used for diagnostics, preparation, surgery, therapies, nursing, and patient education. In medicine, this technology makes patients wholly engulfed in a situation that is almost similar to a regular procedure. It has now been used widely in many hospitals around the world for teaching purposes (Ammanuel et al., 2019; Dyer et al., 2018).

Extended Reality for Healthcare Systems, Volume 1
ISBN 978-0-323-98381-5, https://doi.org/10.1016/B978-0-323-98381-5.00016-7

Body mapping is another effective way to incorporate VR into the analysis process. VR is feedback based and mimics real-life scenarios with learning capabilities involving AI and machine learning to guide it through its complexities. When the condition is complex or the doctor is physically unable to attend the test, a complete reconstruction of the patient's anatomy can be helpful. It is carried out in bursts of intensive preparation and evaluation of different solutions, resulting in the gradual realization of the plan step by step over brief periods. Many procedures in modern clinics and hospitals are performed using a variety of instruments and high-tech methods. Surgeons have abilities to closely link to their familiarity with instruments within the operating room. It promotes teamwork skills and correct workflow, and intervention coherence necessitates a variety of teaching and practice. An instructor or administrator may use VR technologies to monitor their clinicians and employees as they exercise their communication skills. Leaders may introduce additional scenarios or meetings that will result in change (de Faria et al., 2016; Huang et al., 2016).

Relevant medical knowledge is one of the most critical components of successful care. However, even in the most sophisticated data collection programs, the scope of collected knowledge can be impenetrable or misleading to physicians. VR can display statistics directly on the patient's body, highlighting troublesome areas and displaying relevant data. The whole system can be housed in a pair of smart glasses and accessible with a smartphone. This digital reality is hastening the advancement of education and preparation.

In contrast, educators searched to locate rooms, funds, and services that met their needs. VR training can be completed from a distance, which is very useful in COVID time. Educators are often concerned with the amount of space used to teach and the health and welfare of their students. Using VR in healthcare education also ensures that the content can be easily modified and applied (Labovitz and Hubbard, 2020; Zhao et al., 2021).

The adoption of VR technologies can improve the quality of healthcare facilities. VR headphones are used to distract children in disagreeable processes such as blood screening in many hospitals. Another use of VR in medicine is virtual models of the world view and experience of people with mental health conditions. Instead of sending any expert interested guidance or presentations individually, they use VR to show their items. This technology has a long way to go to be completely incorporated, but it is fair that it is popular one day. It produces a wholly rendered digital environment that replaces the actual environment of the user. It can measure the dosage delivery in 3D intuitively, saving preparation time for future higher accuracy and fewer drug side effects (Vozenilek et al., 2004; Roy et al., 2006).

This chapter is structured as Section 6.1 introducing the application of VR technologies in healthcare education. Section 6.2 provides an overview of VR. Section 6.3 highlights the advancement of VR technologies in the past decade. Section 6.4 discusses

the role of VR for healthcare education. Section 6.5 provides a discussion. Section 6.6 highlights the future of VR in healthcare. Finally, Section 6.7 concludes the study.

6.2 Background

VR is a computer-generated simulation that produces a virtual 3D environment with various devices such as special goggles, screens, gloves, and other innovative electronic devices. VR technology has several applications in various industries ranging from real state to tourism. The application of VR could find out in the automobile industry, architecture, automation, sports, journalism, education, marketing, training, healthcare, and many more. For instance, VR is used as marketing by providing the customers feel through these technologies. Coca-Cola become the first companies to explore the VR experiences in their marketing when they transformed their Christmas advert into VR in Poland. Some universities also used the VR technologies for the campus tour of the university. Similar to this, tourism sector is also exploring the VR application by providing a virtual experience about the holiday before a customer buy it through virtual tours of hotels, restaurants, and tourist landmarks. Healthcare is not lagging behind in the applications of VR, and it applies in the several domains of the healthcare such as dentistry, cardiology, orthopedics, surgery and medical education and training, etc.

VR is being used to assist and treat people suffering from various problems, including anxiety, depression, sensory disability, phobias, and psychiatric illnesses. Physicians are now playing with VR, with some tangible improvement for their patients. VR environments would necessitate including patients, parents, community groups, and advocates in the content creation process. VR has many advantages and many promises in relationships with educators, students, professionals, supervisors, and employees. Patients may use VR to engage in outpatient programs in the privacy of their own homes rather than going to a hospital. Progress can be tracked with the program that allows for at-home care. Patients will access the same treatment and more convenience in a one-stop service supervised by a healthcare provider (Winkler-Schwartz et al., 2019; Dimitropoulos et al., 2008; Hamilton et al., 2021).

VR aimed at the healthcare industry is an undeniably profitable growth path. Healthcare professionals, clinicians, academics, patients, and chemists all have great opportunities for this technology. Aside from the shortage and unequal allocation of healthcare professionals, the inadequacy of educational facilities affects the implementation of consistent healthcare services globally. Healthcare institutions have been working on implementing policies to expand the number of healthcare professionals while improving the quality and importance of medical services. This technology is essential for patients with multiple diseases that prevent them from leaving their homes and those with reduced mobility. The use of VR in physical therapy treatment centers

demonstrates the potential to speed up the recovery process and encourage activity. VR in healthcare has already yielded many benefits and will continue to evolve and grow. Healthcare providers, medical students, educators, employers, and workers can determine real-time status with virtual images (Kuehn, 2018; Dunne and McDonald, 2010; Almousa et al., 2021).

Hospital and traumatic emergencies can be frightening and exhausting, particularly for medical professionals in their early careers. It is challenging to prepare inexperienced doctors to respond successfully to medical emergencies until they are faced with a real-life situation. Recipients in VR-based medical training record improved their understanding of anatomical roles, and reduced surgical time in the real world. It increases protection for both surgeon and patient, positive psychological impacts on participants, and overall progress and reduces training costs and efforts,. The opportunity for VR in nursing education is endless and will continue to grow with the release of new hardware technologies. VR is intended to immerse a student in a practical world, which might be to see a patient. Instead of relying on the actual ability, the primary goal of VR is to show the student when and where a choice should be taken. Students will go through their tests based on the case that a teacher creates, making choices for each point of the way (Sultan et al., 2019; Falah et al., 2014a,b).

VR technology provides full integration in a virtual world with the use of specialized devices. In this case, the patient or doctor gets the entire image, which is only constrained by the virtual world's boundaries. The immersive surgery with the aid of VR games such as touch surgery without actually injuring a patient. This technology also allows the medical college during the practical study, which helps avoid the moral and ethical problems that may arise during this period. VR can be used to model surgery before it is done, allowing the surgeon or student to consider different situations. Doctors and students can quickly determine risk during an operation. There are video streaming and patient navigation applications that assist doctors in controlling both real-time and remote surgeries. VR in healthcare applications precisely expresses with the most incredible precision. This system may be used to quickly collect sensitive information and enabling treatments (Nicholson et al., 2006; Arents et al., 2021; Javaid and Khan, 2021).

VR provides tremendous help in the medical evaluation process. It offers tactile aids for getting input, putting the patient in a better place to assess and treat symptoms. VR technology has the potential to help both patients and clinicians. It has the ability to provide customised care that is not purely based on a clinician's own experience. Healthcare practitioners can use VR medical devices to support their patients better and assist physicians in a more customised manner. Medical VR was shown to effectively stop the brain in hospitalized patients from processing and healing pain. This is also helpful to reduce the patient's hospital stay and thus reduces the cost of the whole care process. A range of software is developed to distract the patients' minds in difficulty concentrating

more on VR worlds, which help reduce tension. The app uses immersive VR technology that provides movement and sensory experiences to help correct disorders in the brain (Djukic et al., 2013; Fertleman et al., 2018).

6.3 Significant advancements of virtual reality for healthcare

Health professionals quickly realized the advantages of VR innovations. Healthcare staff can read a lot about anatomy and the way the body works. VR applications allow doctors to visualize and engage with three-dimensional corporate representations. It also proves extremely useful as a guide to help patients understand the operation of surgery and how drugs work. This helps for better understanding for physicians as compared with conventional approaches. The progress made by VR in studying and learning makes it easier and more engaging for medical professionals to practice. VR simulations enable the human body to be explored without endangering patients. This will also lead surgeons to conduct a proper operation. Virtual worlds allow students to experiment and create errors and learn from them (Gao et al., 2021).

VR allows enhancing patient protection by reducing their radiation exposure. A very time-consuming, complicated exercise is the way a patient radiation therapy schedule is designed. 3D modeling by surgeons for operation planning is one of the most exciting things where VR applications improve the experience. Surgeons can communicate with the organ from any desired angle and transform between 3D and the actual CT picture in a three-dimensional environment. In a controlled approach, VR is also able to isolate anxiety-induced stimuli. Other psychological illnesses and childhood disabilities go very well, where VR can play a significant role. Today, businesses worldwide are using VR to improve their learning environment, including healthcare, retail, property, and automobile, in schools and sectors. The blackboards and textbooks are replaced by technologically enabled learning and smart boards (Sales et al., 2011; Lilly et al., 2019; Harden and Hart, 2002).

VR apps allow medical students and doctors to perform surgery and build registered experience to study and continue learning in a healthy and risk-free atmosphere. VR wellness preparation inefficiency and implementation continue to increase. Furthermore, the incorporation in hospitals and health facilities of VR technologies also successfully resolves issues related to training systems focused on simulation. It enables actual knowledge to be practiced at a reduced cost. Advanced VR training platforms allow more frequent and routine practice of different surgical procedures with flexible settings and present scenarios (Samadbeik et al., 2018; Haowen et al., 2021). In difficult situations, VR can address the availability problem. The VR world can be linked to a remote-controlled robot that will handle the actual surgery, while the surgeon does the procedure elsewhere. One of the most severe global problems confronting the healthcare industry is a chronic shortage of skilled practitioners. Learning about normal body

functions and processes in an immersive and demonstrative manner can help relieve discomfort while both setting and managing goals. It assists doctors ensuring that their patients appreciate the treatments they are consenting to. This technology opens up new avenues for addressing neuropsychological disorders in which the brain and nervous system affect a patient's perception and behavior. VR is an excellent way to learn human anatomy layer by layer and the physiology of the human body. This offers interactivity, which offers a much deeper glimpse into the dynamics of body functions (Haleem et al., 2020a,b; Lai and Zou, 2018; Mariani and Pêgo-Fernandes, 2011).

It is crucial for people suffering from phobias to learn coping skills as part of their treatment and exercise in a healthy atmosphere using VR healthcare. Clinics have also started to effectively use VR to construct worlds that reduce a patient's illness in a healthy and regulated manner. Based on the patient's success, therapy sessions can be quickly interrupted. Patients' chances of successfully treating their disorder when confronted with their phobia in everyday life are increasing gradually. Many diseases and traumas inflict severe damage to the head, muscles, and nerve system. It allows them to cope with their trauma and the emotions that emerge as a result positively. VR can be used effectively in psychotherapy to treat phobias and other mental health conditions. This technology enables doctors to build a controllable atmosphere and place patients in stressful situations predetermined for each specific case to train them to resolve their fears and develop trust (Lopes and Jorge, 2019; Levinson et al., 2007; Zack off et al., 2021).

Dentists benefit from medical VR systems using 3D versions of teeth using special drills that simulate real-life tactile feedback. Doctors may study these models from various angles, cross sections, and views to prepare potential operation scenarios. They can also perform procedures in simulated operating rooms before using actual scalpels on live patients to measure and evaluate the optimum sequence of procedure stages. VR allows for the development of personalized situations and simulations, such as those used in paramedic training under specific and optionally challenging circumstances. This technology can also aid in learning how to use different medical devices and equipment by allowing them to be used in a secure simulated environment (Rizzetto et al., 2020; Bradley, 2006; Baniasadi et al., 2020).

6.4 Impact of virtual reality for healthcare education

Visual reality provides an entirely artificial and interactive 3D interface that allows users to imagine like the virtual world is real. VR has a significant effect on primary education because it improves essential learning in a fun and stimulating manner. It helps students get to know each other better and learn quicker and especially helpful for providing scenario-based decision-making material and teaching skills that can be risky. VR teaching is adaptable, compact, and affordable to the healthcare community. With VR patient training, one can visit the hospital on a simulated basis. Students can

incorporate VR into lectures and use VR apps to reach templates and get them excited about learning. The presenter should engage students and reinforce lessons taught in lectures (Javaid et al., 2020; Scalese et al., 2008; Ye et al., 2020).

A high-quality VR training program shows the fundamentals and specifics of different procedures. This technology can move a learner inside a human body so that they can adequately view and train in environments that are not readily available. As a result, VR is handy for developing surgical training simulators. VR has been shown to have its therapeutic powers, allowing for a reduction in medicine, making it an essential adjunct to other approaches. VR applications have an incredible learning environment as well as limitless training opportunities. The use of VR in the educational process is incredible, which transforms and engages the learning atmosphere. It can describe complex and challenging ideas (McGrath et al., 2018; Akaike et al., 2012). Table 6.1 discusses the positive role of VR in healthcare education.

Higher education often categorizes advanced subjects using this immersive technology. It can also boost professional education. Training on VR has become and is a component of several ecosystems around the world. This will replicate the training in a comfortable, wholly immersive, fun, and risk-free manner. VR is being used to help people learn what to expect before and after surgery. This approach is benificial for healthcare organizations, delivering clinician preparation and patient engagement services (McGaghie et al., 2011; Haixu et al., 2021; Saxena et al., 2018). It has also shown interest in the field of mental health. VR not only frees up doctors to do what they do best, but it can also make better doctors. This allows physicians and other medical practitioners to view CT and MRI images before and after surgical procedures. VR teaching modules have many advantages, including reducing medical complications and creating a more practical setting for training doctors and students (Fiani et al., 2020; Li et al., 2017).

There are several challenges while adopting the VR in healthcare education such as resource constraints, explicit technical knowledge, physical safety risks, and possibility to skip the real situation. The VR requires a significant investment for the development of VR setup and maintains them at the operational. Further, a technical knowledge is demanded for the development of the VR environment. Some VR tools are very bulky, and it could threaten the safety of the user. Due to extensive use of VR technologies, the health professional might lose the interest in real-world experiment. This could pose a threat for the healthcare education. These limitations create a challenge for the VR applications in healthcare education system.

These challenges need to be mitigated through the identification of explicit application of VR in healthcare education. The financial challenge could be reduced through the development of laboratory that could be shared with other similar organizations. Government financial aid could also support investment required for the development of the VR infrastructure. Research and development are also required for the process

Table 6.1 Positive Role of Virtual Reality (VR) for improving healthcare education.

S No	Role of VR	Description
1.	Teaching and training	Currently, VR applications in healthcare focus on providing modern and immersive ways to teach and train medical professionals. It is developing novel approaches to treat people with neuropsychological problems. One of the most successful applications of VR in the medical field is education and patient care. Doctors and nurses use VR as a modern way of learning and teaching, providing the closest thing to real-world experience (Tang et al., 2021; Antoniou et al., 2020; Fabris et al., 2019).
2.	Recovery plan	Treatments and therapy are much more successful when the patient understands just what to expect. VR technology aids in the creation of realistic goals at different phases of a recovery plan. Patients may use VR to understand what will happen to their body before and during treatment and allow them to give consent more knowledgeably and manage their perceptions of the healing process (Alfalah et al., 2019; Jensen and Konradsen, 2018).
3.	The practice of social behaviors	VR builds opportunities for people to practice their social behavior. Teaching positive coping skills in various simulated contexts assists young adults with autism practice and understands socially healthy activities. Patients can overcome any of the psychological consequences of phantom limb suffering by using VR. It constructs a three-dimensional universe in which the phantom limb can be controlled, assisting them in managing this disorder. VR has been used to help individuals with disabilities or chronic illnesses (Schmidt et al., 2021; Chang et al., 2018; Shi, 2014).
4.	E-learning	Several forms of E-Learning have been used in recent years to disseminate knowledge and provide instruction to medical students with VR applications. VR environments allow users to interact with simulated and real-life experiences and learn practical information. VR increases patient education standards and allows inexperienced medical professionals to study concepts in settings that mimic real-life situations (Hammoud et al., 2008; Javaid and Haleem, 2020).
5.	Emergency	VR uses an effective teaching technique in emergency rooms. This is particularly true when it comes to teaching emergency responders. Such lifesaving treatments and techniques are only undertaken in real life, despite being vital to a patient's safety. VR tools allow

(*Continued*)

Table 6.1 Positive Role of Virtual Reality (VR) for improving healthcare education.—cont'd

S No	Role of VR	Description
		clinicians to mimic scenarios that infrequently exist to apply their theoretical experience and maintain their lifesaving skills sharp, even though they are just required occasionally. Healthcare providers sometimes have minutes or seconds to intervene to save a patient's life. As a result, providers need to keep their professional expertise and skills up to date and effectively educated (Alharbi et al., 2020; Klemm et al., 2021; Meyer, 2020).
6.	Treatment of burn patient	VR can be used as a distraction treatment technique for burn patients. Pain affects the senses through pathways in the brain and nerves. VR can be used to distract a patient, alleviating some of the pain experienced during excruciating activities such as wound care or physical therapy. Using a simulated plane to test neuropsychological impairments will further increase patient outcomes accuracy. Care professionals gain a more accurate view of a patient's actions and overall health by using VR to more closely depict real-life conditions during an evaluation (De Ponti et al., 2020; Haleem et al., 2020a,b; Papadopoulou et al., 2019).
7.	View situation from different angles	The most crucial benefit of VR is that it helps people to view any situation from any angle. VR experience is an essential educational tool for assisting medical students in developing empathy for patients. Medical practitioners may use this technology to simulate the human body's interior and understand more about human anatomy. VR preparation is also helpful in mechanical ventilation and intrusive hemodynamic control (Falah et al., 2013; Yongjie et al., 2021).
8.	Medical care	Patients and doctors use VR to become more knowledgeable of their medical needs and care principles. Clinical practitioners may use artificial scenarios to convey the effects of undesirable lifestyle habits such as noxious substance use, metabolic dysfunctions, obesity, tumor formation, the influence of smoking and alcohol on lung and liver functions, and so on. Patients who are chronically ill or injured will use VR goggles or headphones to view their homes in an immersive world (Mantovani et al., 2003; Singh et al., 2020).
9.	Explanation of surgical procedure	VR allows the surgeon or lecturer to present a case and then explains surgical procedures and approaches. Medical students will wander around inside a pounding heart, stimulate it, and observe how a defect affects the organ's

(*Continued*)

Table 6.1 Positive Role of Virtual Reality (VR) for improving healthcare education.—cont'd

S No	Role of VR	Description
		normal functions. VR can also improve empathy between healthcare providers and families by allowing them to immerse themselves in the presence of a patient or family member (Pensieri and Pennacchini, 2016; Izard et al., 2017; Yu et al., 2019).
10.	Improve real-life experience	Via its display-based method, this technology blends physical and simulated images. By rendering the exhibition interactive in real time, these immersive images improve the real-world experience. Health education, research, and surgical preparation have all benefited from the digital convergence of VR. It is extremely capable of guiding complicated surgical procedures. This dupes consumers and gives them the idea of truth. VR is one of the most exciting emerging health innovations. It effectively assists doctors in performing advanced procedures (Pierce et al., 2008; Plotzky et al., 2021; Al-Hiyari and Jusoh, 2020).
11.	Better imagination	It allows students to imagine and perceive any muscle and vein in the human body in three dimensions. In this way, VR has the potential to transform medical education. Pharma firms benefit from VR by supplying critical drug knowledge. With the aid of a VR technology-enabled kit, lab staff can observe and understand the function within the human body in three dimensions (Yao et al., 2020; Javaid and Haleem, 2020).
12.	Medication knowledge	Pharmacy students can develop a better knowledge of the medications with VR learning. Students can engage in multistudent operating room exercises, which will help them learn better and exercise their positions. The most critical explanation is that VR equipment is much less expensive and simpler to maintain lab equipment (Hilty et al., 2006; Maresky et al., 2019; Bogomolova et al., 2021).
13.	Improve skill	Healthcare professionals can be quickly and efficiently improve their skills using VR. When training resources are running short, VR will assist in showing workers the techniques and methods used to complete any mission. Students can learn the routines of both new and performed activities by using the hands-on experience of a simulated world. VR can be used to teach hands-on skills. It can also be used to build situations that teach healthcare practitioners and help them engage with patients. This is particularly relevant when dealing with people suffering from mental health issues or chronic pain (Zare Bidaki, 2018; Mathur, 2015; Sarma, 2021).

(Continued)

Table 6.1 Positive Role of Virtual Reality (VR) for improving healthcare education.—cont'd

S No	Role of VR	Description
14.	Distance learning	Distance learning can be done effectively through VR systems. Training would be almost as valuable as it will be in a face-to-face presentation. The machine will trace reactions, head turns, and hand motions in VR. Programs shared with VR can be used repeatedly. This allows healthcare providers to continue working while still furthering their education (Letterie, 2002; Zhao et al., 2020; Kron et al., 2010; Zhu et al., 2018).
15.	Procedure planning	VR surgery can provide planning a procedure and carry out different situations to refine the sequence and formulate a course of action for each situation. VR world will be able to play out any aspect of the planned scenario anytime quickly. The possibility of this technology enhances the planning phase. Surgeons may use VR to plan the procedure ahead of time and simulate future scenarios without dealing with them in real life. This will help to improve the overall precision and control of the process (Moro et al., 2021; Andreatta et al., 2010).
16.	Relieving tension	VR is currently being investigated to assist individuals in relieving tension, combating fear, and achieving a more peaceful and relaxed emotional state. VR immerses the user in the same world in which he or she would meet a patient. It provides a thorough understanding of how and why provider-based decisions are made and receive real-time input in the event (Shao et al., 2020; Ustun et al., 2020; Sattar et al., 2019).
17.	Virtual classroom	In a virtual classroom environment, students can use VR devices. Additionally, professors and students can use the Web-based tool to view VR scenarios. In any case, professors will configure their cases or build personalized cases depending on the result they want their students to achieve. Students can exercise these cases as many times as they choose because they are accessible over the internet. The framework offers a self-assessment in the context that students may use to measure their success and discover which places they will need to spend more time on (Lessick and Kraft, 2017; Lange et al., 2000).
18.	VR therapy	VR therapy is more than just a sophisticated diversion strategy. It provides an engaging, soothing, and immersive experience in which an immersed patient entirely consumes their consciousness. VR immersion therapy may aid in the prevention or reduction of panic attacks and other types of anxiety symptoms. VR healthcare solutions may help distract children from their discomfort during blood checks, flu vaccines, and other medical procedures. VR therapy is a promising treatment

(Continued)

Table 6.1 Positive Role of Virtual Reality (VR) for improving healthcare education.—cont'd

S No	Role of VR	Description
		option that has shown to be highly successful for many patients (Beverly et al., 2021; Berg and Steinsbekk, 2020; Palanica et al., 2019).
19.	3D scanning	In VR, 3D scans of patients' organs can be explored in a manner that conventional approaches cannot do. In tandem with counseling, they use VR to help people resolve phobias by simulating a real-world environment. 3D computer graphics help customers face anxiety in a managed and secure environment. Surgeons use many tools to visualize their operation and project 3D images of the patient's anatomy into the region of vision of the surgeon. Today, it is crucial for those experts to be up-to-date with medical instruments that undergo technological updates more quickly than ever before (Konge et al., 2014; Ruthenbeck and Reynolds, 2015; Moro et al., 2017; Harpham-Lockyer et al., 2015).
20.	Presenting complex ideas	It can present complex ideas through simulation and interaction and offer valuable insights into the real world. VR helps to maintain formal health training, providing stable consistency and structured qualifications for medical experts. VR is an exciting tool for learning, such as rehearsal of dynamic operations, hard news and other frequent in-hospital scenarios. Specialists will also be best trained to confront dynamic and complicated medical conditions (Górski et al., 2016; Abbas et al., 2020; Iwanaga et al., 2021).

development of the explicit technical knowledge about VR so that they could easily apply the VR technology in their respected filed.

6.5 Discussion

VR enable students to provide several opportunities to learn vital skills. It delivers sensory opportunities and offers anxiety reduction treatment. This often eliminates distractions for disordered pupils. It offers freedom to experiment in an atmosphere that is protected and regulated. The use of VR in science provides an engaging, immersive learning atmosphere for students. This helps improve student access, making the whole learning experience more cost-effective. VR is a low-cost way for surgeons to study and perform different operations and their related instruments. For multiple reasons, such as health concerns, information and privacy problems, have prevented the widespread use

of VR headsets. It provides a quantitative segmentation of the healthcare market. This technology is being used to develop innovative VR educational systems for dental students and professionals. To replicate an actual patient in VR, a 3D model is first obtained. VR healthcare applications are now extensive and evolving exponentially. In the context of education, it appears to be an outstanding asset. Its popularity is based on a foundation that assumes people recall more than reading, see, or hear something from direct experience. The use of these technological solutions provides success in the healthcare system. These innovative technologies provide feasible alternatives to the healthcare system.

Medical VR has found another effective distractor with many similar applications related to patient pain and irritation. By directing the patients' attention to the simulated world, the researchers demonstrated the efficacy of VR in treating pain. The excellent visual influence of VR can be used to inform people, especially young people, about the adverse consequences of behaviors. Patients will relax and abstract away from the treatment by wearing VR headphones instead of protective glasses. The technology has contributed to the advancement of safer solutions and customised medical records. It makes diagnostics possible for the patient to see detailed photographs of body configurations without invasive procedures. This technology enhances the patient experience, therapeutics, training, pharmacy managers, and medical processes in the healthcare industry.

6.6 Future of virtual reality in healthcare

In future, VR will provide promising applications in healthcare, which can be able to visualize customised patient-specific 3D models. Students will practice their surgical operations in a VR environment using VR software that simulates in an operating room. It will help advanced surgeons to practice new procedures. This will assist in conducting interactive medical seminars, medical interns, reducing pain and fear among patients, and accelerating physical therapy by tailoring activities to patients' clinical needs. Researchers, surgeons, and nurse educators are developing novel approaches to use immersive technology to improve healthcare education and practice. Medications, therapies, medical research, patient safety, and empathy will be highlighted by using this technology. This enhances collaborations between universities and academics, and hospitals. The video captures are converted into holograms that can be put in any location to enable students to assess patients and make diagnoses. Students will be able to communicate with holographic materials. By mixing physical and simulated images, VR will assist future healthcare professionals. It aids them in gaining a more detailed and impressive understanding of human anatomy.

The advancement of VR in healthcare education requires a huge investment that could be developed with the government support in terms of finance. Several

technological developments are also required for the efficient adoption of the VR technologies in healthcare. This opens the door for the researchers, scientists, and academicians to conduct research in this domain. Furthermore, the bulky equipment is needed to reduce their weight for efficient utilization of VR applications. The VR implementations framework is also required for the efficient application of the VR technologies in healthcare, and this type of framework is not available. These areas could be explored in the future studies.

6.7 Conclusion

Healthcare education needs to stay up with the new technologies. VR is an innovative approach that proves to be very effective in saving patients' lives and ensuring a quick recovery. This technology allows computer-generated data, facts to interact in real time with physical objects. It provides adequate instruction and education to caregivers in the healthcare industry. This dramatically increases decision-making abilities under challenging circumstances, as well as motivation and expertise. The proper use of this technology increases the efficiency of healthcare organizations. It has shown to be successful in providing patients with improved treatment. This makes substantial and steady progress in the healthcare field by serving vital functions. The knowledge provided by VR assists people in carrying out activities in the real world by broadening the users' perspectives. With the advent of VR, there is a change in the understanding of healthcare education. It is used to improve every feature, from educating students to planning specific procedures. VR can improve the healthcare system for better patient service. In future, this will help patients to keep track of the status of health innovatively.

References

Abbas, J.R., Kenth, J.J., Bruce, I.A., 2020. The role of virtual reality in the changing landscape of surgical training. Journal of Laryngology & Otology 134 (10), 863–866.

Akaike, M., Fukutomi, M., Nagamune, M., Fujimoto, A., Tsuji, A., Ishida, K., Iwata, T., 2012. Simulation-based medical education in clinical skills laboratory. Journal of Medical Investigation 59 (1, 2), 28–35.

Al-Hiyari, N., Jusoh, S., 2020. The current trends of virtual reality applications in medical education. In: 2020 12th International Conference on Electronics, Computers and Artificial Intelligence (ECAI). IEEE, pp. 1–6.

Alfalah, S.F., Falah, J.F., Alfalah, T., Elfalah, M., Muhaidat, N., Falah, O., 2019. A comparative study between a virtual reality heart anatomy system and traditional medical teaching modalities. Virtual Reality 23 (3), 229–234.

Alharbi, Y., Al-Mansour, M., Al-Saffar, R., Garman, A., Alraddadi, A., 2020. Three-dimensional virtual reality as an innovative teaching and learning tool for human anatomy courses in medical education: a mixed methods study. Cureus 12 (2).

Almousa, O., Zhang, R., Dimma, M., Yao, J., Allen, A., Chen, L., Qayumi, K., 2021. Virtual reality technology and remote digital application for tele-simulation and global medical education: an innovative hybrid system for clinical training. Simulation & Gaming 52, 614–634, 10468781211008258.

Ammanuel, S., Brown, I., Uribe, J., Rehani, B., 2019. Creating 3D models from radiologic images for virtual reality medical education modules. Journal of Medical Systems 43 (6), 1–3.

Andreatta, P.B., Maslowski, E., Petty, S., Shim, W., Marsh, M., Hall, T., Frankel, J., 2010. Virtual reality triage training provides a viable solution for disaster-preparedness. Academic Emergency Medicine 17 (8), 870–876.

Antoniou, P.E., Arfaras, G., Pandria, N., Athanasiou, A., Ntakakis, G., Babatsikos, E., Bamidis, P., 2020. Biosensor real-time affective analytics in virtual and mixed reality medical education serious games: cohort study. JMIR Serious Games 8 (3), e17823.

Arents, V., de Groot, P.C., Struben, V.M., van Stralen, K.J., 2021. Use of 360° virtual reality video in medical obstetrical education: a quasi-experimental design. BMC Medical Education 21 (1), 1–9.

Baniasadi, T., Ayyoubzadeh, S.M., Mohammadzadeh, N., 2020. Challenges and practical considerations in applying virtual reality in medical education and treatment. Oman Medical Journal 35 (3), e125.

Berg, H., Steinsbekk, A., 2020. Is individual practice in an immersive and interactive virtual reality application non-inferior to practicing with traditional equipment in learning systematic clinical observation? A randomised controlled trial. BMC Medical Education 20, 1–10.

Beverly, E.A., Love, C., Love, M., Williams, E., Bowditch, J., 2021. Using virtual reality to improve health care providers' cultural self-efficacy and diabetes attitudes: pilot questionnaire study. JMIR Diabetes 6 (1), e23708.

Bogomolova, K., Hierck, B.P., Looijen, A.E., Pilon, J.N., Putter, H., Wainman, B., van der Hage, J.A., 2021. Stereoscopic three-dimensional visualisation technology in anatomy learning: a meta-analysis. Medical Education 55 (3), 317–327.

Bradley, P., 2006. The history of simulation in medical education and possible future directions. Medical Education 40 (3), 254–262.

Chang, P.W., Chen, B.C., Jones, C.E., Bunting, K., Chakraborti, C., Kahn, M.J., 2018. Virtual reality supplemental teaching at low-cost (VRSTL) as a medical education adjunct for increasing early patient exposure. Medical Science Educator 28 (1), 3–4.

de Faria, J.W.V., Teixeira, M.J., Júnior, L.D.M.S., Otoch, J.P., Figueiredo, E.G., 2016. Virtual and stereoscopic anatomy: when virtual reality meets medical education. Journal of Neurosurgery 125 (5), 1105–1111.

De Ponti, R., Marazzato, J., Maresca, A.M., Rovera, F., Carcano, G., Ferrario, M.M., 2020. Pre-graduation medical training including virtual reality during COVID-19 pandemic: a report on students' perception. BMC Medical Education 20 (1), 1–7.

Dimitropoulos, K., Manitsaris, A., Mavridis, I., 2008. Building virtual reality environments for distance education on the web: a case study in medical education. International Journal of Social Sciences 2 (1), 62–70.

Djukic, T., Mandic, V., Filipovic, N., 2013. Virtual reality aided visualisation of fluid flow simulations with application in medical education and diagnostics. Computers in Biology and Medicine 43 (12), 2046–2052.

Dunne, J.R., McDonald, C.L., 2010. Pulse!!: a model for research and development of virtual-reality learning in military medical education and training. Military Medicine 175 (Suppl. 7), 25–27.

Dyer, E., Swartzlander, B.J., Gugliucci, M.R., 2018. Using virtual reality in medical education to teach empathy. Journal of the Medical Library Association: JMLA 106 (4), 498.

Fabris, C.P., Rathner, J.A., Fong, A.Y., Sevigny, C.P., 2019. Virtual reality in higher education. International Journal of Innovative Science and Modern Engineering 27 (8).

Falah, J., Harrison, D.K., Charissis, V., Wood, B.M., 2013. July). The characterisation of a virtual reality system to improve the quality and to reduce the gap between information technology and medical education. In: International Conference on Virtual, Augmented and Mixed Reality. Springer, Berlin, Heidelberg, pp. 122–131.

Falah, J., Khan, S., Alfalah, T., Alfalah, S.F., Chan, W., Harrison, D.K., Charissis, V., August 2014. Virtual Reality medical training system for anatomy education. In: 2014 Science and Information Conference. IEEE, pp. 752–758.

Falah, J., Charissis, V., Khan, S., Chan, W., Alfalah, S.F., Harrison, D.K., August 2014. Development and evaluation of virtual reality medical training system for anatomy education. In: Science and Information Conference. Springer, Cham, pp. 369–383.

Fertleman, C., Aubugeau-Williams, P., Sher, C., Lim, A.N., Lumley, S., Delacroix, S., Pan, X., 2018. A discussion of virtual reality as a new tool for training healthcare professionals. Frontiers in Public Health 6, 44.

Fiani, B., De Stefano, F., Kondilis, A., Covarrubias, C., Reier, L., Sarhadi, K., 2020. Virtual reality in neurosurgery: can you see it?. In: A Review of the Current Applications and Future Potential. World Neurosurgery, pp. 291–298.

Gao, Y., Zhao, Q.P., Zhou, X.D., Guo, Q.M., Xi, T., 2021. The role of virtual reality technology in medical education in the context of emerging medical discipline. Sichuan da xue xue bao. Yi xue ban= Journal of Sichuan University. Medical Science Edition 52 (2), 182–187.

Górski, F., Buń, P., Wichniarek, R., Zawadzki, P., Hamrol, A., 2016. Effective design of educational virtual reality applications for medicine using knowledge-engineering techniques. Eurasia Journal of Mathematics, Science and Technology Education 13 (2), 395–416.

Haixu, W., Ribo, C., Jing, L., 2021. The application of virtual reality technology based on optical motion capture technique in the teaching of cardiovascular medicine. Chinese Journal of Medical Education 41 (6), 528.

Haleem, A., Javaid, M., Khan, I.H., 2020a. Virtual reality (VR) applications in dentistry: an innovative technology to embrace. Indian Journal of Dental Research 31 (4), 666.

Haleem, A., Javaid, M., Khan, I.H., 2020b. Holography applications toward medical field: an overview. Indian Journal of Radiology and Imaging 30 (3), 354.

Hamilton, D., McKechnie, J., Edgerton, E., Wilson, C., 2021. Immersive virtual reality as a pedagogical tool in education: a systematic literature review of quantitative learning outcomes and experimental design. Journal of Computers in Education 8 (1), 1–32.

Hammoud, M.M., Nuthalapaty, F.S., Goepfert, A.R., Casey, P.M., Emmons, S., Espey, E.L., Peskin, E.G., 2008. To the point: medical education review of the role of simulators in surgical training. American Journal of Obstetrics and Gynecology 199 (4), 338–343.

Haowen, J., Vimalesvaran, S., Kyaw, B.M., Car, L.T., 2021. Virtual reality in medical student's education: a scoping review protocol. BMJ Open 11 (5), e046986.

Harden, R.M., Hart, I.R., 2002. An international virtual medical school (IVIMEDS): the future for medical education? Medical Teacher 24 (3), 261–267.

Harpham-Lockyer, L., Laskaratos, F.M., Berlingieri, P., Epstein, O., 2015. Role of virtual reality simulation in endoscopy training. World Journal of Gastrointestinal Endoscopy 7 (18), 1287.

Hilty, D.M., Alverson, D.C., Alpert, J.E., Tong, L., Sagduyu, K., Boland, R.J., Yellowlees, P.M., 2006. Virtual reality, telemedicine, web and data processing innovations in medical and psychiatric education and clinical care. Academic Psychiatry 30 (6), 528–533.

Huang, H.M., Liaw, S.S., Lai, C.M., 2016. Exploring learner acceptance of the use of virtual reality in medical education: a case study of desktop and projection-based display systems. Interactive Learning Environments 24 (1), 3–19.

Iwanaga, J., Kamura, Y., Nishimura, Y., Terada, S., Kishimoto, N., Tanaka, T., Tubbs, R.S., 2021. A new option for education during surgical procedures and related clinical anatomy in a virtual reality workspace. Clinical Anatomy 34 (3), 496–503.

Izard, S.G., Méndez, J.A.J., Palomera, P.R., 2017. Virtual reality educational tool for human anatomy. Journal of Medical Systems 41 (5), 76.

Javaid, M., Haleem, A., 2020a. Virtual reality applications toward the medical field. Clinical Epidemiology and Global Health 8 (2), 600–605.

Javaid, M., Haleem, A., 2020b. Critical components of Industry 5.0 towards a successful adoption in the field of manufacturing. Journal of Industrial Integration and Management 5 (03), 327–348.

Javaid, M., Khan, I.H., 2021. Virtual reality (VR) applications in cardiology: a review. Journal of Industrial Integration and Management 2, 100117, 2130001.

Javaid, M., Haleem, A., Vaishya, R., Bahl, S., Suman, R., Vaish, A., 2020. Industry 4.0 technologies and their applications in fighting COVID-19 pandemic. Diabetes & Metabolic Syndrome: Clinical Research Reviews 14 (4), 419–422.

Jensen, L., Konradsen, F., 2018. A review of the use of virtual reality head-mounted displays in education and training. Education and Information Technologies 23 (4), 1515–1529.

Klemm, P., Kleyer, A., Tascilar, K., Schuster, L., Meinderink, T., Steiger, F., Simon, D., 2021. A virtual reality-based app to educate health care professionals and medical students about inflammatory arthritis: feasibility study. JMIR Serious Games 9 (2), e23835.

Konge, L., Albrecht-Beste, E., Nielsen, M.B., 2014. Virtual-reality simulation-based training in ultrasound. Ultraschall in der Medizin-European Journal of Ultrasound 35 (02), 95–97.

Kron, F.W., Gjerde, C.L., Sen, A., Fetters, M.D., 2010. Medical student attitudes toward video games and related new media technologies in medical education. BMC Medical Education 10 (1), 1–11.

Kuehn, B.M., 2018. Virtual and augmented reality put a twist on medical education. JAMA 319 (8), 756–758.

Labovitz, J., Hubbard, C., 2020. The use of virtual reality in podiatric medical education. Clinics in Podiatric Medicine and Surgery 37 (2), 409–420.

Lai, P., Zou, W., 2018. The application of virtual reality technology in medical education and training. Global Journal of Information Technology: Emerging Technologies 8 (1), 10–15.

Lange, T., Indelicato, D.J., Rosen, J.M., 2000. Virtual reality in surgical training. Surgical Oncology Clinics of North America 9 (1), 61–79.

Lessick, S., Kraft, M., 2017. Facing reality: the growth of virtual reality and health sciences libraries. Journal of the Medical Library Association: JMLA 105 (4), 407.

Letterie, G.S., 2002. How virtual reality may enhance training in obstetrics and gynecology. American Journal of Obstetrics and Gynecology 187 (3), S37–S40.

Levinson, A.J., Weaver, B., Garside, S., McGinn, H., Norman, G.R., 2007. Virtual reality and brain anatomy: a randomised trial of e-learning instructional designs. Medical Education 41 (5), 495–501.

Li, L., Yu, F., Shi, D., Shi, J., Tian, Z., Yang, J., Jiang, Q., 2017. Application of virtual reality technology in clinical medicine. American Journal of Tourism Research 9 (9), 3867.

Lilly, J., Kaneshiro, K.N., Misquith, C., Dennett, B., 2019. Creating a new "reality" for medical education: the Nexus Reality Lab for virtual reality. Journal of the Medical Library Association: JMLA 107 (4), 609.

Lopes, D.S., Jorge, J.A., 2019. Extending medical interfaces towards virtual reality and augmented reality. Annals of Medicine 51 (Suppl. 1), 29-29.

Mantovani, F., Castelnuovo, G., Gaggioli, A., Riva, G., 2003. Virtual reality training for health-care professionals. CyberPsychology & Behavior 6 (4), 389–395.

Maresky, H.S., Oikonomou, A., Ali, I., Ditkofsky, N., Pakkal, M., Ballyk, B., 2019. Virtual reality and cardiac anatomy: exploring immersive three-dimensional cardiac imaging, a pilot study in undergraduate medical anatomy education. Clinical Anatomy 32 (2), 238–243.

Mariani, A.W., Pêgo-Fernandes, P.M., 2011. Medical education: simulation and virtual reality. Sao Paulo Medical Journal 129 (6), 369–370.

Mathur, A.S., 2015. Low cost virtual reality for medical training. In: 2015 IEEE Virtual Reality (VR). IEEE, pp. 345–346.

McGaghie, W.C., Issenberg, S.B., Cohen, M.E.R., Barsuk, J.H., Wayne, D.B., 2011. Does simulation -based medical education with deliberate practise yield better results than traditional clinical education? A meta-analytic comparative review of the evidence. Academic Medicine: Journal of the Association of American Medical Colleges 86 (6), 706.

McGrath, J.L., Taekman, J.M., Dev, P., Danforth, D.R., Mohan, D., Kman, N., Won, K., 2018. Using virtual reality simulation environments to assess competence for emergency medicine learners. Academic Emergency Medicine 25 (2), 186–195.

Meyer, B., 2020. A shift in reality: virtual and augmented systems in higher and medical education. Current Issues in Emerging eLearning 6 (1), 7.

Moro, C., Štromberga, Z., Stirling, A., 2017. Virtualisation devices for student learning: comparison between desktop-based (Oculus Rift) and mobile-based (Gear VR) virtual reality in medical and health science education. Australasian Journal of Educational Technology 33 (6).

Moro, C., Birt, J., Stromberga, Z., Phelps, C., Clark, J., Glasziou, P., Scott, A.M., 2021. Virtual and augmented reality enhancements to medical and science student physiology and anatomy test performance: a systematic review and meta-analysis. Anatomical Sciences Education 14 (3), 368−376.

Nicholson, D.T., Chalk, C., Funnell, W.R.J., Daniel, S.J., 2006. Can virtual reality improve anatomy education? A randomised controlled study of a computer-generated three-dimensional anatomical ear model. Medical Education 40 (11), 1081−1087.

O'Sullivan, D.M., Foley, R., Proctor, K., Gallagher, S., Deery, A., Eidem, B.W., McMahon, C.J., 2021. The use of virtual reality echocardiography in medical education. Pediatric Cardiology 1−4.

Palanica, A., Docktor, M.J., Lee, A., Fossat, Y., 2019. Using mobile virtual reality to enhance medical comprehension and satisfaction in patients and their families. Perspectives on Medical Education 8 (2), 123−127.

Papadopoulou, P., Chui, K.T., Daniela, L., Lytras, M.D., 2019. Virtual and augmented reality in medical education and training: innovative ways for transforming medical education in the 21st century. In: Cognitive Computing in Technology-Enhanced Learning. IGI Global, pp. 109−150.

Pensieri, C., Pennacchini, M., 2016. Virtual reality in medicine. In: Handbook on 3D3C Platforms. Springer, Cham, pp. 353−401.

Pierce, J., Gutiérrez, F., Vergara, V.M., Alverson, D.C., Qualls, C., Saland, L., Caudell, T.P., 2008. Comparative usability studies of full vs. partial immersive virtual reality simulation for medical education and training. Studies in Health Technology and Informatics 132, 372−377.

Plotzky, C., Lindwedel, U., Sorber, M., Loessl, B., König, P., Kunze, C., Meng, M., 2021. Virtual reality simulations in nurse education: a systematic mapping review. Nurse Education Today 104868.

Pottle, J., 2019. Virtual reality and the transformation of medical education. Future Healthcare Journal 6 (3), 181.

Rizzetto, F., Bernareggi, A., Rantas, S., Vanzulli, A., Vertemati, M., 2020. Immersive Virtual Reality in surgery and medical education: diving into the future. The American Journal of Surgery 220 (4), 856−857.

Roy, M.J., Sticha, D.L., Kraus, P.L., Olsen, D.E., 2006. Simulation and virtual reality in medical education and therapy: a protocol. CyberPsychology & Behavior 9 (2), 245−247.

Ruthenbeck, G.S., Reynolds, K.J., 2015. Virtual reality for medical training: the state-of-the-art. Journal of Simulation 9 (1), 16−26.

Sales, B.R.A., Machado, L.S., Moraes, R.M., 2011. Interactive collaboration for virtual reality systems related to medical education and training. Technology and Medical Sciences 157−162, 2011.

Samadbeik, M., Yaaghobi, D., Bastani, P., Abhari, S., Rezaee, R., Garavand, A., 2018. The applications of virtual reality technology in medical groups teaching. Journal of Advances in Medical Education & Professionalism 6 (3), 123.

Sarma, V.P., 2021. The era of virtual reality in medical education: do we still need the dissection table? International Surgery Journal 8 (2), 771−772.

Sattar, M.U., Palaniappan, S., Lokman, A., Hassan, A., Shah, N., Riaz, Z., 2019. Effects of Virtual Reality training on medical students' learning motivation and competency. Pakistan Journal of Medical Sciences 35 (3), 852.

Saxena, N., Kyaw, B.M., Vseteckova, J., Dev, P., Paul, P., Lim, K.T.K., Car, J., 2018. Virtual reality environments for health professional education. Cochrane Database of Systematic Reviews 2018 (10).

Scalese, R.J., Obeso, V.T., Issenberg, S.B., 2008. Simulation technology for skills training and competency assessment in medical education. Journal of General Internal Medicine 23 (1), 46−49.

Schmidt, M.W., Köppinger, K.F., Fan, C., Kowalewski, K.F., Schmidt, L.P., Vey, J., Nickel, F., 2021. Virtual reality simulation in robot-assisted surgery: meta-analysis of skill transfer and predictability of skill. BJS Open 5 (2), zraa066.

Shao, X., Yuan, Q., Qian, D., Ye, Z., Chen, G., le Zhuang, K., Qiang, D., 2020. Virtual reality technology for teaching neurosurgery of skull base tumor. BMC Medical Education 20 (1), 1−7.

Shi, Y.L., 2014. Application of virtual reality technology in medical education. In: Frontier and Future Development of Information Technology in Medicine and Education. Springer, Dordrecht, pp. 467–476.

Singh, R.P., Javaid, M., Kataria, R., Tyagi, M., Haleem, A., Suman, R., 2020. Significant applications of virtual reality for COVID-19 pandemic. Diabetes & Metabolic Syndrome: Clinical Research Reviews 14 (4), 661–664.

Sultan, L., Abuznadah, W., Al-Jifree, H., Khan, M.A., Alsaywid, B., Ashour, F., 2019. An experimental study on usefulness of virtual reality 360 in undergraduate medical education. Advances in Medical Education and Practice 10, 907.

Tang, Y.M., Ng, G.W.Y., Chia, N.H., So, E.H.K., Wu, C.H., Ip, W.H., 2021. Application of virtual reality (VR) technology for medical practitioners in type and screen (T&S) training. Journal of Computer Assisted Learning 37 (2), 359–369.

Ustun, A.B., Yilmaz, R., Yilmaz, F.G.K., 2020. Virtual reality in medical education. In: Mobile Devices and Smart Gadgets in Medical Sciences. IGI Global, pp. 56–73.

Vozenilek, J., Huff, J.S., Reznek, M., Gordon, J.A., 2004. See one, do one, teach one: advanced technology in medical education. Academic Emergency Medicine 11 (11), 1149–1154.

Winkler-Schwartz, A., Bissonnette, V., Mirchi, N., Ponnudurai, N., Yilmaz, R., Ledwos, N., Del Maestro, R.F., 2019. Artificial intelligence in medical education: best practices using machine learning to assess surgical expertise in virtual reality simulation. Journal of Surgical Education 76 (6), 1681–1690.

Yao, P., Challen, C., Caves, C., 2020. An experimental study on usefulness of virtual reality 360° in undergraduate medical education. Advances in Medical Education and Practice 10, 1103–1104.

Ye, Q., Abdessalem, H.B., Boukadida, M., 2020. Hypocrates+: a virtual reality medical education platform with intelligent real-time help system. In: International Conference on Brain Function Assessment in Learning. Springer, Cham, pp. 96–101.

Yongjie, W., Libin, C., Xin, Y., Xueming, C., 2021. Application of virtual reality technology in clinical medical education. Medical Education Management 7 (1), 73.

Yu, W., Wen, L., Zhao, L.A., Liu, X., Wang, B., Yang, H., 2019. March). The applications of virtual reality technology in medical education: a review and mini-research. In: Journal of Physics: Conference Series, vol. 1176. IOP Publishing, p. 022055. No. 2.

Zack off, M.W., Young, D., Sahay, R.D., Fei, L., Real, F.J., Guiot, A., Klein, M., 2021. Establishing objective measures of clinical competence in undergraduate medical education through immersive virtual reality. Academic Pediatrics 21 (3), 575–579.

Zare Bidaki, M., 2018. Application of virtual reality simulators and virtual labs in medical education. Interdisciplinary Journal of Virtual Learning in Medical Sciences 9 (1).

Zhao, J., Xu, X., Jiang, H., Ding, Y., 2020. The effectiveness of virtual reality-based technology on anatomy teaching: a meta-analysis of randomised controlled studies. BMC Medical Education 20, 1–10.

Zhao, G., Fan, M., Yuan, Y., Zhao, F., Huang, H., 2021. The comparison of teaching efficiency between virtual reality and traditional education in medical education: a systematic review and meta-analysis. Annals of Translational Medicine 9 (3).

Zhu, J.W., Pan, Z.X., Chen, S., Wang, Q., Shen, Z., Zhu, H.J., Pan, H., 2018. Application and prospect of virtual reality in the medical field. Basic & Clinical Medicine 38 (3), 422–425.

Zito, F.A., Marzullo, F., D'Errico, D., Salvatore, C., Digirolamo, R., Labriola, A., Pellecchia, A., 2004. Quicktime virtual reality technology in light microscopy to support medical education in pathology. Modern Pathology 17 (6), 728–731.

CHAPTER SEVEN

Extended reality for development of clinical skills

Samiya Khan
Faculty of Science & Engineering, University of Wolverhampton, Wolverhampton, United Kingdom

7.1 Introduction

Healthcare organizations typically employ a large chunk of a country's workforce. For instance, NHS England is the largest employer in Europe, having 1.3 million people employed and 16,000 students studying to get employed in its healthcare workforce.[1] Education and training are essential facets of the healthcare sector owing to the evolving need for its staff to remain responsive to changes in healthcare delivery practices and needs of the patients.

One of the most profound consequences of the COVID-19 pandemic is the grave shortage of healthcare staff. Moreover, the pandemic has also caused considerable delays and disruptions in training, particularly when limited physical sessions for simulation were performed. Therefore, students were not able to get exposure to diverse use cases and patient categories. It is imperative to find alternate methods of training delivery as we move into the post-COVID era.

Health and care education is delivered around the work with the help of a well-established technique called simulation-based education (SBE). The efficacy of this technique has been realized by students and teachers alike. XR has seamlessly integrated with this form of education as it provides students easy access to varied environments and use cases, at lower costs and higher convenience, thus contributing to continuing education and workforce training even in times when conventional education is halted.

The healthcare sector is usually under immense financial pressure. Therefore, there is an evolving need to lower the costs of education and training. As a result, more accessible and cost-effective solutions are the need of the hour. Moreover, this is also the reason why XR technologies have garnered so much attention (Bowyer et al., 2008; Haluck, 2000; Logeswaran et al., 2020). In other words, XR has a wide range of applications in education, which is provisioned at lower costs, in a scalable manner and with higher accessibility, making XR an excellent prospect for healthcare education settings.

1 https://www.xrhealthuk.org/the-growing-value-of-xr-in-healthcare.

Extended Reality for Healthcare Systems, Volume 1
ISBN 978-0-323-98381-5, https://doi.org/10.1016/B978-0-323-98381-5.00012-X

However, like any transformative technology, XR adoption is also not short of challenges. This chapter discusses the different facets of XR usage for clinical skill development and training.

The rest of the chapter is organized in the following manner: Section 7.2 provides an overview on the use of XR for medical education and training. Section 7.3 discusses how this technology can also be used for patient education and improving the efficiency of service delivery. Section 7.4 elaborates on the challenges and opportunities in this sector. Lastly, Section 7.5 summarizes the findings and provides insights on future scope for research.

7.2 Medical education and training

XR has changed the way medical students learn, providing a viable alternative to conventional clinical education and surgical training. The use of surgical training simulators with visual and haptic feedback has demonstrated good prospect, particularly with respect to training students for diverse settings, which improves their capability to deal with real-world challenges (Caccianiga et al., 2021). A study by Harvard Business Review suggested that the use of virtual reality (VR) for surgical training is known to boost overall performance by as much as 230%.[2] Other findings of the study evidenced that VR-trained surgeons demonstrated higher surgical accuracy with lower time to perform the surgery. Fig. 7.1 shows a medical student practicing surgery using a VR-based simulation tool.

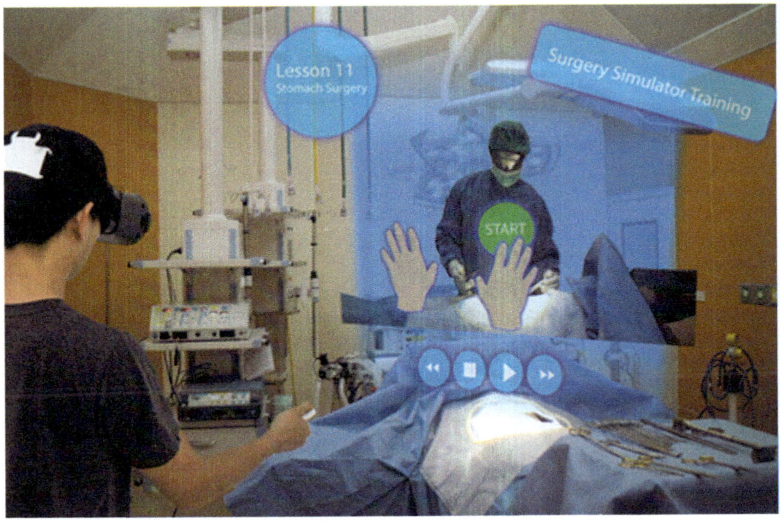

Figure 7.1 Medical student using VR simulation tool for practicing surgery. *VR*, virtual reality.

2 https://medicalfuturist.com/5-ways-medical-vr-is-changing-healthcare/.

From an educational perspective, VR-based training involves collaborative and experiential learning, both of which are supported by constructivist learning model. Existing literature suggests that the constructivist learning model is more effective than other digital delivery methods or traditional learning models (Logeswaran et al., 2020; Kyaw et al., 2019). Besides this, VR is also known to drive contextual learning, which has a direct impact on the motivation and learning growth of students.

The landscape of scenarios and settings supported by XR-based training is the strength of this delivery method. With AR (augmented reality) and VR, students can train on risk-free environments that cover scenarios such as urgent aid, emergency care, and complex surgeries. Moreover, from a clinical perspective, they can be supported with training on both the physiological and anatomical aspects of these settings. XR has also found applications in nurse training where nurses can be trained to achieve precise hand—eye coordination for equipment handling and performing procedures such as MRIs.

Using XR, it is possible to merge information coming from angiograms and CT scans to develop 3D models for radiography specialists.[3] The fact that XR-based training follows "learning-by-doing" approach where practitioners can learn step-by-step procedures in real time collaboratively gives them a conducive learning environment. This setting shall also give them higher confidence as learning from looking at static images leaves a learning void as they are still unsure of whether they will be able to replicate the steps in a real scenario. Fig. 7.2 shows medical students using AR/VR-based tool to analyze data related to a patient's heart. It is practically not possible to train practitioners

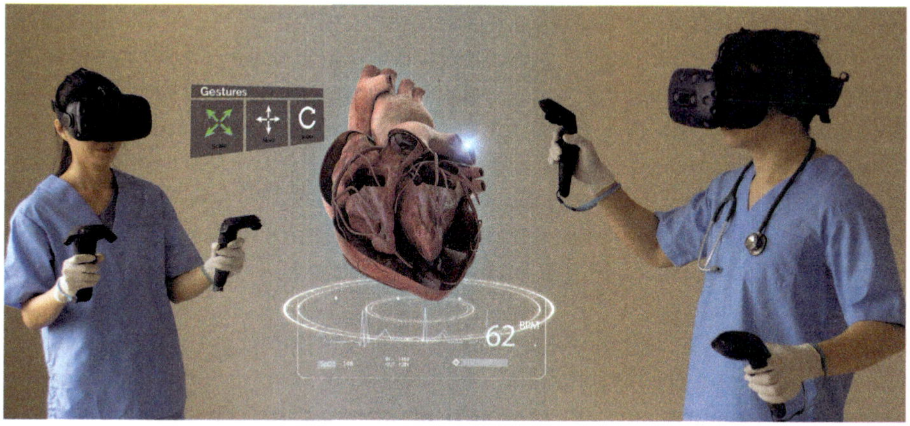

Figure 7.2 Medical students using AR/VR tool to analyze patient health data. *AR*, augmented reality; *VR*, virtual reality.

3 https://med.stanford.edu/news/all-news/2017/07/virtual-reality-system-helps-surgeons-reassures-patients.html.

on high-risk situations or at inconvenient locations. XR allows remote access of such scenarios without risking the lives of patients, providing highly accessible, risk-free setup.

Remote surgery is a futuristic use case of XR. However, recent research has performed experimentation to prove the viability of remote surgeries and their broadcast via high-speed 5G networks. PwC's report on the use of 5G for healthcare[4] suggested that the emergence of tactile Internet will lead to an unimaginable breakthrough. The ultralow latency features of 5G will enable remote surgeries as it will then be possible to duplicate the actions of the surgeon present in one demographic location at a remote location, in real time. This use case is expected to bring benefits to patients who reside in smaller and remote areas, giving them access to best medical resources within the confines of their home or location. Another significant finding shared by PwC in this report is that XR-based training costs 52% lesser that traditional classroom in a study performed for 3000 students.

7.3 Patient-focused education

XR technology provides opportunities and potential for use in not just medical education for healthcare practitioners, but it can also be used for patient education and engagement. There are several use cases identified in this domain, which include visualization of patient care pathways before clinical intervention, personalizing interventions for varied illnesses and patient perspectives, and sharing insights with them. In more ways than one, VR-based solutions are known to facilitate effective communication and interaction between patients, caregivers, and medical practitioners. Human connectivity is an indispensable aspect of providing medical care, and technologies such as VR can play an instrumental role in boosting it.

Providing VR experience of a preintervention visualization of the pathway of care has multiperspective implications and impact. For instance, this visualization can help in anxiety reduction for the patient (Gold et al., 2021). A better understanding of the procedure will allow the patient to calm down, allowing smoother intervention and postintervention recovery. From the clinician's perspective, this VR experience will also allow them to review surroundings, settings, and interactions from multiple perspectives. Therefore, it can trigger a positive change in them while they support the patient through the care journey.

There is a prospect of adding value to both basic VR applications such as virtual tour of operating theater in the hospital and complex computer-generated imagery (CGI). Equipment that supports creation of 360-degree videos have been adopted by many medical facilities to support in-house training that can suit the varying needs of diverse

4 https://www.pwc.com/gx/en/industries/tmt/5g/pwc-5g-in-healthcare.pdf.

categories of individuals. One of the driving factors in its adoption is its low-cost and easy availability.

From the user's perspective, the key benefit of VR-based applications is that it allows them to test, rehearse, and experience training material in such a manner that they get access to best services remotely without feeling "not there." Moreover, healthcare professionals can practice their skills on varied scenarios without worrying about any real-life consequences or risks associated with those situations. Since immersive technologies facilitate a more realistic experience, the responses to scenarios are also intuitive and reflexive, which provides a better assessment of skill development.

7.3.1 Empathy-inclusive training

One of the most crucial aspects of medical training and education is comprehension of patients' perspective. It is only when the healthcare providers understand the emotions and conditions of the patient that they can empathize with them. Empathy and understanding patient situation go a long way in establishing connectivity and effective communication between the patient and the healthcare provider. The ability of healthcare providers to "look through patient's eyes" provides them an opportunity to consider and reflect onto how engagement and healthcare delivery can be improved for the patient, which will have positive impact on healthcare delivery and care provisioning. Empathy training is a fundamental aspect of healthcare education. However, due to challenges such as inconsistent training opportunities and work pressure, in addition to trials involved in providing humanities-focused teaching, healthcare professionals usually do not find the time to reflect on their experiences and empathize with patients.

A service available in the sector of healthy aging is provided by Embodied Labs.[5] They provide training and education to all the people involved in aging spectrum through immersive experiences. Currently, they are the market leaders in the aging sector as far as cutting-edge immersive training is concerned. The immersive experiences provided as part of this service improve patient outcomes and allow experiential engagement between users, to mitigate several challenges in communication, interaction, and engagement. Their services are in line with the vision to provide the best possible life to aging individuals irrespective of their background, ethnicity, status, physical capacity, and cognitive ability.

VR experiences can be developed from the first person's perspective. Thus, an experience of this kind will enable other people involved in the setting to understand patient's conditions in a better way. Such experiences should be provided as part of healthcare education and training, which should be followed by discussions, reflective sessions, and formulation of all-inclusive strategies that can improve patient experience.

5 https://www.embodiedlabs.com/.

Empathy education is a soft skills training that must be provided to healthcare practitioners and support workers, in addition to medical students. Examples of VR experiences in this domain include a patient's ICU journey through interactions with nurses, paramedics, and doctors. One of the recent experiences in this domain provided insights into the impact of PPE kit on patient involvement.

7.3.2 Patient education for differently abled

The brightest prospects of XR for therapeutic intervention are expected for the differently abled. XR can support several applications such as development of skills for employment, independent living, and well-being (Smith et al., 2014). The benefit of using VR for simulating contexts and real-life scenarios is that there are no risks or unforeseen consequences attached. Therefore, VR simulations are predictable and safe in that respect (Bradley and Newbutt, 2018). Moreover, the content can be customized to suit individual ability, capacity, and needs. Fig. 7.3 shows a child with disabilities wearing a VR headset under adult supervision to develop skills for disability management.

It is because of these reasons that VR is considered as a breakthrough technology for the differently abled sector. Research is actively being conducted to test the feasibility and viability of immersive technologies to support autistic groups (Newbutt et al., 2020). The landscape of support includes educational training, travel training, social skills' development, and skill development for independent living. Another facet of support required and yet usually ignored for differently abled individuals is anxiety and well-being. An individual with social anxiety needs therapy for the condition, but he or she also requires anxiety management prior to getting into a social situation so that impact of therapeutic intervention can be maximized.

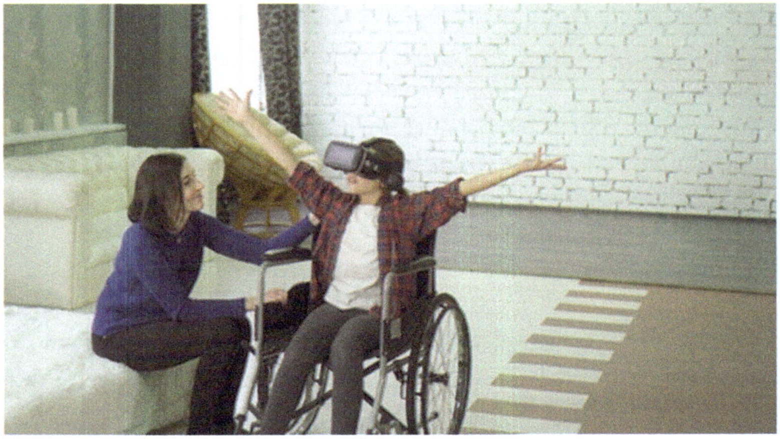

Figure 7.3 Child with disabilities wearing VR headset for therapeutic intervention. *VR*, virtual reality.

Other research works in the domain include Brossonboek et al. (2019) that proposed the use of breath-controlled VR experience for management of disruptive class behavior. This research work also evidenced that VR could potentially be used by neurodiverse young population to deal with anxiety or anger or even to virtually travel places for entertainment. Gold et al. (2021) attested the effects of VR on preoperative anxiety among autistic patients, caregivers, and medical practitioners. The results of this study are not only confined to scare of preoperative procedures, but they also attest that VR can reduce the fear associated with visiting hospitals in children, particularly those who must do so regularly.

One of the services available in this domain is Talking Sense,[6] which is an AR-based conversational training tool. This tool was developed through the collaborative efforts between Neon and Ulster University. The service uses artificial intelligence to support parents of autistic children to better understand their behavior and adopt the best possible coping strategy. This tool is majorly of use for parents who can train themselves by talking to a character on the app in real time who represents their child. They can practice interventions and strategies on this character to get the best results for their child. It also helps in relieving parental stress associated with raising children with disabilities.

7.3.3 Preintervention visualizations

The communication and interaction benefits of XR are evident and demonstrated in several scenarios. Technologies such as VR can allow health workers to get answers to questions about the patient that may otherwise be extremely stressful for the patient to answer (Fig. 7.2). An example of how this works is that patient anatomy is visually represented in an immersive experience. VR is increasingly being used to inform patients and allow them to virtually experience the pathways of care so that they know what they are getting into and can find solutions to problems that they may face, in advance. Fig. 7.4 shows a patient wearing VR headset before surgery for educating him about the procedure and reducing anxiety. Research has proven the effectiveness of VR in reducing preoperation anxiety in epileptic patients (House et al., 2020). In addition to preoperative education, VR has successfully been used to manage depressive symptoms in patients (Migoya-Borja et al., 2020).

This is a significant tool in view of that fact that anxiety can have a substantial impact on the outcome of procedures. The consequences of anxiety can range from delays, halts, need for extra resources such as general anesthetics and can even cause rescheduling of the procedure in extreme cases. Furthermore, issues such as these for one patient can have a rollover effect on other scheduled procedures. Subsequently, such issues, no matter how menial they appear upfront, can cost healthcare facilities significantly.

[6] https://talking-sense.org/.

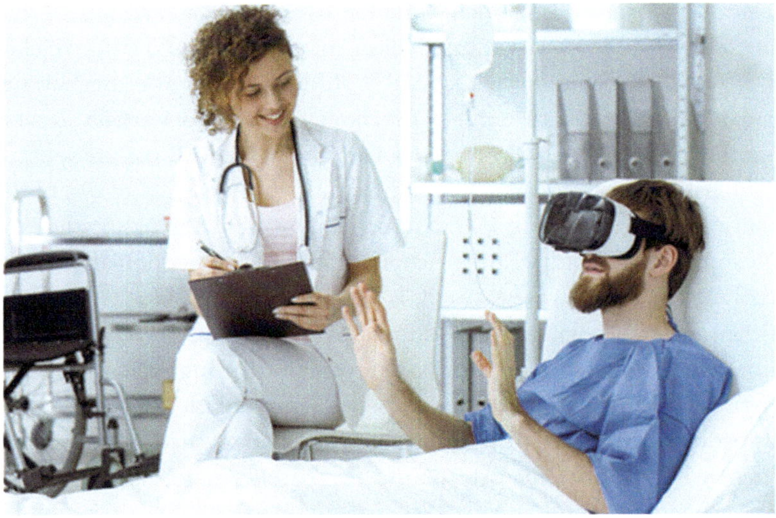

Figure 7.4 Patient being informed about procedure prior to surgery for anxiety management.

Research efforts in this domain include development of interactive video to explain clinical pathways for anxiety management (Koo et al., 2020), instructional video to reduce fear among patient and their families before MRI scan (Ashmore et al., 2019), and preoperative anxiety management for patients undergoing arthroscopy (Yang et al., 2019).

7.4 Challenges and opportunities

While the benefits of XR adoption are apparent, it is critical to acknowledge the challenges and roadblocks that exist in its adoption. Mitigating these risks and challenges will allow development of solutions that can maximize benefits and ensure that no harm is done in the process. An education tool has a direct impact on the skills and knowledge of the learners, who in this case are medical practitioners. Any pitfalls in the training and education structure will ultimately compromise patient safety. Therefore, the design and development of these solutions must be performed after careful assessment and intensive research. For instance, a 360-degree video of a location may bring back memories for some elderly patients, which may trigger varied physical, emotional, and psychological responses. To deal with situations like these, appropriate support mechanisms will need to be crafted and deployed.

Immersive experience may have some negative outcomes on people such as fatigue and postural stress. The design of immersive experiences must follow a learning-driven approach by avoiding these outcomes if they are negligibly reported. Another important aspect of using technology is that its infrastructure can fail and will need maintenance and

support with time. Therefore, proper technical support must be made available to deal with any technical issues that may arise during sessions. Besides these, there are some specific challenges that exist because data related to patient cases and stories are confidential and personal. Therefore, proper permissions may have to be sought before usage to ensure learning data security and information governance.

On similar lines, issues related to health conditions, disabilities, and accessibilities may also have to be considered. For instance, there are multiple levels of dyslexia, and it must be determined prior to widespread usage as to which levels of such conditions may be eligible for using the XR solution. The procedures, approvals, and governance-related issues are a deterrent in XR adoption and will require clear definitions and frameworks for commercial and widespread use of XR-healthcare for education. Quality of medical education and training can be significantly enhanced with the inclusion of XR in the education structure. Besides this, preintervention VR experiences have proved helpful in improving patient experience and outcome. Thus, there is unlimited opportunity in this space for local healthcare settings to create customized content and educational tools for their patients and staff.

7.5 Conclusion

Undoubtedly, XR holds a lot of potential for healthcare education and training. There are several factors that drive exploration in the field of XR healthcare for medical education. Firstly, the number of training hours is decreasing because of increasing work pressure. However, excessive work pressure increases the chances of error, which in turn puts patient safety at risk. Secondly, healthcare service demand outnumbers the number of healthcare professionals available to provision it, limiting training time. Lastly, for countries such as the United Kingdom, a huge chunk of their healthcare workforce is due to retire in the coming years. According to a study,[7] it was estimated that 30%—50% of the workforce working in the healthcare sector of the United Kingdom will retire by 2030.

There are significant investments required on the part of the government and healthcare organizations to provide training, putting significant financial burden on them. As a result, solutions that are accessible and can provide training remotely are sought. However, the delays caused in approvals and compliance of regulations take so long that the demand is not met in time, which is also the reason why research and development in XR is slowing down.

XR is a beneficial extension to the traditional methods of medical education. Since most of the health careers are professional in nature, XR allows trainees to get a real-life experience before he or she actually sees the patient, providing them unique yet valuable

7 https://www.kingsfund.org.uk/publications/health-care-workforce-england.

practice and training. XR-based simulations for training are beneficial in supporting experiential education. However, to start with, it is better to consider them as supplemental to traditional teaching methodology. XR's ability to take quality education to a completely remote setup shall allow clinicians and medical practitioners to learn in their own time and pace, irrespective of their location. While XR makes training more accessible, it also makes training risk-free and minimizes inequalities in training. These benefits are magnified in situations such as the COVID-19 pandemic when regular education faced considerable delays and halts. In addition to education of medical practitioners, XR can be equally effective for educating patients and their families to reduce preintervention stress.

One of the most critical steps for widespread XR adoption is that the design and implementation of developed solutions must undertake after collaborative engagement between stakeholders and businesses. The adoption of this approach will require real-life use cases, evidence-based practice, and compliance with industry standards and best practices. As a recommendation, the design of XR solutions for medical education and training must cater to the learning needs; include assessment, simulation, and evaluation; and have the technological capabilities to support these functional elements for seamless operation. Evidently, XR is the future of medical education and training and is expected to make a considerable impact in this sector over the next decade, particularly outside the hospital setup.

References

Ashmore, J., Di Pietro, J., Williams, K., Stokes, E., Symons, A., Smith, M., Clegg, L., McGrath, C., 2019. A free virtual reality experience to prepare pediatric patients for magnetic resonance imaging: cross-sectional questionnaire study. JMIR Pediatrics and Parenting 2 (1), e11684. https://doi.org/10.2196/116843.

Bossenbroek, R., Wols, A., Weerdmeester, J., Lichtwarck-Aschoff, A., van Rooij, M.J.W., Granic, I., 2019. Efficacy of a virtual reality biofeedback game (DEEP) to reduce anxiety and disruptive classroom behaviour: a single-case study. JMIR Mental Health 7 (3). https://doi.org/10.2196/16066.

Bowyer, M.W., Streete, K.A., Muniz, G.M., Liu, A.V., 2008. Immersive virtual environments for medical training. Seminars in Colon and Rectal Surgery 19 (2), 90–97. https://doi.org/10.1053/j.scrs.2008.02.005.

Bradley, R., Newbutt, N., 2018. Autism and virtual reality head-mounted displays: a state-of-the-art systematic review. Journal of Enabling Technologies 12 (3), 101–113. www.emerald.com/insight/content/doi/10.1108/JET-01-2018-0004)/full/html.

Caccianiga, G., Mariani, A., de Paratesi, C.G., Menciassi, A., De Momi, E., 2021. Multi-sensory guidance and feedback for simulation-based training in robot assisted surgery: a preliminary comparison of visual, haptic, and visuo-haptic. IEEE Robotics and Automation Letters 6 (2), 3801–3808.

Gold, J.I., Annick, E.T., Lane, A.S., Ho, K., Marty, R.T., Espinoza, J.C., 2021. "Doc McStuffins: doctor for a day" virtual reality (DocVR) for pediatric preoperative anxiety and satisfaction-A feasibility study. Journal of Medical Internet Research 23 (4).

Haluck, R.S., Thomas, M.K., 2000. Computers and virtual reality for surgical education in the 21st century. Archives of Surgery 135 (7), 786–792. https://doi.org/10.1001/archsurg.135.7.786.

House, P.M., et al., 2020. Use of the mixed reality tool "VSI Patient Education" for more comprehensible and imaginable patient educations before epilepsy surgery and stereotactic implantation of DBS or stereo-EEG electrodes. Epilepsy Research 159, 106247. https://doi.org/10.1016/j.eplepsyres.2019.106247.

Koo, C.H., Park, J.W., Ryu, J.H., Han, S.H., 2020. The effect of virtual reality on preoperative anxiety: a meta-analysis of randomized controlled trials. Journal of Clinical Medicine 9 (10), 3151. https://doi.org/10.3390/jcm9103151. Published 2020 Sep. 29.

Kyaw, B.M., Saxena, N., Posadzki, P., Vseteckova, J., Nikolaou, C.K., George, P.P., Divakar, U., Masiello, I., Kononowicz, A.A., Zary, N., Tudor Car, L., 2019. Virtual reality for health professions education: systematic review and meta-analysis by the digital health education collaboration. Journal of Medical Internet Research 21 (1), e12959. https://doi.org/10.2196/12959.

Logeswaran, A., Munsch, C., Chong, Y.J., Ralph, N., McCrossnan, J., 2020. The role of extended reality technology in healthcare education: towards a learner-centred approach. Future Healthcare Journal 8, e79–e84. https://doi.org/10.7861/fhj.2020-0112.

Migoya-Borja, M. et al. (2020). Feasibility of a virtual reality-based psychoeducational tool (VRight) for depressive patients. p. 7. www.liebertpub.com/doi/full/10.1089/cyber.2019.0497.

Newbutt, N., Bradley, R., Conley, I., 2020. Using Virtual reality head-mounted display in schools for pupils on the autism spectrum; views, experiences, and future directions. Cyberpsychology, Behavior, and Social Networking 23 (1), 22–33. www.liebertpub.com/doi/full/10.1089/cyber.2019.0206.

Smith, M.J., Ginger, E.J., Wright, K., Wright, M.A., Taylor, J.L., Humm, L.B., Fleming, M.F., 2014. Virtual reality job interview training in adults with autism spectrum disorder. Journal of Autism and Developmental Disorders 44 (10), 2450–2463.

Yang, J.-H., et al., 2019. 'Effects of preoperative virtual reality magnetic resonance imaging on preoperative anxiety in patients undergoing arthroscopic knee surgery: a randomized controlled study', arthroscopy. In: The Journal of Arthroscopic & Related Surgery, vol. 35. Official Publication of the Arthroscopy Association of North America and the International Arthroscopy Association, pp. 2394–2399. https://doi.org/10.1016/j.arthro.2019.02.037, 8.

CHAPTER EIGHT

AR/VR telehealth platforms for remote procedural training

Hema Garg[1] **and Viraj Uttamrao Somkuwar**[2]
[1]School of Interdisciplinary Research, Indian Institute of Technology Delhi, New Delhi, India
[2]Department of Textile and Fibre Engineering, Indian Institute of Technology Delhi, New Delhi, India

8.1 Introduction

Telemedicine involves telecommunication and information technology for delivering medical services at places of need. One of the typical advantages of telemedicine is the provision of medical facilities on-site and on-demand, irrespective of the geographical location restriction. The most vulnerable population is the elderly people and children, which require healthcare accessibility at home. The various features of telemedicine in the healthcare sector are depicted in Fig. 8.1. Telemedicine is quite popular in providing mental healthcare around the globe (Hand, 2021). Specific chronic

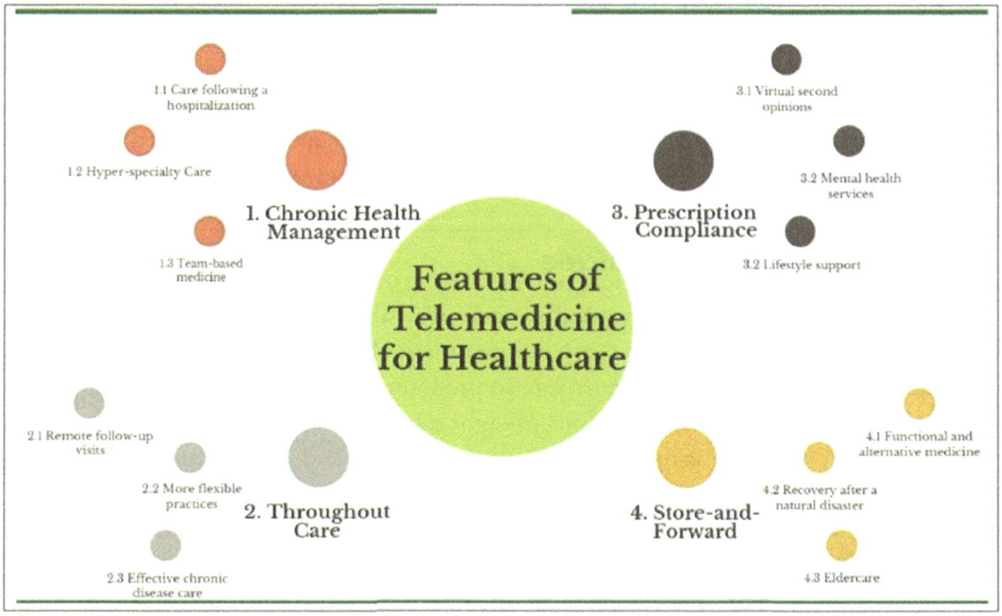

Figure 8.1 Features of telemedicine healthcare (Haleem et al., 2021).

Extended Reality for Healthcare Systems, Volume 1
ISBN 978-0-323-98381-5, https://doi.org/10.1016/B978-0-323-98381-5.00002-7

ailments such as dementia demand in-home care. The merit of delivering in-house medical treatment is the reduction in the hospitalization duration and antibiotic therapy (Pappalardo et al., 2021). Patients suffering from complex trauma injuries require immediate medical help. In countries such as Africa, the infant mortality rate is high primarily due to acute respiratory and diarrheal infections, which require immediate healthcare attention. However, life expectancy deteriorates drastically due to the lack of facilities in underdeveloped nations and rural areas. Delay in transportation and effective treatment decreases the life expectancy as well in such cases.

Telemedicine also plays a crucial role in disease surveillance and upskilling the medical staff in rural areas. The conventional learning method includes online platforms of e-learning, textbooks, and grasping knowledge through journals/articles. However, practical training has turned out to be more effective. Therefore, training techniques that are short, cost-effective, interactive, user friendly, and easily grasping become important. Digital interventions that could be made available 24/7 access and provide flexibility at the workplace are in demand. Simulation strategies such as games, augmented reality (AR), virtual reality (VR), and mixed reality are much more productive than traditional teaching. There is better retention of knowledge, higher satisfaction, and motivation among the medical trainees. The participants can frequently practice without harming the patients, which avoids distressing them and facing negative consequences (Gasteiger et al., 2021). Computer-driven technologies such as AR/VR enable independent learning without the need of an instructor.

The AR/VR technology is found to be valuable and effective in combination with telementoring. Telementoring conveys instructions by an expert from a remote location to an inexperienced practitioner in a rural area. This approach is beneficial for both the patient and the community practitioner as it enables expert advice to a completely unfamiliar clinical procedure (Bui et al., 2021). This chapter explores the AR/VR medium for medical diagnosis, its merits and demerits, and scope in the near future.

8.2 History of telemedicine

For millennia, people have used sounds or visible signals to communicate over long distances. Drums, horns, and other instruments have been used to deliver signals using specific sound patterns that correlate to predefined codes, and they are still utilized in some places. After so many years, line-of-sight transmission remains crucial as a critical parameter for communicating information. In the 17 and 18 centuries, a series of electrical inventions empowers the subsequent availability of instantaneous communications across a vast distance (Masys, 1997). This invention begins to apply for the aid of medical facilities using the telegraph during the 1860 Civil War. The telegraph transmissions were utilized to send casualty lists, medical aids, and clinical diagnostic data from the doctor's situated miles away to the soldier (Zundel, 1996). The transmission of

medical data through media was explored in 1905 when the first electrocardiograph developed. Waller and William, two physiologists, were the first to discover that the heart's beating produces a weak electric potential that can be detected with an electrode placed to the skin. In 1860, Waller recorded a first electrocardiogram (ECG) of a human being and successfully transmitted the ECG from his lab to an academic hospital (Hjelm and Julius, 2005). Following the telegraph, major technologies have evolved, such as the radio, telephone, smartphone, and advancements in connectivity, such as morse code to the Internet, satellite, and cellular network for medical consultation and communication (Fig. 8.2). As technology and connectivity progressed, the growth of telemedicine has grown immense popularity for medical diagnosis.

At the time of World War I, radio communications were utilized to transmit medical data from remote areas such as Alaska and Australia. During several wars, radio communications were used to send the medical teams and supplies (Zundel, 1996). By the 1940s, Gerson Cohen formulated the term "telegnosis," which describes the procedure of diagnosing using radiographs broadcast over the telephone or the radio. He transmits the radiograph data between two clinics at a remote location in Philadelphia at a distance of 60 miles. The research published by Gerson concluded that the utility of telemedicine could help train the young radiologist through expert teaching and supervision and provide specialist services in small hospitals at remote locations(Gershon-Cohen et al., 1952). After achieving success in short-distance clinical data transmission, transmission across continents was explored. Briskier claimed transmitting clinical data and heart sounds from New York to Paris and Rome via radio and radio photo media (Briskier, 1959).

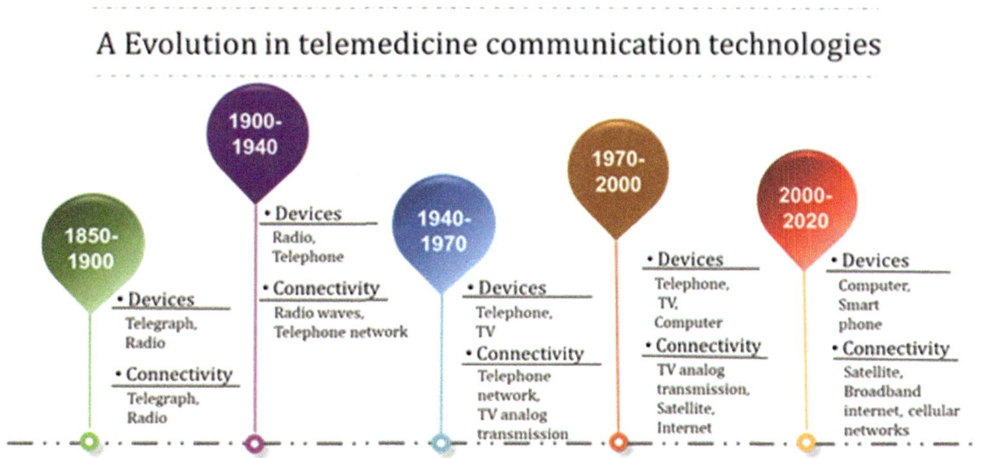

Figure 8.2 Advancement in devices and connectivity used for telemedicine communication technologies.

However, ordinary telephones are inexpensive, widely available, and perfect for quick consultations and various elements of patient care. Skin conditions, radiography, pathology images, and fundoscopic findings can be viewed via transmitted still images. A two-way interactive television (IATV) approach was used for real-time audio—visual (AV) signal transmission over the two geographically distant sites following long-distance communications. In 1959, Wittson et al. used the IATV approach for telepsychiatry diagnosis from Omaha's Nebraska Psychiatric Institute and a state mental institution 112 miles apart (Wittson et al., 1961). A similar approach was found by Raymond et al. at Massachusetts general hospital in 1960, giving medical treatment to the patient 2.7 miles away using a two-way AV microwave circuit. The whole diagnosis procedure was observed and evaluated by expert physicians. The rapid interpretation of X-rays, despite the distance, demonstrated the feasibility of telediagnosis could enhance medical care. The treatment was done across 1000 patient encounters as the largest telediagnosis at the time, and Bird, one of the team's scientists, invented the term "telemedicine" (Bird and Murphy, 1974; Murphy and Bird, 1974). Grundy et al. (1977) assessed the implementation of telemedicine by hypothesizing a few medical problems (Fig. 8.3). The study focuses on delivering instructions to patients in an intensive care unit (ICU) in a small hospital with limited resources, the feasibility of AV technology and associated hardware failures, the use of telemedicine as a valuable educational resource for nurses,

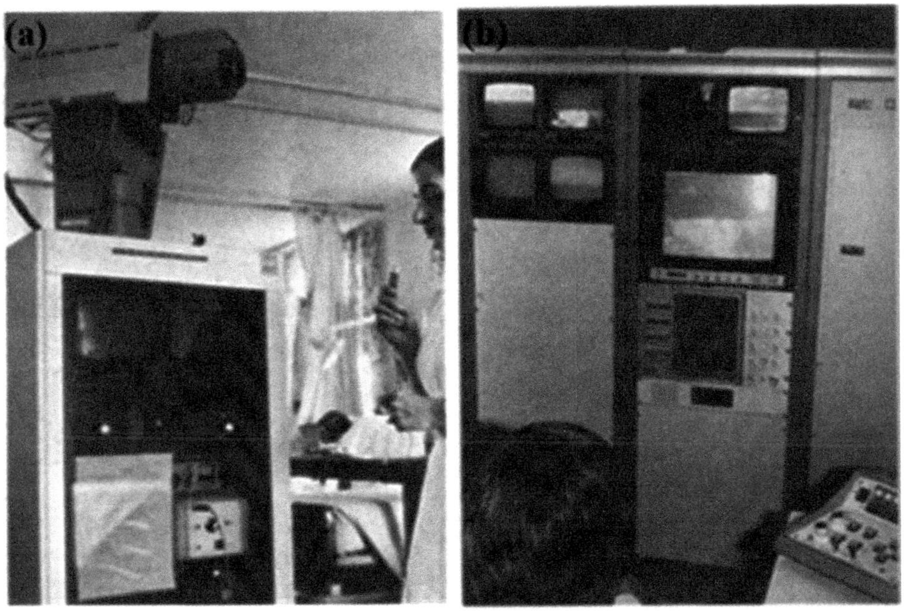

Figure 8.3 (A) Mobile camera unit used by the physician for telemedicine; (B) monitoring station for remote access and control (Grundy et al., 1977).

doctors, and physicians, and the improvement in treatment following the use of telemedicine. He concluded that AV technology is preferable to telephone and adequate to provide critical care but expensive for small hospitals. Also, telemedicine may be an excellent resource for medical personnel education and can act as a vital link between small hospitals in distant areas and more extensive medical facilities (Grundy et al., 1977).

Telemedicine gained a further boost from the application of space technology for promoting healthcare services in rural areas. In 1975, the NASA (National Aeronautics and Space Administration), in cooperation with Lockheed Missiles and Space Corporation, developed a project named STARPHAC (Space Technology Applied to Rural Papago Advanced Health Care) that provides remote healthcare facilities to the population of southern Arizona. The system uses space technology to monitor the astronaut's vital parameters and deliver instructions remotely from the control station. The design featured a two-way connection between a Health Service hospital and a mobile health unit staffed by nurses and advanced practice providers using television radio and remote telemetry (Fig. 8.4). The STARPHAC system provided significant results in applying the telemedicine technique and has represented as a "first generation" of telemedicine in adapting remote healthcare delivery (Freiburger et al., 2007). Video conferencing was also made to provide the physician and the medical officer with customized video communication. The Maritime Health Services (MHS) created a telemedicine project Medical consultation network (MedNet) in 1960 to provide occupational health services. The system indicates the possibility of an AV communication between the ongoing ship in the ocean and the physician from MHS. The MedNet technology was initially put to the test onboard the Golden Alaska, a big fishing trawler. The MHS

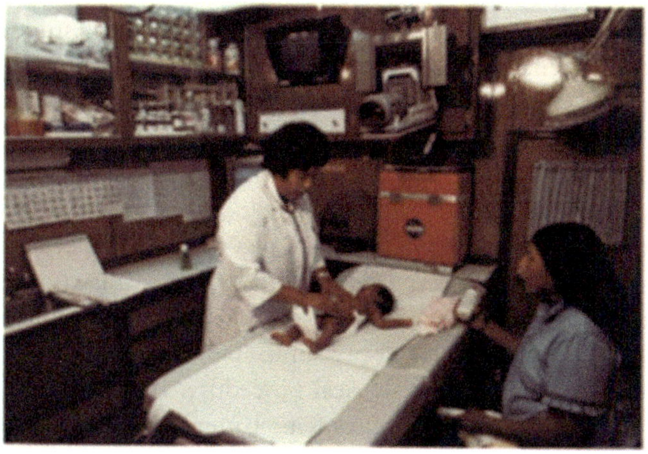

Figure 8.4 Examination of an infant by the specialist using a STARPHAC remote telemedicine unit (Freiburger et al., 2007).

system shows the feasibility of expert medical attention on remote sites in the ocean, saving lives and money (Jarris, 1994).

The era of 2000–20 brings a rapid transformation in information technology with computers, smartphones, telemedicine devices, broadband, and Internet connectivity. The revolution in smartphone and cellular network technology has brought telemedicine devices on a handheld gadgets to access personalized healthcare data. Telemedicine technology has evolved manifold, and companies in telemedicine businesses have introduced a model such as "TelaDOC Health" for direct consumer interaction (Jagarapu and Savani, 2021).

8.3 Conventional versus modern telemedicine medical treatment

Teleconferencing is the most sought-after conventional telemedicine medium in the early decades. A range of situations such as medical help during travel by aircraft, boat, etc., military operations, in remote locations, or reinforcing home care by family members demands interaction between an untrained individual with complicated medical devices. Also, most countries lack specialized facilities such as the ICU and the emergency division professionals. A 2D approach or telecommunication could not achieve this. Instead, it requires a sophisticated, interactive 3D medium to effectively communicate the physician and practitioner's knowledge. AR and VR are two such platforms for providing.

8.4 Augmented and virtual reality in telemedicine

AR is a technology that enables new insight into human–computer interaction by assimilating computer-generated virtual objects with the real environment. However, it is a well-known technology that gained popularity after imparting student education and medical support to patients and medical training to nurses or untrained professionals in remote areas through Google glasses and Microsoft HoloLens. AR was first discovered and implemented by Ivan Sutherland at Harvard University in the 1960s (Herron, 2016). In a nutshell, the AR system entails a target activating a virtual trigger that introduces a virtual presence into a user's reality. The target includes patterned marking, geolocation, object identification, or voice activation. The display of virtual objects is realized through either head-mounted displays or hand-held devices such as smartphones and tablets. A significant advantage of AR is the visualization of the virtual element in place and does not require any external monitor or display unit. This is important to realize the convenience for the practitioner in the surgical procedure. It also suffers from drawbacks such as the inconvenience of wearing headsets, battery life, limited field of view, the latency of network availability, and video and audio signals.

While AR and VR are closely related technologies, it becomes crucial to understand the difference between them to grasp their contribution to the medical field better. AR involves the superimposition of an interactive, digital layer onto the physical environment. A typical example includes AccuVein, which displays the veins when projected onto a body surface. In contrast, VR, often termed the sister technology of AR, involves the complete formation of the computer-simulated 3D figure with which a user can communicate in real space (Khor et al., 2016). In AR, only certain elements are overlayed and do not involve complete immersion in the virtual environment.

8.5 Medical visualization and diagnosis

AR technology has great potential to aid in medical education—grasping anatomy and conveying surgical procedures in real time, particularly for laparoscopic, cardiology, and neurosurgery. In medical education, anatomy is a subject that requires 3D visualization of organs and becomes difficult to understand in 2D textbooks, as shown in Fig. 8.5. The AR platform greatly assists in studying the particular orientation of organs with better understanding. However, AR is still at its infancy stage, with only some countries developing such implementation. With its jaw-dropping features of Dynamic 365 Remote Assist and HoloLens 2, Microsoft has enabled a real-time view of the operating room. Mount Sinai Health System Hospital in New York can provide intensive medical care with the help of their leading-edge expertise and techniques around the globe. A country like Uganda, deprived of such facilities, is connected to the

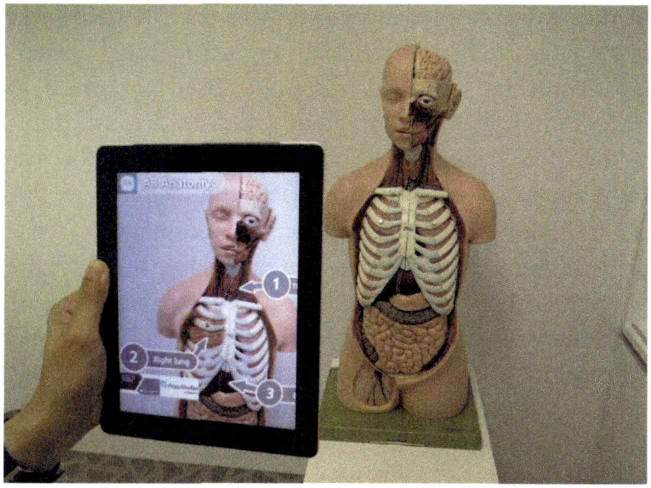

Figure 8.5 Anatomy study using AR technology (Anonymous). *AR*, augmented reality.

hospital thousands of miles away. Doctors can add mixed reality annotations to the local doctor's field of view to demonstrate their advice in real time. This concept of "shared surgery" has revolutionized the healthcare systems in the rural and underprivileged regions worldwide with an increased success rate in complicated surgery (Anonymous). In rural Mississippi, the United States, a Tel-Emergency program was also initiated where specially trained nurses in real-time collaboration with a physician are responsible for treating patients.

Several studies proved AR to increase the quality and speed of the treatment. Ponce et al. introduced a Virtual Interactive Presence (VIP), whereby a patient and the surgeon could interact, facilitating verbal, visual, and manual interactions (Ponce et al., 2016). The VIP system is essential in postoperative care, and the utmost care time of healing and recovery involves regular doctor follow-ups.

8.6 Applications of augmented reality/virtual reality

8.6.1 Infectious pandemic

There has been an outbreak of infectious disease such as bubonic plague, Spanish flu, severe acute respiratory syndrome (SARS), and Middle East respiratory syndrome (MERS), which has caused a significant impact on public health, economic, and social problems (Asadzadeh and Samad-Soltani, 2021). Recently, there has been an outbreak of highly infectious SARS coronavirus-2 (SARS-COV-2), which has spread from Wuhan, China, worldwide. The necessary preparedness steps to stop the transmission involve social distancing, quarantine, and complete shutdown of many countries. In such situations, medical staff is always at a higher risk of transmission, which could lead to the collapse of the medical system of a nation. AR/VR has emerged as advanced technologies to cater to the emergency of restricted physical contact among the medical fraternity and the patient (Fig. 8.6). VR system is used to spread awareness on the transmission ways and improve human hygiene (washing hands, disinfecting frequently touched surfaces) by giving training and interactive displays for visualization of the warning signs and symptoms of the infection (Pawar and Bansode, 2021).

An excellent example of teleconsultation implementation in the COVID-19 pandemic is Peru. After reporting the first case on March 6, 2020, the country implemented the telemedicine model on March 20, 2021, to cater to the situation (Alvarez-Risco et al., 2021). The various stages of implementation included teleconsultation and telemonitoring of patients, which extended to providing online drug prescriptions and online transfer of medical records between the medical institution, insurance companies, and pharmacies. The Peruvian government too launched the teleconsultation website and app to monitor the suspected cases better. This whole integrated approach helped Peru fight the pandemic.

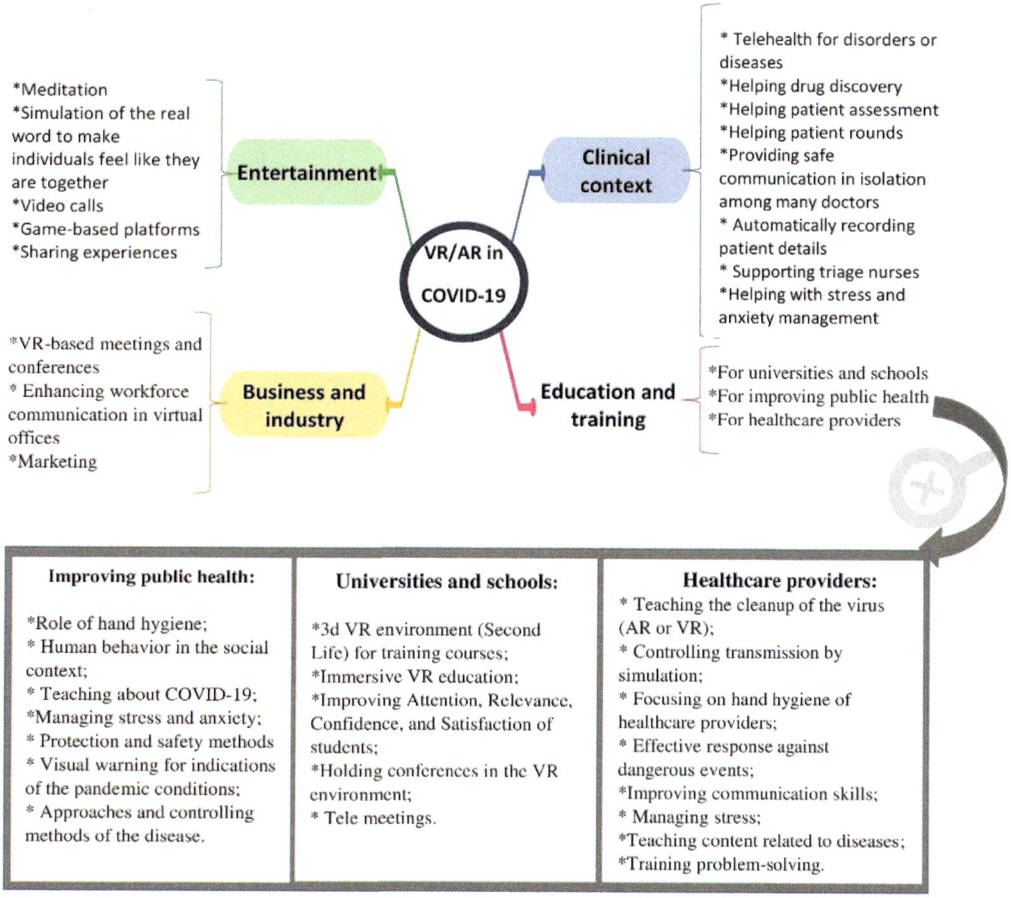

Figure 8.6 AR/VR applications in COVID-19 outbreak (Asadzadeh and Samad-Soltani, 2021). *AR*, augmented reality; *VR*, virtual reality.

8.6.2 Treatment of behavioral health conditions

Due to changing lifestyles, depression is becoming a challenging issue in the healthcare domain. According to a study, a person dies every 40 s due to mental health disorder. Identification of the disorder in the early stages is quite essential for avoiding critical situations. Internet of things (IoT) and VR concepts are becoming popular in this respect (Katkuri and Mantri, 2021). AR/VR has been designed to treat certain disorders such as autism, drug/alcohol addiction, phobia-related problems, and stress/anxiety management (Berenguer et al., 2020; Ghiţă and Gutiérrez-Maldonado, 2018; Lam et al., 2020; Oing and Prescott, 2018). Various healthcare and therapeutic games are built to enhance user interface and attractiveness to elevate the user's motivation (Fig. 8.7).

Figure 8.7 Treatment of behavioral health conditions with the aid of exercise via VR technique (Anonymous). *VR*, virtual reality.

8.7 Challenges in implementation of telemedicine

Telemedicine has recently been implemented in several developed and developing countries. Despite the rapid acceptance of telemedicine services and their immense potential, the clinical and treatment advantages associated with their use are limited. The telemedicine implementation has faced two significant challenges: the cost-effectiveness and the research data presented limited to case studies observed on a few patients. The effectiveness of implementation is also affected by the type of treatment. For example, studies have shown a reduction in mortality among patients by 15% through telemedicine in the ICU. In contrast, some treatments have significantly less or no improvement in the mortality and recovery of the patient (Higgins and Thompson, 2002; McCambridge et al., 2010; Wilcox and Adhikari, 2012; Young et al., 2011). The literature offers a limited justification for this diversity (Hersh et al., 2001; Friedman and Wyatt, 1997). The research mainly focuses on outcomes of patient results without accounting for the user's acceptability, morality, and feasibility (Kahn, 2015; Lustig, 2012; McLean et al., 2013). Telemedicine data storage and communication have raised the doubt of ethical and regulatory practices involved in implementing the technology (Nittari et al., 2020). The infrastructure of this technology needs to be designed to handle the regulatory and legal aspects. In traditional approaches, physicians and nurses collaborate and work together in person, while telemedicine forces them to work and receive orders virtually. Similarly, patients have to take advice from a person they never met results in hesitation and lack of trust (Wilcox and Adhikari, 2012; Nittari et al., 2020; Ash et al., 2007; Harrison et al., 2007). The various challenges that arise in the implementation of telemedicine technology are discussed in this section (Fig. 8.8).

8.7.1 Development and implementation cost

Apart from the trained medical personal, the technology and the infrastructure remain the prime hurdle in the complete implementation of telemedicine. The necessary software, computer equipment, and developing tools for the core infrastructure need to be installed at the required location. Also, the installation and demonstration require trained personal for maintenance and training the staff. Such a requirement from the information and communication technology demands an additional cost, which needs support from the critical policy driven by the government (Khemapech, 2019; Telematics, 1997). As the telemedicine services are based on existing digital technologies, development teams must typically understand the underlying infrastructure and the needs of the consumers. System developers are expected to appropriately pick and use the tools and devices to design the future generation of telemedicine solutions that meet the requirements (Peng et al., 2012). The aforementioned conditions become more challenging in India, where the rural population is high, with minimal access to technology and digital resources. The lack of broadband and a high-speed network infrastructure becomes the major technological hurdle in rural areas. The new multibillion-rupee "Digital India" initiative in India is an excellent first step in overcoming such technical impediments (Chandwani and Dwivedi, 2015).

8.7.2 Digital competence

According to experts, the most positive impact of telemedicine implementation will be to improve the quality of life of elderly people by offering more autonomous and active care in hospitals (Peng et al., 2012). However, the large population of elderly people is not familiar with the current digital technologies such as smartphone and application software interaction, particularly old people in long-term care facilities (LTCFs). People at LTCFs frequently choose not to use the Internet and cannot afford Internet connectivity or ICT devices. The inadequate technical solutions for connecting virtually with doctors and other healthcare professionals via telemedicine have physical or cognitive constraints that may restrict them from using telemedicine and lack the skills to use telemedicine even if they have access (Batsis et al., 2019; Beaunoyer et al., 2020; Seifert et al., 2019; Zhai, 2020). Hence, unlike the current generation of people, the elderly have to learn to use these devices to become more independent. A Nielsen Norman research team of computer use in various age groups revealed that users over 65 had a success score of just 53% in performing a set of prescribed activities (such as accessing information and making an online purchase), compared with a success rate of 78% for users younger than 65 (Bujnowska-Fedak et al., 2015). The results suggest that individuals over 65 years old have a poorer success rate and make more errors on computer-based tasks than younger users (Telematics, 1997).

8.7.3 Digital accuracy

Even while digital technology has become an integral part of many aspects of our everyday life, some individuals, particularly the elderly, refuse to accept it due to concerns about privacy and security. Even though banks and financial institutions are establishing digital services such as online banking and mobile application-based services, customers in the healthcare sector have had different reactions. First and foremost, health-related personal data must be safely sent across digital networks and precisely delivered to predetermined recipients. Second, additional sensors, such as wearable devices that can monitor heart rate and blood pressure, are used to detect and transmit readings to the receiving unit. The accuracy of telemedicine medical devices compared with that of standard medical equipment is the primary concern for telemedicine implementation (Bujnowska-Fedak et al., 2015). The telemedicine hardware requirements such as computer, tablet, and Internet access capability also become the hurdle in remote locations. According to one research, participants declined telemedicine for a variety of reasons, including a lack of access to computers and bandwidth and minimal computer skills (Moo et al., 2020). Connection and audio issues were the most often cited concerns in studies that revealed technological challenges. Images freezing, sound delay, echoing, and static sounds were among the connectivity concerns (Azad et al., 2012; Burton and O'Connell, 2018; Morgan et al., 2011). The acceptability and successful implementation of telemedicine are primarily reliant on user orientation, training, equipment provision, and a preceding test run to enable the user to become more familiar with the technology.

8.7.4 Technological adaptation

While face-to-face diagnostics build confidence and trust in patients, telemedicine raises questions about diagnosis accuracy when patients are treated from a distance. Because older individuals prefer personal engagement with health professionals, telemedicine-based treatments offered from a distance are often not preferred. In a survey of the elderly aged 60 and above conducted by Bujnowska-Fedak and Mastalerz-Migas, 61% of seniors stated that direct contact with health professionals was very beneficial. This may be due to the perception that the new technology may reduce social interaction and personal touch. The need for social interaction may be so profound that older people would turn down technologies that will improve their health and quality of life in the long term (Bujnowska-Fedak et al., 2015). The acceptability and successful implementation of telemedicine are primarily reliant on user orientation, training, equipment provision, and an initial test run to enable the user to become more familiar with the technology.

8.7.5 Privacy and security

The importance of privacy and security in developing trust in telemedicine consumers cannot be overstated. It is vital to secure the right to privacy and protection of personal

data throughout data collection and processing, particularly regarding health data. Telemedicine entails concerns because of the digital transmission of personal data. Disclosure of confidential material, such as a patient's health condition or medical reports, can negatively affect the patient's personal and professional life. Furthermore, even for insurance and pharmaceutical businesses, patient data is an important marketing tool. Privacy and security remain the prime concern among telemedicine users due to the potential of privacy exploitation with high-resolution photographs and videos and enhanced technology that allows real-time surveillance of personal location and activities in real time (Hanson et al., 2007; Percival and Hanson, 2006). Patients who use telemedicine are concerned about the privacy of their transmitted electronic medical record (EMR), which might disclose their identity. Any interference with an EMR's integrity can result in profound implications, including the death of a patient. As a result, every e-healthcare network must meet the C-I-A (confidentiality, integrity, and availability) criteria for patient health data. Several legislative, financial, and technical solutions have been implemented to counter security and privacy threats in the telemedicine network. The design and implementation of the Health Insurance Portability and Accountability Act (HIPAA) in 1996 was one such deterrence action (Das and Mukhopadhyay, 2011). Even though several precautions have been implemented to prevent possible exploitation, a minor vulnerability always remains that may cause immense damage; therefore, a continual risk assessment plan should be followed (Fig. 8.9).

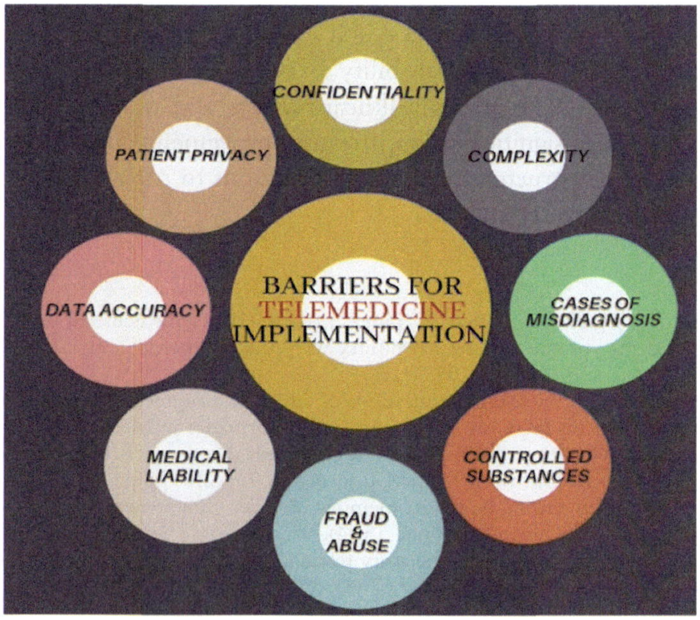

Figure 8.8 Barriers to the implementation of telemedicine (Haleem et al., 2021).

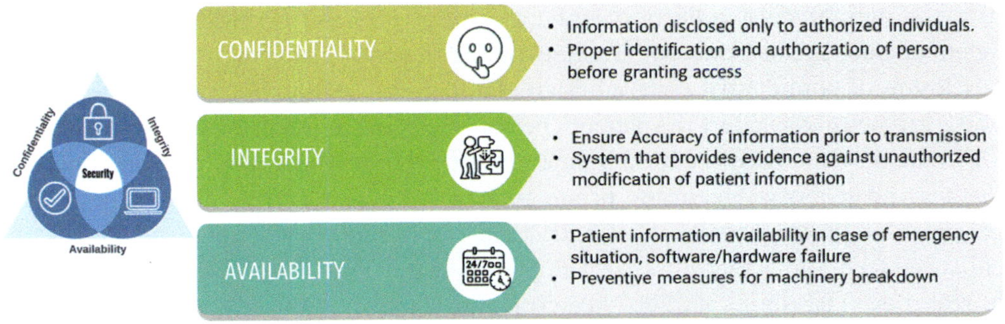

Figure 8.9 Challenges to the implementation of telemedicine.

8.8 Conclusion and future outlook

In this chapter, we reviewed telemedicine as a medium for remote procedural training in rural areas via various technologies such as AR and VR. Telemedicine has proven to be a vital technology in linking an expert on-demand with the patient via a clinical practitioner for immediate medical assistance, tackling pandemic situations effectively, and providing elderly care. People with limited mobility receive doctor's advice and medications more swiftly. The most extensive application of telemedicine is to provide medical care in rural areas and underdeveloped nations, where quality treatment is else not possible to achieve. The skills of medical practitioners are uplifted without geographical location and time restrictions. It has become a great source of imparting knowledge among the medical fraternity. However, technological advancement also suffers from various drawbacks such as maintaining privacy and confidentiality of patient information and massive reliability on the gadgets, headsets, and networks. In the future perspective, to improve healthcare access and outcomes for people with disabilities in the future, significant, long-term improvements in technological, regulatory, and legislative infrastructure and specific solutions to particular patient and health system demands are necessary.

References

Alvarez-Risco, A., Del-Aguila-Arcentales, S., Yáñez, J.A., 2021. Telemedicine in Peru as a result of the COVID-19 pandemic: perspective from a country with limited internet access. The American Journal of Tropical Medicine and Hygiene 105 (1), 6.

Asadzadeh, A., Samad-Soltani, T., 2021. Applications of virtual and augmented reality in infectious disease epidemics with a focus on the COVID-19 outbreak. Informatics in Medicine Unlocked 100579.

Ash, J.S., et al., 2007. Categorizing the unintended sociotechnical consequences of computerized provider order entry. International Journal of Medical Informatics 76, S21−S27.

Azad, N., et al., 2012. Telemedicine in a rural memory disorder clinic—remote management of patients with dementia. Canadian Geriatrics Journal 15 (4), 96.

Batsis, J.A., et al., 2019. Effectiveness of ambulatory telemedicine care in older adults: a systematic review. Journal of the American Geriatrics Society 67 (8), 1737−1749.

Beaunoyer, E., Dupéré, S., Guitton, M.J., 2020. COVID-19 and Digital Inequalities: Reciprocal Impacts and Mitigation Strategies, vol. 111, p. 106424.

Berenguer, C., et al., 2020. Exploring the impact of augmented reality in children and adolescents with autism spectrum disorder: a systematic review. International Journal of Environmental Research and Public Health 17 (17), 6143.

Bird, K.T., Murphy Jr., R.L., 1974. Telediagnosis: a new community health resource. American Journal of Public Health 64 (2).

Briskier, A., 1959. Heart examination and consultation by radio and radio-photo transmission. Journal of the American Medical Association 169 (17), 1981–1983.

Bui, D.T., et al., 2021. Tele-mentoring using augmented reality technology in healthcare: a systematic review. Australasian Journal of Educational Technology 81–101.

Bujnowska-Fedak, M.M., Grata-Borkowska, U.J.S.H.T., 2015. Use of telemedicine-based care for the aging and elderly: promises and pitfalls. Smart Homecare Technology & Telehealth 3, 91–105.

Burton, R.L., O'Connell, M.E., 2018. Telehealth rehabilitation for cognitive impairment: randomized controlled feasibility trial. JMIR Research Protocols 7 (2), e9420.

Chandwani, R.K., Dwivedi, Y.K., 2015. Process, and policy, telemedicine in india: current state, challenges and opportunities. Transforming Government: People, Process and Policy.

Das, S., Mukhopadhyay, A., 2011. Security and privacy challenges in telemedicine. CSI Communications 35, 20–22.

Freiburger, G., Holcomb, M., Piper, D., 2007. The STARPAHC collection: part of an archive of the history of telemedicine. Journal of Telemedicine and Telecare 13 (5), 221–223.

Friedman, C.P., Wyatt, J.C., 1997. Subjective approaches to evaluations. In: Friedman, C.P., Wyatt, J. (Eds.), Evaluation Methods in Medical Informatics. Springer, New York, pp. 205–222.

Gasteiger, N., et al., 2021. Upskilling health and care workers with augmented and virtual reality: protocol for a realist review to develop an evidence-informed programme theory. BMJ Open 11 (7), e050033.

Gershon-Cohen, J., et al., 1952. Telognosis: three years of experience with diagnosis by telephone-transmitted roentgenograms. Journal of the American Medical Association 148 (9), 731–732.

Ghiţă, A., Gutiérrez-Maldonado, J., 2018. Applications of virtual reality in individuals with alcohol misuse: a systematic review. Addictive Behaviors 81, 1–11.

Grundy, B.L., et al., 1977. Telemedicine in critical care: an experiment in health care delivery. Journal of the American College of Emergency Physicians 6 (10), 439–444.

Haleem, A., et al., 2021. Telemedicine for Healthcare: capabilities, features, barriers, and applications. Sensors International 100117.

Hand, L.J., 2021. The Role of Telemedicine in Rural Mental Health Care Around the Globe. Telemedicine and e-Health.

Hanson, J., et al., 2007. Attitudes to telecare among older people, professional care workers and informal carers: a preventative strategy or crisis management? Universal Access in the Information Society 6 (2), 193–205.

Harrison, M.I., Koppel, R., Bar-Lev, S., 2007. Unintended consequences of information technologies in health care—an interactive sociotechnical analysis. Journal of the American Medical Informatics Association 14 (5), 542–549.

Herron, J., 2016. Augmented reality in medical education and training. Journal of Electronic Resources in Medical Libraries 13 (2), 51–55.

Hersh, W.R., et al., 2001. Clinical outcomes resulting from telemedicine interventions: a systematic review. BMC Medical Informatics and Decision Making 1 (1), 1–8.

Higgins, J.P., Thompson, S.G., 2002. Quantifying heterogeneity in a meta-analysis. Statistics in Medicine 21 (11), 1539–1558.

Hjelm, N., Julius, H., 2005. Centenary of tele-electrocardiography and telephonocardiography. Journal of Telemedicine and Telecare 11 (7), 336–338.

Jagarapu, J., Savani, R.C., 2021. A brief history of telemedicine and the evolution of teleneonatology. In: Seminars in Perinatology. Elsevier.

Jarris Jr., R.F., 1994. Satellite video system aids offshore treatment. Health Management Technology 15 (4), 32–34.

Kahn, J.M., 2015. Virtual visits—confronting the challenges of telemedicine. New England Journal of Medicine 372 (18), 1684–1685.

Katkuri, P.K., Mantri, A., 2021. Study of behavioral changes and depression control mechanism using IoT and VR. In: Innovations in Computer Science and Engineering. Springer, pp. 605–613.

Khemapech, I., 2019. Telemedicine – meaning, challenges and opportunities. Siriraj Medical Journal 71, 246–252.

Khor, W.S., et al., 2016. Augmented and virtual reality in surgery—the digital surgical environment: applications, limitations and legal pitfalls. Annals of Translational Medicine 4 (23).

Lam, P.P., et al., 2020. Oral health status of children and adolescents with autism spectrum disorder: a systematic review of case-control studies and meta-analysis. Autism 24 (5), 1047–1066.

Lustig, T.A., 2012. The Role of Telehealth in an Evolving Health Care Environment: Workshop Summary. National Academies Press.

Masys, D.R., 1997. Telemedicine: A Guide to Assessing Telecommunications in Health Care. BMJ Group BMA House, Tavistock Square, London, WC1H 9JR.

McCambridge, M., et al., 2010. Association of health information technology and teleintensivist coverage with decreased mortality and ventilator use in critically ill patients. Archives of Internal Medicine 170 (7), 648–653.

McLean, S., et al., 2013. The impact of telehealthcare on the quality and safety of care: a systematic overview. PLoS One 8 (8), e71238.

Microsoft. Customer Stories. Mount Sinai Hospital Customer Stories 2021 [cited 2021 March 2, 2021]; Available from: https://customers.microsoft.com/en-us/story/858292-mount-sinai-health-system-health-provider-teams-hololens-remote-assist-dynamics-365.

Moo, L.R., et al., 2020. Home-based video telemedicine for dementia management. Clinical Gerontologist 43 (2), 193–203.

Morgan, D.G., et al., 2011. Evaluation of telehealth for preclinic assessment and follow-up in an interprofessional rural and remote memory clinic. Journal of Applied Gerontology 30 (3), 304–331.

Murphy Jr., R., Bird, K.T., 1974. Telediagnosis: a new community health resource. Observations on the feasibility of telediagnosis based on 1000 patient transactions. American Journal of Public Health 64 (2), 113–119.

Nittari, G., et al., 2020. Telemedicine practice: review of the current ethical and legal challenges. Telemedicine and e-Health 26 (12), 1427–1437.

Oing, T., Prescott, J., 2018. Implementations of virtual reality for anxiety-related disorders: systematic review. JMIR Serious Games 6 (4), e10965.

Pappalardo, M., et al., 2021. Telemedicine in pediatric infectious diseases. Children 8 (4), 260.

Pawar, J., Bansode, T., 2021. AR for maintenance training during COVID-19 pandemic. In: Techno-Societal 2020. Springer, pp. 459–467.

Peng, H., Yang, C.Q., Wang, S., 2012. Nonformaldehyde durable press finishing of cotton fabrics using the combination of maleic acid and sodium hypophosphite. Carbohydrate Polymers 87 (1), 491–499.

Percival, J., Hanson, J., 2006. Big brother or brave new world? Telecare and its implications for older people's independence and social inclusion. Critical Social Policy 26 (4), 888–909.

Ponce, B.A., et al., 2016. Telemedicine with mobile devices and augmented reality for early postoperative care. 38th Annual International Conference of the IEEE Engineering in Medicine and Biology Society (EMBC). IEEE.

Seifert, A. et al., 2019. Digital participation. 10, 978–983.

Telematics, 1997. W.H.O.G.C.o.H., A Health Telematics Policy in Support of WHO's Health-For-All Strategy for Global Health Development : Report of the WHO Group Consultation on Health Telematics. World Health Organization, Geneva, 11–16 December.

Wilcox, M.E., Adhikari, N.K., 2012. The effect of telemedicine in critically ill patients: systematic review and meta-analysis. Critical Care 16 (4), 1–12.

Wittson, C.L., Affleck, D.C., Johnson, V., 1961. Two-way television in group therapy. Psychiatric Services 12 (11), 22–23.

Young, L.B., et al., 2011. Impact of telemedicine intensive care unit coverage on patient outcomes: a systematic review and meta-analysis. Archives of Internal Medicine 171 (6), 498–506.

Zhai, Y.J.P., 2020. A call for addressing barriers to telemedicine: health disparities during the COVID-19 pandemic. Psychotherapy and Psychosomatics 1.

Zundel, K.M., 1996. Telemedicine: history, applications, and impact on librarianship. Bulletin of the Medical Library Association 84 (1), 71.

Further reading

EdZbarzhyvetsky. Available from: https://depositphotos.com/category/holidays-events.htmlGirl stretching body with virtual reality headset on yoga mat at home-depositphotos.com.

piqsels.com. Available from: https://www.piqsels.com/en/public-domain-photo-sbytn.

CHAPTER NINE

Envisioning big data in IoT with augmented and virtual reality: challenges,opportunities, and potential solutions

Ab Rouf Khan
VIT Bhopal University, Ashta, Madhya Pradesh, India

9.1 Introduction to the Internet of Things and augmented/virtual reality

The Internet's pervasiveness has allowed it to reach practically every corner of the planet and have an unparalleled impact on humans, yet it still has a long road ahead to go. We are entering the "Internet of Things-IoT" paradigm as more and more heterogenous devices are connected to the Internet. As per the literature, Kevin Ashton is considered to have coined the term-Internet of things (Kevin Ashton, 2009). As per Gubbi et al. (2013), IoT is "interactive, stimulating network wherein actuators and sensors blend impeccably with the environment around us, sharing the information across the various platforms." These days, the IoT is quite a buzzword to describe the situations in which computing competence and Internet connectivity outspread to a diverse set of devices, objects, sensors, and other devices capable of sensing data. To interconnect several devices quickly and cheaply, there has to be the convergence of several technologies and support from the market trends. The technologies (Rose et al., 2015; Al-Fuqaha et al., 2015; Lee, 2019) that make IoT a reality can be listed as follows:

- Ubiquitous connectivity
- The extensive embracing of IP-based networking
- Progress in data analytics
- The upsurge of cloud computing
- Miniaturization-manufacturing advances in computing and communication technologies
- Computing economics

The paradigm of the IoT is the integration of various visions rather than a single technology. IoT should not be only a global EPC scheme in the broader context, in which RFIDs are the only objects. They are only part of the whole story, and the same

Extended Reality for Healthcare Systems, Volume 1
ISBN 978-0-323-98381-5, https://doi.org/10.1016/B978-0-323-98381-5.00005-2

extends to the unique/universal/ubiquitous ID (uID) alternative architecture, whose key idea is still (middleware-based) solutions to community resource object transparency in an IoT ecosystem (Turchet et al., 2020).

IoT has already had a significant influence on practically every aspect of life. The World Economic Forum predicts that 26 to 30 billion devices will be connected to the Internet by 2025 in homes and workplaces (WEF, 2021). The Gartner Inc. estimates that in the year 2025, the market for IT services for IoT will be worth 58 billion dollars, growing at a 34% CAGR from 2020. The build phase will provide the most money, while the run period will see the most growth (Gartner, no date). The Business Insider Intelligence predicts to have over 41 billion IoT devices by 2027, rising from 8 billion in 2019 (Newman, 2020). The dominant IoT applications' expected market share by 2025 (Global Building Automation Market Expected to Grow 60% in Next Decade | HPAC Engineering, no date) is demonstrated in Fig. 9.1.

To build a favorable atmosphere for enterprises to produce quality goods, an architecture plan can be perceived as a backbone for the IoT. In connecting billions of devices, architecture is characterized as a structure for describing the network's physical components and serviceable association and structure, its operational doctrines and procedures, and the data setups used. An assembly of physical objects, sensors, actuators, users, cloud services, communication layers designers, business strata, and IoT practices are part of the IoT architecture (Sobin, 2020; Chang et al., 2019).

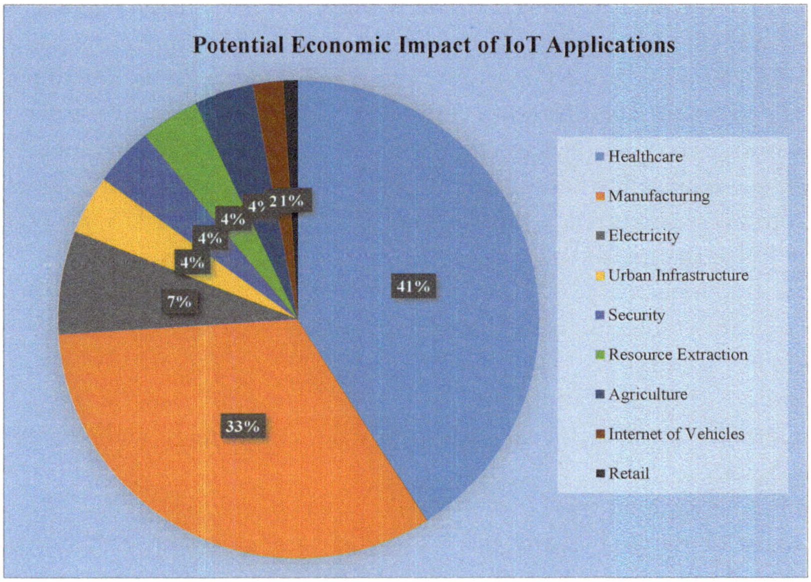

Figure 9.1 The dominant IoT applications' expected market share by the year 2025.

The prominent architectures for the IoT paradigm can be classified as follows:

- three-layered architecture,
- five-layered architecture,
- middleware-based architecture, and
- service-oriented architecture (SOA).

The most basic architecture developed for IoT is a three-layer architecture. It is considered the conventional architecture for IoT. The three-layered architecture comprises perception, network, and application layers. The three-layer architecture specifies the fundamental IoT definition, but it is not sufficient. For IoT analysis, since the finer aspects of IoT are also the focus of research (Sethi and Sarangi, 2017; Uviase and Kotonya, 2018). Fig. 9.2 demonstrates the three- and five-layer architectures of IoT.

Merging information with the physical environment would make it simpler to understand and interact with it. AR improves the way we acquire, comprehend, and disseminate facts by overlaying virtual material on top of our current perspective of the world around us as a mode of display. In regard of technology and applications, AR has advanced by great strides in recent years. AR is a data-centric application that needs more data compared with any other application IoT provides. Other than application-specific contents that users view and interact with, AR apps require information about the world as well as representations of how they connect to system data (Huang et al., 2013; MacIntyre et al., 2013). VR allows the users to have a truly interactive experience by displaying information in several dimensions and allowing them to easily alter viewing positions and seeing what they wish (Hu et al., 2021).

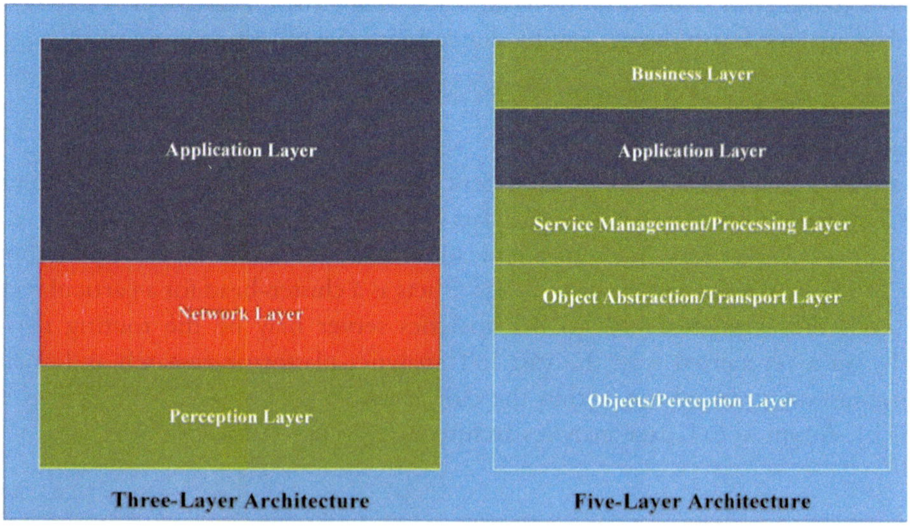

Figure 9.2 Three-layer versus a five-layer architecture for the Internet of Things.

9.2 Classification of mammoth data in the Internet of Things

Connected IoT systems will see a compound annual growth rate (CAGR) of 28.7% over the 2018–25 projected timeframe. It has been predicted that there will be 41.6 billion IoT-based devices, generating 79.4 zettabytes (ZB) of data in 2025, as per the International Data Corporation (*IDC*) annual report of the year 2019 (International Data Corporation, 2019). Based on the intrinsic characteristics of data in the IoT paradigm, the IoT data can be categorized into two broad categories (Yannuzzi et al., 2014):

i. Data generation

ii. Data quality

Beholding deep into the various aspects of data generation and quality, the various aspects related to the two broader categories can be tabulated, as shown in Table 9.1 (Qin et al., 2016).

It is essential to analyze massive data produced by IoT applications to fetch decision-making information. IoT analytics can be characterized by analyzing vast chunks of big data as a process to monitor and optimize decision-making in real time (Rathore et al., 2018; Ahmed et al., 2017; Strohbach et al., 2015). IoT analytics are needed to analyze every massive IoT data segment to obtain critical data stream patterns. It can assist industries in proactive equipment maintenance and other infrastructural facilities, enable them to follow new products and services, expedite operating processes, and generate new products and services for consumers. Specifically, the imminent paradigm of IoT analytics is more likely to impact three industries—smart lifestyle, smart cities, and smart transport (Bibri, 2018; Kim and Jung, 2019; Sharma and Wang, 2017).

To analyze the massive data generated in IoT, a state-of-art analytical framework is required (Jiang et al., 2018; Andres-Maldonado et al., 2019; Marjani et al., 2017). The IoT data analytical architecture can be divided into the following:

- sensor networks,
- routing network, and
- data analytics.

The sensor networks sense the environment and produce readings. It is also the role of these networks to optimally exchange data with the routing system or gateway concerning power usage, overhead communication, and the shortest distance path (Sharma et al., 2018). A gateway device or IoT device acts as a cluster-head for a particular cluster and sends the aggregated data to the analytics server may be the routing network. Usually, in a centralized way, the vast IoT analytics platform's research and decision-making tasks are carried out within the cloud servers (Liu et al., 2019; Firouzi et al., 2018a,b). A typical IoT data analytics architecture is depicted in Fig. 9.3.

Table 9.1 Classification of data in Internet of Things (IoT).

category	Attributes
Data generation	**i.** *Velocity*: Generation of data at diverse rates. Handling very high sampling data rates poses a challenge here. **ii.** *Scalability*: The scale at which the data in IoT networks are going to be tremendously huge. It is challenging to design scalable IoT systems. **iii.** *Dynamics*: The IoT elements are mobile and fragile and possess intermittent connections. Handling dynamic data are always a challenge. **iv.** *Heterogeneity*: IoT inherently is heterogenous due to the presence of a diverse set of devices. Dealing with heterogenous data is a challenge.
Data quality	**i.** *Uncertainty*: Uncertainty in IoT exists from different IoT devices, including missing readings, reading a tag multiple times, accuracy difference between the actual and indicated value, and even the sensing precision. **ii.** *Redundancy*: Redundant sensing data can easily be generated in IoT environments because of various factors, including reading the same tag at a particular location or multiple locations. Sensors of the same type may generate similar readings due to high sampling rates in a particular area. **iii.** *Ambiguity*: Data generated from various IoT devices can be perceived differently depending upon the need for data. Interpreting the data correctly is a significant challenge to deal with. **iv.** *Inconsistency*: Missing tag readings, monitoring of the same environment by different sensors, and packet losses during the data transmission lead to inconsistency in IoT.

9.3 Augmented/virtual reality and big data in Internet of Things

For a few years, AR/VR and big data have been developing their own paradigms in numerous sectors. But the convergence of two emerging technologies, which seem to be unrelated, has received little attention thus far. The merging of AR/VR and big data

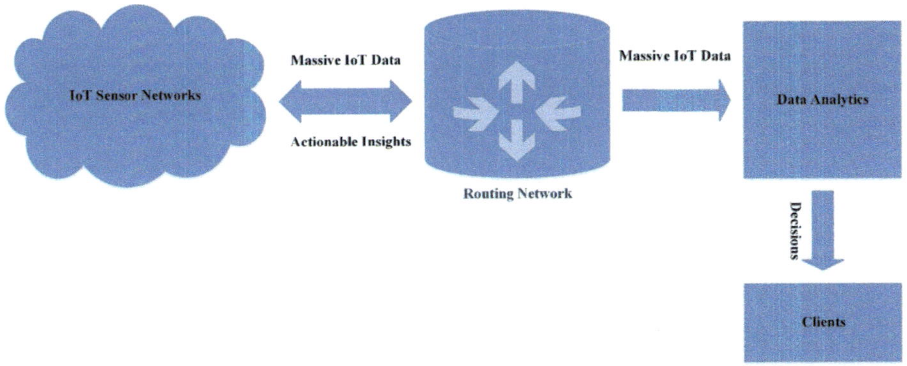

Figure 9.3 IoT data analytics architecture.

is being aided by the extensive perception of big data and the unique presentation modality of AR/VR. In terms of visualization and interaction, AR/VR offers a lot of potential for bringing innovation to big data.

9.3.1 Challenges associated with the big data in Internet of Things with augmented and virtual reality

The big data IoT analytics from the AR/VR viewpoint is only in its infancy, amid all the speculation around data analytics in real time. There are many challenges to the real-time AR/VR big data IoT analytics platform. The system's effective operation requires high-quality decision-making data, effectively reduced power routing schemes, high energy efficiency overall, optimum overhead communication, and fair privacy. The resource-constrained existence of IoT networks further aggravates the situation. As shown in Fig. 9.3, the overall data analytics architecture does not have the competence to support constructive analytical applications involving analysis in real time. An apparent issue with the paradigm of dominant data analytics is that the layer of analytics is discrete. Whereas this framework has performed well enough for longitudinal data analysis, contemporary-day comprehensive analysis needs online decision-making within a relatively short-time budget. This chapter will classify the open challenges related to AR/VR big data analytics in three categories: open challenges associated with the sensor network, associated with a routing network, and associated with data analytics (Jawhar et al., 2020; Vodyaho et al., 2020).

9.3.1.1 Open challenges associated with augmented reality/virtual reality Internet of things sensing networks

IoT devices are small-sized, and the battery is even smaller. Most of the IoT devices are disposed of as they are out of charge due to this design limitation. IoT devices, therefore, have significant resource constraints concerning power, processing, and storage

capacities. A key issue is to increase the battery life of IoT devices, and it is yet to be solved. Hence, IoT devices need a completely different method when making decisions based on machine learning techniques. In contrast, conventional machine learning procedures need sophisticated resources that are not accessible to these small IoT devices. In a resource-constrained scenario, the issue of performing IoT analytics poses a significant challenge.

9.3.1.2 Open challenges associated with the routing network

Inefficient device-to-device (D2D) communications in AR/VR-enabled IoT networks lead to many problems such as more packet loss, high overhead in transmission, and more significant power loss. Furthermore, the routing network experiences immense power loss and packet loss in complex scenarios where AR/VR-enabled IoT devices are dynamic, further aggravating AR/VR enabled IoT data routing. Hence, better systems that can provide superior efficiency for static and dynamic IoT scenarios are needed for the IoT routing network.

9.3.1.3 Open challenges associated with Internet of Things data analytics

The raw AR/VR-enabled IoT sensor data are exceptionally uncertain. Uncertainty is induced by the inclusion of outliers, redundancy, imprecision, missing values, and biased interpretations. Redundancy poses a challenge to system performance. Usually, the machine overwhelms redundant data losses a significant portion of its computing energy to process duplicate data on multiple occasions. Inaccuracy and bias are often very difficult to identify in a large IoT data set because it is context-dependent. The artifact of the system or entity requires considerable domain knowledge.

Apart from the challenges mentioned before, issues related to AR/VR display and viewer devices and AR/VR interfaces and interaction methods pose a major challenge as well.

9.3.2 Big data in Internet of things with augmented and virtual reality: opportunities

Business Insider Intelligence predicts to have over 41 billion IoT devices by 2027, rising from 8 billion in 2019 (Newman, 2020). Another estimate predicts that the international IoT market is projected to hit a $1256.1 billion value by 2025 from $690 billion in 2019 at a CAGR of 10.53% during 2020–25 (Mordor Intelligence, 2020).

Applications for healthcare alone are projected to produce around 1.1 USD to 2.5 USD trillion annually in international economic progression by 2025 (I-SCOOP, 2019). Instances comprise portable health and telecare for successful medical well-being, prevention, diagnosis, treatment, and monitoring provided through electronic media. Such devices send and collect data via a wireless network using a specific IP address. On average, these technologies are low cost, low power, compact, and tranquil to use

(Firouzi et al., 2018a,b; Elhoseny et al., 2018). Bluetooth-compatible smart watches working on Bluetooth and Wi-Fi technologies and testing toolkits for COVID-19 screening are examples that already have observable results (Michael, 2020; Khan and Chishti, 2020; Bermejo et al., 2017).

Advisory tools in manufacturing expect the global Internet to generate 1.3 USD trillion in value by 2020, with a 149% rise in reappearance on investment. Navigant recently announced that it is projected that the Building Automation Systems (BAS) market will grow by 60%, from $58.1 billion in 2013 to $100.8 billion by 2021 (Global Building Automation Market Expected to Grow 60% in Next Decade | HPAC Engineering, no date). This growth will provide telecommunications organizations with a chance to link modern facilities with energy efficiency, security, and data protection through optimized networks. ISPs are expected to offer their networks to deliver a high quality of service for machine-to-machine (M2M), person-to-person (P2P), and person-to-machine (P2M) movement flows to distribute IoT and related services globally.

The dominant IoT applications' expected market share by 2025 is demonstrated in Fig. 9.1. We could exploit the advantages and features AR/VR-enabled IoT systems to provide in all of the major applications the IoT market with big data provides (Cirillo et al., 2019).

9.3.2.1 Augmented reality/virtual reality–enabled healthcare

Prompt disclosure of relevant and required information is critical while making healthcare-related choices. The relevance of AR/VR in the health sector may be ascribed to its capacity to provide important information in real time when needed. By projecting CT scans or medicinal pictures on a present view, AR/VR's remarkable capacity of X-ray visualization is leveraged to give contextual signals for identifying patients and training materials for the medical fraternity. While early instances have demonstrated AR/VR's ability to revolutionize the healthcare sector, AR's huge life-saving potential for the health sector cannot be completely exhibited without the backing of big data. AR/VR is only useful for medical training and as a medical expertise evaluation if there are not enough data sets. Rather than the facts, doctors' experiences are heavily weighted in making decisions. Patient records are becoming digitized, resulting in a deluge of digital information from which therapeutic choices that were formerly dependent on guessing and experience may now be made using big data analytics. The physician can rapidly obtain vital interpretation of patient records by creating an AR/VR scenario to present a pertinent health record. As wearable gadgets that can measure blood oxygen level, heart rate, etc., become more widely available, AR/VR can play an increasingly crucial role in self-tracking health. AR/VR provides real-time information to help comprehend individual health status and provide recommendations based on medical data and nutrition (Jo and Kim, 2019).

9.3.2.2 Augmented reality/virtual reality—enabled retail industry

AR/VR has been used in the retail industry for some time now. To provide broad information, the majority of AR/VR apps overlay geospatial-related statistics on the existent display. Virtual advertisements have become a new normal in all the new product launch events, whether Apple Inc. or any other tech giant. To make AR/VR more penetrative in the retail industry, the preferences of the end users should be taken into account. Exploring vast volumes of data to comprehend consumers' buying interests and behavior is beneficial, as it aids in tailoring promotions and suggestions to customers.

Moreover, when integrated with AR/VR technology, it can be much more exciting. AR encourages vertical commerce to a potential customer using big data. Customers' mental presentations are influenced by an aware and activated buying setting, making them more engaged, aware, and forceful in their purchasing selections. AR/VR removes physical constraints to improve the digital shopping experience everywhere at any anytime, thanks to the benefits of high mobility and dependable virtual content given by big data (Shafique et al., 2020; Huang et al., 2013).

9.3.2.3 Augmented reality/virtual reality—enabled public services

The public services are provided by the government, which generate and use vast amounts of data. The regime has numerous avenues for accumulating statistics from community services, including national security, retail, transportation, manufacturing, social services, etc. Big data can help the government deliver better services to the people. When public services, such as healthcare and manufacturing, are offered directly to people in their specific environments via AR/VR technology, they will be more efficient and productive. Big data from video scrutiny, smartphone and sensor networks, and digital media will make the AR/VR approach for public service delivery more promising. As gadgets and network devices continue to encompass vehicles, it will be increasingly easier to collect large amounts of traffic data to keep us informed about the situation. The data from the internet of vehicles could be used for deriving better insights in view of transportation public services. AR/VR can project information to the drivers, examining threads and forecasting probable automobile accidents.

9.3.2.4 Augmented reality/virtual reality—enabled tourism industry

AR/VR focuses on providing ambient information and supporting in daily tasks in terms of ecological consciousness, which is especially useful when users are unaware of the world around them. AR/VR provides an intuitive way to boost a sightseeing adventure by highlighting intriguing sights or bringing the past to life. Since travel is often connected with geospatial discovery, several AR/VR travel guide applications use geospatial data. As the world becomes more intelligent because of big data, it becomes possible to follow and measure travelers' demands and behaviors to provide a proactive

and smart vacation experience. According to the demands of travelers, information, for example, the locations of adjoining rest stops and eateries, can be provided in a much interactive way by merging the AR/VR-enabled tourism industry with big data analytics (Rathore et al., 2018; Farid et al., 2021).

9.3.3 Big data in Internet of Things with augmented and virtual reality: potential solutions

A novel and efficient three-tier architecture is proposed to deal with the challenges posed to the efficient implementation of AR/VR systems in the IoT framework. The proposed architecture is given in Fig. 9.4. The overall process of handling the challenges posed will consist of collecting the data at the edge of the network using the embedded sensors in

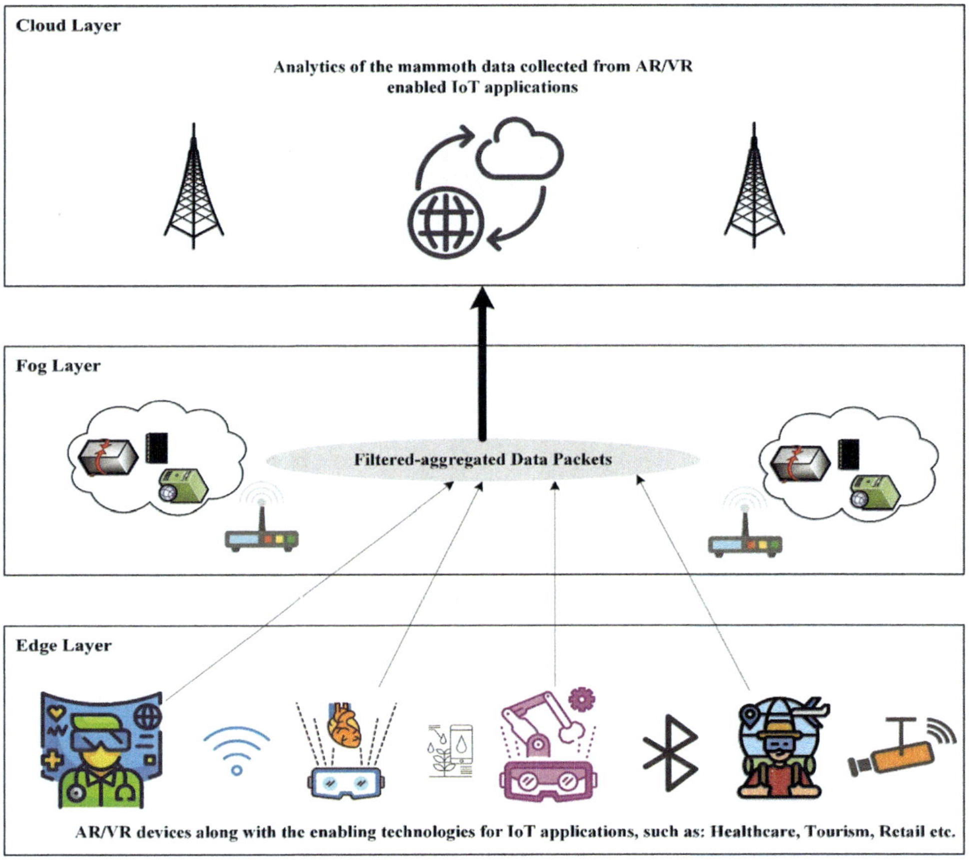

Figure 9.4 The proposed architecture of the envisioning big data in IoT with augmented and virtual reality. *IoT*, Internet of things.

the various AR/VR-enabled applications and sending the same to the cloud for processing and analytics in an efficient manner. The sensors embedded in the AR/VR devices corresponding to different applications—healthcare, manufacturing, retail, etc. will collect the data at the network's edge. This layer is also referred to as the sensor layer in the literature. Since the data collected from numerous AR/VR devices are mammoth, we need to efficiently transmit the data to the higher layers—fog/cloud. Aggregating the data at the edge layer of the proposed architecture and sending the aggregated data rather than the individual data packets could be a game-changer in the AR/VR industry. To aggregate the data collected from the AR/VR devices, Beta-Dominating Set Centered Cluster-Based Data Aggregation Mechanism (βDSC^2DAM) (Khan and Chishti, 2022) is used. The fog nodes owing to their lesser latency and better efficiency will be used to aggregate the data rather than the normal IoT nodes. Aggregating the data at the edge layer will decrease the total number of data transmissions.

By reducing the number of communications between devices, bandwidth and energy consumption limits are naturally reduced. By and large, this improves the overall efficiency, data freshness, and security of the IoT framework. At the edge of the network, we will filter the false data to improve efficiency. False data consume a lot of bandwidth. Besides, approximately packet loss of 25% occurs due to retransmissions (Cheng et al., 2017). Thus, we need to filter the false data at the aggregator itself to improve the efficiency of the network. The process of intelligent data filtering includes filtering the data, which is practically nonexistent but may be added to the sensor data due to some noise or other anomaly (González-Manzano et al., 2016; Santamaria et al., 2018). The data filtration process starts with discarding the values, which cannot exist, or we can call it the data sensed, which is out of range (Kim et al., 2017; Incorvia, 2015; Khattab et al., 2016).

9.4 Conclusion

The merging of AR/VR and big data is aided by the extensive perception of big data and the unique presentation modality of AR/VR. This chapter presented the challenges, opportunities, and potential solutions to envision IoT big data with AR/VR techniques. The challenges are mainly associated with the sensor network, routing network, and data analytics. The opportunities of envisioning big data in the AR/VR enabled IoT applications lie in healthcare, manufacturing, retail, tourism, etc. The data collected from the AR/VR-enabled IoT applications were aggregated and filtered before sending to the cloud layer for analytics. The fog nodes carried out the data aggregation and filtration at the fog layer. The aggregation and filtration of the data in the AR/VR-enabled IoT applications assisted with the big data analytics improve the processes of AR/VR in the IoT framework.

References

Ahmed, E., et al., 2017. The role of big data analytics in Internet of Things. Computer Networks 129, 459–471. https://doi.org/10.1016/j.comnet.2017.06.013.

Al-Fuqaha, A., et al., 2015. Internet of things: a survey on enabling technologies, protocols, and applications. IEEE Communications Surveys and Tutorials 17 (4), 2347–2376. https://doi.org/10.1109/COMST.2015.2444095.

Andres-Maldonado, P., et al., 2019. An analytical performance evaluation framework for NB-IoT. IEEE Internet of Things Journal 6 (4), 7232–7240. https://doi.org/10.1109/JIOT.2019.2915349.

Ashton, K., 2009. That "internet of things" thing. RFID Journal. https://doi.org/10.1038/nature03475.

Bermejo, C., et al., 2017. When augmented reality meets big data. In: Proceedings - IEEE 37th International Conference on Distributed Computing Systems Workshops. ICDCSW, pp. 169–174. https://doi.org/10.1109/ICDCSW.2017.62, 2017.

Bibri, S.E., 2018. The IoT for smart sustainable cities of the future: an analytical framework for sensor-based big data applications for environmental sustainability. Sustainable Cities and Society 38, 230–253. https://doi.org/10.1016/j.scs.2017.12.034. October 2017.

Chang, C., Srirama, S.N., Buyya, R., 2019. Internet of Things (IoT) and New Computing Paradigms. Fog and Edge Computing, pp. 1–23. https://doi.org/10.1002/9781119525080.ch1.

Cheng, C.T., Ganganath, N., Fok, K.Y., 2017. Concurrent data collection trees for IoT applications. IEEE Transactions on Industrial Informatics 13 (2), 793–799. https://doi.org/10.1109/TII.2016.2610139.

Cirillo, F., et al., 2019. Embracing the future internet of things. Sensors 19 (2), 1–6. https://doi.org/10.3390/s19020351.

Elhoseny, M., et al., 2018. Secure medical data transmission model for IoT-based healthcare systems. IEEE Access 6, 20596–20608. https://doi.org/10.1109/ACCESS.2018.2817615.

Farid, A.M., et al., 2021. Smart city drivers and challenges in urban-mobility, health-care, and interdependent infrastructure systems. IEEE Potentials 11–16. https://doi.org/10.1109/MPOT.2020.3011399, 1 January.

Firouzi, F., Rahmani, A.M., Mankodiya, K., Badaroglu, M., Merrett, G.V., Wong, P., Farahani, B., 2018a. Internet-of-Things and big data for smarter healthcare: from device to architecture, applications and analytics. Future Generation Computer Systems 78, 583–586. https://doi.org/10.1016/j.future.2017.09.016.

Firouzi, F., Rahmani, A.M., Mankodiya, K., Badaroglu, M., Merrett, G.V., Wong, P., Farahani, B., et al., 2018b. Secure medical data transmission model for IoT-based healthcare systems. IEEE Access 6, 20596–20608. https://doi.org/10.1109/ACCESS.2018.2817615.

Gartner (no date) Forecast: IT Services for IoT, Worldwide, 2019-2025. Available at: https://www.gartner.com/en/documents/4004741/forecast-it-services-for-iot-worldwide-2019-2025 (Accessed: 13 January 2022).

Global Building Automation Market Expected to Grow 60% in Next Decade | HPAC Engineering (no date). Available at: https://www.hpac.com/building-automation/article/20927495/global-building-automation-market-expected-to-grow-60-in-next-decade (Accessed: 29 October 2020).

González-Manzano, L., et al., 2016. PAgIoT - privacy-preserving aggregation protocol for internet of things. Journal of Network and Computer Applications 71, 59–71. https://doi.org/10.1016/j.jnca.2016.06.001.

Gubbi, J., et al., 2013. Internet of Things (IoT): a vision, architectural elements, and future directions. Future Generation Computer Systems 29 (7), 1645–1660. https://doi.org/10.1016/j.future.2013.01.010.

Hu, M., et al., 2021. Virtual reality: a survey of enabling technologies and its applications in IoT. Journal of Network and Computer Applications 178, 1–19. https://doi.org/10.1016/j.jnca.2020.102970.

Huang, Z., et al., 2013. Mobile Augmented Reality Survey: A Bottom-Up Approach. Available at: http://arxiv.org/abs/1309.4413.

Incorvia, C., 2015. Data Filtering in the Internet of Things (12) United States Patent.

International Data Corporation, 2019. The Growth in Connected IoT Devices is Expected to Generate 79.4ZB of Data in 2025, According to a New IDC Forecast, International Data Corporation. Available at: https://www.idc.com/getdoc.jsp?containerId=prUS45213219. Accessed: 10 December 2020.

Jawhar, Q., Thakur, K., Singh, K.J., 2020. Recent advances in handling big data for wireless sensor networks. IEEE Potentials 39 (6), 22–27. https://doi.org/10.1109/MPOT.2019.2959086.

Jiang, N., et al., 2018. Random access analysis for massive IoT networks under a new spatio-temporal model: a stochastic geometry approach. IEEE Transactions on Communications 66 (11), 5788–5803. https://doi.org/10.1109/TCOMM.2018.2854275.

Jo, D., Kim, G.J., 2019. AR enabled IoT for a smart and interactive environment: a survey and future directions. Sensors. https://doi.org/10.3390/s19194330.

Khan, A.R., Chishti, M.A., 2020. On contact tracing in COVID-19 (SARS-CoV-2) pandemic using lowest common ancestor in m-ary data aggregation tree in the fog-computing enhanced internet of things. International Journal of Pervasive Computing and Communications 19. https://doi.org/10.1108/IJPCC-08-2020-0110.

Khan, A.R., Chishti, M.A., 2022. βDSC 2 DAM : beta - dominating set centered cluster - based data aggregation mechanism for the internet of things. Journal of Ambient Intelligence and Humanized Computing 12 (12). https://doi.org/10.1007/s12652-021-03692-x.

Khattab, A., Abdelgawad, A., Yelmarthi, K., 2016. Design and implementation of a cloud-based IoT scheme for precision agriculture. Proceedings of the International Conference on Microelectronics, ICM 0, 201–204. https://doi.org/10.1109/ICM.2016.7847850.

Kim, T., Jung, B.C., 2019. Performance analysis of grant-free multiple access for supporting sporadic traffic in massive IoT networks. IEEE Access 7, 166648–166656. https://doi.org/10.1109/ACCESS.2019.2954203.

Kim, D.Y., Jeong, Y.S., Kim, S., 2017. Data-filtering system to avoid total data distortion in iot networking. Symmetry 9 (1), 1–13. https://doi.org/10.3390/sym9010016.

Lee, I., 2019. The Internet of Things for enterprises: an ecosystem, architecture, and IoT service business model. Internet of Things 7 (2019), 100078. https://doi.org/10.1016/j.iot.2019.100078.

Liu, D., et al., 2019. A survey on secure data analytics in edge computing. IEEE Internet of Things Journal 6 (3), 4946–4967. https://doi.org/10.1109/JIOT.2019.2897619.

MacIntyre, B., Rouzati, H., Lechner, M., 2013. Walled gardens: apps and data as barriers to augmenting reality. IEEE Computer Graphics and Applications 33 (3), 77–81. https://doi.org/10.1109/MCG.2013.51.

Marjani, M., et al., 2017. Big IoT data analytics: architecture, opportunities, and open research challenges. IEEE Access 5, 5247–5261. https://doi.org/10.1109/ACCESS.2017.2689040.

Michael, K., 2020. Getting behind COVID-19 contact trace apps : the google- Apple partnership. IEEE Consumer Elctronics Magzine. https://doi.org/10.1109/MCE.2020.3002492.

Mordor Intelligence, 2020. The Global IoT Market Is Expected to Reach a Value of USD 1256.1 Billion by 2025 from USD 690 Billion in 2019 at a CAGR of 10.53%, during the Period 2020-2025, Global News Wire. Available at: https://www.globenewswire.com/news-release/2020/05/13/2033070/0/en/The-global-IoT-market-is-expected-to-reach-a-value-of-USD-1256-1-billion-by-2025-from-USD-690-billion-in-2019-at-a-CAGR-of-10-53-during-the-period-2020-2025.html. Accessed: 29 October 2020.

Newman, P., 2020. Internet of things report - business insider. Available at: https://www.businessinsider.com/internet-of-things-report?IR=T. Accessed: 29 October 2020.

Qin, Y., et al., 2016. When things matter: a survey on data-centric internet of things. Journal of Network and Computer Applications 64, 137–153. https://doi.org/10.1016/j.jnca.2015.12.016.

Rathore, M.M., et al., 2018. Exploiting IoT and big data analytics: defining Smart Digital City using real-time urban data. Sustainable Cities and Society 40, 600–610. https://doi.org/10.1016/j.scs.2017.12.022. August 2017.

Rose, K., Eldridge, S., Chapin, L., 2015. The Internet of Things: An Overview - Understanding the Issues and Challenges of a More Connected World. The Internet Society (ISOC), p. 80. https://doi.org/10.1017/CBO9781107415324.004 (October).

Santamaria, A.F., et al., 2018. A real IoT device deployment for e-Health applications under lightweight communication protocols, activity classifier and edge data filtering. Computer Communications 128 (June), 60–73. https://doi.org/10.1016/j.comcom.2018.06.010.

SCOOP, I.-, 2019. 'The Internet of Things in Manufacturing: Benefits, Use Cases, and Trends', towards a More Integrated and Mature IoT-Enabled eHealth Reality, pp. 1–13. Available at: https://www.i-scoop.eu/internet-of-things-guide/internet-things-healthcare/. Accessed: 29 October 2020.

Sethi, P., Sarangi, S.R., 2017. Internet of things: architecture,Issues and applications. International Journal of Engineering Research in Africa 07 (06), 85–88. https://doi.org/10.9790/9622-0706048588.

Shafique, K., et al., 2020. Internet of things (IoT) for next-generation smart systems: a review of current challenges, future trends and prospects for emerging 5G-IoT Scenarios. IEEE Access 8, 23022–23040. https://doi.org/10.1109/ACCESS.2020.2970118.

Sharma, S.K., Wang, X., 2017. Live data analytics with collaborative edge and cloud processing in wireless IoT networks. IEEE Access 5, 4621–4635. https://doi.org/10.1109/ACCESS.2017.2682640.

Sharma, S., Chen, K., Sheth, A., 2018. Toward practical privacy-preserving analytics for IoT and cloud-based healthcare systems. IEEE Internet Computing 22 (2), 42–51. https://doi.org/10.1109/MIC.2018.112102519.

Sobin, C.C., 2020. A Survey on Architecture, Protocols and Challenges in IoT, Wireless Personal Communications. Springer US. https://doi.org/10.1007/s11277-020-07108-5.

Strohbach, M., et al., 2015. Towards a big data analytics framework for IoT and smart city applications. Modeling and Optimization in Science and Technologies 4, 257–282. https://doi.org/10.1007/978-3-319-09177-8-11.

Turchet, L., et al., 2020. The Internet of Audio Things: state-of-the-art, vision, and challenges. IEEE Internet of Things Journal XX (X). https://doi.org/10.1109/jiot.2020.2997047, 1–1.

Uviase, O., Kotonya, G., 2018. IoT architectural framework: connection and integration framework for IoT systems. In: Electronic Proceedings in Theoretical Computer Science, vol 264. EPTCS, pp. 1–17. https://doi.org/10.4204/EPTCS.264.1.

Vodyaho, A., et al., 2020. Data collection technology for ambient intelligence systems in internet of things. Electronics 9 (11), 1–26. https://doi.org/10.3390/electronics9111846.

WEF, 2021. Digital Transformation - Reports - World Economic Forum, World Economic Forum. Available at: http://reports.weforum.org/digital-transformation/the-race-to-meet-customer-expectations/. Accessed: 13 January 2022.

Yannuzzi, M., et al., 2014. 03- key ingredients in an IoT recipe fog computing. Pdf', (CAMAD) Computer Aided Modeling and Design of Communication Links and Networks 5. https://doi.org/10.1016/S0039-9140(99)00178-2.

CHAPTER TEN

Evolution and contribution of extended reality in smart healthcare systems: toward a data-centric intelligent healthcare approach

Tawseef Ayoub Shaikh[1], Tabasum Rasool Dar[2] and Shabir Sofi[3]
[1]Department of Computer Science & Engineering, Baba Ghulam Shah Badshah University, Rajouri, Jammu & Kashmir, India
[2]Interdisciplinary Centre for Water Research (ICWaR), Indian Institute of Science, Bangalore, Karnataka, India
[3]ITE Department, National Institute of Technology, Srinagar, Jammu & Kashmir, India

10.1 Introduction

With recent technological advancements, how people experience both physical and virtual environments is beginning to change. Virtual reality (VR), augmented reality (AR), and mixed reality (MR) technologies have revolutionized their respective fields by changing how their members' collaborations integrate real and virtual objects. Combining VR, AR, and MR has given birth to the much-hyped extended reality (XR) technology. On the VR spectrum, VR is the furthest on the virtual side, as it gives people the sensation of being in a computer-generated virtual environment. Big tech companies, such as Google, Microsoft, and Facebook, are looking to get involved in the XR space, by developing their own hardware and software and targeting the medical industry. A sense of immersion can be created using a head-mounted display (HMD) or a Cave Automatic Virtual Environment (CAVE) technique. Virtual world interactions are performed with movements that are identical to how we express emotions and ideas with our bodies and have always been the primary means of getting things done. In a nutshell, AR technology blends virtual and real elements to enrich our lives. Placing virtual elements such as information and the parameters of an object is popular, especially among designers and creativity researchers. With the majority of AR techniques relying on mobile devices, the user's interaction with the AR system is mostly restricted to a certain area or object. MR does a better job of mixing the virtual and real-world aspects. To comprehend this concept, many people consider MR to feature two particular attributes of VR and AR: heightened immersion and realism. MR and VR environments entirely cover their users, but MR and AR each blend real-world and virtual elements (Hu et al., 2021).

Extended Reality for Healthcare Systems, Volume 1
ISBN 978-0-323-98381-5, https://doi.org/10.1016/B978-0-323-98381-5.00008-8

The state-of-the-art literature suggests the largest benefits for businesses and consumers will likely be in industries such as retail (Berg and Vance, 2016; Bonetti et al., 2018), tourism (Griffin et al., 2017), education (Merchant et al., 2014), healthcare (Freeman et al., 2017), entertainment (Lin et al., 2017), and research (Bigné et al., 2016), where VR is poised to have a big impact. An astonishing increase in interest for VR devices shows that sales of VR headsets HMD have just exceeded 1 million in a quarter for the first time, and VR device sales are estimated to increase from $1.5 billion in 2017 to $9.1 billion by 2021 (Canalys Media Alert, 2017). The decreasing prices of VR headsets, as well as new models coming to market (such as the Oculus GO and HTC Vive Focus, FastCompany), will cause the use of VR to drastically increase in the future. Apart from VR, AR and MR are also included in the top 10 strategic trends for 2018 (Canalys Media Alert, 2017). The estimated year-over-year revenue increase from these technologies between 2016 and 2020 is huge (rising from $2.9 billion to $61.3 billion (Canalys Media Alert, 2017)). These results suggest that these technologies will have a bright future.

10.1.1 Definitions

XR technologies act as connection bridge between people, places, and information. The innovation could also help reduce the frustration many people feel about the limitations of healthcare, giving them the ability to stay healthy while performing their work remotely. XR means the range of experiences that take the boundary between the real world and the simulated world and smears it into obscurity as depicted in Table 10.1. To bring people into the experience, the technology can include visuals, audio, and potentially even scents and physical sensation. XR, a term given by Paul, blurs the lines between physical and simulated worlds. This is due to VR and AR technology. XR technology is making immersive experiences more commonplace; it has done so by reducing the importance of distance, which was a big barrier in the past. The spectrum of realities, "reality—virtuality continuum," has opened up for researchers to discover and

Table 10.1 The extended reality spectrum.

Virtual reality	Merged reality	Mixed reality	Augmented reality
Interactive virtual objects	Interactive virtual objects	Interactive virtual objects	Virtual objects
Virtual background	True background	True background	True background
Immersive display	Immersive display	See-through display	See-through display
⟵ Fully Immersive Experience			Unobstructed Experience ⟶

drive deep into this VR environment (Milgram and Kishino, 1994). The continuum ranges from reality to virtuality at the two extremes. The whole world is captured in a real environment (RE). Direct or indirect (seen on a video display) views of a real scene are both included. "Virtual environments" (VEs) are computer-generated environments that use virtual objects (which do not exist in the real world) and an interface that makes it seem like they are appearing on the device in real time. Virtual worlds (VWs) such as Second Life (a type of VW) provide open virtual environments for their users to interact in real time with other users, who are represented by avatars.

VR: It is a digital space in which a person can move around and interact with the surroundings, creating a real sensory experience in real time. Fig. 10.1 shows the VR system located at the far-right end of the continuum, which contains entirely computer-generated content. Users are completely surrounded by the virtual environment, and there is no chance they will see or interact with the real world. A strong sense of presence and immersion in VR simulations allow people to play hypothetical scenarios to their hearts' content. The PlayStation VR is an excellent example. VR system setups can be grouped into three categories. One of the first examples is an integrated VR headset that integrates with either a smartphone-based system or an overall system that is based on a cardboard device. One other system is CAVE, where the walls and floors are made of multiple large screens, and users are completely immersed in it. The final configuration is done with the HMD, which is linked to a separate computer. The setup has become popular recently because it is getting cheaper and better at VR experience, 360 films, and video mapping.

AR: To augment something means to add to it. Digital enhancements are added to real-world experiences, with a focus on making them more interesting, in AR. Everything happens in a single instant. The Snapchat filters are the most common AR apps almost used by everybody. Systems that use computers to overlay virtual information in the real world are defined by these systems. AR environments have a great potential to give rise to exciting tools in many application areas. The wildly popular smartphone game Pokémon Go is an excellent example of AR (ICRC Innovation Board, 2018). The ability of AR to create 3D medical models in real time and then project them into remote locations in medical training and simulation boosts the potential of the technology. Furthermore, real-time deformable medical models increase the application of the simulation. A medical 3D AR app for

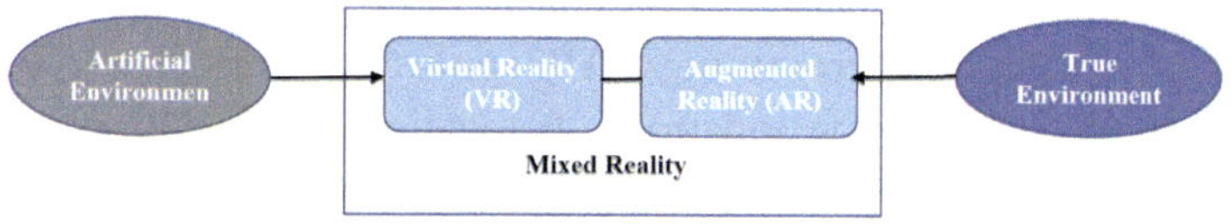

Figure 10.1 Relation between real and virtual worlds in different proportions.

smartphone-equipped medical professionals exists to quickly train and have them administer the procedure for common but critical procedures in a safe and rapid sequence. When it comes to AR systems, they are capable of overlaying digital contents, such as information and objects, onto the real world. The end result is an enhanced experience where the user can see and interact with the surrounding environment while having the ability to use text, imagery, and animation. Smart glasses and handheld devices both improve the user experience. One will never get to race an Olympic 200 m, travel the world on airplane wings, or go on a trip to Mars. The IKEA Place app is a prime example of an AR application. Customers use the app on their smartphones to see products they have placed into their homes. For the first time, in 2011, Crisis Commons and OpenStreetMap responded to the Haiti earthquake with their respective applications (ICRC Innovation Board, 2018).

MR: The extremes are called virtual, mixed, merged, hybrid, or augmented virtuality environments. This environment is a blend of both virtual and real environments. MR came about as the junction where physical and virtual objects overlapped. MR systems are ahead of AR because they give users the ability to interact with virtual objects as if they were actually present in the real world. To make an MR headset, you need to put together a computer glass that is clear so the sensors can see, and an integrated computer in a headset. To allow virtual objects to interact with the user's actual environment, the real-world space is usually mapped in real time using integrated sensors. MR offers more interactive and immersive AR if you think about it. The Microsoft HoloLens, known for its MR applications, is one example of a widely used MR headset.

Three main criteria (immersion, interaction, and information) help explain the distinctions between AR, VR, and MR (Venkatesan et al., 2021). The immersion of the technology is defined by the experience the user has. Though AR enhances real-world views with virtual information, VR provides an entirely virtual immersive experience. MR is able to translate between the virtual and real worlds in real time, which gives it a spatial mapping capability. Interaction is used to describe the type of interactions possible with technology. Virtual objects can be interacted with via VR, while AR makes it possible to interact with real-world objects. MR allows people to interact with virtual and physical things. Data refer to the kind of information being processed during visualization. A virtual 3D space is used to keep track of virtual objects in VR. AR provides an interactive annotation, while the user is performing an activity. MR takes place in 3D space and time, with correlation to the user's surroundings. One can argue that every extremely immersive XR experience depends on the effortless interaction between the real and virtual realms. It is vitally important to consider user's context, including his or her physical surroundings. AR applications emphasize this importance well, thanks to their contextual foundations. A location-based AR experience may be triggered when the customer arrives at a specific store. In the same way, a marker-based AR experience can happen when the customer finds a special AR marker on a display,

Table 10.2 Selected XR technology.

Technology name	Type (VR, AR/MR)	Estimated cost (USD)	Description
Microsoft Hololens	MR	3000	Head-mounted wireless computer system including AR display
Google Cardboard	AR	5–10	Compact inexpensive cardboard adapter to use smartphones as VR glasses
Oculus Rift	VR	350	High-fidelity VR headset display; requires a powerful connected computer
Oculus Go	VR	200	Stand-alone VR headset; works wirelessly without
HTC Vive	VR	500	High-fidelity VR headset display; requires a powerful connected computer
Magic leap	MR	2295	Standalone AR headset; works wirelessly without a computer
Google Glass	AR	1500	Standalone AR headset; works wirelessly without a computer
zSpace		4000	AR desktop display monitor; can be viewed by multiple users at once; requires a powerful connected computer
Vuzix Blade	AR	\$899	Projects information in both eyes
PlayStation VR	VR	\$350	Mostly for gaming with play station. Needs PlayStation and PlayStation Camera

point-of-sale unit, or packaging. Table 10.2 gives a description of various XR tools and techniques in the market with the corresponding price range.

10.1.2 Extended reality in healthcare applications

The healthcare industry has gained significant benefits in recent years from VR and AR. There exist multiple ways that VR is helping to improve life for many people in need of

additional assistance: teaching social skills to children with autism, helping patients with PTSD and depression, identifying early signs of schizophrenia and Alzheimer's, and improving the lives of patients with brain injuries. People are also taking advantage of the new VR therapy programs to teach relaxation and control anxiety through VR experience, particularly during the COVID-19 quarantine protocol. Today, the video game industry brings in the most revenue from XR applications, but that does not mean that it will stay that way. AR has made advances in visualizations that can be helpful for patients and healthcare providers in surgeries and in assisting them to find patients' veins. Doctors are now able to put images on a patient's body in real time by projecting them with AR. The healthcare field is innovating in many different ways, from experimenting with new learning styles to discovering new medical devices that require no physical contact. Many believe that the 2025 mark will have the entertainment industry alone exceeding USD 5 billion, making it the second-largest and most rapidly expanding sector to use XR (Goldman Sachs Global Investment Research, 2021). XR will enable patients to watch surgical procedures before their own surgeries, which will be helpful for their post-operative recovery and for ensuring surgeons give them a successful procedure. It will also help surgeons train and practice by providing them with better surgical guides and experiences that reduce the cost of surgical care (Forbes Inc, 2021). To make things clearer, this can let doctors see data about surgeries as they are happening. It can also be used to treat mental illness and to control pain medication use. XR may play into the USD 16 billion patient monitoring device market by providing a new way to monitor patient progress remotely (Grand View Research Inc, 2021).

The AR paradigm is widely adopted because it is needed for medical imaging. Image-guided is one of the most common uses of medical software surgery (Lorensen et al., 1993; Grimson et al., 1995). The surgeon has access to an inside look at the patient's body thanks to imaging studies, such as CT or MRI scans, which the surgeon performs before surgery. A surgical plan is made with these images. To understand the surgical route to the target area, imaging software creates a 3D model by assembling a complete image of all the relevant views and slices. The AR system can be implemented to the extent that the surgical team will be able to view patient CT or MRI scans that are correctly positioned in the operating room and visible during the surgery (State et al., 1994). Beyond those two examples, AR is also in imaging ultrasound and optical diagnostics (Argotti et al., 2002). The healthcare industry has embraced any new technological innovation in recent years. Innovative VR has finally been brought into the healthcare field, first in surgeries, then medical training, patient care, and more. The multidimensional uses of XR tools and techniques in smart healthcare are given as follows:

- Doctors currently have to rely on X-rays, ultrasounds, MRIs, and other similarly sophisticated imaging tools to understand a patient's organs. For hard-to-diagnose cases, however, more advanced imaging techniques come up short. Every patient has

a distinct body and organ type, which means that doctors have to tailor treatment plans to each individual. And with XR tools and techniques, it has gotten significantly easier. Surgeons can observe organs in three dimensions and then make more accurate incisions. A successful conjoined twin separation procedure took place thanks to XR.

- Doctors who have decades of experience always feel time-pressed. Even with newbies, they do not have time to train, and it is just barely left. XR can do tremendous work here. Using holograms, these experienced doctors can now train students or new doctors, with guidance, without actually being present.
- Stroke patients have trouble moving and are therefore more likely to fall. The patient's recovery is supported by XR's virtual rehabilitation environment, which is motivational, task-oriented, and controlled (Wurst, 2020).
- To prove that XR supports the recovery from surgical trauma and muscle weakness, Stanford is conducting a clinical trial on a physical therapy system that offers treatment following injury and surgery. Patients will be assessed for their mobility, but XR's benefits will be explored through pain management.
- The UCLA study revealed that VR-goggle-trained surgeons are faster, more accurate, and complete more steps than their non-VR counterparts. The surgeons utilized Osso VR to assist with their work. The improvement will result in thousands of lives saved in the long run.
- Abbvie created a VR environment to simulate the effects of Parkinson's disease as a means of increasing Parkinson's disease awareness. Family members of Parkinson's patients may find this useful for getting a better understanding of the disease and learning how to cope with it (Wurst, 2020).
- GSK created a migraine simulator to increase migraine awareness and understanding. They came up with this for an advertising project for one of their pharmaceuticals.
- Bayer set up XR at a booth for a medical meeting, where attendees averaged 10 min at a time over an initial 2 min. In areas such as product demonstrations and eDetailing, similar approaches could be effective.
- Cedars-Sinai ran a study to look into the effects of VR-based pain relief treatments. 21% of patients who received regular immersive VR relaxation therapy were able to perform the treatment in their homes regularly and noticed significant improvements in their pain levels (Safavi and Kalis, 2018).
- To treat lazy eyes, the Jessenius Faculty of Medicine (Slovakia) has used an Oculus Rift device loaded with a custom game designed to strengthen patients' better eyes and condition their weaker eyes to treat amblyopia, a condition commonly referred to as lazy eye.
- Medical animation studio Random has broad effects on pharma manufacturing plans to provide insight on how manufacturing processes could be visualized and monitor process adherence (Safavi and Kalis, 2018).

- Using VR headsets to experience molecule structures, scientists at Novartis are doing "molecule walking" to analyze protein structures and functions perfectly.
- VR-assisted clinical trials—The research suggests that the VR experiences increased patient compliance with treatment supporting its enlarged potential implications for clinical trials.
- Patients who have surgery have to face the fear of potential complications, such as pain and complications following surgery. Doctors can easily locate veins for frequent injections in XR, which will lead to better treatment quality and greater patient satisfaction.

Broadly, the XR applications in transforming healthcare services can be interpreted from the following three categories.

10.1.2.1 Distance to people

Think about a sick old person in a remote location that gets to consult with an internationally renowned physician without ever leaving their home. To prepare for operations in the real world, surgical residents perform procedures using a simulator instead of operating on cadavers. Nurses use vein finders to help insert IVs in patients' veins on the first try. A VR cognitive treatment helps a PTSD sufferer, who is a veteran, rebound. One such example would be a condition requiring treatment by someone who is not a specialist and a patient who lives far away. XR will provide a realistic experience with a virtual physician. 82% of health executives agree that XR is removing the hurdle of distance in access to people, information, and experiences (Florida Hospital Tampa, 2016). For people patients and their families, Florida Hospital Tampa is using VR models to view inside a patient's brain tumor or aneurysm. Neurosurgeons can view the same model during the operation to create short-term fixes on the spot, instead of performing last-minute, drastic changes. The ability to model a surgical procedure using a three-dimensional computer simulation and the understanding of the medical condition in question will allow patients to make more educated medical decisions and understand their current medical status better (Florida Hospital Tampa, 2016). XR can get healthcare providers and their students in sync with each other by providing them with tools that work. For example, an internationally recognized subspecialist could be working with a medical resident in another country on teaching them a new method. So, the research says that "XR" is helping health professionals get to people, information, and experiences that were once out of reach and diminish the distance gap between delivery of value-based care to the patients.

10.1.2.2 Distance to information

XR is cutting down the gap between consumers and clinicians, which, in turn, eliminates the information that providers have to gather. A doctor can access detailed information with it, which removes obstacles to crucial decisions. For example, a surgeon

could use AR glasses to look at digital content that is projected over the patient without taking their focus off of them. Doctors could gain greater accuracy and leverage outcomes that were impossible previously. They do this by having their information appear directly over their physical movements. XR is improving data accessibility, as well as making new discoveries easier to find. With the advent of new XR tools, information is being conveyed in 3D environments, just as humans do, so things look and feel more familiar. This makes it possible for new visualizations, which can lead to fresh discoveries in healthcare. Surgical procedures are becoming more precise, thanks to the use of 3D mapping and imagery that serves as a "GPS system" for navigating complex anatomy. Doctors used minimally invasive sinus surgery to treat a patient with the recent application of this technology. The system can help surgeons learn how to perform this procedure and the surgical planning (MobiHealthNews, 2018). To better understand the extent of various diseases, it is important to view medical scans in a 3D form, which is what The Body VR allows with interactive 3D scans (The Body VR Website). Oxford researchers developed VR models of genetic data to improve understanding of what goes on inside living cells (Virtual Reality Headsets, 2017). Drishti is an AI-powered solution from Accenture that helps the visually impaired enhance their experience of the world around them and better adapt to their working environment. The solution includes an app that is capable of notifying users about the number of people in a room, their ages, genders, and emotions. In addition, the user can use it to read text out loud, such as from books and documents, or to identify doorways (e.g., glass doors) that could pose a threat to safety (Accenture Press Release, 2017).

10.1.2.3 Distance to experiences

Perhaps the greatest potential to the healthcare industry by XR is delivered by providing shared and communal experiences. In the past, medical professionals could not relate to their patients' medical problems, as they could not be the victims of these illnesses. XR can change this by allowing medical personnel to attempt to get an idea of how it feels to be afflicted with an illness. For example, Embodied Labs makes VR labs that train seniors on services like assisted living. A medical research lab enables medical students to experience what it is like to be 74-years-old with various health issues through the "We Are Alfred" lab. The Beatriz laboratory takes users on a progressive Alzheimer's disease journey (Embodied Labs website). Using XR, clinicians will be able to see how debilitating mental illness is, and they will also be able to see the necessity of providing care to people struggling with mental illness. In this case, one example researchers have found compelling is the use of VR therapy to help military veterans confront post-traumatic stress disorder. This allows patients to see images or relive experiences while discussing responses with therapists in real time. Bravemind, a VR-based exposure therapy tool, was developed by the Institute for Creative Technologies at the University of Southern California in collaboration with the US government. It allows

psychologically scarred veterans to face their triggers and conquer their PTSD by exposing them to environments that cause stress but are not dangerous. After receiving treatment, 80% of the patients reported a decrease in symptoms, including depression (USC Institute for Creative Technologies). Younger patients can benefit even more by using advanced technologies. XR is being used in hospitals to help children cope with painful experiences such as injections and changing clothes. The child about to receive an IV has a chance to visit a virtual ocean right before the procedure (CNET, 2018). To teach healthcare professionals CPR techniques, Nicklaus Children's Hospital in Miami has created VR training content (Next Galaxy, 2015).

10.1.3 Organization

The rest of the paper is sorted out as: Section 10.2 talks about extensively the state-of-the-art methods, tools, and techniques of XR implemented in revolutionizing the healthcare sector. The related survey is divided into seven sections, and each section is again divided into subsections. A detailed review with proper interpretation and discussion is carried out in each section. The challenges, potentials, and future directions of the XR in the healthcare field are debated in Section 10.4. The summary of the entire paper is carried out in Section 10.5 followed by Reference section.

10.2 Previous related work

The potential of AR is being harnessed in several fields of medicine such as surgical workplace robotic surgery, neurosurgery endoscope-assisted microsurgery, pediatric surgery, and obstetrics and gynecology. The given section describes the state-of-the-art XR tools and techniques in medicinal services. The section is categorized into seven subsections, and each subsection carries a detailed survey with corresponding discussions about XR in healthcare facilities (Fig. 10.2).

10.2.1 Extended reality in medical education

XR solutions could revolutionize medical students' and trainees' anatomical learning as well as their invasive procedure training. Students can use XR to examine internal human systems and their mechanisms of action. XR simulation can easily accommodate additional patient-specific data, such as CT scans and MRIs. XR allows trainees to make and revise their practice sessions many times, in a setting without fear of real-world harm. There are tons of learning and training options available for learners using XR. Applications can use VR to simulate everything that an environment, complete with any needed instructional materials, has to offer. The second class of applications introduces medical simulations to VR to trainees as the next available platform. It is possible to use VR in all kinds of applications on the most popular consumer VR headsets. In addition, several MxR applications allow multiple students to collaborate while they interact with

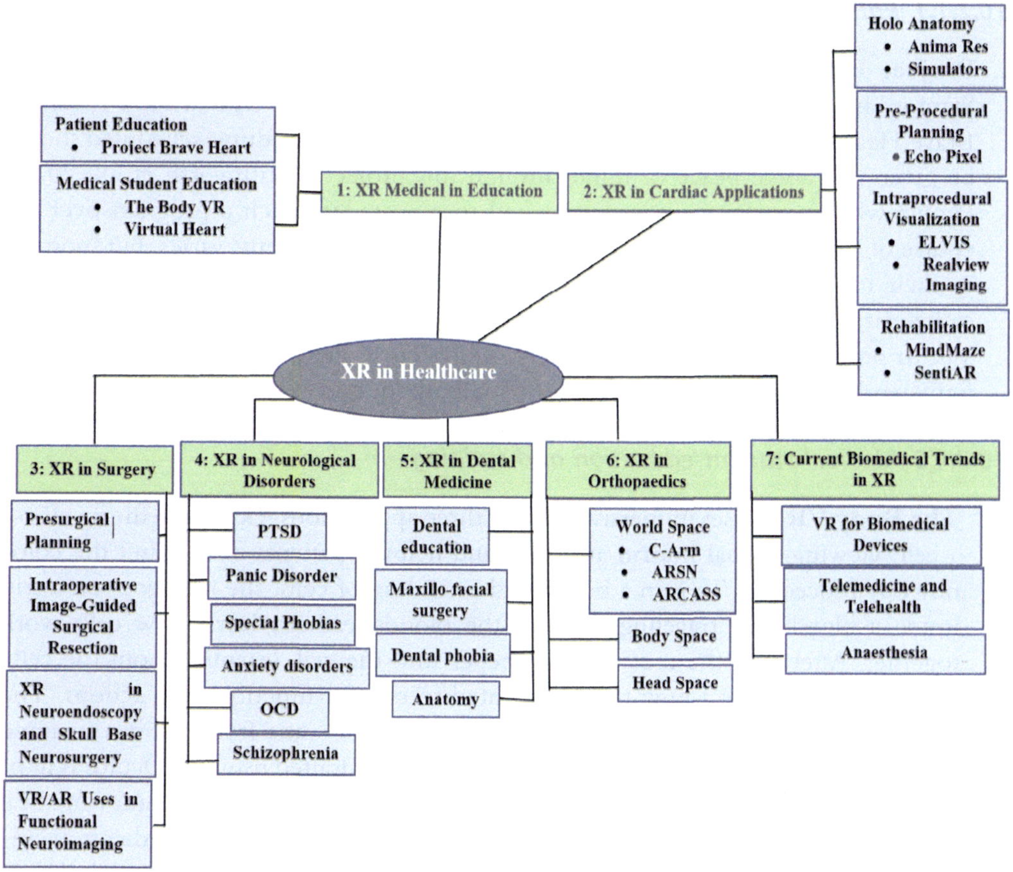

Figure 10.2 Schematic flowchart of the state-of-the-art works in XR in healthcare. *XR*, extended reality.

each other and discuss educational material in a more natural setting. MxR is ideal for this purpose because the software can use the headset's unlimited freedom to walk and communicate freely in the augmented environment. For example, patient-facing applications could be developed to educate cardiac patients about how to handle their illness, while medical personnel-facing applications could be used to help hospital staff better administer treatment. The software may enable separate features for customers on each side of the customer base. Applying the 3D model to learning applications is done to enable users to control the content they are studying, and, in effect, their own education. The XR in medical sciences education can serve the interests of both patients and students.

10.2.1.1 Patient education

i. **Project Brave Heart**: The VR program at Lucile Packard Children's Hospital Stanford has three distinct goals (Hospital SCsHLPCs, 2018). One of these is Project Brave Heart. Patients who have cardiac catheterization procedures scheduled should be given additional peace of mind through this project (Southworth et al., 2020). People who are getting surgery are asked to repeatedly watch a program over the course of a week before their procedure. The number of patients varies, but most are in their teens or early 20s. Prior to the procedure, the patient walks through the catheterization lab, cardiac surgery lab, recovery ward, and their regular hospital via VR to provide them with an immersive experience. Fig. 10.3 outlines the general framework for the relationship between medical education and XR technology.

10.2.1.2 Medical student education and training

i. **The Body VR**: It uses immersive VR for three applications: exploring the insides of a cell, viewing virtual human anatomy, and helping patients understand the colonoscopy procedure. To gain a better understanding of cells, the user goes on a trip inside a blood cell, traveling through the bloodstream to learn how cells work together. After this, the user can dig deeper into the cell, learning about the cell's internal structures and how they are related to cellular functions and actions. The DICOM Viewer can be utilized to look at medical scans from various imaging machines such as MRI, CT, and PET. The results are drafted using an Oculus Rift or HTC Vive VR headset. Gong et al. (2021) designed a training program for medical personnel to learn how to perform intubations with hand—eye coordination utilizing AR concepts to succeed. A novel adaptive synchronization algorithm (ASA) used to maintain the shared state of the collaborative AR environment increases the

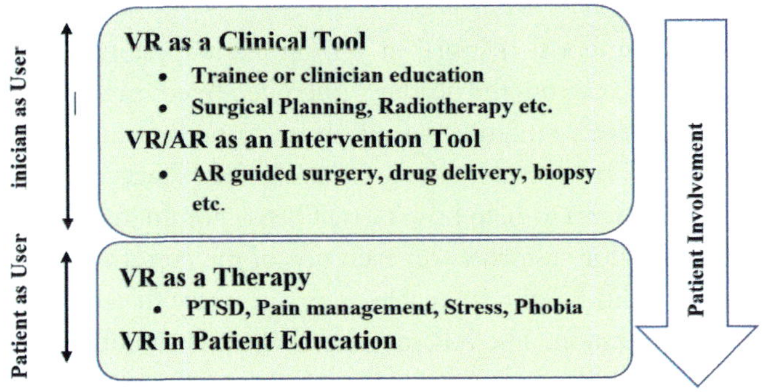

Figure 10.3 Medical virtual reality applications by patient involvement, including whether the clinician or patient will be using the virtual reality.

sense of presence among participants and thus allows them to interact despite any delays caused by the infrastructure.

ii. **Stanford virtual heart**: Lighthaus, Inc. partnered with Stanford University on a project to provide education about the human heart with the help of virtual reality known as the Stanford Virtual Heart Project (Stanford Children's Health). This project has several unique components. The first focuses on educating patients and their families about their child's cardiac anatomy, which at the moment is limited to plastic models and drawings. Stanford medical students and trainees can see what their anatomy looks like and see how abnormalities in it affect the body's functions. The students can walk through, examine, and move the models around to have a better understanding of the body's internal organs and physiology. The trainees have a library of about two dozen common congenital lesions at their disposal. The goal of these training exercises is to increase one's comprehension of these diseases and the changes they produce in body physiology. Echopixel is a 3D monitor that is used as the final step in the cardiothoracic operating room. 3D equipment in the surgical suite might make it possible to do detailed assessments of intracardiac anatomy and geometry after the patient has been placed on cardiopulmonary bypass and the heart has been deflated. This could be difficult because the anatomy may be difficult to see with these procedures. Even though XR is not used regularly in medical education at the moment, there have been a significant number of experiments and pilots involving anatomy. Table 10.3 shows several examples of these pilots.

10.2.2 Cardiac applications of extended reality

Various cardiac applications of virtual reality are depicted in this section.

10.2.2.1 Medical student training cardiac

i. **HoloAnatomy**: Microsoft's HoloLens is being used at Case Western Reserve University to enhance anatomy learning among medical students by tailoring instruction and observation (Sarah et al., 2019; Case Western Reserve). Students who can better comprehend how 3D anatomic relationships work will find that learning becomes less frustrating and more enjoyable because they will be able to "think like a doctor." The team has developed HoloAnatomy, a program that allows medical students to perform holographic dissections to better understand the body's organs and systems. The program is a joint effort of the university and the Cleveland Clinic and has public access to the people for demo.

ii. **Anima Res**: Anima Res is a company that specializes in creating 3D medical animations for AR, MR, and VR (Butler et al., 2018). This team's task is to help medical education become more relevant to doctors, med students, and patients. In particular, "Insight Heart" allows users to get a visceral understanding of the human

Table 10.3 Examples of XR anatomy innovations at medical schools.

Description	XR type	School
Stanford Neurosurgical Simulation and Virtual Reality Center (Stanford Medicine, 2019)	XR type: VR level of learners: medical students, residents, surgeons. Focus: neuroanatomy, neurosurgical procedure training	Stanford
HoloAnatomy with Microsoft Hololens (Workman, 2018)	XR type: MR level of learners: medical students. Focus: general anatomy	Case Western/Cleveland Clinic
Immersive Education at CHLA with Oculus Go (Oculus. Immersive Education)	XR type: VR level of learners: all incoming residents, optional for medical students. Focus: pediatric trauma procedures, pediatric resuscitation training	Children's Hospital of Los Angeles
Virtual Reality Anatomy at USCF with HTC Vive (UCSF VR)	XR type: VR level of learners: first-year medical students. Focus: general anatomy	University of California, San Francisco
Enduvo VR Teaching and Learning Platform using HTC Vive (The University of Illinois)	XR: VR level of learners: medical students, surgeons, faculty. Focus: general anatomy	University of Illinois College of Medicine Peoria

heart by using immersive visual effects to display atrial fibrillation, systemic hypertension, and myocardial infarction, all seen in a three-dimensional space. This encounter is feasible on many different types of advanced reality platforms.

iii. Simulators: Stand-alone training applications have also been developed through hardware integration (Talbot et al., 2017). The Vimedix transesophageal echocardiogram (TEE) and transthoracic echocardiogram (TTE) simulator by CAE demonstrates how MR can be used in a variety of different medical applications. Using the MR simulator gives students insight into their anatomical relationships, and they are able to understand how to position the probe.

10.2.2.2 Preprocedural planning

The use of the systems has been validated, as well, having been proven both clinically effective and able to achieve the task of preprocedural planning.

i. **EchoPixel** True 3D, developed by Echopixel, is an innovative DICOM workstation that includes the first DICOM-certified 3D system to be approved by the FDA (Chan et al., 2013). 3D visualization is accomplished using a technique similar to that of 3D movie theaters and early 3D consumer televisions, which provides two different images to each eye through glasses that are equipped with small liquid crystal displays (LCDs). The image can be manipulated through handheld wands if one Echopixel user is wearing polarized glasses. Echopixel, which can show arteries in people with pulmonary atresia with a significant collateral vessel in the heart, has been utilized in preliminary cardiology research. Doctors using the True 3D display interpreted findings more quickly than those who used a traditional display, finishing in 13 min as opposed to 22 min. The interpretations were just as accurate when compared with catheter angiography.

10.2.2.3 Intraprocedural visualization

i. **Enhanced electrophysiology visualization and interaction system (ELVIS)**: Though better visualization has been a big research and development push for many years, equal advances in interaction have not been made. ELVIS, the interventional electrophysiologist, can see real-time patient-specific 3D cardiac geometry and view catheter locations on that display without breaking sterility, which is critical in this field (Silva et al., 2017). ELVIS shows data from either an electroanatomic mapping system (EAMS), computed tomography, or cardiac magnetic resonance imaging, in which the EAMS is to be used before any procedures are done. A new capability that has been recently discovered allows seeing cases from the past, as well as live ones from the control room. In addition to making it easier to see the content, the system enables sterile control of the display to be done via gestures, gaze, or voice. The new interaction method is ideal for one-off interventions, as it allows the interventionalist to commandeer the unified model in a way that is advantageous for the task at hand. A shared cardiac holographic model is located in the room and lets each user look at the model from his or her perspective while they operate on it. It is possible to pass control to another user at any time while only ever having one person in control at a time.

ii. **Realview Imaging**: The Schneider Children's Medical Center pediatric cardiology group in Yokneam, Israel, was able to successfully use 3D imaging software from Realview Medical Imaging to analyze whether real-time holograms are feasible in a standard cardiac catheterization laboratory in 2016 (Silva et al., 2018). The Realview CGH, a set of computational holograms that was created with a combination of 3D rotational angiography, was coupled with transoesophageal echocardiography. This study included participants with both preexisting heart disease and postsurgery patients. Using a "very easy" image marking process, all

patients were able to use the tool to generate real-time 3D holograms of high accuracy that could be cropped, zoomed, rotated, moved, and even sliced.

iii. **EchoPixel**: The EchoPixel system was used in the intraprocedural procedures. The third arm of Stanford's VR program includes three-dimensional VR Imaging, which aids surgeons in preprocedural planning for various types of cardiac surgery. Before they perform the procedure, the CT surgeons use this arm to perform virtual run-throughs (Wired, 2017). EchoPixel, a California-based company, makes the True3D technology that EchoPixel Tech uses. The technology uses a 1080p, active 3D VR display and stylus, which was built by Hewlett Packard Enterprise. EchoPixel's software paired with a monitor-style display and 3D glasses and a stylus allows a user to interact with data in 3D. The stylus allows the user to rotate, cut into, and measure the parts of the anatomy they are working on (Wired, 2017).

10.2.2.4 Rehabilitation

i. **MindMaze**: The VR space company MindMaze is developing hardware and software together to create neurorehabilitation apps (Chevalley et al., 2015). MindMotion PRO, their solution that is approved by the US Food and Drug Administration, is being used in the poststroke patient population to enhance upper limb mobility by combining VR, brain imaging, and gaming technologies. An interesting study with results that are easy to explain involved poststroke patients who received a 20-to-30-minute training session starting on day 4 of hospitalization, which allowed the patients to return to normal activities while a stroke healed in them. Almost everyone who used the MindMotion PRO reported an improvement in movement capability by about 90% (Chevalley et al., 2015).

ii. **SentiAR**: SentiAR solution is in development for use in electrophysiology labs at the moment, with the option of more work with cardiac interventional procedures down the line. The electrophysiology laboratory's current restrictions include the fact that each piece of equipment in the lab is unique in that it uses its own control panel and display. No equipment interfaces with any other. A 2D screen is used to store the compressed complex 3D data set known as electroanatomic mapping data. The SentiAR system utilizes electroanatomic mapping data from a commercially available system and projects patient-specific, real-time geometries, cardiac and electroanatomic mapping, and catheter locations in stereoscopic 3D on a Microsoft HoloLens 720p.

10.2.3 Presurgical and intraoperative augmented reality in neurooncologic surgery

One of the most-studied applications of 3D models in surgical planning is 3D printing, wherein 3D models of anatomy and pathology derived from medical images are

Table 10.4 Clinical summary of XR in neurooncologic surgery.

Authors	Surgical purpose	Sample size	Article objective	Brain tumor classification	Analysis software	Clinical outcomes
Gerard et al. (2018)	Intraoperative guidance	8 patients	To evaluate the feasibility of combining intraoperative ultrasound and AR in tumor surgery	Meningioma ×2, glioma ×4, metastases ×2	IBIS (Drouin and Kochanowska, Montreal, Canada)	Not reported
Finger et al. (2017)	Neuroendoscopy and presurgical planning of biopsy and other procedures	28 total patients with 14 having the underlying oncologic disease	To evaluate AR-enhanced navigated neuroendoscopy system for intraventricular pathologies	Various periventricular tumors	Nova Plan 2.6.10 (Scopis, Berlin, Germany)	Not reported
Chen et al. (2017)	Presurgical planning	16 patients	To assess an AR system using mobile devices for presurgical planning of supratentorial lesions	Parietal, temporal, and frontal lesions (meningioma × 15, glioma × 1)	3D Slicer 4.0 (Surgical Planning Laboratory, Brigham and Women's Hospital, Boston, Massachusetts, United States)	Not reported
Sun et al. (2016)	Presurgical planning and intraoperative guidance	79 patients with functional neuronavigation and intraoperative MRI and 55 control patients	To investigate the utility of combined VR and AR for intraoperative MRI and neuronavigation in glioma surgery	Glioma	iPlan 2.6 (BrainLab AG, Munich, Germany) SARL, Bernex, Switzerland)	69.6% of the study group achieved complete resection, with an average extent of 95.2%

(Continued)

Table 10.4 Clinical summary of XR in neurooncologic surgery.—cont'd

Authors	Surgical purpose	Sample size	Article objective	Brain tumor classification	Analysis software	Clinical outcomes
						compared with 36.4% and 84.9% in the study group; language, motor, and vision preservation were significantly higher in the study group
Watanabe et al. (2016)	Intraoperative guidance	6 patients	To assess AR-based navigation system with whole operating room tracking	Various tumors	Amira (FEI, Hillsboro, Oregon, United States)	Not reported
Rotariu et al. (2017)	Neuroendoscopy and presurgical planning	22 patients	To evaluate the role and accuracy of virtual endoscopy for presurgical assessment	Pituitary adenomas	OsiriX (Pixmeo)	Not reported
Tabrizi and Mahvash (2015)	Intraoperative guidance	5 patients	To intraoperatively evaluate a novel AR neuronavigation system	3 metastases, 2 glioblastoma	MRIcro (Chris Rorden, Columbia, South Carolina, United States)	All tumors were successfully removed with no complications

Table 10.4 Clinical summary of XR in neurooncologic surgery.—cont'd

Authors	Surgical purpose	Sample size	Article objective	Brain tumor classification	Analysis software	Clinical outcomes
Inoue et al. (2015)	Presurgical planning	99 patients	To assess the utility of a 3D CT model for obtaining preoperative information regarding sphenoidal sinus procedures	Pituitary adenomas	3D Advantage Workstation Volume Share 4 (GE Healthcare, Wauwatosa, Wisconsin, United States)	Not reported
Inoue et al. (2013)	Intraoperative guidance	3 patients	To assess novel AR neuronavigation system using Web cameras	Glioblastoma × 2, meningiomas × 2	3D Slicer	No new neurologic deficits occurred; two of three tumors were successfully removed in their entirety
Stadie and Kockro (2013)	Presurgical planning	208 patients (Dextroscope) and 33 patients (Setred)	To report experiences with two different VR systems	Various tumors	Dextroscope (Volume Interactions Pte. Ltd., Singapore, Singapore) and Setred system (Setred, Stockholm, Sweden)	Not reported
Wang et al. (2012)	Presurgical planning	60 patients	To examine the utility of VR in planning	Various tumors in the sellar region	Dextroscope	Of the selected group of 30 participants,

(Continued)

Table 10.4 Clinical summary of XR in neurooncologic surgery.—cont'd

Authors	Surgical purpose	Sample size	Article objective	Brain tumor classification	Analysis software	Clinical outcomes
			sellar region tumor resections			hormone levels and vision were improved; complications including CSF leakage and diabetes insipidus were noted in five patients
Stadie et al. (2011)	Craniotomy placement	48 patients	To describe the method of defining the placement of the craniotomy for minimally invasive procedures	Various tumors	Dextroscope	Not reported
Low et al. (2010)	Presurgical planning	5 patients	To assess the utility of AR surgical navigation for resection of meningiomas	Parasagittal, falcine, and convexity meningiomas	Dextroscope	Four of five patients had complete resection; 1 patient had near-total excision; all patients had good neurologic recovery

Table 10.4 Clinical summary of XR in neurooncologic surgery.—cont'd

Authors	Surgical purpose	Sample size	Article objective	Brain tumor classification	Analysis software	Clinical outcomes
Qiu et al. (2010)	Presurgical planning	45 patients	To assess the utility of VR presurgical planning using DTI tractography for cerebral gliomas with pyramidal tract involvement	Cerebral gliomas involving pyramidal tracts	Dextroscope	Gross tumor resection in 33 of 45 (73%) patients and subtotal resection in 6 (13%) patients; 7 of 45 (16%) patients had improved motor function, and 30 of 45 (67%) patients had no change
Ferroli et al. (2013)	Presurgical planning	64 patients	To assess clinical experience using stereoscopic virtual reality for surgical planning	Various tumors	Dextroscope	Not reported
Yang et al. (2009)	Presurgical planning	42 patients in VR group and in the control group	To evaluate the outcome of presurgical planning using dextroscope in	Meningioma × 15, schwannoma × 15, other × 12	Dextroscope	Total resection rate was 83% in VR group compared with 71% in the control

(Continued)

Table 10.4 Clinical summary of XR in neurooncologic surgery.—cont'd

Authors	Surgical purpose	Sample size	Article objective	Brain tumor classification	Analysis software	Clinical outcomes
			patients with skull base tumors			group; complication rate, length of postoperative stay, and surgery duration were significantly reduced in VR group
Stadie et al. (2008)	Presurgical planning	106 total cases, including 100 cranial lesions	To report on experiences with 3D virtual reality systems for minimally invasive surgical planning	Various tumors	Dextroscope	Not reported
Anil et al. (2007)	Presurgical planning	1 patient	To report preoperative planning with dextroscope for fourth ventricular ependymoma	Fourth ventricular ependymoma	Dextroscope	Tumor was entirely removed with patient having no immediate postoperative neurologic deficits
Rosahl et al. (2006)	Resurgical planning	110 patients	To investigate the usefulness of VR in image guidance for skull base procedures	Various tumors	Image Guidance Laboratories (Stanford University, Stanford, California, United States	Not reported

manufactured using a 3D printer. One advantage of AR/VR over 3D printing is that it can additionally provide simultaneous displays of real and virtual images. Table 10.4 gives the thorough study of the XR application in the field of surgery.

10.2.3.1 Presurgical planning

The benefits of using VR/AR for surgical planning have been widely recognized, with the chief benefit being seen in surgical cases where navigation systems have failed to properly register. Targeted approaches are crucial in neurooncology surgeries, and their success heavily relies on the proper location of the tumor (Orringer et al., 2012). Stadie et al. (2011) compared the success rates of cranial surgery by simulating each technique using a virtual planning station named Dextroscope and VectorVision. It was discovered that these techniques are equivalent to localizing craniotomy. Nonetheless, neuronavigation resulted in inaccurate or failed results 3% of the time (out of 48 cases). Using preoperative VR planning, minimally invasive surgery was completed successfully in these scenarios. Apart from patient factors, such as brain swelling and fluid leakage, 3D imaging systems can be prone to errors in probe tracking and image-to-patient registration (Orringer et al., 2012; Widmann et al., 2012). Through better visualization of anatomic features, VR/AR has the potential to help neuronavigation processes improve in areas that have novel case-based applications.

10.2.3.2 Intraoperative image-guided surgical resection

The AR-based surgical system can show anatomic and functional imaging altogether. The system, developed by Besharati Tabrizi and Mahvash (Tabrizi and Mahvash, 2015), is being used in many hospitals and helps surgeons by placing a virtual image onto the patient's skull or brain so that it is visible in real time. Despite having all tumors removed, an imaging system showed more discrepancies than did tumor navigation with regard to defining tumor edges and other tumor parameters. A higher rate of complete glioma resection has been achieved by combining VR/AR protocols using functional neuronavigation and intraoperative MRI. The different tools from the Brainlab software suite were used to help better navigate the tumor through 3D visualization and tracing of critical structures. The neurosurgeon used AR to see a 3D view or 2D image section as a visual representation of the virtual image that was superimposed on the neurosurgeon's view. The intraoperative MRI made it possible to perform more complete tumor removal in patients using this technique (69.6% compared with 36.4% in the control group). Most significant among benefits of intraoperative MRI, the technique accounts for brain shift, the deformation of the brain that results from factors such as brain swelling and cerebrospinal fluid loss. Another tool in the investigation is intraoperative ultrasound, which may also enable better measurement of brain shift (Nimsky et al., 2001). A different study used the AR surgical navigation platform DEXRay in combination with presurgical planning with Dextroscope to aid in successful resection of meningiomas in the falcine, convexity, and parasagittal regions.

10.2.3.3 Augmented reality in neuroendoscopy and skull base neurosurgery

Stereotactic biopsies, when used for treating tumors in a patient's brain, pose problems, particularly when the tumor is located in or near the ventricles or at the boundary of the ventricles and cortex. Neuroendoscopic approaches are best suited for these tumors (Chrastina et al., 2012). Finger et al. (2017) analyzed the accuracy and value of an AR-assisted neuroendoscopy system for identifying internal disorders such as those found within the ventricles. The region of interest was superimposed as AR on the endoscopic field of view after the virtual planning of the neuronavigation system. The results of all the biopsies were a 100% diagnostic yield. Surgery on the skull base relies heavily on visualization, which is why stereoscopic visualization is important for imaging deep-seated tumors. By using VR, this method has been possible: this new surgical technique, and the training associated with it, which was difficult to simulate in the past. The accuracy of virtual endoscopy images for surgical planning of transsphenoidal pituitary adenoma surgery is especially great. This method of investigation uncovered a wealth of knowledge about the nasal cavity's structures and landmarks, including the carotid prominence and the sphenoid septa, giving medical professionals the guidance they needed to determine the best surgery plan and strategy (Mikhail et al., 2019). In their prospective, randomized clinical trial, Yang et al. (2009) focused on using VR in skull base tumor resection procedures as part of presurgical planning. Similar use was found with the technology, and it resulted in a significantly reduced surgical time and post-operative stay, as well as fewer complications because of cerebrovascular injury. According to Schwam et al. (2021), AR may be of use in lateral skull base surgery. More research is needed into the potential of AR technology to help with skull base surgery.

10.2.3.4 Virtual reality/augmented reality uses in functional neuroimaging

The use of VR and AR for imaging has increased with the goal of successfully completing all functions while preserving patients' quality of life (Zhao et al., 2012). Lesions in eloquent brain areas, including the primary motor cortex and subcortical motor pathways, which are hard to map and identify, are more likely to be impacted during brain tumor surgery (Amidei and Kushner, 2015). Diffusion tensor imaging tractography can be used to visualize key white matter tracts and also to provide information on the tumor's location. T1-weighted MRI, T2-weighted MRI, and diffusion tensor imaging are some of the imaging techniques that could be used to implement this. Understanding the interrelationship between white matter tracts and glioma lesions can be better done with the help of fiber tracking and glioma segmentation. This method was proved to be very effective, as it was found to increase surgical accuracy, thereby minimizing the potential for damage to motor function, as well as ensure complete tumor removal without worsening neurovascular function for patients with glioma tumors with metastases that were close to or touching the pyramidal tract and one patient who had a tumor located in the corticospinal tract. Using 3D Slicer, Inoue et al. (2013)

designed a 3D AR neuronavigation system capable of adding tumors, vascular structures, and tractography to images obtained via Web cameras.

10.2.4 Extended reality in neurological disorders

10.2.4.1 Posttraumatic stress disorder

Posttraumatic stress disorder (PTSD) is pervasive among veterans of numerous wars, which has prompted the hunt for both VR immersion therapy (TERV) and other methods to be successful treatments (Rothbaum et al., 1999). These sessions were set in a virtual Vietnam, with jungle, rice fields, and rivers, where the patient could move and act on his own. Details such as helicopter sounds and blasts from explosions, gunfire, and flashes of light were added. The battle was made even more real with the addition of fog and soldiers shouting orders. The application of VR for the treatment of PTSD was first studied by Rothbaum et al. (Difede and Hoffman, 2002) in 1999. In the second simulation, which took place in a jungle clearing, the patient had the virtual experience of flying a Huey helicopter over Vietnam. Improvements with this treatment approach lasted more than 6 months. In 2002, a case study was done by Beck et al. (2007) who utilized exposure to VR treatment on a World Trade Center (WTC) terrorist attack survivor suffering from PTSD. It has been established that this patient has failed with exposure therapy in the past (Table 10.5).

Table 10.5 Studies on the treatment of posttraumatic stress disorder (PTSD) using virtual reality.

Authors	Study type	Number of participants	Technique applied	Results
Difede and Hoffman (2002)	Case study	1	VRET	90% reduction in symptoms of PTSD, 83% reduction in symptoms of depression
Beck et al. (2007)	Case study	6	VRET, relaxation techniques, in vivo exposure and in imago exposure	Reduction of symptoms of PTSD. No significant reduction in symptoms was found anxiety and depression
Rothbaum et al. (1999)	Case study	1	VRET	4 5% reduction in symptoms of PTSD retention beyond 6 months
McLay et al. (2011)	Controlled randomized trial	20	Group A: VRET, cognitive restructuring Group B: pharmacotherapy and group therapy	7 0% of those who underwent VRET showed >30% improvement of PTSD symptoms compared with 12.5% of group B

The VR training involved short 1-h sessions where participants were introduced to different virtual plane crashes, including an approximation of the twin towers' collapse. After finishing the treatment, the patient saw their symptoms significantly diminish. Using VR, Beck et al. (Rothbaum et al., 1999) helped six people with PTSD resulting from a car accident in 2007. People who were in the treatment group with 10 sessions in the virtual environment showed a decrease in PTSD symptoms. In 2011, McLay et al. (2011) performed the first controlled randomized trial on the effects of VR and cognitive reconstruction on cognitive rehabilitation. This trial studied the program in active-duty military personnel who had postcombat PTSD, comparing it to standard pharmacotherapy and group therapy. Patients who received no treatment other than sham exposure showed greater improvement than the control group after 10 weeks. 11 September survivors were successfully treated by TERV whose instance was described in an academic study (Difede and Hoffman, 2002). The study concluded with a more significant research effort on a larger participant pool of 13 survivors. This experiment also revealed therapeutic effects beyond what was compared with the control group. Recent veterans have benefitted greatly from the TERV, which helps reduce their anxiety and PTSD symptoms (Difede and Hoffman, 2002). A more recent study completed a clinical trial comparing classic CBT to TERV on 10 subjects with PSTD. The study showed no significant difference between the therapeutic effects of the two but showed a slight preference for TERV (Dayan, 2006). Additionally, treatment for victims of road accidents is another major topic. Using the VR therapeutic system of the Argaman Virtual Reality Software Suite is a great treatment option for PTSD, especially after exposure to extreme terror and traumatic experiences (Dayan, 2006).

10.2.4.2 Panic disorder

North et al. (1996) studied the desensitization process of panic disorder with agoraphobia in 30 students using VR. In 2000, Jang et al. (2000) conducted an open, uncontrolled study to determine the benefits of VR therapy in people with panic disorder with agoraphobia. Panic disorder with agoraphobia can be treated by the experiential—cognitive therapy developed by Vincelli et al. (2003b) (Table 10.6).

A complex cognitive—behavioral therapy strategy uses VR as its integration point. There was a big plaza, a subzero supermarket, and many other large public areas in the virtual environment. Researchers assigned 12 panic disorder patients with agoraphobia to either 8 sessions of experiential—cognitive therapy, 12 session GCS, or a waiting list, all on a random basis. The study found that had fewer sessions, compared to CBT, and in a few months were able to achieve the same results as CBT when it comes to reducing anxiety, depression, and panic attacks. Choi et al. (2005) compared the two groups and found that those who received brief experiential—cognitive therapy fared better. The results of the treatment showed a significant improvement in symptoms, regardless of treatment type. In 2007, Botella et al. handpicked 37 patients who met the DSM-IV panic disorder with agoraphobia criteria and were randomized to three nine-session experiments: exposure to VR, exposure to reality, and no treatment of these patients.

Table 10.6 Studies on the treatment of panic disorder using virtual reality.

Authors	Study type	Number of participants	Technique applied	Results
North et al. (1996)	Controlled study	60	Group A: VRET Group B: None No treatment	Significant reduction in PTSD symptoms of the group receiving VRETcontrol group showed no improvement
Jang et al. (2000)	Open, uncontrolled study	7	VRET	The effectiveness of VRET was not supported VR
Vincelli et al. (2003b)	Controlled randomized study	12	Group A: BHT Group B: GCCS Group C: List Waiting list	Similar reduction in panic attack, anxiety, depression symptoms in both treatments. 33% fewer sessions in the BHT. The superiority of both over the waiting list
Choi et al. (2005)	Controlled randomized study	40	Group A: BHT Group B: GCCS	Significant reduction in seizure symptoms Panic attacks in both treatments
Botella et al. (2007b)	Controlled randomized study	37	Group A: VRET Group B: In vivo report Group C: List Waiting list	Same reduction in panic attack symptoms and superiority over both treatments the waiting list. ΔMaintenance of treatment gains 12 months later

Subjects who were treated with the VR experienced significantly better improvements in their symptoms, compared to the subjects who were only on the control waiting list.

10.2.4.3 Special phobias

Despite having its first clinical trial in the 1990s, VR has been in use as a therapeutic tool to treat anxiety disorders to this day. A string of tests and studies have also been done to evaluate its effectiveness (Gorini and Riva, 2008). For a lot of them, their focus is on anxiety disorders, but a few also expand to eating disorders, substance dependence, and control of pain. Some even go so far as to cover palliative care and rehabilitation (Tarnanas et al., 2009).

10.2.4.3.1 Flight phobia

Many people suffer from the fear of flying, which prevents them from going on trips or they end up being too anxious to travel. Many people with severe anxiety are able to

travel but resort to alcohol and medications to cope with their symptoms (Roberts, 1989). It is believed that between 10% and 25% of the general population has this disorder. A team led by Rothbaumet et al. (Roberts, 1989) in 2000 researched the impact of an exposure-based treatment for 49 patients with extreme fear of flying. This group treatment included one session where the patients worked in an in-house airplane and also visited a local airport. Patients were also placed on a waiting list. The treatment was delivered in a total of 8 sessions of 60—90 min for 8 weeks. The participants were measured in their desire and anxiety as they prepared to get on an actual plane. Exposure to VR, or even real life, turned out to be better than the waitlist, whereas the two treatments proved equally good (Rothbaum et al., 2002). Rothbaum et al. (2006) found similar results in 2006 when they replicated the study.

The first experiment in 2001 involving exposure to VR compared with exposure to fantasy by Wiederhold et al. (2001) studied the results of 30 people with the fear of flying. One group experienced a simulation in VR along with seeing what their bodies were doing (heart rate, sweating, etc.). A second group experienced VR with no knowledge of their bodily status. Finally, the third group experienced their daydreams. In comparison, after treatment ended, of the people who received psychosomatic treatment with the aid of fantasy, 20% were able to fly without any medication, but 80% were able to fly using VR without recording psychosomatic reactions. In contrast, 100% of those who received treatment with VR with psychosomatic recordings could fly without any medication. A 2003 study by Mühlberger et al. (2003) of 45 patients with phobia found that the 6-month evaluation concluded that the virtual environment helped cure symptoms only when the patient was shown both visual and auditory stimuli. The study by Krijn et al. (2007), which examined the effects of VR exposure, exposure in the real world, and bibliotherapy, involved a phobia of flying and involved 83 people. Two days of group cognitive therapy were given to all of the groups. It was discovered that VR exposure therapy and GSCT were more effective treatments in comparison with bibliotherapy because they had the most success in getting their participants to break free of their fears. The results of the study are laid out in Table 10.7.

10.2.4.3.2 Social phobia

Using VR, North et al. (1998) treated social phobia for the first time in 1998. People with social phobia were exposed to a virtual environment with speeches for 6 weeks, and those who received real exposure were compared with those who received exposure to a neutral sham environment. Findings revealed that those who completed the virtual public speaking experience experienced significant improvements. A pilot study done by Pertaub et al. (1999) included 10 participants who speak in front of an audience who was virtual and, importantly, that the audience displayed overtly positive or overtly hostile emotions. The study set out to find out which type of virtual audience would be able to provoke social phobia symptoms. The participants' anxious reactions were mostly

Table 10.7 Studies on the treatment of fear of flying using virtual reality.

Authors	Study type	Number of participants	Technique applied	Results
Rothbaum et al. (2000)	Controlled randomized study	45	Group A: VRET Group B: VRET Group C: Waiting list	VRET and GCS were superior to the list Waiting list. Between them, there was no significant difference. There was no significant difference between the treatment options Gains were maintained 6 months after
Rothbaum et al. (2000, 2002)	Controlled randomized study	83	Group A: VRET Group B: GCS Group C: Waiting list	75 completed treatment. Confirmation of the results of the previous study. The therapeutic benefits were maintained at 6 and 12 months after
Wiederhold et al. (2001)	Controlled randomized study	30	Group A: VRET with feedback of physical stimuli Group B: VRET without VRET feedback of the without somatic stimuli Group C: In imago exposure	VRET with feedback: 100% effectiveness VRET without feedback: 80% effectiveness Both two were superior to in imago exposure (20%)
Mühlberger et al. (2003)	Controlled randomized study	45	Group A: TBI with VRET (flight simulation) Group B: TF with VRET Without motion Group C: TTH	Reduction of symptoms was observed only in VRET with or without motion Therapeutic benefits were maintained 6 months after
Krijn et al. (2007)	Controlled randomized study	83	Group A: VR report, Group B: GCCS Group C: Bibliotherapy Without communication with the therapist All groups: after treatment, they received an additional 2 days of CrCB	59 completed treatment. The VRET and GCS were superior to bibliotherapy GCS after CrCB showed the greatest efficacy

provoked by the virtual audience, which was shown by the results. The study by Harris et al. (2002), which exposed eight individuals with a public speaking phobia to VR and compared the results with a control group, concluded that exposure to VR was effective. Roy et al. (2003) and Klinger et al. (2004a) introduced the concept of treating social phobia with VR in 2003. In this treatment, the virtual environments imitated four major elements of the patients' lives with social phobia: performing, intimacy, control, and empowering behavior. Following this, the same researchers (Klinger et al., 2004b) published an identical study in the controlled nonrandomized form in which they discovered that VR treatments were as effective in aiding recovery as GCS. The relevant results are presented in Table 10.8.

Table 10.8 Studies on the treatment of social phobia using virtual reality.

Authors	Type Study type	Number of participants	Technique applied	Results
North et al. (1998)	Controlled study	16	Group A: VRET public speech Group B: VRET speech in neutral audience	14 people completed the treatment. Significant improvement of symptoms only group A hostile audience
Pertaub et al. (1999)	Pilot study	10	Group A: VRET on neutral emotionally neutral audience Group B: VRET in a hostile/hostile audience/friendly audience	Stressful reactions were primarily elicited by the
Harris et al. (2002)	Controlled randomized trial	14	Group A: VRET Group B: Waiting list	The superiority of VRET over waiting list
Roy et al. (2003)	Clinical trial (Within groups design)	10	VRET in four conditions: Performance, control, familiarity, and confidence	Improvement of symptoms
Klinger et al. (2004a)	Clinical trial	10	VRET in four conditions: performance, control, intimacy, and confidence	Improvement of symptoms
Klinger et al. (2004b)	Controlled trial	36	Group A: VRET Group B: GCS	Significant improvement of symptoms in both treatments

10.2.4.3.3 Acrophobia

TERV was first applied to acrophobia in 1993, which laid a platform of collaboration between computer scientists and psychotherapists. The first test of the TERV occurred on a 19-year-old patient who was suffering from acrophobia in 1995. The researchers discovered that the participants' acrophobia got better (Rothbaum et al., 1995, 2000). The same team made another experiment in the same year with 17 acrophobic individuals, which proved fruitful. Further investigations involved exposing 10 and 33 acrophobes to exposure therapy (Emmelkam et al., 2001). In addition to finding similar outcomes in both treatments, the authors discovered affordable PCs to use in the treatment.

10.2.4.3.4 Arachnophobia

In 1996, the first research on TERV in the context of arachnophobia occurred where a woman in her late 30s who was afflicted with crippling arachnophobia received the treatment. Following 12 sessions, it was concluded that the patient's arachnophobia had reduced enough that she is now able to sleep in a tent. One of the more notable parts of this experiment was the incorporation of a real object, which resembles a hairy spider. This gave participants the chance to practice using feedback from pseudohaptic and tactile sensations (Emmelkam et al., 2001).

10.2.4.3.5 Claustrophobia

Botella et al. (1998) used eight sessions of VR exposure treatment in a patient who met the criteria for claustrophobia. Assessment results showed an improvement in symptoms, which was also found to persist a month later. Researchers in the following year investigated the utility of VR exposure in treating a patient who had multiple phobias, specifically claustrophobia, fear of storms, and panic disorder accompanied by agoraphobia (Botella et al., 1999), using a virtual environment in eight one-on-one sessions. After 3 months, measurements of clinical improvement were found in claustrophobia patients, where significant clinical improvement was found before, during, and after the conclusion of treatment. Agoraphobia and storm phobia even improved though they had not had specific treatment. VR is used for the treatment of 4 claustrophobic patients in eight sessions (Botell et al., 2000). There was a marked reduction in fear and avoidance among patients after having been evaluated at three intervals (before, after, and 3 months later). Table 10.9 contains a compilation of the study results.

10.2.4.3.6 Fear of driving

Two studies were conducted on a single case and 10 cases per team. They looked at the fear of driving, and the results were published (Wald and Taylor, 2000). A significant improvement in their clinical condition was seen after TERV treatment in both of these studies.

Table 10.9 Studies on the treatment of claustrophobia using virtual reality.

Authors	Study type	Number of participants	Technique applied	Results
Botella et al. (1998)	Study Case study	1	VRET	Reduction of symptoms—maintenance of therapeutic gains 1 month after
Botella et al. (1999)	Study Case study	1 of treatment gains 3 months after	VRET	Significant reduction in symptoms of claustrophobia. Reduction of agoraphobia and phobia symptoms storm fears without specific treatment for these phobias. ΔMaintenance
Botell et al. (2000)	Controlled study	4	VRET	Significant reduction in symptoms of fear and avoidance. ΔMaintenance of therapeutic gains 3 months after

10.2.4.3.7 Agoraphobia

Many locations are involved in this phobia that involves many distinct places (plane, subway, cinema, driving, places deserts). Next, many protocols that study agoraphobia include exposure to VR in addition to cognitive therapy (such as cognitive restructuring, psychoeducation, self-instruction, etc.). And with clinical efficacy, the experiments have found that relaxation treatments are effective (Vincelli et al., 2003a). A study comparing classical CBT and TERV with traditional therapy demonstrates that the TERV treatment uses the same amount of time while being more effective. The results were further supported by a later study on a larger sample of 37 patients (Difede and Hoffman, 2002). The therapeutic efficacy of TERV was observed not only in how the environment is cost-effective for a therapist but also in a measurable therapeutic advantage on all scales of 18 agoraphobes (behavioral, physiological, and subjective).

10.2.4.3.8 Anxiety disorders

TERV has been conducted for various anxiety disorders. When treating people with anxiety disorders, it is important to introduce new settings that mimic what they fear the most and expose them to the stressful environment for a long period of time. These VR exposure sessions last an hour and consist of anywhere from 5 to 12 individual sessions, the latter can last 30–60 min (Botella et al., 2007a). The following are different types of tests the subject completes before and after the exposure: questionnaires, the results of which are both subjective and objective. For these types of tests, behavioral testing is used rarely, while physiological testing (including body temperature, skin conductance, etc.) is more frequently used (Malbos et al., 2008). Both subjective and objective tests were used to strengthen the results discovered in the presence of notable correlations.

10.2.4.3.9 Obsessive–compulsive disorder

Little clinical studies exist in this area, as currently, TERV is facing a unique challenge in treating this disorder, and the development of virtual environments is difficult. Despite this, a pilot study found that people could practice being compulsively free in VR, and many people reported having anxiety symptoms following the VR exposure in 33 people (Kim et al., 2009). As TERV sessions progress, psychometric tests reveal decreasing anxiety. This indicates the use of VR for the treatment of obsessive–compulsive disorders.

10.2.4.3.10 Schizophrenia

Schizophrenia exhibits some abnormal perceptions that could lead to mistakes about the boundaries of reality. This illness could therefore seem to discourage therapy involving VR at first glance. A 2014 study on four patients with schizophrenia that used VR technology concluded that patients can tolerate this immersive technology with little or no cognitive detriment (Kim et al., 2009). Additionally, cognitive rehabilitation in VR was then applied to 12 schizophrenics aged 60 or older, to assist them in their recovery of simple mental tasks such as navigating an artificial ocean by using their bodies, catching bullets, etc. (Chan et al., 2010). The cognitive scores and memory abilities improved more than those of the control group ($n = 15$). Finally, a clinical study that focused on cognitive rehabilitation for schizophrenics in regard to social competence employed VR. This study includes 91 schizophrenics, with a group who received VR therapy and a control group who had traditional role-playing therapy (Park et al., 2011). VR appealed to participants more than traditional treatment based on what initially seemed like a lack of incentive.

10.2.4.3.11 Depression

We found only two studies using immersive VR with depression as an explicit focus. A number of smaller studies involving a single treatment technique were performed to evaluate the efficacy of treatment over time, with the depression levels observed to decrease. A nonimmersive VR-type task with a focus on spatial navigation memory assessment was performed in a study on depression (Freeman et al., 2017; Gould et al., 2007).

10.2.4.3.12 Eating disorders

It has been noted that the VR field is plagued by very few well-conducted studies, despite an early application of VR for eating disorders (Riva, 1998). To induce hunger, it is possible to use VR to do it, as demonstrated by experiments showing that people will experience hunger-related reactions when in VR just as they would if the food were actually in front of them (Pallavicini et al., 2016). In a fascinating VR study, Keizer et al. (2016) used VR to help patients with anorexia nervosa experience a healthy BMI body,

which patients with the condition reported was an important first step in overcoming their disease and also an impact on reducing body size overestimation that lasted for at least 2 hours.

10.2.5 Extended reality in dental medicine

The development of AR and VR in dental medicine, and to identify future research needs to accomplish its clinical translation is debated in the current section.

10.2.5.1 Dental education

In the field of dental education, studies have explored various topics, including how to develop patients' 3D vision during oral health procedures, up to complex methods for correcting defects of the facial bones. In 2014, Eve et al. (2014) compared dental undergraduates to prosthodontics residents on a simulated caries removal exercise. Novice and experienced operators achieved significant increases in efficiency: defined as the percentage of the carious lesion removed over drilling time. Al-Saud and Mushtaq (2017) used a haptic VR simulator to investigate the impact of feedback on the rate of motor skill acquisition for tooth preparation. A dental instructor teaching others how to use a haptic device accelerated the learning of basic manual dexterity skills when inexperienced participants were trained in person while receiving feedback on their performance via the device. A study on the subject used haptic VR to test manual dexterity in preclinical dental education is carried out by Urbankova et al. (2013). According to the study, using VR simulators is one way to identify students with learning challenges in dental training in the preclinical stage. Suebnukarn et al. (2014) also performed a similar experiment using a prototype haptic VR dental simulator and assessed its utility for motor skill training. The results proved the VR accuracy simulators by demonstrating that they can distinguish between the performances of experts and nonexperts. The new learning objective explored the acceptance of XR in teaching preparation design, which was found to be successful (Espejo-Trung et al., 2015). de Boer et al. (2015) investigated how the use of virtual learning settings affected the appreciation and success of students. The inferior alveolar nerve block was examined by Correa et al. (2017) in regard to the dental anesthesia training simulator. It was proven to be ideal for needle application because it tested very suitable in the simulation and in all relevant aspects of needle use, including localization, depth of insertion, and virtual tissue resistance. Khelemsky and Hil (2017) investigated the usefulness of a new VR surgical simulator for orbital floor reconstruction in more complex surgical treatment techniques. To evaluate a basic training system that uses VR technology to remove submandibular glands, Miki et al. (2016) ran a study. Training for oral surgeons (novices) was successful thanks to the endoscope-assisted surgery-based VR training system.

10.2.5.2 Maxillofacial surgery

In the oral and maxillofacial surgery field, several types of orthognathic operations (repair of the jaw or the use of distraction osteogenesis) as well as reconstructive surgeries for the mandible and the operation of saliva glands are performed. A report on a case in which a 42-year-old woman's pleomorphic adenoma in the lacrimal gland was successfully removed using a microscope-based AR system is featured in the report of Scolozzi and Bijlenga (2017). In 2016, Yamada et al. (2016) conducted a study of 21 patients who were sent with custom-made titanium mesh trays to fill the gap created from bone and marrow surgery. The 3D-printed skull model was combined with titanium mesh bent around it. VR simulation was completed by using computer software and preoperative radiographic data. In 2014, Qu et al. (2015) employed an AR toolkit for distraction osteogenesis to guide mandibular osteotomy line placement and to aid with distractor positioning in 20 patients with hemifacial microsomia. Using AR technology to produce a new imaging and visualization tool, the research team led by Zinser et al. (2013) published a protocol that integrates orthognathic surgical navigation with a computer-assisted technique for displaying 3D anatomy overlaid with simulated surgical instruments. The goal was to investigate in vivo accuracy and flexibility. In 2014, Fernandez-Alvarez et al. (2014) conducted a study to validate a VR software for anthropometric measurements before preoperative planning and facial graft harvesting began. Using VR, we were able to match the results of the conventional analog method. The 3D reconstructions produced by VR software could help with understanding the donor's face and be beneficial.

10.2.5.3 Dental phobia

One of the most prevalent phobias today is a dental phobia. In 2016, Raghav et al. (2016) tested the effectiveness of a noninvasive virtual reality exposure therapy (VRET) on patients with dental phobia when compared with patients who had only received informational pamphlets. The authors suggested that VRET might be an alternative treatment for dental anxiety and phobia, based on the 6-month follow-up results, which were reported.

10.2.5.4 Anatomy

One must study the 3D anatomy of the cavernous blood sinus to ensure successful skull base surgery for the treatment of lesions in this region. Cadaver dissection, while common, lacks the ability to illustrate the human body's spatial anatomy. In 2018, Qian et al. (2018) used VR to model the cavernous sinus in 3D. VR-based procedures are convenient, noninvasive, time saving, and more accurate than traditional ones.

10.2.6 Extended reality in orthopedics

Since 2000, research on AR has been focusing on using the technology in orthopedics, first on preclinical studies using cadavers, bone phantoms, and models (Bagwe et al., 2021). These are grouped into three areas, the world, body, and head-based position of the display system (Bagwe et al., 2021). The World space encompasses locations where displays exist in a stationary place, such as computers and projector-based screens. HMDs such as HoloLens and AR-augmented microscopes with HUDs such as Pentero 900 make up the Headspace. Smartphones and tablets, among other items, fit into the category of Body Space. Several AR systems have been used in orthopedic surgeries since 2013.

10.2.6.1 World space

i. **Camera-augmented mobile (C-arm) AR system:** A camera-augmented mobile (C-arm), developed by Siemens in 1998, has a camera built into the C-arm device, and that device is present in this image. The image of the patient's nonvisible parts was superimposed over the visible parts, which was easier to see and made the device more effective. von der Heide et al. (2018) tested a novel AR system called CamC on orthopedic surgeries that deal with trauma. The orthopedic surgery community found that CamC's promise was in orthopedic surgery.

ii. **Augmented reality surgical navigation system (ARSN):** ARSN is a novel system for spinal surgery navigation developed by Philips. Using ARSN in the surgical placement of 253 pedicle screws, Elmi-Terander et al. (2018) had high accuracy with acceptable operative time after performing the first series of surgical studies with cadavers. The only issue with the system was that it was hard to use on patients who were obese.

iii. **Augmented reality computer-assisted spine surgery (ARCASS):** Wu et al. (2014) used ARCASS for percutaneous vertebroplasty (PVP) in three patients. It was tested prior to use on humans, on things such as dummy patients, animal models, and 3D models. With the Visible Patient tool, this technology uses 3D images from preoperative CT scans to project images of patients during the intra-operative period via a camera and projector.

10.2.6.2 Body space

i. **Depth camera with an optical marker:** In 2013, Shen et al. (2013) used this system to create a new lightweight AR system for patient-specific plate-based acetabular contouring using optical markers on printed pictures and a video camera, combined with a desktop. A digital plate was created from the fracture pattern of the real plate, which was used to assist in the design of the virtual plate. This implant was utilized in patients using an implantable lightweight AR system. They reported having less invasive surgeries with greater anatomical precision.

ii. **Smartphone camera with QR code:** Ogawa et al. (2018) established an AR-based portable navigation system utilizing the smartphone display for viewing functional pelvic plane and placing acetabular cup during THA.

10.2.6.3 Headspace

i. **Augmented reality with heads up display operating microscope**: Carl et al. (2020) used the Pentero 900 operating microscope with AR for 42 spinal surgeries, integrated with the HUD-enabled operating microscope. To do automatic registration and integration of data, they used a nonlinear registration system that used low-dose intraoperative CT scans. The people interviewed stated that the use of AR in surgery significantly improved anatomical orientation and surgical accuracy.
ii. **Augmented reality using wearable HMD:** Thanks to advances in technology, today exist a whole bunch of new wearable headsets, with firms such as Epson (MOVERIO), Google (Google Glass), Microsoft (HoloLens), and Vuzix in the mix (Smart Glasses M400). The surgical field promises simple and easy use of AR thanks to these devices. Researchers conducted an experiment using virtual protractor with augmented reality (VIPAR), a system composed of an HMD with a tracking camera and a marker sheet, to perform PVP in five patients in 2013 (Abe et al., 2013). First, 40 spine phantom models were used to determine the accuracy and viability of the system. AR was more accurate in positioning the needles.

10.2.7 Current biomedical trends in extended reality

XR helps visualize and analyze 3D data through interactive visualizations by providing a tool for looking at 3D models as opposed to in a 2D representation. Thus, XR technology helps give XR volumetric data a massive boost. For instance, VR and AR have greatly increased the visualization capabilities and ability to interact with microscopic images, molecular data, and anatomical data sets in biomedical engineering. Google's AR Microscope (ARM) diagnoses cancers from microscopic images in real time (Chen et al., 2019). The augmented bright-field microscope, computer, and trained deep learning algorithms were all in the ARM system. The training of the DL model in ARM has been done for the purposes of detection of prostate cancer and lymph node metastasis in breast cancer. Using AR, the output from the DL predictions was mapped as contours, heatmaps, or textural information onto the microscopic sample. The pathologists that utilize this system can conduct faster scans for cancer in huge images. Expansion microscopy was paired with VR to use on small-scale images to render microscopic structures that were previously undetectable (Duffy, 2019). They presented ExMicroVR, a tool that could accommodate up to six scientists working remotely to accomplish more complicated tasks. A key feature of expansion microscopy is that it greatly expands tissue sample volume, which allows for ease of visualization of molecules and interactions between cells. For the 3D VR interface, the project's images of the 2D expansion microscopy were combined with 360 degree view VR.

The cell structure and protein distribution are examined using a VR application called ConfocalVR (Stefani et al., 2018). In addition to cellular images such as confocal microscopy stack images, the software envisioned these cells in red—green—blue (RGB) volumes. Users of the ConfocalVR interface could use controllers to drag and rotate an image to get it into the exact position and size required for a better view of a particular region of interest. One could easily control the appearance of their poster with items such as hue, brightness, and opacity. It also gives options for multiple people to use it at the same time, like many VR programs that have been discussed here. Engineers and doctors use the Microsoft HoloLens, a holographic MR HMD, to look at three-dimensional images, such as anatomical structures, and to interact with them in a way that delivers a much clearer image. HoloLens-based VR for medical procedures gave surgeons and medical personnel the ability to view complex organs during operations (Limonte, 2018). An autopsy is performed while the deceased's brain was navigated via HoloLens, which allowed the pathologist to get a closer look at what they were doing and view WSIs in a more immersive way. The pathologist also used HoloLens for telepathology and telepathology—radiology correlation to improve workflow and enable them to assess patients in a timely manner.

Neuroscientists also use VR to trace neurons in brain images using TeraVR to annotate neurons in teravoxel-scale brain images (Wang et al., 2019). The neurons of mice that were analyzed in VR improved upon existing representations (e.g., 2D or 360 degree views). To maximize the model's ability to adjust to individual preferences, a U-Net-based DL model was trained on reconstructions to adjust its output. The other tool in the VR neuron tracing set discovered and sorted the neuron data's spatial relations. Theart et al. (2017) created a user interface for 3D microscopy data visualization and data colocation within VR. The arivis application VisionVR is open-source software that creates a VR program for viewing three-dimensional microscopic images and offers a set of tools for virtual manipulation and evaluation (Calı et al., 2015).

10.2.7.1 Virtual training for surgeries and biomedical devices

Training a physician to perform operations is difficult for several reasons. For example, doing them well requires skills that must be learned through extreme practice and training. The students are given the chance to practice an easy procedure in VR before moving on to the next level of the program, and that is because it can help medical students perform surgery on a patient safely. Students who have practiced with medical simulators such as RASimsAs, AnatomyX, and SimSurgery are better equipped to handle unexpected medical emergencies. With the increase in medical school students' usage of these simulators, they can train their abilities to think and solve problems on the fly, respond in a high-pressure environment, and perform tasks despite the stress. Because of the isolation headsets that accompany VR programs, some simulators have to employ solutions such as having participants point to various aspects of a virtual world to

stimulate collaboration. In the virtual world, students and educators can work together, allowing them to not say which body part or object they are referring to and instead point to the object in the virtual world so that their classmates can see it.

Advanced simulators, such as RASimAs, created by the University of Aachen, allowed surgeons to be able to practice surgeries with more accurate representations of how the body reacts. These simulators use information from actual surgical cases to help surgeons learn how to correctly plan their procedures. The RASimAs simulator, for example, helped students develop skills for injection accuracy by letting them work with tissue reactions (ThinkMobiles Team, 2016). When the physicians used RASimAs and used an MRI machine to visualize where they were injecting the RASimAs, they would find that when the needle pierced the nerves, they were imitating reactions that the physicians were to look for when trying to insert a needle into a specific nerve in a patient. When mistakes are brought to light immediately after they occur, medical students benefit greatly.

AnatomyX will enable a collaboration-focused AR learning experience for students, that is, hands-on and interactive. Multiple students collaborate to improve surgery by collaborating and working in real time on a shared model (Medivis, 2021). The program is a superb resource with many valuable features, including getting instant access to the latest medical data, which is refreshed with updated information during learning. In addition to virtual learning materials, AnatomyX provides virtual examinations and assessments for educators.

SimSurgery was also available to adjust the exercise difficulty based on the trainee's skills. Also, the use of these tools makes it easier for educators to understand students' experiences since they allow educators to view students' activities firsthand in real time. To train medical students, who need to learn procedures without harm, VR gives students the opportunity to practice difficult procedures in a safe environment.

Researchers found that VR simulation training is inexpensive and easy to replicate. The largest issue in the diagnosis and treatment of cancer and radiology surgery is precision. To adopt Virtual and Augmented Surgical Intelligence is necessary because of this (VASI). Simulators in surgical training have been around longer, such as LapSim and MIST VR, both of which are surgical simulators (Kamarudin and Zary, 2019). VR training will likely become a big part of schools' programs. In the future, VR and MR technology will probably have a big impact on anatomical education. VR is a great alternative to manuals when it comes to training people on the correct way to do things in the biopharma industry and will help to ensure everyone's success.

10.2.7.2 Telemedicine and telehealth screening

In spite of being a completely different domain from XR, telemedicine and telehealth are gaining ground fast. Telehealth via XR offers patients the opportunity to interact with their doctors in a virtual environment where they can experience their remote

consultations in a three-dimensional and immersive fashion. The XRHealth platform provides neurocognitive, physical, and emotional support through VR and AR, offering training for recovering from injury and illness as well as stress and pain management. The VR platform uses virtual environments, games, and movement-tracking exercises to provide clinicians with an environment in which they can provide feedback. XRHealth has begun to treat patients through its new telehealth clinic, which concentrates on rehabilitation (XRHealth VR Telehealth, 2021). The beauty of this form of therapy is that it allows people to exercise at home in peace. People with paralysis, especially those who have trouble getting around, will benefit greatly from this. VR platform helps keep a person engaged as they immerse themselves in environments that resemble what they know so that they can continue recovery in a realistic setting. VR treatments can even be modified based on what the patients see, with the physicians having the ability to see these same views. Participants receive posttraining information on their level of rehabilitation and their progress, so it is easier to keep track of their progress. The XRHealth platform, which includes AI, targets the individual, helping to supplement traditional therapeutic methods such as prescription drugs, tailoring to patient needs.

10.2.7.3 Anesthesia

A system known as AREA, designed to assist with epidural anesthesia, is described (Ashab et al., 2012). The findings indicate that using Micron Tracker with a simple setup can yield a satisfactory level of accuracy for back level and line recognition, all with little training required.

10.3 Challenges, future directions, and potentials

Different types of VR, including immersive VR, computer-assisted reality, AR, and MR, can all be used for educational purposes. However, each type's benefits and shortcomings must be taken into account, with special attention paid to how well VR environments emulate the real world. While AR and MR improve vision, VR and MeR completely obstruct it. When the screen is powered down, it is clear that VR and MeR are entirely opaque, whereas AR and MR are only semitransparent, allowing the user to see underneath digital additions. Students could use VR to practice in a totally immersive simulation, free from any outside distraction. Interventionalists, though, can take advantage of AR, MeR, and MR to stay present in a physical room, where they can conduct procedures and stay engaged with the patient while also working with their team. MeR platforms are a threat to safety if the procedure loses power and then the entire view gets blocked. To make the most visually appealing, mobile, interactive, and interactive VR applications, you need to keep the power, processing speed, and weight of your hardware low while avoiding using lots of money on big, heavy equipment.

Resolution, brightness, focal depth, and FOV impact the appearance of photos. One of the most challenging aspects of XR is the display technology, which is commonly the most expensive and size-constrained part of the system (Kress and Shin, 2013). On top of this, 3D systems need two displays to create the illusion of depth using vergence, which means that every eye sees a different picture. To maximize visual quality, you have to the human visual system (HVS). A normal human visual acuity is about 6/6 or 20/20, with a resolution of 1 arcmin/pixel and an elliptical FOV that is 1500—1700 by 1350—1500. A system with the angular resolution in pixels described before will have pixels known as "retina display" pixels, as defined by Apple. It is equivalent to a total pixel/eye size of roughly 9000 by 8100 pixels to fully immerse the human visual system. One reason for this is that 4K HMDs (high-definition displays) contain 3840*2160 pixels and, often, require workstation-class graphics for their processing. Clearly, today's optical and display technologies do not make this resolution and FOV affordable. To reach their required performance, device manufacturers have to reduce FOV, pixel density, and display brightness. New AR capabilities such as handheld miniaturization and smaller power requirements have just been released, as with the Apple iPhone X (Caelli, 2014), which demonstrates lower power and weight, as well as smaller dimensions. Stereoscopic displays also have trouble showing depth at close distances because of the accommodation of our eyes, which adjusts for distance so that everything is in focus. We need an accommodation to help people focus on surgical tools and digital objects (like surgical guides) at different simulated distances (for example, they can focus on objects inside their "personal space" and "action space"). It is possible for HVS accommodation and HMD vergence to be in conflict (commonly known as VAC vergence and accommodation conflict) and result in discomfort while working at near distances. Most digital display systems only offer a single, static focal plane to all content, though recent technologies such as adaptive optics are capable of creating multiple fixed focal planes.

Challenges remain with AR and VR. The headache, dizziness, and discomfort that come from surgeons using headsets as preferred hardware cannot be overlooked. Learning the basics of AR and VR is challenging. For surgeons, the fact that VR systems currently cannot simulate physical touch is hugely significant. Skin marker tracking systems have a significant drawback: there can be a variation in the position of the tissue relative to the bones. However, these obstacles can be overcome through new technology and increased collaboration between medical experts and engineers. Despite some notable constraints, AR and VR possess unique advantages, low costs, and numerous additional advantages for their potential use with other technologies, thus ensuring that their future in the field of spine surgery will be promising. A fact to note is that AR and VR can serve to provide information to the clinician, students, trainees, patients, and surgical robots by linking all of them together to better facilitate surgical operations.

AR and VR, being linked to so many healthcare technologies, create new capabilities and avenues for the exploration and application of AR. Gaming is a fantastic AR and VR technology that will be used in healthcare to help with pain management, learning, and teaching for patients and students. The merger of AI, wearable sensors, gaming, AR, and VR can lead to the creation of physical rehabilitation environments that will enable the physician to remotely interact with the patient in a virtual space and monitor them in real time. Combining robotic guidance, AI, and AR technologies results in useful, fast, and precise navigation systems. Surgeons can get live feedback during surgeries if they utilize AI, wearables, and AR technologies. Merging VR, AR, and MR with AI and surgical robotics can help advance the implementation of telemedicine and also provide the ability to semiautomate and remotely perform surgeries. Surgeons could use AR glasses for many medical procedures in the near future, and soon doctors will be able to see 3D images during consultations, rehabilitation, training, and surgery. Doctors in training will be exposed to advanced VR and AR simulations. Spine surgery, being a major discipline in orthopedics, should be prepared to accept and implement these new technologies. Clinical teams, industrial designers, and gaming professionals are collaborating on spine surgery and are planning to use everything they are learning from these partnerships to change how all aspects of the surgery are conducted within the next decade. Additionally, the use of these applications in minimally invasive procedures and open surgical procedures will significantly grow. Some hardware solutions are also being developed for extended realities. Many of these solutions have been designed by small businesses, but the advances in the field will come from big businesses, such as Microsoft, constantly improving their products. The newest ideas about HoloLens 2, such as the latest plans, are simply the beginning of a lengthy road. In addition, AR/MR displays that pass beyond HMDs may be developed. In the future, the hardware may open up some new possibilities.

10.4 Conclusion

In the coming years, VR and AR will see an uptick in use in radiologic imaging, surgery, and a host of other therapeutic applications. Previous research has mostly focused on patient-as-user applications in medicine, with VR/AR playing the role of the intervention. New developments have made it possible to expand clinician—user applications, whose acceptance was held back in the past by limitations in the technology's ability to provide a satisfying experience.

The research for the application and success of the extended realities in medical sector is accumulating, and its work encompasses education and training as well as patient rehabilitation. It also runs the gamut from planning ahead of a procedure to using VR in the middle of a procedure on systems with modest resolutions and minimal hardware specifications. New hardware advances have made 3D imaging better, which increases

patient therapy, training comprehension, and patient procedure accuracy. Using 3D data, we will see a quick increase in the ability to quickly complete complicated tasks thanks to increasing improvements in software and hardware. The various medical imaging tools will develop further to offer improved visualization and more options for handling specific problems. Patients can be given control over their own data for better health outcomes, which can be accomplished through various applications. This includes, for example, allowing patients to visualize three-dimensional objects, understand how to use data visualizations, or learn how to use technology they previously could not use. Making these promises a reality is the next big hurdle in present data-driven 21st-century healthcare systems.

References

Abe, Y., Sato, S., Kato, K., 2013. A novel 3D guidance system using augmented reality for percutaneous vertebroplasty. Journal of Neurosurgery: Spine 19 (4), 492–501.

5"Accenture develops artificial intelligence-powered solution to help improve how visually impaired people live and work, July 28, 2017. Accenture Press Release (Last accessed 13 May 2021).

Al-Saud, L.M., Mushtaq, F., Allsop, M.J., 2017. Feedback and motor skill acquisition using a haptic dental simulator. European Journal of Dental Education 21 (4), 240–247.

Amidei, C., Kushner, D.S., 2015. Clinical implications of motor deficits related to brain tumors. Neuro-Oncology Practice 2, 179–184.

Anil, S., Kato, Y., Hayakawa, M., Yoshida, K., Nagahisha, S., Kanno, T., 2007. Virtual 3-dimensional preoperative planning with the dextroscope for excision of a 4th ventricular ependymoma. Minimally Invasive Neurosurgery 50, 65–70.

Argotti, Y., Davis, L., Outters, V., Rolland, J.P., 2002. Dynamic superimposition of synthetic objects on rigid and simple-deformable objects. Computers & Graphics 26, 919–930.

Ashab, H.A.D., Lessoway, A.V., Khallaghi, S., Cheng, A., Rohling, R., Abolmaesumi, P., 2012. AREA: an augmented reality system for epidural anaesthesia. In: Proceedings of 34th Annual International Conference of the IEEE EMBS San Diego, California USA, pp: 2659-2663, 28 August - 1 September 2012.

Bagwe, S., Singh, K., Kashyap, A., Arora, S., Maini, L., 2021. Evolution of augmented reality applications in orthopaedics: a systematic review. Journal of Arthroscopy and Joint Surgery 8, 84–90.

Beck, J.G., Palyo, A.S., Winer, H.E., Schwagler, E.B., Ang, E.J., 2007. Virtual reality exposure therapy for PTSD symptoms after a road accident: an uncontrolled case series. Behavior Therapy 38 (1), 39–48.

Berg, L.P., Vance, J.M., 2016. Industry use of virtual reality in product design and manufacturing: a survey. Virtual Reality 21 (1), 1–17.

Bigné, E., Llinares, C., Torrecilla, C., 2016. Elapsed time on first buying triggers brand choices within a category: a virtual reality-based study. Journal of Business Research 69 (4), 1423–1427.

Bonetti, F., Warnaby, G., Quinn, L., 2018. Augmented reality and virtual reality in physical and online retailing: a review, synthesis and research agenda. In: Jung, T., Tom Dieck, M. (Eds.), Augmented Reality and Virtual Reality, Progress in IS, Cham: Springer, pp. 119–132.

Botell, C., Baños, R.M., Villa, H., Perpiña, C., Garcia-Palacios, A., 2000. Virtual reality in the treatment of claustrophobic fear: a controlled, multiple-baseline design. Behavior Therapy 31 (3), 583–595.

Botella, C., Baños, R.M., Perpiña, C., Villa, H., Alnañiz, M., Rey, A., 1998. Virtual reality treatment of claustrophobia: a case report. Behaviour Research and Therapy 36 (2), 239–246.

Botella, C., Villa, H., Baños, R.M., Perpiña, C., Garcia-Palacios, A., 1999. The treatment of claustrophobia in virtual reality: changes in other phobic behaviors not specifically treated. CyberPsychology and Behavior 2 (2), 135–141.

Botella, C., Garcia-Palacios, A., Villa, H., 2007a. Virtual reality exposure in the treatment of panic disorder and agoraphobia: a controlled study. Clinical Psychology & Psychotherapy 14, 164–175.

Botella, C., García-Palacios, A., Villa, H., Baños, R.M., Quer, S., Alcañiz, M., Riva, G., 2007b. Virtual reality exposure in the treatment of panic disorder and agoraphobia: a controlled study. Clinical Psychology & Psychotherapy 14 (3), 164–175.

Butler, J.F., Holcomb, J.B., Shackelford, S., Montgomery, H.R., Anderson, S., Cain, J.S., 2018. Management of suspected tension pneumothorax in tactical combat casualty care: TCCC guidelines chang. Journal of Special Operations Medicine 18 (2), 19–35.

Caelli, T., 2014. Visual Perception: Theory and Practice: Pergamon International Library Of Science, Technology, Engineering, and Social Studies, pp. 103–114.

Calı, C., Baghabra, J., Boges, D.J., Holst, G.R., Kreshuk, A., Hamprecht, F.A., Srinivasan, M., Lehva slaiho, H., Magistretti, P.J., 2015. Three-dimensional immersive virtual reality for studying cellular compartments in 3D models from EM preparations of neural tissues. Journal of Comparative Neurology 524, 23–38.

Canalys Media Alert, 2017. Virtual Reality Headset Shipments Top 1 Million for the First Time. Retrieved from. https://goo.gl/PvBHjx (Accessed 14 June 2021).

Carl, B., Bopp, M., Sab, B., Pojskic, M., Voellger, B., Nimsky, C., 2020. Spine surgery supported by augmented reality. Global Spine Journal 10 (2), 41–55.

Case Western Reserve, Cleveland Clinic Collaborate with Microsoft on 'Earth-Shattering' Mixed-Reality Technology for Education. Available at: http://case.edu/hololens/. (Last Accessed August 11, 2021).

Chan, C.L.F., Ngai, E.K.Y., Leung, P.K.H., Wong, S., 2010. Effect of the adapted virtual reality cognitive training program among Chinese older adults with chronic schizophrenia: a pilot study. International Journal of Gerjiatric Psychiatry 25 (6), 643–649.

Chan, F., Aguirre, S., Heaton, H.B., Hanley, F., Perry, S., 2013. Head tracked stereoscopic pre-surgical evaluation of major aortopulmonary collateral arteries in the newborns. In: Proceedings of Radiological Society of North America 2013 Scientific Assembly and Annual Meeting; Chicago, Illinois.

Chen, J.G., Han, K.W., Zhang, D.F., Li, Z.X., Li, Y.M., Hou, L.J., 2017. Presurgical planning for supratentorial lesions with free Slicer software and Sina app. World Neurosugery 106, 193–197.

Chen, P.C., Gadepalli, K., MacDonald, R., Liu, Y., Kadowaki, S., Nagpal, K., Kohlberger, T., Dean, J., Corrado, G.S., Hipp, J.D., 2019. An augmented reality microscope with real-time artificial intelligence integration for cancer diagnosis. Nature Medicine 25, 1453–1457.

Chevalley, O.H., Schmidlin, T., Marcos, P.D., 2015. Intensive upper limb neurorehabilitation with virtual reality in chronic stroke: a case report. In: Proceedings of Annual Meeting of American Society of Neurorehabilitation; Chicago, Illinois; October 15–16.

Choi, Y.H., Vincelli, F., Riva, G., Wiederhold, B.K., Lee, J.H., Park, K.H., 2005. Effects of group experiential cognitive therapy for the treatment of panic disorder with agoraphobia. CyberPsychology and Behavior 8 (4), 387–393.

Chrastina, J., Novak, Z., Riha, I., Hermanova, M., Feitova, V., 2012. Diagnostic value of brain tumor neuroendoscopic biopsy and correlation with open tumor resection. Journal of Neurological Surgery Part A: Central European Neurosurgery 75, 110–115.

VR Could Be Your Next Painkiller, March 13, 2018. CNET (Last Accessed 02 June 2021).

Correa, C.G., Machado, M.A.A.M., Ranzini, E., 2017. Virtual reality simulator for dental anesthesia training in the inferior alveolar nerve block. Journal of Applied Oral Science 25 (4), 357–366.

Dayan, E., 2006. ARGAMAN: rapid deployment virtual reality system for PTSD rehabilitation. In: Proceedings of International Conference on Information Technology: Research and Education, pp. 34–38.

de Boer, I.R., Lagerweij, M.D., Wesselink, P.R., 2015. Evaluation of the appreciation of virtual teeth with and without pathology. European Journal of Dental Education 19 (2), 87–94.

Difede, J., Hoffman, G.H., 2002. Virtual reality exposure therapy for world trade center post-traumatic stress disorder: a case report. CyberPsychology and Behavior 5 (6), 529–535.

Duffy, J., 2019. Microscopy and VR Illuminate New Ways to Prevent and Treat Disease. https://www.cmu.edu/news/stories/archives/2019/july/vr-expands-microscopy.html. (Last Accessed 30 June 2021).

Elmi-Terander, A., Nachabe, R., Skulason, H., 2018. Feasibility and accuracy of thoracolumbar minimally invasive pedicle screw placement with augmented reality navigation technology. Spine 43 (14), 1018e1023.

Embodied Labs website. http://www.embodiedlabs.com/labs. (Last accessed 22 May 2021).

Emmelkam, P., Bruynzeel, M., Drost, L., 2001. Virtual reality treatment in acrophobia: a comparison with exposure in vivo. CyberPsychology and Behavior 4, 183–202.

Espejo-Trung, L.C., Elian, S.N., Luz, M.A., 2015. Development and application of a new learning object for teaching operative dentistry using augmented reality. Journal of Dental Education 79 (11), 1356–1.

Eve, E.J., Koo, S., Alshihri, A.A., 2014. Performance of dental students versus prosthodontics residents on a 3D immersive haptic simulator. Journal of Dental Education 78 (4), 630–637.

Fernandez-Alvarez, J.A., Infante-Cossio, P., Barrera-Pulido, F., 2014. Virtual reality AYRA software for preoperative planning in facial allotransplantation. Journal of Craniofacial Surgery 25 (5), 1805–1809.

Ferroli, P., Tringali, G., Acerbi, F., 2013. Advanced 3- dimensional planning in neurosurgery. Neurosurgery 72, A54–A62.

Finger, T., Schaumann, A., Schulz, M., Thomale, U.W., 2017. Augmented reality in intraventricular neuroendoscopy. Acta Neurochirurgica 159, 1033–1041.

Florida hospital Tampa integrates virtual reality into surgical planning and patient education. Florida Hospital website, October 25, 2016 (Last accessed 09 April 2021).

Forbes Inc, 2017. (Last accessed 15 May 2021).

Freeman, D., Reeve, S., Robinson, A., Ehlers, A., Clark, D., Spanlang, B., Slater, M., 2017. Virtual reality in the assessment, understanding, and treatment of mental health disorders. Psychological Medicine 47 (14), 2393–2400.

Gerard, I.J., Oertel, K.M., Drouin, S., 2018. Combining intraoperative ultrasound brain shift correction and augmented reality visualizations: a pilot study of eight cases. Journal of Medical Imaging 5 (021210), 1–13.

Goldman Sachs Global Investment Research, 2016. (Last acceded 15 March 2021).

Gong, L., Berglund, Å., Johansson, B., 2021. A framework for extended reality system development in manufacturing. IEEE Access 9, 24796–24813.

Gorini, A., Riva, G., 2008. Virtual reality in anxiety disorders: the past and the future. Expert Review of Neurotherapeutics 8 (2), 215–233.

Gould, N.F., Holmes, M.K., Fantie, B.D., Luckenbaugh, D.A., Pine, D.S., Gould, T.D., Burgess, N., Manji, H.K., Zarate, C.A., 2007. Performance on a virtual reality spatial memory navigation task in depressed patients. American Journal of Psychiatry 164, 516–519.

Grand View Research Inc, 2017. (Last acceded 15 May 2021).

Griffin, T., Giberson, J., Lee, S.H., Guttentag, D., Kandaurova, M., Sergueeva, K., Dimanche, F., 2017. Virtual reality and implications for destination marketing. In: Proceedings of the 48th Annual Travel and Tourism Research Association (TTRA), Quebec City, Canada, pp. 547–560.

Grimson, W.E.L., Ettinger, G.J., White, S.J., Gleason, P.L., Lozano-Pérez, T., Wells, W.M., Kikinis, R., 1995. Evaluating and validating an automated registration system for enhanced reality visualization in surgery. In: Proceedings of Computer Vision, Virtual Reality, and Robotics in Medicine, Nice, France.

Harris, S.,R., Kemmerling, R.L., North, M.M., 2002. Brief virtual reality therapy for public speaking anxiety. CyberPsychology and Behavior 5, 543–550.

Hospital SCsHLPCs, 2018. Lucile Packard Children's Hospital Stanford pioneers use of VR for patient care, education and experience. Available from: http://https:%20//www.stanfordchildrens.org/en/about/news/releases/2017/virtual-reality-%20program (Last Accessed 09 April 2021).

Hu, X., Nanjappan, V., Georgiev, G.V., 2021. Bursting through the blocks in the human mind: enhancing creativity with extended reality technologies. Interactions 57–61. https://doi.org/10.1145/3460114.

Extended reality-Brief Determining needs, expectations and the future of XR for the ACRC, 2018. A study commissioned by the ICRC Innovation Board, pp. 1–7 (Last acceded 10 March 2021.

Inoue, D., Cho, B., Mori, M., 2013. Preliminary study on the clinical application of augmented reality neuronavigation. Journal of Neurological Surgery Part A: Central European Neurosurgery 74, 71–76.

Inoue, A., Ohnishi, T., Kohno, S., 2015. Utility of three-dimensional computed tomography for anatomical assistance in endoscopic endonasal transsphenoidal surgery. Neurosurgical Review 38, 559–565.

Jang, D.P., Ku, J.H., Shin, M.B., Young, H.C., Kim, I.S., 2000. Objective validation of the effectiveness of virtual reality psychotherapy. CyberPsychology and Behavior 3, 369–374.

Kamarudin, M.F.B., Zary, N., April 2019. Augmented Reality, Virtual Reality and Mixed Reality in Medical Education: A Comparative Web of Science Scoping Review. Preprints, pp. 1–10. www.preprints.org.

Keizer, A., van Elburg, A., Helms, R., Dijkerman, H.C., 2016. A virtual reality full body illusion improves body image disturbance in anorexia nervosa. PLoS One 11, e0163921.

Khelemsky, R., Hil, B., Buchbinder, 2017. Validation of a novel cognitive simulator for orbital floor reconstruction. Journal of Oral and Maxillofacial Surgery 75 (4), 775–785.

Kim, K., Kim, C.H., Kim, S.Y., Roh, D., Kim, S.I., 2009. Virtual reality for obsessive-compulsive disorders: past and future. Psychiatry Investigation 6, 115–121.

Klinger, E., Légeron, P., Roy, S., Chemin, I., Lauer, F., Nugues P, P., 2004a. Virtual reality exposure in the treatment of social phobia. Studies in Health Technology and Informatics 99, 91–119.

Klinger, E., Légeron, P., Ro, S., Chemin, I., Lauer, F., Nugues, P., 2004b. Virtual reality exposure in the treatment of social phobia. In: Riva, G., Botella, C., Légeron, P., Optale, G. (Eds.), Cybertherapy: Internet and Virtual Reality as Assessment and Rehabilitation Tools for Clinical Psychology and Neuroscience. IOS Press, Amsterdam, pp. 91–119.

Kress, B., Shin, M., 2013. Diffractive and holographic optics as optical combiners in head-mounted displays. In: Proceedings of the 2013 ACM Conference on Pervasive and Ubiquitous Computing Adjunct Publication, pp. 1479–1482.

Krijn, M., Emmelkamp, P.M.G., Lafsson, R, O., Bouwman, M., van Gerwen, L.J., Spinhoven, P., 2007. Fear of flying treatment methods: exposure vs cognitive behavioral therapy. Aviation Space & Environmental Medicine 78 (2), 121–128.

Limonte, K., 2018. AI in Healthcare: HoloLens in Surgery. https://cloudblogs.microsoft.com/industry-blog/en-gb/health/2018/12/20/ai-health care-hololens-surgery/ (Last Accessed 22 August 2021).

Lin, J.H.T., Wu, D.Y., Tao, C.C., 2017. So scary, yet so fun: the role of self-efficacy in the enjoyment of a virtual reality horror game. New Media & Society 20 (4), 1–20.

Lorensen, W., Cline, H., Nafis, C., Kikinis, R., Altobelli, D., Gleason, L., 1993. Enhancing reality in the operating room. In: Proceedings of IEEE Visualization Conference, San Jose, CA, USA.

Low, D., Lee, C.K., Dip, L.L.T., Ng, W.H., Ang, B.T., Ng, I., 2010. Augmented reality neurosurgical planning and navigation for surgical excision of parasagittal, falcine and convexity meningiomas. British Journal of Neurosurgery 24, 69–74.

Malbos, E., Mestre, D.R., Note, I.D., Gellato, C., 2008. Virtual reality and claustrophobia: multiple components therapy involving game editor virtual environments exposure. CyberPsychology and Behavior 11 (6), 695–707.

McLay, R.N., Wood, D.P., Webb-Murphy, J.A., Spira, J.L., Wiederhold, D.M., Pyne, J.M., 2011. A randomized, controlled trial of virtual reality-graded exposure therapy for post-traumatic stress disorder in active duty service member with combat-related posttraumatic stress disorder. Cyberpsychology, Behavior, and Social Networking 14 (4), 223–229.

Medivis, 2021. The Augmented Reality Anatomy Lab and Learning Platform. https://www.medivis.com/anatomyx (Last Accessed 28 June 2021).

Merchant, Z., Goetz, E.T., Cifuentes, L., Keeney-Kennicutt, W., Davis, T.J., 2014. Effectiveness of virtual reality-based instruction on students' learning outcomes in K12 and higher education: a meta-analysis. Computers & Education 70, 29–44.

Texas Surgeons Perform First Sinus Surgery Using AR, March 14, 2018. MobiHealthNews (Last accessed 09 April 2021).

Mikhail, M., Mithani, K., Ibrahim, G.M., 2019. Presurgical and intraoperative augmented reality in neuro-oncologic surgery: clinical experiences and limitations. World Neurosurgery 128, 268–276.

Miki, T., Iwai, T., Kotani, K., 2016. Development of a virtual reality training system for endoscope-assisted submandibular gland removal. Journal of Cranio-Maxillo-Facial Surgery 44 (11), 1800–1805.

Milgram, P., Kishino, F., 1994. A taxonomy of mixed reality visual displays. IEICE Transactions on Information and Systems 77 (12), 1321–1329.

Mühlberger, A., Wiederman, G., Pauli, P., 2003. Efficacy of one-session virtual reality exposure treatment of fear of flying. Psychother Research 13 (3), 323–336.

Next Galaxy to Develop Virtual Reality Applications for Miami Children's Hospital, March 2015 (Last accessed 11 July 2021).

Nimsky, C., Ganslandt, O., Hastreiter, P., Fahlbusch, R., 2001. Intraoperative compensation for brain shift. Surgical Neurology 56, 357–364.

North, N.M., North, S.M., Coble, J.R., 1996. Effectiveness of virtual reality environment desensitization in the treatment of agoraphobia. Presence: Teleoperators and Virtual Environments 5, 346–352.

North, M.M., North, S.M., Coble, J.R., 1998. Virtual reality therapy: an effective treatment for the fear of public speaking. International Journal of Virtual Technology 3, 2–8.

Oculus. Immersive Education. CHLA and Oculus Expand VR Medical Training Program to New Institutions, 2018 [cited 2019 Jan 29]. Available from: https://www.oculus.com/blog/immersive-education-chla-andoculus-expand-vr-medical-training-program-to-new-institutions/ (Last Accessed 09 July 2021).

Ogawa, H., Hasegawa, S., Tsukada, S., Matsubara, M., 2018. A pilot study of augmented reality technology applied to the acetabular cup placement during total hip arthroplasty. The Journal of Arthroplasty 33 (6), 1833e1837.

Orringer, D.A., Golby, A., Jolesz, F., 2012. Neuronavigation in the surgical management of brain tumors: current and future trends. Expert Review of Medical Devices 9, 491–500.

Pallavicini, F., Serino, S., Cipresso, P., Pedroli, E., Chicchi Giglioli, I.A., Chirico, A., Manzoni, G.M., Castelnuovo, G., Molinari, E., Riva, G., 2016. Testing augmented reality for cue exposure in obese patients: an exploratory study. Cyberpsychology. Behavior, and Social Networking 19, 107–114.

Park, K.M., Ku, J., Choi, S.H., Jang, H.J., Park, J.Y., 2011. A virtual reality application in role-plays of social skills training for schizophrenia: a randomized, controlled trial. Psychiatry Res 189, 166–172.

Pertaub, D.P., Slater, M., Barker, C., 1999. Public speaking in virtual reality: facing an audience of avatars. IEEE Computer Graphics and Applications 19 (2), 6–9.

Qian, Z.H., Feng, X., Li, Y., 2018. Virtual reality model of the three-dimensional anatomy of the cavernous sinus based on a cadaveric image and dissection. Journal of Craniofacial Surgery 29 (1), 163–166.

Qiu, T.M., Zhang, Y., Wu, J.S., 2010. Virtual reality presurgical planning for cerebral gliomas adjacent to motor pathways in an integrated 3-D stereoscopic visualization of structural MRI and DTI tractography. Acta Neurochirurgica 152, 1847–1857.

Qu, M., Hou, Y., Xu, Y., 2015. Precise positioning of an intraoral distractor using augmented reality in patients with hemifacial microsomia. Journal of Cranio-Maxillo-Facial Surgery 43 (1), 106–112.

Raghav, K., van Wijk, A.J., Abdullah, F., 2016. Efficacy of virtual reality exposure therapy for the treatment of dental phobia: a randomized control trial. BMC Oral Health 25, 16.

Riva, G., 1998. Virtual environment for body image modification: virtual reality system for the treatment of body image disturbances. Computers in Human Behavior 14, 477–490.

Roberts, R., 1989. Passenger fear of flying: behavioral treatment with extensive in-vivo exposure and group support. Aviation Space & Environmental Medicine 60, 342–348.

Rosahl, S., Gharabaghi, A., Hubbe, U., Shahidi, R., Samii, M., 2006. Virtual reality augmentation in skull base surgery. Skull Base 16, 59–66.

Rotariu, DiL., Ziyad, F., Budu, A., Poeata, I., 2017. The role of OsiriX based virtual endoscopy in planning endoscopic transsphenoidal surgery for pituitary adenoma. Turkish Neurosurgery 27, 339–345.

Rothbaum, B.O., Hodge, L., Kooper, R., 1995. Effectiveness of computer-generated (virtual reality) graded exposure in the treatment of acrophobia. American Journal of Psychiatry 152 (4), 626–628.

Rothbaum, B.O., Hodges, L., Alarcon, K., Ready, D., Shahar1, F., Graap, K., 1999. Virtual reality exposure therapy for PTSD Vietnam veterans: a case study. Journal of Traumatic Stress 12 pp:263–227.

Rothbaum, B., Hodges, L., Smith, S., 2000. A controlled study of virtual reality exposure therapy for fear of flying. Journal of Consulting and Clinical Psychology 68 (6), 1020–1026.

Rothbaum, B.O., Hodge, L., Smith, S., Lee, J.H., Price, L., 2000. A controlled study of virtual reality exposure therapy for the fear of flying. Journal of Consulting and Clinical Psychology 68 (6), 1020–1026.

Rothbaum, B.O., Hodges, L., Anderson, P.L., Price, L., Smith, S., 2002. Twelve-month follow up of virtual reality and standard exposure therapies for the fear of flying. Journal of Consulting and Clinical Psychology 70 (2), 428–432.

Rothbaum, B.O., Anderso, P.L., Zimand, E., Hodges, L., Lang, D., Wilson, J., 2006. Virtual reality exposure therapy and standard (in vivo) exposure therapy in the treatment of fear of flying. Behavior Therapy 37 (1), 80–90.

Roy, S., Klinger, E., Légeron, P., Lauer, F., Chemin, I., Nugues, P., 2003. Definition of a VR-based protocol to treat social phobia. CyberPsychology and Behavior 6, 411–420.

Safavi, K., Kalis, B., 2018. Extended reality trend 2 the end of distance. Digital Health Technology Vision 1–7.

Sarah, M., Zweifacha, M., Triolab, M., 2019. Extended reality in medical education: driving adoption through provider-centered design. Digit Biomark 3, 14–21.

Schwam, G.Z., Kaul, V.F., Bu, D.D., Iloreta, A.M.C., Bederson, J.B., E. P., Cosetti, M.K., Wanna, G.B., 2021. The utility of augmented reality in lateral skull base surgery: a preliminary report. American Journal of Otolaryngology-Head and Neck Medicine and Surgery 42 (102942), 34–53.

Scolozzi, P., Bijlenga, P., 2017. Removal of recurrent intraorbital tumour using a system of augmented reality. British Journal of Oral and Maxillofacial Surgery 55 (9), 962–964.

Shen, F., Chen, B., Guo, Q., Qi, Y., Shen, Y., 2013. Augmented reality patient-specific reconstruction plate design for pelvic and acetabular fracture surgery. International Journal of Computer Assisted Radiology and Surgery 8 (2), 169e179.

Silva, J.N., Southworth, M.K., Dalal, A., Van Hare, G.F., Silva, J.R., 2017. Improving visualization and interaction during transcatheter ablation using an augmented reality system: first-in-human experience. Circulation 136, A15358.

Silva, N.A.J., Southworth, M., Raptis, C., Silva, J., 2018. Emerging applications of virtual reality in cardiovascular medicine. Journal of the American College of Cardiology: Basic to Translational Science 3 (3), 420–430.

Southworth, K.M., Silva, J.R., Silva, J.N.A., 2020. Use of extended realities in cardiology. Trends in Cardiovascular Medicine 30, 143–148.

Stadie, A.T., Kockro, R.A., 2013. Mono-stereo-auto stereo: the evolution of 3-dimensional neurosurgical planning. Neurosurgery 72, A63–A77.

Stadie, A.T., Kockro, R.A., Reisch, R., 2008. Virtual reality system for planning minimally invasive neurosurgery. Journal of Neurosurgery 108, 382–394.

Stadie, A.T., Kockro, R.A., Serra, L., 2011. Neurosurgical craniotomy localization using a virtual reality planning system versus intraoperative image-guided navigation. International Journal of Computer Assisted Radiology and Surgery 6, 565–572.

Stanford Children's Health, Stanford LPCsH, Lucile Packard Children's Hospital Stanford pioneers Use of VR for Patient Care, Education, and Experience. Available at: http://www.stanfordchildrens.org/en/about/news/releases/2017/virtual-reality-program. (Last Accessed December 11, 2020).

Stanford Medicine, Neurosurgery. Stanford Neurosurgical Simulation and Virtual Reality Center, 2019 [cited 2019 Jan 29]. Available from: http://med.stanford.edu/neurosurgery/divisions/vr-lab.html. (Accessed 15 August 2021).

State, A., Chen, D.T., Tector, C., Brandt, A., Chen, H., Ohbuchi, R., Bajura, M., Fuchs, H., 1994. Case study: observing a volume rendered fetus within a pregnant patient. In: In Proceedings of the 1994 IEEE Visualization Conference, Washington, DC.

Stefani, C., Lacy-Hulbert, A., Skillman, T., 2018. ConfocalVR: immersive visualization for confocal microscopy. Journal of Molecular Biology 430, pp: 4028–403.

Suebnukarn, S., Chaisombat, M., Kongpunwijit, T., 2014. Construct validity and expert benchmarking of the haptic virtual reality dental simulator. Journal of Dental Education 78 (10), 1442–1450.

Sun, G.C., Wang, F., Chen, X.L., 2016. Impact of virtual and augmented reality-based on intraoperative magnetic resonance imaging and functional neuronavigation in glioma surgery involving eloquent areas. World Neurosurgery 96, 375–382.

Tabrizi, L.B., Mahvash, M., 2015. Augmented reality guided neurosurgery: accuracy and intraoperative application of an image projection technique. Journal of Neurosurgery 123, 206–211.

Talbot, H., Spadoni, F., Duriez, C., Sermesant, M., O'Neill, M., Jais, P., 2017. Interactive training system for interventional electrocardiology procedures. Medical Image Analysis 35, 225–237.

Tarnanas, I., Wasserstrom, J., Giotakos, O., 2009. Using virtual reality emotional human agents as a relative - scored personality measure. Journal of CyberTherapy & Rehabilitation 2 (1), 155–158.

The Body VR website. http://thebodyvr.com/anatomy-viewer/. (Last Accessed 11 April 2021).

The University of Illinois Helps Develop Revolutionary Virtual Reality for Learning. http://www.govtech.com/education/higher-ed/University-of-Illinois-Helps-Develop-Revolutionary-Virtual-Reality-for-Learning.html. [Last Accessed January 29, 2021].

Theart, R.P., Loos, B., Niesler, T.R., 2017. Virtual reality assisted microscopy data visualization and colocalization analysis. BMC Bioinformatics 18 (2), 64–85.

ThinkMobiles Team, 2016. VR Apps in Medicine Transforming Healthcare We Once Knew. https://thinkmobiles.com/blog/virtual-reality-applicationsmedicine/ [Last Accessed 22 July 2021].

UCSF VR. https://www.ucsf.edu/news/2017/09/408301/how-vr-revolutionizing-way-future-doctors-arelearning-about-our-bodies. [Last Accessed 09 May 2021].

Urbankova, A., Eber, M., Engebretson, S.P., 2013. A complex haptic exercise to predict preclinical operative dentistry performance: a retrospective study. Journal of Dental Education 77 (11), 1443–1450.

USC Institute for Creative Technologies Website, Medical Virtual Reality. (Last accessed 05 Jun 2021).

Venkatesan, M., Mohan, H., Ryan, J.R., Schurch, C.M., Nolan, G.P., Frakes, D.H., Coskun, A.F., 2021. Review virtual and augmented reality for biomedical applications. Cell Reports Medicine 2 (100348), 1–13.

Vincelli, F., Anolli, L., Bouchard, S., 2003a. Experiential cognitive therapy in the treatment of panic disorders with agoraphobia: a controlled study. CyberPsychology and Behavior 6 (3), 321–328.

Vincelli, F., Anolli, L., Bouchard, S., Wiederhold, B.K., Zurloni, V., Riva, G., 2003b. Experiential cognitive therapy in the treatment of panic disorders with agoraphobia: a controlled study. CyberPsychology and Behavior 6 (3), 3210328.

Virtual Reality Headsets Might Help Cure Genetic Diseases, September 22, 2017. Futurism (Last accessed 21 April 2021).

von der Heide, A.M., Fallavollita, P., Wang, L., 2018. Camera-augmented mobile Carm (CamC): a feasibility study of augmented reality imaging in the operating room. International Journal of Medical Robotics 14 (2), 1e8.

Wald, J., Taylor, S., 2000. Efficacy of virtual reality exposure therapy to treat driving phobia: a case report. Journal of Behavior Therapy and Experimental Psychiatry 31, 249–257.

Wang, S.S., Zhang, S.M., Jing, J.J., 2012. Stereoscopic virtual reality models for planning tumor resection in the sellar region. BMC Neurology 12, 146–172.

Wang, Y., Li, Q., Liu, L., Zhou, Z., Ruan, Z., Kong, L., Li, Y., Wang, Y., Zhong, N., Chai, R., 2019. TeraVR empowers precise reconstruction of complete 3-D neuronal morphology in the whole brain. Nature Communications 10, 3474–3487.

Watanabe, E., Konno, S.M., Hirai, T.M., Yamaguchi, T., 2016. The trans-visible navigator: a seethrough neuronavigation system using augmented reality. World Neurosurgery 87, 399–405.

Widmann, G., Schullian, P., Ortler, M., Bale, R., 2012. Frameless stereotactic targeting devices: technical features, targeting errors and clinical results. International Journal of Medical Robotics and Computer Assisted Surgery 8, 1–16.

Wiederhold, B.K., Gevirtz, R., Spira, J.L., 2001. Virtual reality exposure therapy vs imagery desensitization therapy in the treatment of flying phobia. Cyberpsychology: Journal of Psychosocial Research on Cyberspace: Mind, Cognition, and Society 253–272.

Wired, M., 2017. EchoPixel Announces Progress in the Clinical Adoption of Inter- Active Virtual Reality for Pediatric Surgery. Available from: http://www.marketwired.com/press-release/echopixel-

announces-progress-clinical-adoption-interactive-virtual-reality-pediatric-2202796.htm (Last Accessed 23 April 2021).

Workman, S., 2018. Mixed reality: a revolutionary breakthrough in teaching and learning. EDUCAUSE Review. Available from: https://er.educause.edu/articles/2018/7/mixed-reality-a-revolutionary-breakthrough-inteaching-and-learning.

Wu, J.R., Wang, M.L., Liu, K.C., Hu, M.H., Lee, P.Y., 2014. Real-time advanced spinal surgery via visible patient model and augmented reality system. Computer Methods and Programs in Biomedicine 113 (3), 869e881.

Wurst, S., 2020. "Extended Reality in Life Sciences and Healthcare", Solutions, Market Moves, and Opportunities, 1, pp. 1–13.

XRHealth, VR Telehealth. https://www.xr.health/, 2021. (Last Accessed 23 May 2021).

Yamada, H., Nakaoka, K., Sonoyama, T., 2016. Clinical usefulness of mandibular reconstruction using custom-made titanium mesh tray and autogenous particulate cancellous bone and marrow harvested from tibia and/or ilia. Journal of Craniofacial Surgery 27 (3), 586–592.

Yang, D.L., Xu, Q.W., Che, X.M., Wu, J.S., Sun, B., 2009. Clinical evaluation and follow-up outcome of the presurgical plan by Dextroscope: a prospective controlled study in patients with skull base tumors. Surgical Neurology 72, 682–689.

Zhao, Y., Chen, X., Wang, F., 2012. Integration of diffusion tensor-based arcuate fasciculus fiber navigation and intraoperative MRI into glioma surgery. Journal of Clinical Neuroscience 19, 255–261.

Zinser, M.J., Mischkowski, R.A., Dreiseidler, T., 2013. Computer-assisted orthognathic surgery: waferless maxillary positioning, versatility, and accuracy of an image-guided visualization display. British Journal of Oral and Maxillofacial Surgery 51 (8), 827–833.

Further reading

A.G. arivis, arivis VisionVR. https://imaging.arivis.com/en/imagingscience/arivis-invie. (Last Accessed 09 August 2021).

Strasburger, H., Rentschler, I., Jüttner, M., 2011. Peripheral vision and pattern recognition: a review. Journal of Vision 11, 13–25.

CHAPTER ELEVEN

Heart disease prediction with machine learning and virtual reality: from future perspective

Ashima Arya[1], Mitu Sehgal[1], Neha Bhatia[1], Sapna Juneja[2] and Deepika Koundal[3]
[1]Department of Information Technology, Samalkha, Haryana, India
[2]Department of Computer Science, KIET Group of Institutions, Ghaziabad, Uttar Pradesh, India
[3]School of Computer Science, University of Petroleum & Energy Studies, Dehradun, Uttarakhand, India

11.1 Introduction

In a normal daily routine, everyone is undergoing the busy schedule, which results into stress and anxiety. This leads to multiple diseases within a human body. Heart disease, cancer, and brain issues are only a few of the major ailments. The most common cause of reduce lives is due to sustainability of heart disease (Mohan et al., 2019). Every year, 17.5 million people die from heart disease. It not only affects heart functioning but also affect others part of body. The problem exits when it comes to correctly predict the cause of disease and to know how it can be treated (Khourdifi and Mohamed, 2019). Multivariate regression analysis can be used to create a prediction model. As we all know, digital technology is rapidly evolving. As a result, healthcare facilities retain a large quantity of data in their databases, which can be difficult to analyze. With the machine learning (ML) algorithm, analysis of data and predictions of data becomes easier with better accuracy. A data set can be collected from the huge database for creating the model (Rajdhan et al., 2020).

The types of cardiac disorders are as follows: The word "cardio" is derived from the word "heart." All cardiac disorders fall under the umbrella of cardiovascular diseases such as coronary heart disease, angina pectoris, congestive heart failure, cardiomyopathy, congenital heart disorders, and coronary artery diseases (narrowing of the coronary arteries), which are the numerous types of heart disease. The coronary arteries supply the heart with blood and oxygen. A high number of persons fall ill or die as a result of it. It is well known as the most common type of heart disease (Krishnan and Geetha, 2019). High blood glucose disturbs the blood vessels and nerves that run the heart and its working as a result of diabetes. If a person has diabetes for a long time, there is a considerable possibility that they will develop heart disease in the future. Additional reasons that add to heart disease in people with diabetes consist of smoking, which surges the risk of heart disease, which makes it challenging for the heart to perform well and sets

Extended Reality for Healthcare Systems, Volume 1
ISBN 978-0-323-98381-5, https://doi.org/10.1016/B978-0-323-98381-5.00011-8

stress on the heart, fading blood vessels (Barik et al., 2020). Heart disease, high blood pressure, and obesity are also interrelated to irregular cholesterol levels. Family history can also be the reason to have a heart disease in someone. The other risk factors that must be included are age, gender, stress, and unhealthy diet. From the study, it has been observed that men have a more risk of heart disease than women.

From this study, it can be expected that the predictions of heart disease based on the factors presented earlier using ML techniques. The chosen algorithms have been taken into account and implemented onto the data set according to its requirements. The objective of this study is to increase precision and create the system well organized in foretelling outcomes. The data set, process model and implementation, and results will be discussed in Section 11.5—11.7.

11.2 Literature review

The healthcare industry is one of the vast application areas in which there are a huge number of resources that hold a tedious job to be handled and operated manually. A great amount of research and study has been done on the heart disease prediction systems using various noninvasive methods such as algorithms based on ML.

Sahaya and Murugeshwari (2018) focused on data mining techniques to foretell many heart diseases that are prevailing in the world. The core purpose behind this study is to extract and analyze the unknown patterns in the data and to focus on the enhancement of accuracy status of any data set. They also have provided descriptive information about the databases and tools used such as Rapid Miner, Weka, Datamelt, Apache Mahout, etc. Based on the results, it proves that combination of two or more algorithm produces a better and accurate output than a single algorithm use. Sharmila and Chellammal (2018) proposed an efficient heart disease system that predicts the heart-related issues using data mining. Nowadays, it plays a major role in predicting various diseases in comparatively reduced number of tests. It recommended using big data tools such as SVM and HDFS for a better and improved result. A comparative study is done on parallel and sequential SVM, which yielded that better computation time is achieved in parallel SVM. Beyene and Kamat (2018) suggested predicting the prevalence of any heart-related disease using mining algorithms. The major objective of this research is to sense and treat heart problems initial on and in a relatively short period of time. Various attributes are taken into account for better evaluation of the person extent of sufferings from the heart disease. Data set analysis has been done using Weka Software. Kaur and Arora (2018) emphasized on approaches that are suitable and helpful in forecasting diseases by taking out old data records of patients using several algorithms such as K-nearest neighbor (K-NN), artificial neural network (ANN), naïve Bayes (NB), and support vector machine (SVM). From this, we were able to reach to a conclusion that SVM provided the highest accuracy rate as compared with other mining algorithms. Tithi and Akhtar (2019)

proved that mining techniques can be used to predict cardiac disease. This aids the medical practitioner's study and decision-making. ECG is a medical diagnosis of the activity of the heartbeat of a human being as the conventional methods are very time-consuming and tedious to work on. The apt algorithm can be used to classify ECG signals. This recommended approach increases ECG classification accurateness, letting for extra particular finding. Hosh et al. (2021) used Cleveland Heart Disease data set and generated an automatic diagnostic system for heart disease prediction. They used different sets of information with three ML approaches: decision tree (DT), K-NN, and random forest (RF). The three approaches were applied to the entire set of features, a set of 10 features selected using "Pearson's correlation," and a group of six features selected using the Relief algorithm. The accuracy, precision, compassion, and a variety of other catalogs were used to evaluate the results. The blend of the RF classifier and the features preferred by Relief made the best results, with an accuracy of 98.36%. This could be improved even more by using a fivefold cross validation (CV) technique, which would end result in 99.337% accuracy.

According to Xiao-Yan Gao et al. (2021) in the medical field, ML is becoming gradually significant. Ensemble learning approaches are employed to increase the recital of forecasting heart disease in this article. To pick serious characteristics from the data set, two types of mining procedures are used: linear discriminant analysis (LDA) and principal component analysis (PCA). On identified features, a relationship among ML algorithms and ensemble learning approaches is implemented. Models are valued using a range of methodologies, including exactitude, ability to remember, exactness, F-measure, and ROC. In the study by Lokaiah et al. (2021), the issue of heart ailments among pregnant women in India has been discussed by determine the fetal electrocardiogram (FECG) signals, which specify how healthy the heart is functioning. Numerous heart problems can be familiar built on the individual signs identified by the FECG scan. It is dangerous to discover and analyze FECG signals through labour. The mixing of impulse wave patterns produced by the several cardiac tissues is identified as a heartbeat. The identification of waveforms reasons the pregnant woman's situation to change. As a result, FECG signal categorization is critical for noticing cardiac disease.The arrangement of FECG using ML techniques as input of waveforms (features) is the matter of this study. When the data set is too enormous, the ML techniques are simplified using Python. For refining the accuracy rate for disease identification, the recommended method uses two methods: DT and K-NN algorithms. When the two systems are related on the FECG heartbeat categorization data set, the high-level accuracy in binary level classification is succeeded.

Manoj described one of the causes of heart disease has emerged due to poor eating habits, tension, absence of workout, high blood pressure, smoking, liquor, drug misuse, cholesterol, and high blood sugar. The blood arteries of the human body have become feeble as a result of oily foods, which can lead to a variety of cardiac ailments. Increased

artery pressure can cause the heart's walls to congeal. It can reason a blood clot, which can lead to heart problems. Identifying such diseases at an initial period is the only mode to save lives. This research talks over a trusting strategy that extensively recovers the prediction accuracy rate of cardiac disorders to cover this gap. Tata Balaji et al. used one-dimensional and multidimensional signals. Computer-aided diagnosis supports in detecting heart diseases at an initial period, saving human life (Balaji et al., 2021). The objective of this script is to propose a general idea of HDs, indications, and the role of ML in HD likelihood, trailed by numerous state-of-the-art ML algorithms that aid in the early detection and prediction of HD to save human lives.

In the study by Saiyed and Koteeswaran, for early heart disease prediction, an upgraded form of the K-means neighbor classifier is utilized, which promises higher accuracy than the original K-means neighbor classifier and other comparable classifiers. The expressions that have a substantial impact on heart disease include smoking, eating habits, diabetes, blood pressure, and other associated phrases. In such cases, a particular strategy or a hybrid combination of approaches to forecast cardiac disease at an early stage is necessary. The classifier is the fourth stage in the phases of expecting cardiac disease, and it is a crucial step for reaching accurateness, thoughtfulness, and specificity. When compared with the actual K-means neighbor classifier, the enriched K-means neighbor in the Python surroundings produces more accurate output with less information sets.

In the study by Manoj Diwakar et al., the authors include a discussion of ML and image fusion classification algorithms that have been shown to aid healthcare practitioners in detecting cardiac disease. Cainelli focused on the relationship between cognitive and psychopathological traits in kids with inherited heart disease (CHD) and clinical characteristics. A prospective observational research was undertaken in kids with CHD who underwent heart surgical procedure before the age of 4. They conducted a comprehensive neuropsychological (cleverness, linguistic, care, supervisory purpose, reminiscence, and community expertise) and psychopathological examination at minimum 18 months following heart operation, using a machine learning method for clustering and influencing variable classification. They looked at 74 kids (37 with CHD and 37 age-matched controls).

In the study by Swathy and Saruladha (2021), cardiovascular diseases (CVDs) are discovered to be widespread in the population, often resulting in death. Fatness, cholesterol, high blood pressure, and tobacco use are all increasing the death rate, according to the results of a recent poll. Because of the aforementioned variables, severity of the disease is increasing. The need of the hour is to research the various variants of these components and their impact on CVD. This needs the application of sophisticated technology to detect the disease early on and aid in the reduction of mortality rates. With their vast techniques, the artificial intelligence (AI) and data mining areas offer a study potential in assisting in the forecast of CVD monastery and identifying their social designs in massive volumes of records. The outcomes of these forecasts will aid clinicians in

creating decisions and initial identification, lowering the likelihood of patients dying. This study links and reports on the several categorizations, data mining, ML, and deep learning models utilized for CVD prediction. The survey is divided into three sections: CVD classification and data mining methods, CVD ML models, and CVD deep learning models. This survey also compiles and reports the performance measures used to report precision, the data set utilized for estimate and grouping, and the apparatuses used for each one set of these methods.

In the study by Rubini et al. (2021), the authors have taken basic symptoms such as age, sex, pulse rate, latent blood pressure, saturated fat, abstaining blood sugar, resting electrocardiographic results, exercise-induced angina, ST depression, ST section that rise at peak during exercise, the number of major vessels colored by fluoroscopy, and extreme heart rate reached; an application was developed that can envisage the susceptibility of heart disease. Doctors can use this to double-check and confirm their patient's condition. In previous surveys, only 10 features were evaluated for prediction; however, in this proposed research endeavor, 14 essential features were considered. In addition, this paper compares the classification of cardiovascular illness using ML approaches such as RF, logistic regression, SVM, and NB. According to the results of the comparison research, the ML algorithm RF has verified to be the utmost perfect and trusty algorithm, which is why it was chosen for usage in the suggested system. This approach also shows the link between diabetes and the degree to which it affects heart disease.

According to the study by Kumar et al. (2020), the human body's blood circulation is necessary for survival. As a result, it is not incorrect to claim that the heart is the most energetic element of our physique because it pushes (or circulates) blood to all parts of the body. Getting a condition related to the heart means getting a disease related to the most vital portion of our bodies. According to the WHO, cardiac ailments claim the lives of 17 million people each year, accounting for more than a third of all deaths. ML algorithms are now being used to solve difficulties in a variety of medical sectors, and we may use them to anticipate heart disease in this case as well. Over the "heart.csv"" data set, we will examine the accuracy of various ML techniques and determine which method produces the best results.

According to Archana Singh and Rakesh Kumar, in living creatures, the heart serves an important role. Diagnosis and forecast of cardiac illnesses necessitate better precision, perfection, and accuracy because even a minor error can result in exhaustion or death. There are frequent demise cases associated to the heart, and the figure is growing exponentially day by day. To report the issue, a predictive system for disease awareness is required. ML is a branch of AI, which offers significant assistance in anticipating any type of occurrence using natural events as training. Using the UCI repository data set for training and testing, they analyze the accurateness of ML methods for expecting cardiac disease, including K-NN, DT, linear regression, and SVM. Anaconda (Jupyter) notebook is the finest tool for implementing Python programming since it has various types

of libraries and header files that make the task more exact and precise. According to Kataria and Srinivas, predicting and detecting cardiac disease has always been a difficult and time-consuming undertaking for doctors. To treat cardiac disorders, clinics and hospitals are giving costly treatments and operations. As a result, anticipating cardiac disease in its initial phases will be beneficial to individuals all round the world, allowing them to take required treatment before it becomes serious. Heart disease has been a major issue in current years, with the primary causes being extreme alcohol habit, tobacco usage, and absence of physical doings. ML has revealed to be useful in making decisions and estimates from a huge set of information created by the healthcare business over time. ANNs, DTs, RF, SVM, NB), and the K-NN methods are a number of the supervised ML approaches utilized in this forecast of heart disease. In addition, the results of various algorithms are summarized.

11.3 Virtual reality

Virtual reality (VR) is a skill that generates 3D reality for the user built on sights and objects chronicled from a computer-generated and technical atmosphere using scenes and things, which replicate genuineness for the user. Ivan Sutherland coined the term "virtual reality" in the 1960s, relating it as "a frame over which a user visions the simulated world as if it observed, touched, and expressed actual, and in which the user might perform genuinely" (Juneja et al., 2020). "The essential notion after the three-dimensional demonstration is to show the user with a viewpoint appearance that fluctuates as he travels," Sutherland (Li et al., 2011) explains. "... We can put suitable 2D images on the spectator's retinas, forming delusion that he is nearsighted 3D object. The partaker can turn into a character in the scene perceived over the VR glasses or headdress. In the beginning, VR technology was first developed for entertainment (Juneja, 2021). However, in current years, it has been used in a wide variety of scientific settings, together with pain controlling, physical restoration, and psychiatric disorders management (e.g., fears, posttraumatic hassle ailment, and nervousness ailment) (Fig. 11.1).

11.3.1 Platform for virtual reality

With an increasing presence across numerous facets, 3D visualization technologies have grown to become a standard in the care of congenital heart disease (CHD). Printed and virtual 3D models provide for a more thorough approach to training trainees and members of the care team. By predicting postprocedural outcomes and assisting in surgical approach, computational fluid dynamics can take 3D modeling to the next level. 3D printing and augmented reality are developing tools for preprocedural planning and intraprocedural guidance that have the potential to transform decision-making and procedural success.

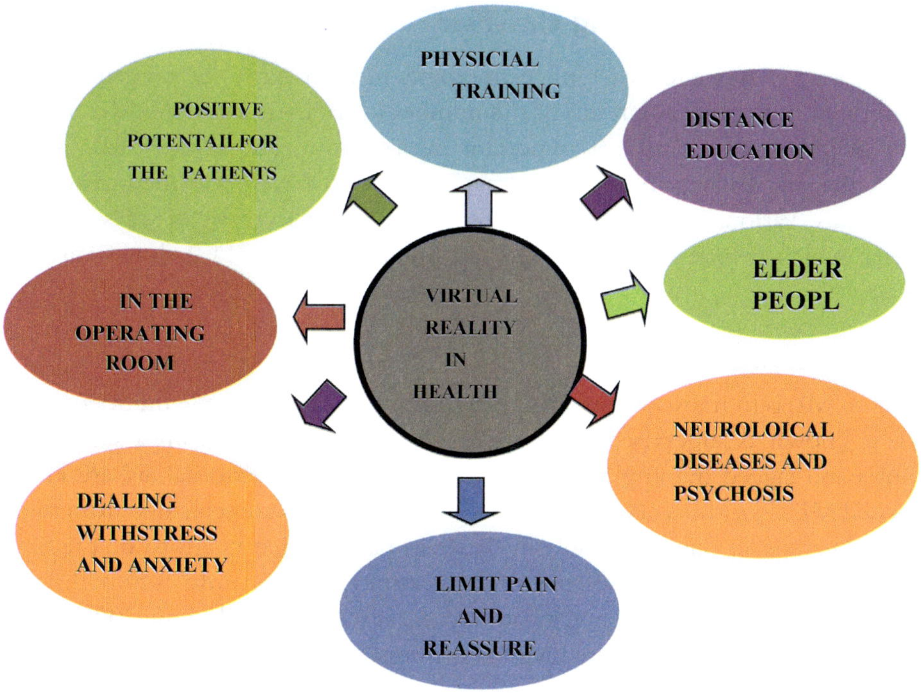

Figure 11.1 Virtual reality future with healthcare.

11.3.2 Virtual reality using three-dimensional technology

VheaRts is a virtual reality platform created with Unity (Unity Technologies) for use with the commercial Oculus Rift/Quest headsets (Facebook Technologies). The Oculus Rift was connected to an Alienware 17 R5 gaming laptop for the duration of the study. A VheaRts module containing the six cases was created. The user was presented with two menus when entering the virtual room, one for activating digital models and the other for accessing the following tools:•Handling and rotating the 3D heart models: the user can 'grab' models in 3D space and move them around freely using the controllers. • Slicing 3D heart models: a slicing tool allows the user to freely 'crop' the mesh in real time. Clipping with a plane or sphere is an option (Fig. 11.2). • Displaying structure labels: primary anatomical parts of each case are highlighted and labeled, and each structure can be grasped and manipulated separately (Fig. 11.2). • Ultrasound probe simulator: a 2D ultrasound projection is presented on a screen when the probe is moved inside the 3D model. • Marking and measuring points in 3D space: the user can mark points in 3D space to highlight structures and measure distances.

11.4 Machine learning classifier

ML is a technique for teaching technologies in what way to better grip data. ML is becoming more popular as a result of the number of data sets available. ML is used in a variety of businesses to abstract important information. The goal of ML is to gain knowledge from statistics. Numerous experiments must have performed to acquire the better results. Various statisticians and system analysts use a variety of ways to solve this problem, which involves large data sets (Mahesh, 2018).Fig. 11.2 shows various ML classifiers.

11.4.1 Supervised learning

The job of ML is to translates any idea to an output based on sample input—output sets, which is known as supervised learning. It assumes a function from a set of training examples and tagged training data. Algorithms that require external aid are known as supervised ML algorithms. The training and testing data sets are separated from the input data set. There is an output variable in the train data set that needs to be predicted or categorized. For prediction or sorting, each algorithm acquires outlines from the training data set and relates them to the investigation data set (Cauana et al., June 2006). Fig. 11.3 depicts the workflow of supervised ML algorithms.

11.4.2 Unsupervised learning

Unsupervised learning is distinct as supervised learning, there are not even one right response and not any instructor. Algorithms are leftward to their specific devices when it

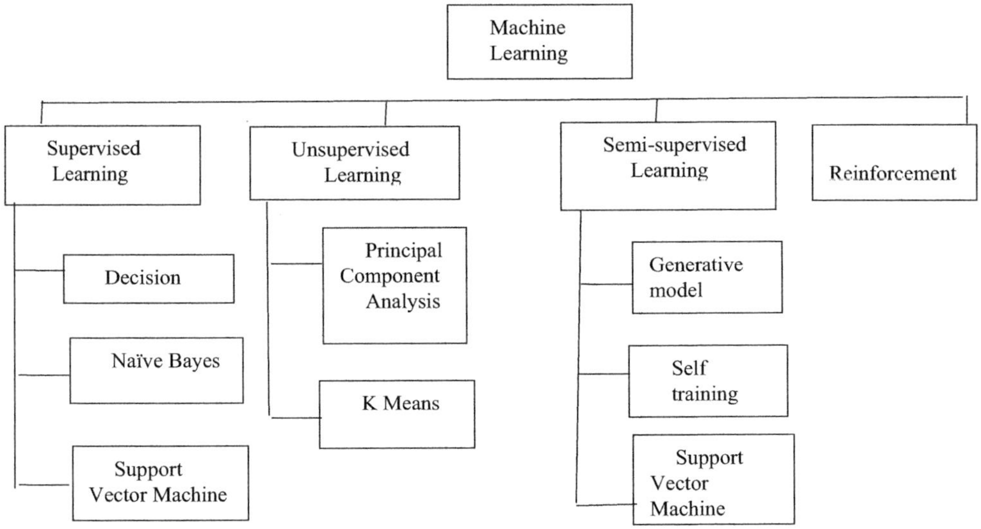

Figure 11.2 Machine learning classifiers.

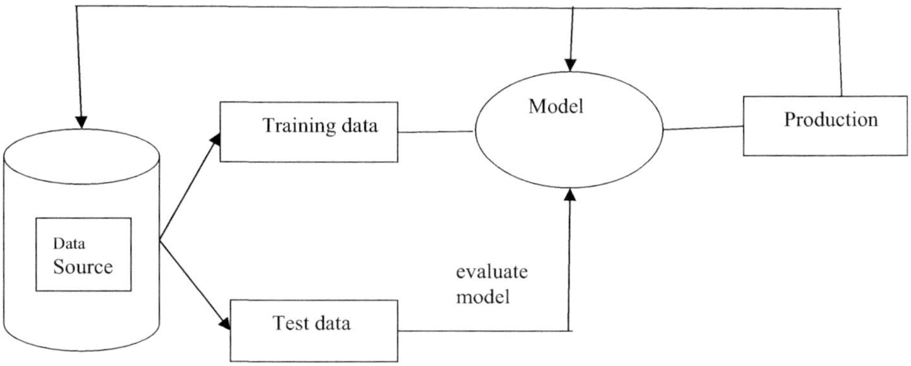

Figure 11.3 Workflow of supervised algorithm.

comes to discovering and presenting the intriguing configuration of the data. Unsupervised learning algorithms extract a small number of features from the information. When fresh information is presented, it recognizes the class of the data by earlier learned structures. It is mostly utilized for feature reduction and clustering.

11.4.3 Semisupervised machine

Semisupervised ML combines the benefits of both supervised and unsupervised ML methods. It can be useful in fields such as ML and data mining if there are existing unlabeled data, and collecting the labeled data is a time-consuming procedure. In supervised ML, you train an ML algorithm using a "labeled" data set, where every record contains the conclusion facts (Zhu and Goldberg, 2009).

11.4.4 Reinforcement learning

Learning through reinforcement learning is a division of ML which studies how software managers have to behave in a specified situation to maximize a metric of increasing recompense. Reinforcement learning, besides supervised and unsupervised learning, is one and only of the three central ML models (Fig. 11.4).

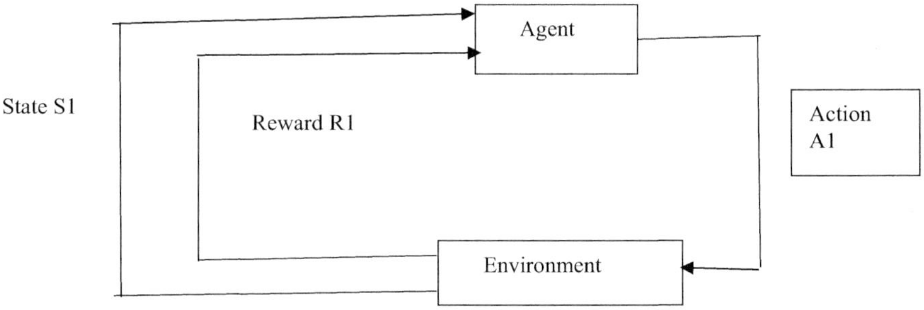

Figure 11.4 Reinforcement learning.

11.4.5 Multitask learning

Multitask learning is a subfield of ML that seeks to resolve numerous diverse odd jobs at the similar period by exploiting commonalities between them. This can help to increase book learning efficacy while implemeting the same different data sets. If there are n tasks then according to traditional deep learning algorithms, only one task will be handle by single model, and these n jobs (or a subset of them) are associated on the other hand not alike. Multitask learning (MTL) will aid in the upgrading of a model's learning by utilizing the information confined in all n tasks learning in a group.

11.4.6 Ensemble learning

Ensemble learning is the practice of scientifically producing and linking many models, such as classifiers or experts, to hold a precise computational intelligence problem. Ensemble learning is normally used to mend a model's presentation or lower the risk of an unintended poor model choice. Ensemble learning is also used for conveying a confidence level to the model's decision, selecting optimal structures, data fusion, incremental learning, nonstationary learning, and error correction.

11.4.7 Neural networks

A neural network is a set of algorithms which attempts to recognize essential connections in a batch of data using a method that impersonates how the humanoid mind works. Neural networks, in this context, refer to arrangements of neurons that can be biological or artificial in nature. Because neural networks can adjust to varying input, they can produce the greatest likely outcome deprived of requiring the output standards to be redesigned. The AI -based neural network can create the trading schemes very fast. In the same way, an ANN works. It operates on three levels. The input layer accepts data. The input is processed by the hidden layer. Finally, the determined output is sent by the output layer (Wang et al., 2018) (Fig. 11.5).

11.5 Proposed methodology

The approach for predicting if a patient has heart disease was carried out using the three algorithms listed in the following, and the results were compared.

1. RF (Sutherland, 1968)
2. Logistic regression (Li et al., 2011)
3. NB algorithm (Juneja, 2021) (Fig. 11.6)

11.5.1 Collection of data set

Application of any algorithm depends on the nature of data set. The data set which has been used consists of 14 attributes which are as follows (Table 11.1):

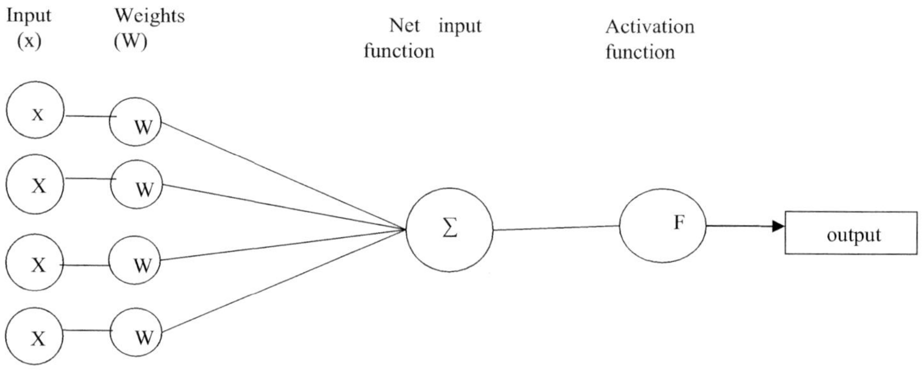

Figure 11.5 Neural network.

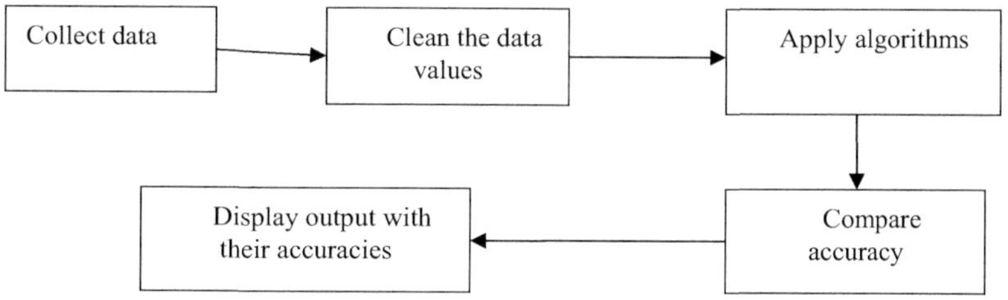

Figure 11.6 Methodology.

11.5.2 Algorithms

11.5.2.1 Logistic regression

An ML classification methodology called logistic regression is used to calculate the possibility of a clear-cut dependent variable (Juneja et al., 2021). The dependent variable in logistic regression is a binary variable that involves data coded as 1 (yes, success, etc.) or 0 (no) (no, failure). In other words, as a function of X, the logistic regression model guesses $P(Y = 1)$.

In data modeling, logistic regression is one and only of the finest used methods for suitable models for categorical data, mostly for binary response data. It goes to a class of models identified as generalized linear models, and it is the most essential (and possibly most commonly used) of them (Upadhyay et al., 2021). The minimal possibilities of the training data are conserved using logistic regression. The coefficients of the model also provide some vision into the relative relevance of each input variable, for example, to conclude whether an email is spam (1) or not (0) and whether the tumor is cancerous (1) or not (0).

11.5.2.1.1 Types of logistic regression

1 .Logistic regression (binary): There are just two possible results for the categorical response. For example: Is it spam or not?

Table 11.1 Attribute taken in data set for prediction.

1. Age: The number of years between one's birth and one's death
2. Gender: Male = 1, Female = 0
3. Cp: Types of chest pain: a) Type 1: typical angina, b) type 2: atypical angina
d) Type 4: asymptomatic, c) Type 3: nonanginal pain
4. Trestbps: Vital sign at rest (in torr on admittance to the hospital)
5. Chol: cholesterol serum concentration in milligrammes per deciliter
6. Fbs: (1 = true; 0 = false) (fasting blood glucose >120 mg/dL)
7. Thalach: reached maximum heart rate
8. Restecg: electrocardiographic data at rest
Type 0: normal; type 1: aberrant STT wave (T wave inversions and/or ST elevation or depression of >0.05 mV)
c) Type 2: left ventricular hypertrophy is suspected or confirmed
9. Exang: exercise-induced angina (yes or no)
10. Oldpeak: Exercise-induced ST depression compared to rest
11. Slope: the slope of the ST segment of the height exercise
i. Type 1 is upsloping; ii. Type 2 is flat; and iii. Type 3 is downsloping
12. Ca: number of large vessels (03) that are fluoroscopy colored
13. Thal: 3 denotes normal, 6 denotes a permanent defect, and 7 denotes a reversible defect
14. Num: a cardiac problem diagnosis (angiographic disease status)
Type 0 (zero): a 50% reduction in diameter
Type 1: a diameter narrowing of more than 50%

2 .Logistic multinomial regression: Without any sorting, there are three or more categories. Predicting which food is preferred more, for example, veg, nonveg, vegan.

3 .Logistic ordinal regression: Three or more categories, each having its own order (Juneja et al., 2021).

11.5.2.1.2 Advantages of logistic regression

1. When the data set is split by a linear line, logistic regression performs well.
2. While logistic regression is less prone to overfitting, it can overfit in data sets with a high number of dimensions. In these cases, regularization (L1 and L2) approaches should be used to minimize overfitting.
3. Logistic regression provides not only a measure of a predictor's relevance (coefficient size) but also the direction of relationship (positive or negative).
4. Logistic regression is more straightforward to implement, analyze, and train (Uppal et al., 2021).

11.5.2.1.3 Limitations of logistic regression

1. The assumption of linearity between the dependent and independent variables is the main drawback of logistic regression. Data are rarely linearly separable in the actual world. Majority of the time, data are a muddled mess (Juneja et al., 2021).

2. Logistic regression should not be employed if the amount of observations is less than the number of features; otherwise, it may result in overfitting. Only discrete functions may be predicted using logistic regression. As a result, dependent variable of logistic regression is limited to the discrete number set. This limitation is troublesome in and of itself, as it makes continuous data prediction impossible (Fig. 11.7).

11.5.2.2 Naïve Bayes

The Bayes theorem is used to create a collection of classification algorithms known as NB classifiers. It is a family of algorithms that share a similar idea, namely that each pair of features being classified is independent of the others. It is a dependable and adaptable tool that has endured the test of time (Shao et al., 2022). The NB classifier approach is built on the Bayesian formula and is especially useful when the inputs have a high dimensionality (Juneja et al., 2021).

Everything in statistics relies around hypotheses. We establish a hypothesis (an educated guess) about how the world works and then set about gathering evidence to test it. Assuming an occurrence and predicting the corresponding outcome is how NB works.

The Bayes theorem calculates the likelihood of an event occurring given the probability of a previous event. The following equation expresses Bayes theorem mathematically:

$$P(c\,|\,x) = \frac{P(x|c)P(c)}{P(x)}. \text{ Here,}$$

$$P(c|x) = P(x1|c)XP(x2|c)X....P(xn|c)XP(c)$$

```
LogisticRegression(C=1.0, class_weight=None, dual=False, fit_intercept=True,
                   intercept_scaling=1, l1_ratio=None, max_iter=100,
                   multi_class='auto', n_jobs=None, penalty='l2',
                   random_state=None, solver='lbfgs', tol=0.0001, verbose=0,
                   warm_start=False)

[[ 95  71]
 [ 35 264]]
              precision    recall  f1-score   support

           0       0.57      0.73      0.64       130
           1       0.88      0.79      0.83       335

    accuracy                           0.77       465
   macro avg       0.73      0.76      0.74       465
weighted avg       0.80      0.77      0.78       465

0.7720430107526882
```

Figure 11.7 Function and output of logistic regression.

- The posterior possibility of class (target) given predictor is $P(c|x)$ (attribute).
- $P(c)$ is the class prior possibility.
- The likelihood is $P(x|c)$, which is the likelihood of a predictor given a class.
- P(x) is the predictor's prior possibility.

11.5.2.2.1 Advantages of NB

1. A NB classifier outperforms other models when the assumption of independent predictors holds true.
2. To estimate the test data, NB requires a little amount of training data. As a result, the training duration is shorter.
3. NB is similarly simple to use.

11.5.2.2.2 Limitations of Naïve Bayes

1. The assumption of independent predictors is the main premise of NB. All of the qualities in NB are assumed to be mutually independent. In actual life, getting a collection of predictors that are totally independent is nearly impossible.
2. If a categorical variable in the test data set has a category that is not included in the training data set, the model will give the possibility of 0 (zero) and will be not capable to mark a likelihood. This is commonly mentioned to as zero occurrence. We can apply the smoothing approach to remedy this. Laplace estimation is one of the best simple smoothing techniques (Fig. 11.8).

11.5.2.3 Random forest

A group of DTs is referred to as an RF. We have a pool of DTs in RF. Every single tree provides a grouping for a new object based on attributes, which we refer to as the tree—votes for that class. The classification with the highest votes is chosen by the forest (over all the trees in the forest).

As the name entails, a random forest is prepared from a large number of decision trees that find out the decision by working in parallel with each other. Every tree in the RF creates class likelihood, and the class with the maximum polls turns into the likelihood for our model.

The following is how each tree is planted and grown:

1. If the amount of cases in the training set is N, a random sample of N cases with replacement is occupied. This sample will serve as the tree's training set.
2. If there are M input variables, a number mM is provided so that m variables are randomly picked from the M at each one node, and the top split on this m is used to split the node. Throughout the development of the forest, the value of m is kept constant.
3. Every single tree is matured to its supreme potential. Snipping is not an option.

```
GaussianNB(priors=None, var_smoothing=1e-09)
[[107  59]
 [ 50 249]]
              precision    recall  f1-score   support

           0       0.64      0.68      0.66       157
           1       0.83      0.81      0.82       308

    accuracy                           0.77       465
   macro avg       0.74      0.74      0.74       465
weighted avg       0.77      0.77      0.77       465

0.7655913978494624
```

Figure 11.8 Ready with function and output of naïve Bayes.

Scikit-learn (sklearn) is used to design a DT and train (fit) it on data. We feed the model both the features and the labels during training so that it may learn to classify points using the features. On the training data, we can check the accuracy of our model. Because we gave it the answers (y) for training and did not limit the depth of the tree, we notice that it gets 100% accuracy, which is what we expected. It turns out that a DT's ability to completely understand the training data can be a drawback because it can lead to overfitting.

11.5.2.3.1 Advantages of random forest

1. The RF algorithm is created on the bagging algorithm and employs ensemble learning. It grows as many trees as possible on a subgroup of the data and then merges the outcomes of all the trees. As a result, the overfitting problem in DTs is reduced, as is the variance, which increases accuracy.
2. RF can be used to address difficulties comprising classification and regression.
3. Both categorical and continuous variables function well with RF.
4. RF can handle missing values automatically.
5. RF does not require feature scaling (standardization and normalization) because it uses a rule-based method rather than distance calculation.

11.5.2.3.2 Limitations of random forest

1. Complexity: RF produces a large number of trees (as opposed to a single tree in a DT) and then mixes their outputs. In the Python sklearn library, it builds 100 trees by default. This approach necessitates a significant increase in processing power and resources. On the other hand, a DT is straightforward and does not necessitate a large amount of computer power.

```
RandomForestClassifier(bootstrap=True, ccp_alpha=0.0, class_weight=None,
                       criterion='gini', max_depth=None, max_features='auto',
                       max_leaf_nodes=None, max_samples=None,
                       min_impurity_decrease=0.0, min_impurity_split=None,
                       min_samples_leaf=1, min_samples_split=2,
                       min_weight_fraction_leaf=0.0, n_estimators=100,
                       n_jobs=None, oob_score=False, random_state=None,
                       verbose=0, warm_start=False)

[[127  39]
 [ 49 250]]
              precision    recall  f1-score   support

           0       0.77      0.72      0.74       176
           1       0.84      0.87      0.85       289

    accuracy                           0.81       465
   macro avg       0.80      0.79      0.80       465
weighted avg       0.81      0.81      0.81       465

0.810752688172043
```

Figure 11.9 Function and output of random forest.

2. Longer training period: RF takes significantly longer to train than DTs since it generates several trees (instead of just one) and makes decisions based on the majority of votes (Fig. 11.9).

11.5.2.4 Interface outputs

Fig. 11.10

11.6 Result

(Table 11.2)

11.7 Conclusion

This chapter includes an overview of the present method as well as a synopsis of the previous work. In this chapter, we have explored numerous classification algorithms. Also we have studied different technologies such as NB, logistic regression, SVM, K-NN, RF, and so on to compare the graph of accuracies obtained by each one of them. We have worked on three different data sets to analyze and experience a better result with more precise and accurate results.

From our study, we chose some combinations of techniques to study the data sets. Our approach leads us to the result that RF provides the highest accuracy of all the approaches we have applied individually on the data set.

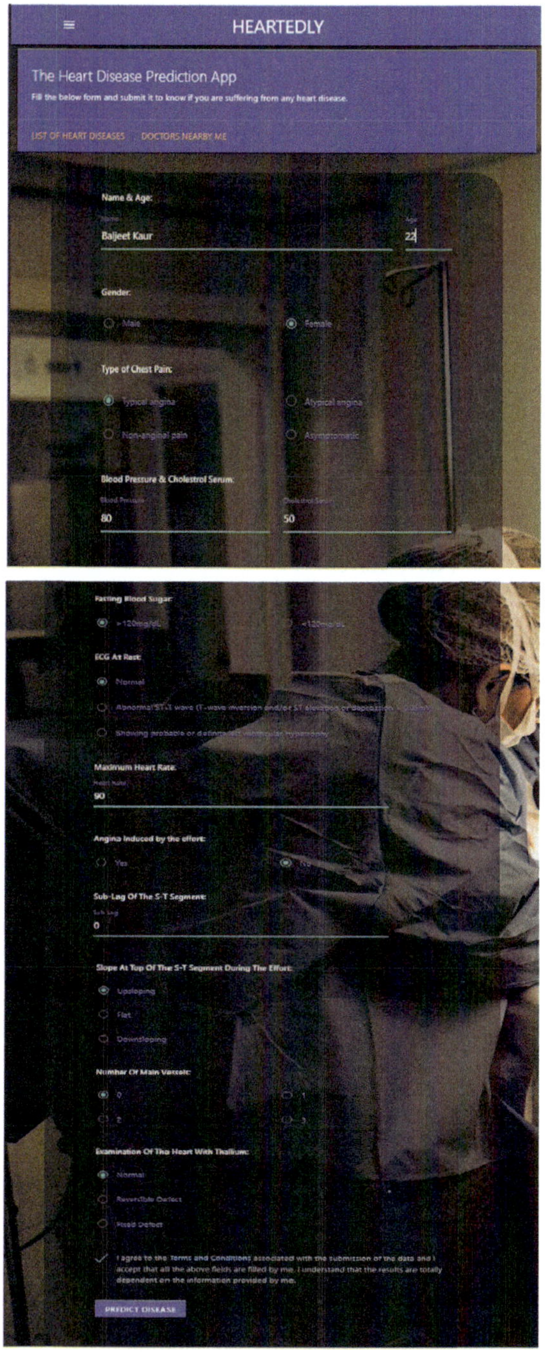

Figure 11.10 Interface for taking input and shows result.

Table 11.2 Details of patients having high chances of heart disease.

$>$ 35 years old
Man = gender
Pain in the chest = 0
Blood pressure is greater than 100
Cholesterol in the blood >131
Fasting 0 for blood sugar
Result of electrocardiography = 0
The maximum heart rate is more than 71 beats per minute
Depression about 1.6
Slope $\geq$ 1
Vessels about 1.1
Thal> 2.5

S.No	Algorithms used	Accuracy achieved
1.	Logistic regression	74.8%
2.	Naïve Bayes	74.6%
3.	Random forest	80.2%

Therefore, in a nutshell, it seems very useful and beneficial to adopt this approach that will help in diagnosing the disease quickly and easily and lessen the medical costs of the patient.

References

Balaji, T., et al., 2021. An insight on machine learning algorithms for predicting heart Diseases. Turkish Journal of Computer and Mathematics Education 12 (No.10), 5867–5877.

Barik, S., et al., 2020. Heart Disease Prediction Using Machine Learning Techniques.

Beyene, M.C., Kamat, P., 2018. Survey on prediction and analysis the occurrence of heart disease using data mining techniques. International Journal of Pure and Applied Mathematics 118 (8), 165–174. ISSN: 1311-8080 (printed version); ISSN: 1314-3395 (on-line version).

E. Cainelli Detecting neurodevelopmental Trajectories in Congenital Heart Diseases with a Machine-Learning Approach Scientific Report, https://doi.org/10.1038/s41598-021-82328-8.

Cauana, R., et al., June 2006. An empirical comparison of supervised learning algorithms. In: ICML '06: Proceedings of the 23rd International Conference on Machine Learning, pp. 161–168. https://doi.org/10.1145/1143844.1143865.

M. Diwakar et. al Latest trends on heart disease prediction using machine learning and image fusion "Emerging Trends in Materials Science, Technology and Engineering, Science Direct, Elsevier, https://doi.org/10.1016/j.matpr.2020.09.078.

Gao, X.-Y., et al., 2021. Improving the Accuracy for Analyzing Heart Diseases Prediction Based on the Ensemble Method. Hindawi Complexity, p. 10. https://doi.org/10.1155/2021/6663455. Article ID 6663455.

Hosh, P., et al., 2021. Implementation of Machine Learning Algorithms to Detect the Prognosis Rate of Kidney Disease. IEEE International Conference for Innovation in Technology.

Juneja, A., 2021. Multi-Criterion decision making for wireless communication technologies adoption in IoT, 10 (1), 1–15. https://doi.org/10.4018/IJSDA.2021010101.

Juneja, A., Bajaj, S., Anand, R., Sindhwani, N., 2020. Improvising green computing using multi-criterion decision making. Journal of Advanced Research in Dynamical and Control 12 (3). https://doi.org/10.5373/JARDCS/V12SP3/20201362.

Juneja, S., Juneja, A., Dhiman, G., Behl, S., Kautish, S., 2021. An Approach for Thoracic Syndrome Classification with Convolutional Neural Networks, vol 2021.

Juneja, A., Juneja, S., Soneja, A., Jain, S., 2021. Real time object detection using CNN based single shot detector model. Journal of Information Technology Management 13 (1), 62–80. https://doi.org/10.22059/jitm.2021.80025.

Juneja, A., Juneja, S., Kaur, S., Kumar, V., 2021. Predicting diabetes mellitus with machine learning techniques using multi-criteria decision making. International Journal of Information Retrieval Research 11 (2), 38–52. https://doi.org/10.4018/ijirr.2021040103.

Juneja, S., Gahlan, M., Dhiman, G., Kautish, S., 2021. Futuristic Cyber-Twin Architecture for 6G Technology to Support Internet of Everything, 2021, pp. 1–7.

R. Kataria, P. Srinivas, "Predicting Heart Disease at Early Stages Using Machine Learning: A Survey", International Conference on Electronics and Sustainable Communication Systems (ICESC), DOI: 10.1109/ICESC48915.2020.9155586.

Kaur, A., Arora, J., 2018. Heart Disease Prediction Using Data Mining Techniques: Asurvey".

Khourdifi, Y., Mohamed, B., 2019. Heart Disease Prediction and Classification Using Machine Learning Algorithms Optimized by Particle Swarm Optimization and Ant Colony Optimization.

Krishnan, S., Geetha, S., 2019. Prediction of Heart Disease Using Machine Learning Algorithms.

Kumar, S., vithula, J., et al., 2020. Role of Machine Learning Algorithms over Heart Diseases Prediction, vol 2292. AIP Conference Proceedings, p. 040013. https://doi.org/10.1063/5.0030743.

Li, A., Montaño, Z., Chen, V.J., Gold, J.I., 2011. Virtual reality and pain management: current trends and future directions. Pain Management 1 (2), 147–157. https://doi.org/10.2217/pmt.10.15.

Lokaiah, D., Madhusudhan, P., Dontha, R., Kakumanu, S., 2021. Recognition of Fetal heart diseases through machine learning techniques" Annals of. R.S.C.B., ISSN:1583-6258 25 (Issue 6), 2601–2615.

B. Mahesh Machine learning algorithms - a review, International Journal of Science and Research (IJSR) ISSN: 2319-7064 ResearchGate Impact Factor (2018): 0.28 | SJIF (2018): 7.426.

Manoj R. Efficient method for prediction accuracy of heart diseases using machine learning" Proceedings of International Conference on Communication and Computational Technologies'pp 113-121.

Mohan, S., Chandrasegar, T., Gautam, S., 2019. Effective Heart Disease Prediction Using Hybrid Machine Learning Techniques.

Rajdhan, A., et al., 2020. Heart Disease Prediction Using Machine Learning.

Rubini, P.E., et al., 2021. A cardiovascular disease prediction using machine learning algorithms. Annals of R.S.C.B., ISSN:1583-6258 25 (Issue 2), 904–912.

Sahaya, A., Murugeshwari, G., April 2018. A Survey on Heart Disease Prediction Using Data Mining Techniques. Conference on Emerging Devices and Smart Systems (ICEDSS), INSPEC Accession Number: 18274810.

Saiyed F., W.S. Koteeswaran "Heart disease early prediction using a novel machine learning method called improved K-means neighbor classifier in python Emerging Trends in Materials Science, Technology and Engineering, Science Direct,Elsevier ,https://doi.org/10.1016/j.matpr.2021.01.570.

Shao, C., Yang, Y., Juneja, S., GSeetharam, T., 2022. IoT data visualization for business intelligence in corporate finance. Information Processing & Management 59 (1), 102736. https://doi.org/10.1016/j.ipm.2021.102736.

Sharmila, R., Chellammal, S., May 2018. A conceptual method to enhance the prediction of heart diseases using the data techniques. International Journal of Computer Science and Engineering.

A. Singh, R. Kumar "Heart Disease Prediction Using Machine Learning Algorithms, International conference on Electrical and Electronics Engineering (ICE3), DOI: 10.1109/ICE348803.2020.9122958.

Sutherland, I.E., 1968. A head-mounted three-dimensional display. In: AFIPS Conference Proceedings Volume 33, Part One: 1968 Fall Joint Computer Conference. Thompson Book Company, pp. 757–764. https://doi.org/10.1145/1476589.1476686.

Swathy, M., Saruladha, K., 2021. A comparative study of classification and prediction of Cardio-Vascular Diseases (CVD) using Machine Learning and Deep Learning techniques. Science Direct ,ICT Express. https://doi.org/10.1016/j.icte.2021.08.021.

Tithi, S.R., Akhtar, A., 2019. Dept "ECG Data Analysis and Heart Disease Prediction Using Machine Learning Algorithms".

Upadhyay, H., Juneja, S., Juneja, A., Dhiman, G., Kautish, S., 2021. Evaluation of ergonomics-related disorders in online education using fuzzy AHP. Computational Intelligence and Neuroscience 2021, 1−11. https://doi.org/10.1155/2021/2214971.

Uppal, M., Gupta, D., Juneja, S., Dhiman, G., Kautish, S., 2021. Cloud-Based Fault Prediction Using IoT in Office Automation for Improvisation of Health of Employees, vol 2021.

S.-C. Wang. Artificial neural network. The Springer International Series in Engineering and Computer Science book series (SECS, vol. 743). In: Cipresso, P., Giglioli, I. A. C., Raya, M. A., & Riva, G. (2018). The past, present, and future of virtual and augmented reality research: A network and cluster analysis of the literature. Frontiers in Psychology, vol. 9, Article 2086. https://dx.doi.org/10.3389%2Ffpsyg.2018.02086.

Zhu, X., Goldberg, A.B., 2009. Introduction to semi-supervised learning. In: Synthesis Lectures on Artificial Intelligence and Machine Learning. Morgan and Claypool publisher, p. 130. https://doi.org/10.2200/S00196ED1V01Y200906AIM006.

CHAPTER TWELVE

Extended reality and edge AI for healthcare 4.0: systematic study

Sonali Vyas
University of Petroleum and Energy Studies, Dehradun, Uttarakhand, India

12.1 Introduction

Healthcare requirements for living beings are continuously changing; almost daily there is a new disease-causing harm to the life. Moreover, pandemic, epidemic, and endemics are affecting the survival of the living beings. At the times, technology always stands out to support human beings in the battle with these diseases. Using Edge AI (artificial intelligence) and extended reality (XR) as an integration forms an ecosystem, which efficiently mitigates the effect of the diseases spread. XR refers to a real or virtual system of human—machine interactions with computer-generated technology, where the 'X' signifies the spatial computer technology (Andrews et al., 2019). Healthcare industry adapts new technologies and creates more room for the technology, e.g., telehealth. When there is no possibility of presence of doctor, treatments occur remotely although it is not the most preferable way. Optimization for both patient and the doctor as per their requirement becomes difficult some or the other times, but a primary care can be given by the use of remotely configured devices. The potential for reality-changing advancements in far-off medical care has been proclaimed since the soonest long stretches of Edge AI and XR for healthcare industry. Early VR required confounding and excessive figuring equipment that kept the innovation restricted, for use by a solitary individual, and restrictively costly until the multiplication of the computing power and hardware (Ong et al., 2021).

12.1.1 Extended reality

XR captures tools, which merge virtual environment with physical world (Lee et al., 2013). XR has human and machine interactions to increase the usability drive and has potential benefits including scalability, enhanced results, and cost-effectiveness (Suh and Prophet, 2018).

XR is a global terminology that exemplifies recent developments in the major divisions, i.e., augmented reality (AR), mixed reality (MR), and virtual reality (VR) (Logeswaran et al., 2021).

Extended Reality for Healthcare Systems, Volume 1
ISBN 978-0-323-98381-5, https://doi.org/10.1016/B978-0-323-98381-5.00010-6

12.1.1.1 Augmented reality

This is the superimposition of advanced information onto this present reality.

12.1.1.2 Mixed reality

This is a consistent hybrid of virtual and innovative situations.

12.1.1.3 Augmented reality

This is an interactive experience of the real world.

12.1.2 Basic architecture

Since a long time, AI's offspring (as shown in Fig. 12.1)—AR, VR, and MR—however, another element—XR—has been currently in great usage. Such idea consolidates three main aces, their expansion and assembly are the motivation behind the rise of XR. Simultaneously, XR is additionally acquiring stream and influencing various cycles and occasions for the coming technologies (Shaptunova, 2019).

Main benefits of this technology are as follows:

- Increased user engagement
- Higher knowledge of retention
- Low operational cost
- Safe virtual environment for efficient training

XR delivers a gainful method to train within a virtual environment. The interactive and immersive method enables trainers to deliver a large quantity of information in a visually appealing way. XR environment will significantly improve many factors, such as knowledge retention levels, user engagement, cost-effectiveness, and performance (Shaptunova, 2019).

Main challenges include the following:

- Compromising privacy: XR is vulnerable to cyberattacks, especially, hacking of data. It can cause grave harm, as XR-based resolutions can access huge confidential information.
- Restricted societal participation: XR offers a wide range of entertainment options, which can focus on people's minds completely and also eliminates the communication need. XR enables communication in various ways with restricted interaction.

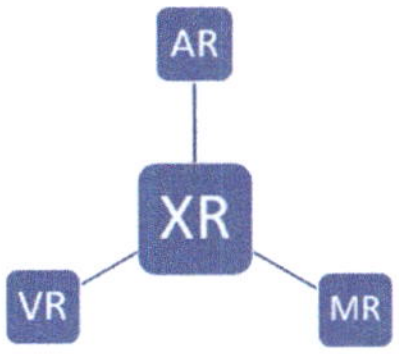

Figure 12.1 Elements for XR (extended reality).

- Impact on health: The prolonged use of AR/VR-enabled devices may result into headaches, eye disorders, nausea, etc.
- Implementation costs: Implementation and maintenance cost of XR is highly expensive.

12.1.3 Edge AI

Edge AI is an amalgamation of edge computing with AI. AI algorithms are executed at local level, either on the device or on the server around the device. Algorithms use data obtained by the devices. Independent decision-making is done by these devices in milliseconds despite Internet connection or cloud systems. Edge AI has no limits in case of potential use cases. Edge AI applications diverge from smart gadgets to production lines and ranging from logistics to smart buildings (Greengard, 2020).

The main features of Edge AI include the following:

- Reduction in costs and time for more improved experience of user.
- Facilitating the combination of wearable technology based on experience of user.
- Increasing data privacy protection by local processing as data are no longer shared over centralized cloud.
- The decrease in bandwidth leads to cost reduction of limited Internet service.
- Devices based on edge technology do not need any special maintenance by the developers.
- The monitoring is done with the help of graphic data flow automation.
- Providing intelligence to the security camera detection process.
- Edge AI enhances the ability of autonomous vehicles to process data and images in real time to detect traffic signals, pedestrians, other vehicles, and roads, to improve transport safety standards.
- Edge AI is used for analyzing image and video, producing responses to audio—visual systems, or recognition of scenes in real time.
- Reducing costs and improving safety for industrial Internet of things (IoT). AI monitors equipment to detect possible errors related to production chain, while machine learning compiles real-time data for the entire process.
- It is utilized for analyzing medical images in emergency healthcare.

Major challenges related to Edge AI (Varghese et al., 2016) include the following:

- Edge node computing: Edge computing can be simplified to several nodes positioned amid the edge devices and the cloud, which includes base station, access points, etc. Standard channels, such as digital signal processors (DSPs), customize the workload they manage. Practically, basic channels may not be able to handle analytical tasks basically as DSPs are not planned for general computing. In addition, it is not easily identified whether these nodes can accomplish calculations with their prevailing workloads.

- Determining edge nodes: Determining resources in a distributed environment is a well-researched area. This is simplified in close and open integrated situations using various techniques embedded in monitoring tools and service brokers. Nevertheless, misusing edge network entails detection methods finding suitable nodes that can be used in decentralized cloud configurations.
- Division and unburdening tasks: Due to the changes in the distributed computer mode in the development of multiple strategies that facilitate the separation of tasks that can be performed in multiple areas. Nevertheless, the use of edge nodes to extract statistics poses the challenge of not only categorizing computer functions correctly but also doing this in an automated manner without having to clearly define the capabilities or location of the edge nodes.
- Rigid service quality: The quality carried by the edge nodes can be determined by QoS and quality delivered to the user by QoE. Only rule required to be accepted in edge computing is to not overload nodes with heavy computer load. The challenge is to make sure the nodes reach the maximum throughput and reliability when distributing their planned workload when additional workloads from the data center or edge devices can be allowed.
- Optimization with the compatible technologies: Devices such as routers and basic channels are required to act as a platform that is publicly accessible edge nodes. Many related challenges need to be addressed. Firstly, the risks linked with public and private administrations who own such devices and devices using them will need to be exposed. Secondly, the planned purpose of the device cannot be negotiated if utilized as edge computing nodes. Lastly, multiple tenancy on edge nodes will happen with technology that puts security in place as a major concern.

12.1.4 Healthcare 4.0

Healthcare 4.0 enables handling and managing a huge volume of patient's real-time data and provides the facility to take accurate and better decisions for treating patients based on that data. It also helped doctors and medical practitioners to carry out predictive analysis of patient's diseases in a better and efficient way (Rehman et al., 2019). Healthcare 4.0 focuses on streamlining the medical processes to provide quick and efficient solutions to healthcare problems. It has introduced concepts of telemedicine and precision medicine, which can improve the availability of care in terms of space and time (Rehman et al., 2019). The current marketing trends and also the old scientific literature witness the importance of healthcare for living.

Healthcare 4.0 features include the following:

- Real-time data collection (Healthcare 4.0)
- Increased use of AI for prediction of results
- Enabling informed decision systems (enabling a check upon treatments)

- Big data storage (medical data records)
- Computing-based treatments (advanced treatments with efficient technology)

Healthcare 4.0 Architecture:

Healthcare 4.0 comprises smart industry and smart engineering as the basic blocks to build the smart healthcare systems. As per Fig. 12.2, the usability of the smart devices precisely in the healthcare system broadens the scope of treatment availability for the patients. Monitoring the real-world processes and producing the corresponding output can assist in treating the diseases (Aceto et al., 2020). Cloud computing enhances the storage capability for the healthcare system and makes the analyzation of the data at ease. The main key element in the healthcare 4.0 system architecture is the use of IoT devices, which is a modern technology including the smart objects, sensors, and devices to track, monitor spectate, and supervise the patients. There are only medical-related IoT systems considered in the Healthcare 4.0, which are there to continuously develop the healthcare (Aceto et al., 2020).

12.2 Revolution of healthcare from 1.0 to 4.0

Healthcare 1.0 signifies the first industrial revolution, first in the 18th century to the middle of the late 19th century. Over time with the emergent technologies, the transformation of machines has facilitated the movement of people from agriculture to industry. In the late 19th and early 20th centuries, a second industrial revolution, namely

Figure 12.2 Basic architecture of Healthcare 4.0 (Aceto et al., 2020).

Industrial 2.0, presented a standard shift from individual- or small-scale production to large-scale production, due to high energy consumption. Henry Ford's integration lines have permitted the systematic distribution and connection of multiple operations and equipment effectively, which completely transformed production processes (Li and Carayon, 2021). In the past decades of the 20th century, rapid advances in computing technologies have resulted into large use of automation systems, digitalized systems, and network communication production and business processes. Integrated and flexible production systems, advanced execution systems, and resource planning have been extensively accepted, which manifested the third industrial revolution, namely, Industry 3.0, and resulted in emergence of the fourth industrial evolution, Industry 4.0 (Fig. 12.3).

Striding into the 21st century, Industry 4.0 signifies the fourth revolution of industry and aims to combine cyberphysical systems with data, process, equipment, and operational skills. The IoT and related services are still widely disseminated and utilized in accordance with big data and techniques of AI. With cloud-based diagnostics, performance, control, and storage of resources, products, and business processes, one can attain sustainable, smart, and affordable integration of supply and service production.

12.3 Trends in Healthcare 4.0

➢ Healthcare with IoT technology

Healthcare 4.0 is focused on monitoring the technological advancements that can provide consistent data of patients active for effective medication. This highlights the requirements of equipment such as sensors, monitors, and other tools that record and display the data (Yuehong et al., 2016). IoT is foundation to become a new trend in healthcare technology.

➢ Exploration through VR

VR has contributed in a superb method when it comes to advancement in healthcare technology during the Healthcare 4.0 evolution.

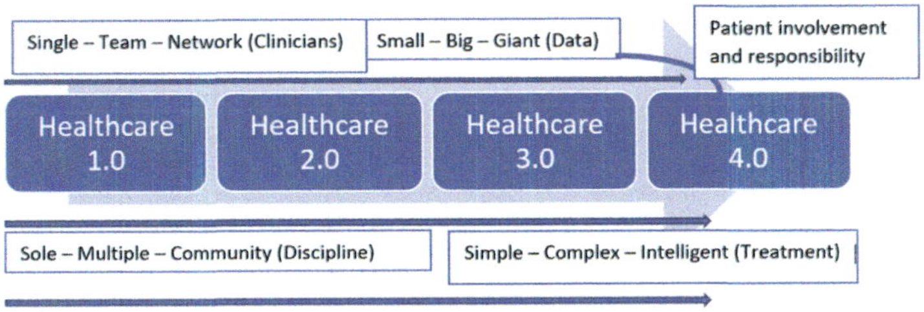

Figure 12.3 Healthcare 1.0 to 4.0 (Li and Carayon, 2021).

➢ Communication through AI

The adaptation of AI is becoming more personalized and advanced with circumstance-based scenarios each day (Yuehong et al., 2016). The implementation of AI is playing a pivotal role in making the patient's diagnosis clearer by allowing them to communicate their mind space that can keep them content and safe while getting treatment.

➢ A virtual treatment is a new treatment style

The adoption of IoT devices in healthcare such as wearable fitness bands, pulsometer, digital thermometer, and any such similar health equipment is glorifying the steps of self-healthcare for individuals. These devices are allowing the users to measure their respective health vitals, record them, and share them with the doctor at the time of requirement of immediate diagnosis (Yuehong et al., 2016).

12.4 Extended reality and Healthcare 4.0

The main area of benefit by XR is medical imaging such as "Modern MRIs and CT scans" that extensively utilizes this technology in Healthcare 4.0. It also allows 3D illustration of human body in spite of the standard 2D image, which improves diagnostic efficiency. Likewise, these technologies greatly increase the chances of surgical training for students so that they can rehearse on patients virtually.

Major areas where extended reality and healthcare integrate are as follows:

➢ Clinical XR for evaluation in clinical neuroscience
➢ Simulation-enabled neuropsychological calculations (Parsons et al., 2020)
➢ Organizing and investigating multisensory physical contents and recognizing all probable cures related to the recovery by utilizing XR (Parsons et al., 2020)
➢ Social XR: Social XR is defined as the research that has been done on the brain, which includes all the stimuli and cognitive approaches to support the results (Parsons et al., 2020)

XR denotes integration of real and virtual systems, where human–machine interaction happens over communications produced by computer and hardware technology. XR technology entails "VR," "MR," and "AR." In Fig. 12.4, an outline of the XR network and relation of VR/AR/MR with real and virtual environment is shown (Doolani et al., 2020).

Integration of XR and Healthcare 4.0 (Elstner, 2020):

➢ XR can help in visualizing medical data more efficiently.
➢ XR can improve therapeutic treatments and even surgeries.
➢ XR can improve well-being in the healthcare environment.
➢ XR can enable to build surgery simulator for training medical practitioners.
➢ XR can help to understand patients' condition better and relieve their pain.

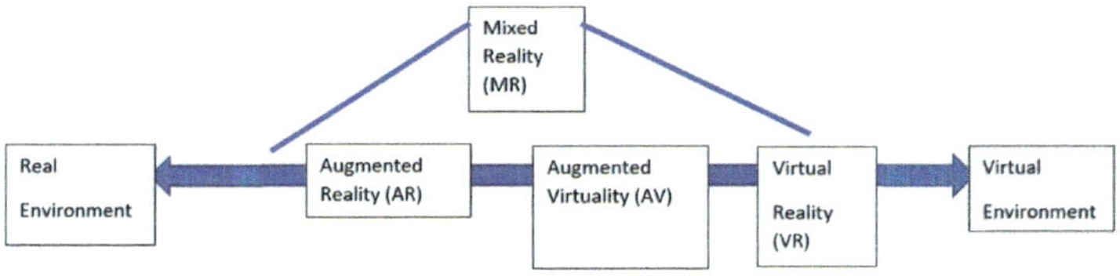

Figure 12.4 XR and Healthcare 4.0. XR, extended reality.

- ➢ XR can prepare surgeons before actual surgeries.
- ➢ XR can conduct educational training for medical students.
- ➢ XR can diagnose health conditions of patients.

XR for remote healthcare applications: XR is a common term for technologies such as VR, AR, and MR, which helps in developing new experience in areas like healthcare (healthcareitnews). XR gives healthcare providers the opportunity to test solutions past the 2D technology. XR helps in enabling patient communication in virtual mode irrespective of their physical distance. Also, XR devices made it possible for taking sensor-based measurements, permitting delicate measurements can be done, and adds insight to patient diagnosis. With the introduction of XR technology in healthcare, the following facilities are being improved:

- Trainers/instructors: They promote training anytime and anywhere, as users can contact their trainers in virtual environment, and also provide proper monitoring of training schedules and use of tools and techniques.
- Healthy food habits: These enable healthy eating by giving users insights on information related to nutrition. It motivates users to keep track of their progress on suggested diets and intakes.
- Physical healing: This provides user-interactive AI-enabled virtual environment for motivating users to follow their physical exercise regime and gives capability to therapists for patient data collection related to recovery.
- Virtual surgery: It generates digitized models for planning surgeries and permitting physicians for sequence optimizing, making course of action details and preparation for exigencies.
- Patient's visions: With the help of virtual environment, patients can easily take insights of their medical situations and actions, keeping records of their medicines and proper dosages, motivating speedy recovery and keeping them at ease.
- Medical training: XR permits interactive skillful experience like human anatomy study and body mechanics to provide depth knowledge of human body functions in 3D virtual mode.
- Telemedicine: Utilization of decentralized specialized care systems such as dermatology and pediatrics enables transferring from huge setting services to community-oriented and cost-effective solutions and provides extensive investments for patients.

- Global pandemic response: It encompasses the speediness, accessibility, and alliance of benefactors without generating difficulties related to resources frequently triggered by epidemics. XR permits professionals to perform remote tests, visualize what is important, and provide immediate treatment

The application of XR with VR and AR is greatly helping healthcare workers in the battleground in saving their lives, offering support in rescue circumstances, providing better care facilities to the patients, ensuring lifesaving medical equipment is properly cared for, and making healthcare available and open to a huge number of unmerited individuals.

12.4.1 Augmented reality in Healthcare 4.0

AR utilizes display, camera, and sensors for edging information in the real world digitally. AR is embraced at huge level by healthcare industry. Actually, nowadays, many doctors are working daily on AR applications to improvise patient health information and results. Healthcare workers quickly saw the benefits of AR technology. Education is the obvious use of AR in the healthcare sector. The basic requirement of a health professionals is to acquire a great amount of information related to human body anatomy and its working. Many AR-based applications provide the ability to envision and communicate with 3D representation of human body (Gerup et al., 2020).

12.4.2 Virtual reality in Healthcare 4.0

Around 85% of medical experts approve that the VR provides an easy way to access and learn the healthcare knowledge for health specialists and medical students. The organization also quotes that approximately 68% of patients will accept VR-based healthcare services as an alternative to traditional healthcare (Mazurek et al., 2019).

The acceptance of VR in healthcare is because of the following reasons:

- Rising requests for improved healthcare service quality
- Reduction in healthcare costs
- Augmented part of connected devices in the healthcare segment

12.4.3 Augmented reality—virtual reality opportunity in Healthcare 4.0

AR permits to display certain information, videos, and photos on smart devices in real time. Hence, it expands the real world by adding more data to it. VR creates simulation by computer technology where a person is completely involved in a digital environment (scand).

Areas of implementation of AR—VR in Healthcare 4.0 are as follows:

- Medical training
- Robotic surgery
- Physical therapy

- Posttraumatic stress management
- Anxiety and depression management
- Emergency treatment

Virtual and augmented authenticities gain acceptance in healthcare systems. They offer top-quality services to medical professionals in treating a patient and to teach healthcare clients about making their healthcare better (scand) (Fig. 12.5).

12.5 Edge AI and Healthcare 4.0

Edge AI enables intelligent solutions to be deployed on edge devices, reducing latency, allowing offline execution, and providing strong privacy guarantees. Unfortunately, achieving efficient and accurate execution of AI algorithms on edge devices, with limited power and computational resources, raises several deployment challenges (Kamruzzaman, 2021).

1. Autonomous monitoring of hospital rooms

Edge AI algorithms make use of huge set of host sensors for data collection and analyzing them for better response. Through utilization of computer vision, as well as data from other sensors, independent monitoring of hospital rooms can be made possible (Efthymiou et al., 2020).

2. New applications in radiology
   - Identification of cardiovascular abnormalities
   - Detection of fractures and other musculoskeletal injuries
   - Supporting the diagnosis of neurological diseases

3. Rural medicine

With the advent of telemedicine and easily accessible health information, medical providers have struggled in providing immediate, high-quality care to people who live far from hospitals and have inadequate Internet access. General healthcare databases face major challenges because of connectivity issues, but the integration of IoT medical devices and edge-of-the-line computing applications has made it easier to resolve such problems (Efthymiou et al., 2020).

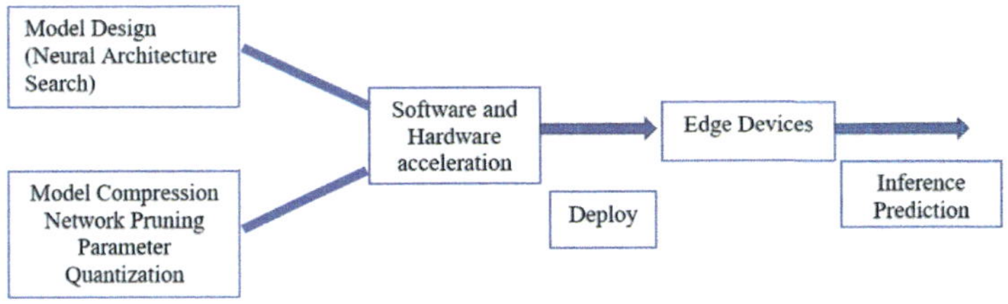

Figure 12.5 Edge AI architecture for Healthcare 4.0 (Kamruzzaman, 2021).

12.6 Future of edge AI and extended reality for Healthcare 4.0

- Edge AI coupled with XR commits to decentralized healthcare services and provides supportive remote healthcare to disabled and elderly patients.
- Its deployment may expand the accessibility of cancer screening organizations and expand effective technology for patient monitoring, such as cardiac pacemakers, in distant areas.
- Edge AI and XR can be utilized in worldwide healthcare applications including visualizations for surgeries and diagnosis and various related studies more efficiently (Gerup et al., 2020).
- Edge AI and XR can also improve advanced therapeutic treatments such as cancer and tumor treatments, and utilization of AR and VR helps in carrying out critical surgeries with safety and understanding, such as heart transplant, etc.
- Edge AI and XR in integration can solve the pain management problem. Physicians and researchers are researching the implementation of VR as an effective alternative of opioids in managing pain. VR impacts emotional state of patients and track intensity of pain and also aids to prevent symptoms of pain to reach to the brain. According to a study of patients having neuropathic pain, they experienced 70% decrease in pain throughout VR sessions and 55% subsequent rapid reduction (Mazurek et al., 2019).

12.7 Conclusion

Edge AI and XR are most powerful tools when we consider their integration in the Healthcare 4.0. In this chapter, we have discussed about the Edge AI and XR with respect to the Healthcare 4.0. We have discussed about the different trends in the Healthcare 4.0 and advanced models for implementing the Edge AI and XR. As there is always a major concern with the data security and storage, the latest industrial revolution and the healthcare evolution to 4.0 might sound promising with all the latest technologies. However, managing, securing, and monitoring the device fleets with reliable and efficient technology in healthcare can be an upcoming challenge. Healthcare 4.0 offers many opportunities and challenges in the engineering of healthcare systems. Both Edge AI and XR are proved to be beneficial in taking advantage of opportunities and facing challenges in obtaining smart and connected healthcare.

References

https://www.healthcareitnews.com/news/extended-reality-holds-incredible-potential-remote-care-applications.

Aceto, G., Persico, V., Pescapé, A., 2020. Industry 4.0 and health: internet of things, big data, and cloud computing for healthcare 4.0. Journal of Industrial Information Integration 18, 100129.

Andrews, C., Southworth, M.K., Silva, J.N., Silva, J.R., 2019. Extended reality in medical practice. Current Treatment Options in Cardiovascular Medicine 21 (4), 1−12.

Doolani, S., Wessels, C., Kanal, V., Sevastopoulos, C., Jaiswal, A., Nambiappan, H., Makedon, F., 2020. A review of extended reality (xr) technologies for manufacturing training. Technologies 8 (4), 77.

Efthymiou, I.P., Sidiropoulos, S., Kritas, D., Rapti, P., Vozikis, A., Souliotis, K., 2020. AI transforming healthcare management during covid-19 pandemic. HAPSc Policy Briefs Series 1 (1), 130−138.

Elstner, M., 2020. Use Cases of Extended Reality in the Construction Industry.

Gerup, J., Soerensen, C.B., Dieckmann, P., 2020. Augmented reality and mixed reality for healthcare education beyond surgery: an integrative review. International Journal of Medical Education 11, 1.

Greengard, S., 2020. Ai on edge. Communications of the ACM 63 (9), 18−20.

Kamruzzaman, M.M., 2021. New Opportunities, Challenges, and Applications of Edge-AI for Connected Healthcare in Smart Cities. In: *2021 IEEE Globecom Workshops (GC Wkshps)*. IEEE, pp. 1−6.

Lee, H.G., Chung, S., Lee, W.H., 2013. Presence in virtual golf simulators: the effects of presence on perceived enjoyment, perceived value, and behavioral intention. New Media & Society 15 (6), 930−946.

Li, J., Carayon, P., 2021. Health Care 4.0: a vision for smart and connected health care. IISE Transactions on Healthcare Systems Engineering 1−10.

Logeswaran, A., Munsch, C., Chong, Y.J., Ralph, N., McCrossnan, J., 2021. The role of extended reality technology in healthcare education: towards a learner-centred approach. Future Healthcare Journal 8 (1), e79.

Mazurek, J., Kiper, P., Cieślik, B., Rutkowski, S., Mehlich, K., Turolla, A., Szczepańska-Gieracha, J., 2019. Virtual reality in medicine: a brief overview and future research directions. Human Movement 20 (3), 16−22.

Ong, T., Wilczewski, H., Paige, S.R., Soni, H., Welch, B.M., Bunnell, B.E., 2021. Extended reality for enhanced telehealth during and beyond COVID-19. JMIR Serious Games 9 (3), e26520.

Parsons, T.D., Gaggioli, A., Riva, G., 2020. Extended reality for the clinical, affective, and social neurosciences. Brain Sciences 10 (12), 922.

Rehman, M.U., Andargoli, A.E., Pousti, H., 2019. Healthcare 4.0: Trends, Challenges and Benefits.

https://scand.com/company/blog/what-does-the-future-hold-for-ar-and-vr-in-healthcare/.

Shaptunova, Y., 2019. What Is Extended Reality and what Can We Do with it? SaM Solutions.

Suh, A., Prophet, J., 2018. The state of immersive technology research: a literature analysis. Computers in Human Behavior 86, 77−90.

Varghese, B., Wang, N., Barbhuiya, S., Kilpatrick, P., Nikolopoulos, D.S., 2016. November). Challenges and opportunities in edge computing. In: 2016 IEEE International Conference on Smart Cloud (SmartCloud). IEEE, pp. 20−26.

Yuehong, Y.I.N., Zeng, Y., Chen, X., Fan, Y., 2016. The internet of things in healthcare: an overview. Journal of Industrial Information Integration 1, 3−13.

CHAPTER THIRTEEN

A miniaturized multilayer triband off-body antenna for heterogenous applications in Internet of Medical Things

Umar Farooq[1], Hushmat Amin Kar[2] and Shoaib Amin Banday[1]
[1]Department of Electronics and Communication, Islamic University of Science and Technology, Awantipora, Jammu and Kashmir, India
[2]Department of Information Technology, National Institute of Technology, Srinagar, Jammu and Kashmir, India

13.1 Introduction

Internet of medical things (IoMT) is a key technology for modern healthcare infrastructure. It comprises medical devices capable of obtaining, analyzing, and sharing the data over Internet. These devices should essentially support high data rate and low latency requirements of 5G-based IoMT applications (Manogaran et al., 2018; Khan and Alam, 2021). To get maximum benefits from IoMT technology, it needs to be integrated with "extended reality (XR)," an umbrella term that covers all of "augmented reality, virtual reality (VR), and mixed reality (MR)." The medical information from the IoMT devices can be integrated with XR devices. These XR devices in turn offer incredible flexibility for 3D/2D displaying and visualizing the medical data obtained from IoMT devices. Unlike IoMT where information is displayed simply on the screen, integrated XR and IOMT has the capability of displaying the same information in full 3D/2D "virtual window." This gives rise to a plethora of modern medical applications including robotic surgeries, remote healthcare, and 3D medical imaging (Andrade and Bastos, 2019; Andrews et al., 2019). IoMT also needs to coexist with other communication systems such as 4G, Wi-Fi, and WiMax to support low-frequency applications (Adarsh et al., 2021). This signifies that IoMT systems are inherently heterogenous and require the multiband antennas capable of operating at mmWave band and L/S/C bands also to support heterogenous network applications (Chung and Chang, 2020). A number of multiband antenna design techniques have been presented in the literature. Chaung et al. have proposed a multiband antenna using a parasitic element for 1.8/2.4/5.2-GHz applications (Chuang et al., 2012). Gu et al. (2017) have proposed a frequency selective method for 2.5/5.3-GHz dual-band antenna operations. M. Saad et al. have investigated a 94/140-GHz dual-band antenna using the triangular fractal structure with a bowtie slot (Khan and Cheema, 2017). Wing Chi et al. have proposed a multiband patch antenna by

Extended Reality for Healthcare Systems, Volume 1
ISBN 978-0-323-98381-5, https://doi.org/10.1016/B978-0-323-98381-5.00004-0

cutting the U-slots at the appropriate position for 3.7/4.95/5.55-GHz operations (Mok et al., 2013). Sharma et al. (2016) have used the stub loading technique to design the multifrequency antenna in S band. Shafai et al. (2000) have investigated the dual L/C band arrays using the single shared aperture technique at resonant frequencies 1.2 and 5.3 GHz, respectively. Mathur and Kumar (2019) have investigated a dual S/X band antenna using a layered structure. However, none of these proposed multiband antennas can operate simultaneously at mmWave band and low-frequency band. Hence, the multiband antennas which can work for mmWave band as well as low-frequency band such as L band or S band or C band are still subject of research. Syeda et al. have also proposed an integrated antenna system for both 4G and 5G applications operating in C band and Ka band (Naqvi et al., 2019). However, the dimensions of the antenna are very high (110 × 75 mm), and hence, it is not feasible for most of the futuristic handheld IoMT devices. As such, there is the need of novel miniaturized multiband antenna system, which can operate at mmWave frequency band as well as lower frequencies. In this context, a triband miniaturized off-body antenna has been designed at C, Ka, and V bands for IoMT applications. The antenna can be used for heterogenous healthcare applications, or the same concept can be utilized to design a multiband band antenna at interested frequencies as per the application requirements according to the FCC/OFCOM regulations of 5G (Wang et al., 2017).

Including the introduction and conclusion, this chapter is organized into six different sections. Section 13.2 describes the proposed antenna configuration and design. Section 13.3 discusses the simulation results. Section 13.4 discusses the experimental results of the proposed antenna. A comparative study to show the effectiveness and novelty of the proposed antenna is given in Section 13.5. Finally, Section 13.6 concludes the chapter.

13.2 Antenna configuration and design

A miniaturized triband antenna is proposed for "C/Ka/V" band operations. The proposed antenna has a multilayered configuration as shown in Fig. 13.1. Layer 1 to layer 5 correspond to electromagnetically coupled C band RMSA, and layer 6 to layer 8 correspond to Ka band and V band RMSA antenna array. Layer 5 acts as patch for C band antenna and also as ground plane for Ka band antenna array. Two symmetrical slots are incorporated on each element of the Ka band array to excite third frequency band, i.e., V band. The detailed description of each layer of the antenna structure is given in Table 13.1.

The proposed triband antenna has been designed in CST Microwave Suite in a schematic manner. First an electromagnetically coupled RMSA antenna at C band (5.5 GHz) is designed as shown in Fig. 13.2. The antenna uses Fr4 material ($\epsilon_r = 4.3$) as antenna substrate. The feed network and the patch are separated by foam material. The design parameters of the antenna have been calculated using the standard mathematical relations (Farooq and Ghulam Mohammad, 2019) and are given in Table 13.2.

Figure 13.1 Configuration of the proposed multilayered multiband antenna.

Table 13.1 Description of each layer of the designed multiband antenna.

Layer	Description
L1	Ground for antenna at C band
L2	Substrate 1 (Fr−4)
L3	Feed of C band antenna
L4	Airgap1/Foam
L5	Antenna patch at C band and also ground for Ka band
L6	Substrate 2 (Rogers RT 5880)
L7	Feed network for Ka band
L8	Airgap 2/foam
L9	Antenna array at Ka band

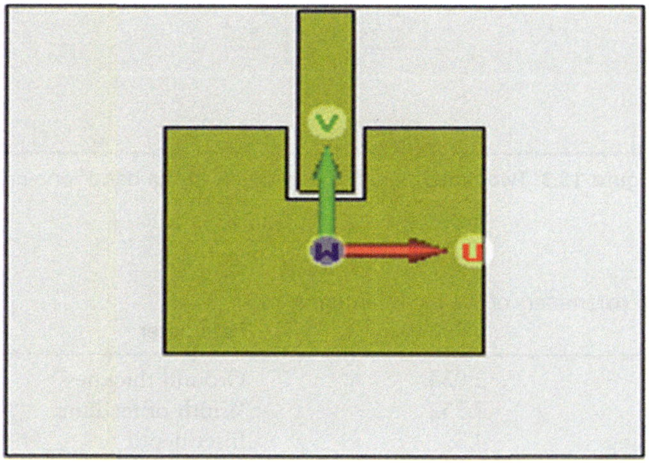

Figure 13.2 Two-dimensional configuration of C band antenna.

Table 13.2 Design parameters of the C band antenna.

Parameter	Value (mm)	Parameter	Value (mm)
Patch length	13	Substrate thickness	1.25
Patch width	17	Length of feedline	10.4
Length of ground	26	Width of feedline	2.72
Width of ground	2.72	Inset depth	4.16
Thickness of ground	0.001	Foam thickness	0.54

Using the C band antenna as a ground plane, a 2×2 "Ka band" antenna array is designed. Duriod 5880 ($\epsilon_r = 2$) is used as substrate material for "Ka band" due to its best suitability for higher-frequency operations (Rogerscorp). The elements of the array are separated by spacing of 4.5 mm (greater than $\lambda/2$) to evade coupling effects. The array is fed using electromagnetic coupled corporate feeding technique so that power is distributed equally to each patch as shown in Fig. 13.3. The feed network and patch elements are separated by foam material. The physical dimensions of each element of the "Ka band" antenna array are given in Table 13.3.

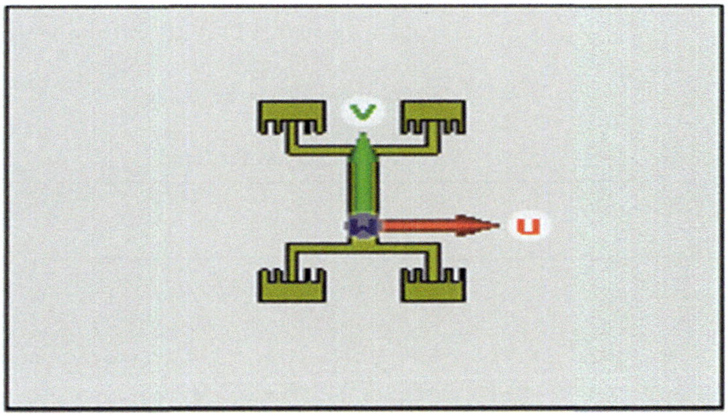

Figure 13.3 Two-dimensional configuration of "Ka band" antenna.

Table 13.3 Design parameters of "Ka band" antenna array.

Parameter	Value (mm)	Parameter	Value (mm)
Patch length	2.034	Ground thickness	0.001
Patch Width	2.752	Width of feedline	0.52
Interelement spacing	4.5	Inset depth	0.79
Substrate thickness	0.45	Foam thickness	0.26

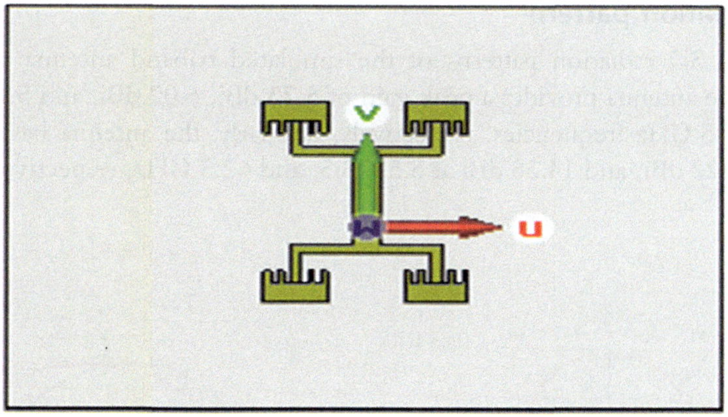

Figure 13.4 Configuration of "Ka and V band" antenna with symmetrical slots.

13.2.1 Excitation of "V band" operation

To excite the "V band" operation, two symmetrically placed rectangular slots with dimension 0.86 × 0.45 mm are introduced at optimum positions in each array element as shown in Fig. 13.4. The slot dimensions are computed using well-defined relations from literature (Farooq and Ghulam Mohammad, 2020):

$$\text{Slot length: } L_s = \frac{0.34 \times c}{f \times \sqrt{\varepsilon_{eff}}} - \frac{h - 1.6}{0.6} \quad (2.1)$$

$$\text{Slot width: } W_s = \frac{0.18 \times c}{f \times \sqrt{\varepsilon_{eff}}} \quad (2.2)$$

where "h" = height of Ka band dielectric material (0.45 mm)
"c" = velocity of light (3×10^8 m/s)
"f" = resonant frequency of higher band
"ε_{eff}" = effective permittivity of Ka band dielectric material (2.02)

13.3 Simulation study

The designed triband antenna is simulated for "off-body" scenarios in CST Microwave Suite, and its performance characteristics such as radiation pattern, return loss, SAR, peak gain, directivity, bandwidth, and beam width are examined. These are presented in the following sections.

13.3.1 Radiation pattern

The 2D and 3D radiation patterns of the simulated triband antenna are shown in Fig. 13.5. The antenna provides a peak gain of 5.23 dBi, 6.02 dBi, and 5.18 dBi at 5.5, 33.5, and 62.5 GHz frequencies, respectively. Similarly, the antenna has directivity of 6.46 dBi, 10.22 dBi, and 14.38 dBi at 5.5, 33.5, and 62.5 GHz, respectively. The beam

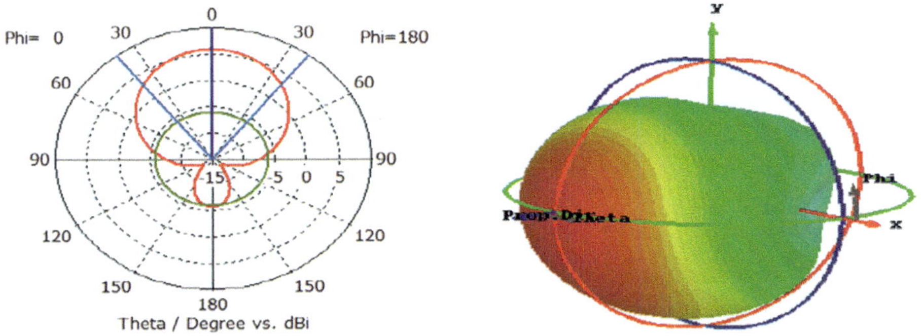

(a) Radiation pattern at 5.5 GHz

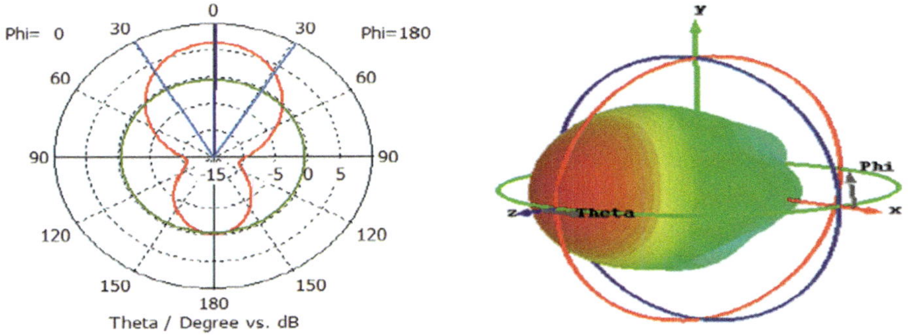

(b) Radiation pattern at 33.5 GHz

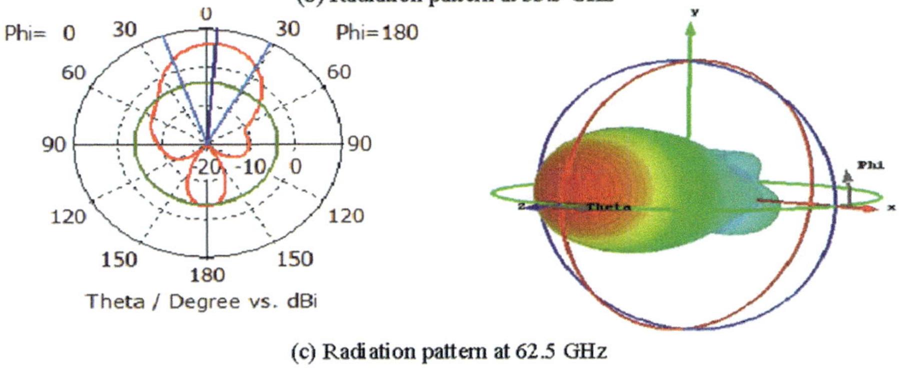

(c) Radiation pattern at 62.5 GHz

Figure 13.5 2D and 3D radiation pattern of the proposed triband antenna.

width as calculated from the radiation pattern plot is 76.43°, 59.53°, and 45.62° at all three predefined resonant frequencies.

13.3.2 Return loss and Voltage Standing Wave Ratio

The return loss plot of the designed triband antenna is given in Fig. 13.6. The figure shows that antenna achieves a minimum return loss of −15.63 dB at 5.5 GHz, −19.35 dB at 33.5 GHz, and −21.05 dB at 62.5 GHz. The voltage standing wave ratio (VSWR) plot is given in Fig. 13.7. The figure shows that the VSWR of the designed antenna is well below

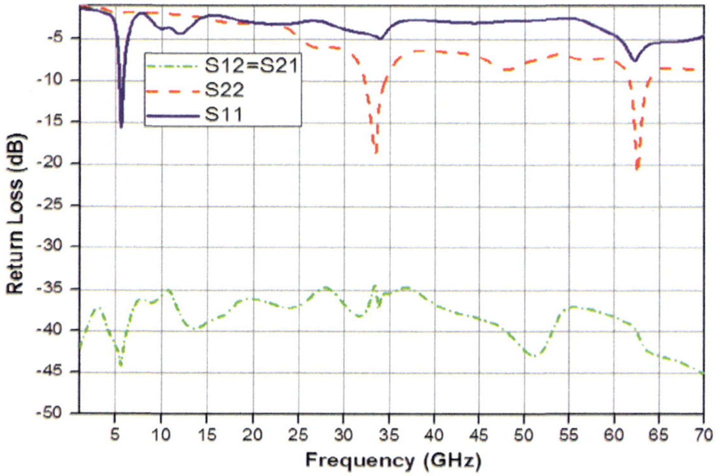

Figure 13.6 Return loss plot of the proposed triband antenna.

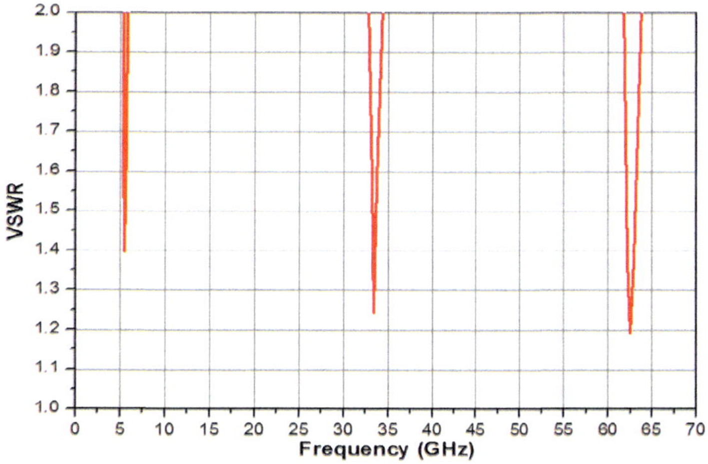

Figure 13.7 VSWR plot of the proposed triband antenna. *VSWR*, voltage standing wave ratio.

Table 13.4 Performance parameters of the designed "C/Ka/V" triband antenna.

	Resonant frequency (GHz)		
Parameter	**5.5**	**33.5**	**62.5**
Return loss	−15.63 dB	−19.35 dB	−21.05 dB
VSWR	1.398	1.242	1.194
Band width	0.36 GHz	1.6 GHz	2.12 GHz
Directivity	6.46 dBi	10.22 dBi	14.38 dBi
Peak gain	5.23 dBi	6.02 dBi	5.18 dBi
SAR (W/kg)	1.56	1.42	1.28
Beam width	76.43	59.53	45.62

the standard value of 2.0 for all resonant frequencies. The low return loss and VSWR values of the designed antenna indicate that the reflections are very less. The various performance parameters of the designed triband antenna obtained from the simulations are also summarized in Table 13.4.

13.4 Experimental study

To test the validity of simulated antenna, the proposed triband antenna has been fabricated, and its return loss results are measured using the vector network analyzer (VNA) for off-body scenarios of IoMT networks. A V-type connector and PTFE semirigid cables have been used in the experimental setup of the designed antenna. The fabricated prototype of the proposed antenna along with the measured results is given in Figs. 13.8 and 13.9. It can be seen that there is a small variation in simulation and experimental results. This small variation can, however, be attributed to the slight differences between the conductivities and loss tangents used in the simulation study and the actual values of the fabricated antenna.

13.5 Comparative study

A comparative study to show the novelty and effectiveness of the proposed antenna with respect to the state-of-the-art multiband mmWave antennas (Shafai et al., 2000; Rogerscorp; Farooq and Ghulam Mohammad, 2020; Zhang et al., 2018; Zhou et al., 2013) is summarized in Table 13.5. The most attractive feature of the proposed antenna is that it operates at three bands, i.e., "C/Ka/V" bands, while most of the referenced antennas support only dual-band applications. The table also shows that the proposed antenna performs comparatively better having smaller size and very low return loss and also provides the reasonable gain and bandwidth compared with the related work.

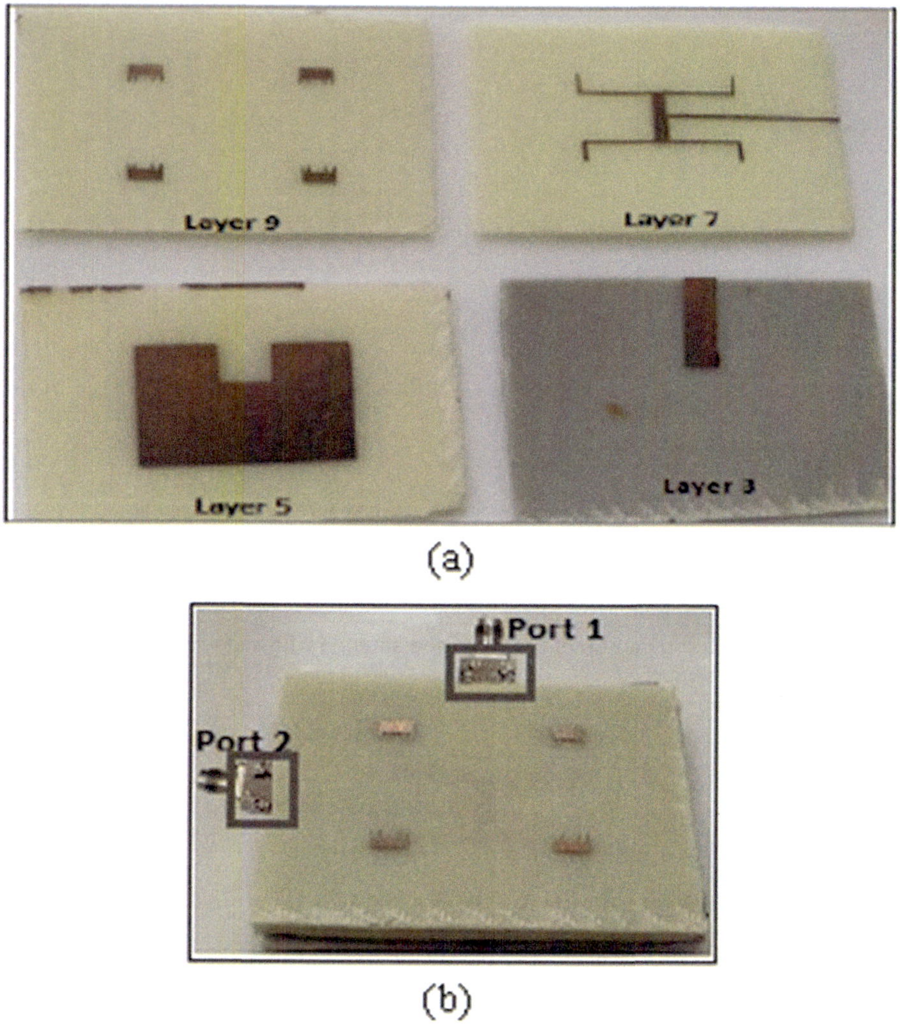

Figure 13.8 (A) Fabricated prototype of multiple layers. (B) Final antenna prototype.

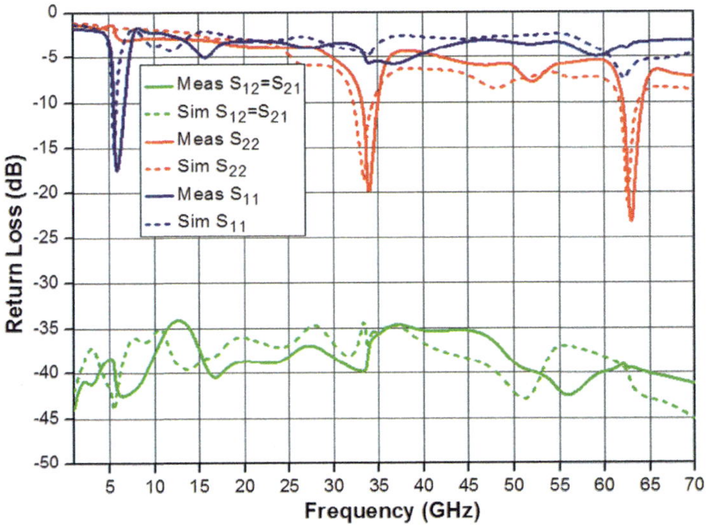

Figure 13.9 Experimental and simulated return loss results.

Table 13.5 Performance comparison of the proposed antenna with state-of-the-art multiband mmWave antennas.

Reference Antenna	Frequency (GHz)	Size (mm^3)	Return loss (dB)	Gain (dBi)	Bandwidth (GHz)
Naqvi et al. (2019)	3.8, 5.5, 28	110 × 75 × 0.51	−16.5, −15.5, −18	3.2, 5.4, 10.29	0.16, 0.45, 4.9
Zhang et al. (2018)	3.5, 60	80 × 86 × 2.6	−20, −13	7.3, 24	0.09, 3.9
Zhou et al. (2013)	0.33, 15.1	350 × 350 × 80	−17, −24	6.5, 22.5	0.09, 1.93
Aliakbari et al. (2016)	28, 38	6.8 × 6.8 × 0.25	−14.25, −19.50	4, 4.5	0.85, 0.75
Xiang et al. (2017)	5.8, 30	85 × 85 × 1.04	−18, −16	10.4, 8.51	0.16, 1.9
Present work	5.5, 33.5, 62.5	26 × 30 × 2.78	−15.63, −19.35, 21.05	5.23, 6.02, 5.18	0.36, 1.6, 2.12

13.6 Conclusion

A miniaturized trifrequency antenna intended for "C/Ka/V" band heterogenous applications of IoMT has been designed and analyzed for various performance parameters such as radiation pattern, return loss, VSWR, gain, directivity, bandwidth, and beam width. The designed antenna resonates at 5.5 GHz, 33.5 GHz, and 62.5 GHz

frequencies. The simulation results show the feasibility and robustness of the designed antenna with return loss, SAR, peak gain, bandwidth, and beam width values in the acceptable range. The performance of the proposed triband antenna has also been tested by the experimental study for off-body scenarios. The simulation results show good agreement with that of the experimental results. The antenna is best suited for the heterogenous networking applications of IoMT wherein mmWave-based IoMT systems can coexist with current low-frequency 4G/Wi-Fi/WiMax systems.

References

Adarsh, A., Pathak, S., Kumar, B., 2021. Design and analysis of a reliable, prioritized and cognitive radio-controlled telemedicine network architecture for Internet of healthcare Things. International Journal of Computer Networks and Applications 8 (1), 54–66.

Aliakbari, H., et al., 2016. A single feed dual-band circularly polarized millimeter-wave antenna for 5G communication. In: 10th European Conference on Antennas and Propagation (EuCAP). IEEE.

Andrade, T., Bastos, D., 2019. Extended reality in IoT scenarios: concepts, applications and future trends. In: 5th Experiment International Conference (Exp. At'19). IEEE.

Andrews, C., et al., 2019. Extended reality in medical practice. Current treatment options in cardiovascular medicine 21 (4), 1–12.

Chuang, C.-S., Jhang, Y.-J., Ku, T.-T., 2012. Compact multi-broadband monopole antenna for integrated mobile broadband wireless radio access system application. In: Asia Pacific Microwave Conference Proceedings. IEEE.

Chung, M.-A., Chang, W.-H., 2020. Low-cost, low-profile and miniaturized single-plane antenna design for an Internet of Thing device applications operating in 5G, 4G, V2X, DSRC, WiFi 6 band, WLAN, and WiMAX communication systems. Microwave and Optical Technology Letters 62 (4), 1765–1773.

Farooq, U., Ghulam Mohammad, R., 2019. Design and analysis of rectangular microstrip antenna (RMSA) for millimeter wave communication applications. Traitement du Signal 36 (5).

Farooq, U., Ghulam Mohammad, R., 2020. Design and analysis of dual band microstrip antenna for millimeter wave communication applications. International Journal of Computing and Digital Systems 9 (4), 607–614.

Gu, C., et al. Dual-band electronically beam-switched antenna using slot active frequency selective surface. IEEE Transactions on Antennas and Propagation 65(3), 1393-1398.

Khan, S., Alam, M., 2021. Wearable Internet of Things for personalized healthcare: study of trends and latent research. In: Health Informatics: A Computational Perspective in Healthcare. Springer, Singapore, pp. 43–60.

Khan, M.S., Farooq, A.T., Cheema, H.M., 2017. A multiband on-chip antenna for 94 and 140 GHz applications. In: 11th European Conference on Antennas and Propagation (EUCAP). IEEE.

Manogaran, G., Chilamkurti, N., Hsu, C.-H., 2018. Emerging trends, issues, and challenges in Internet of Medical Things and wireless networks. Personal and Ubiquitous Computing 22 (5), 879–882.

Mathur, P., Kumar, G., 2019. Dual-frequency dual-polarised shared-aperture microstrip antenna array with suppressed higher order modes. IET Microwaves, Antennas & Propagation 13 (9), 1300–1305.

Mok, W.C., et al., 2013. Single-layer single-patch dual-band and triple-band patch antennas. IEEE Transactions on Antennas and Propagation 61 (8), 4341–4344.

Naqvi, S.I., et al., 2019. An integrated antenna system for 4G and millimeter-wave 5G future handheld devices. IEEE Access 7, 116555–116566.

Available from: http://rogerscorp.com/acs/products/32/rt-duroid-5880-laminates.aspx.

Shafai, L.L., et al., 2000. Dual-band dual-polarized perforated microstrip antennas for SAR applications. IEEE Transactions on Antennas and Propagation 48 (1), 58–66.

Sharma, W.C.S., Kumar, H., Kumar, G., 2016. Single feed dual band circularly polarized stub loaded tunable microstrip patch antenna. In: Asia-Pacific Microwave Conference (APMC). IEEE.

Wang, T., et al., 2017. Spectrum analysis and regulations for 5G. In: 5G Mobile Communications. Springer, Cham, pp. 27–50.

Xiang, B.J., et al., 2017. A flexible dual-band antenna with large frequency ratio and different radiation properties over the two bands. IEEE Transactions on Antennas and Propagation 66 (2),), 657–667.

Zhang, J.F., et al., 2018. A dual-band shared-aperture antenna with large frequency ratio, high aperture reuse efficiency, and high channel isolation. IEEE Transactions on Antennas and Propagation 67 (2), 853–860.

Zhou, S.-G., Tan, P.-K., Chio, T.-H., 2013. Wideband, low profile P-and Ku band shared aperture antenna with high isolation and low cross-polarisation. IET Microwaves, Antennas & Propagation 7 (4), 223–229.

CHAPTER FOURTEEN

Economic impact of XR adoption on healthcare services

Samiya Khan
Faculty of Science & Engineering, University of Wolverhampton, Wolverhampton, United Kingdom

14.1 Introduction

Extended reality (XR) has the potential to impact healthcare in a big way, particularly domains such as medical education. While basic healthcare services can utilize XR for enhancing quality of service and management of increasing demands, medical education is expected to broaden its domain of learning opportunities with XR. At the society level, these aspects will have a cumulative impact on treatment planning and prevention strategies.

XR has been around for quite some time now, but it is garnering immense attention by usage in varied sectors because of the reduced prices of headsets and equipment required for this technology. Reduced costs mean that this technology can now be accessed and made available to mass sections of the society, particularly for trusts and care homes that cannot invest in expensive technology.

Virtual reality (VR) has found its way with reduction in equipment costs, but augmented reality (AR) and mixed reality (MR) are comparatively lagging behind in adoption. Industry collaborations and research community are actively working toward making this technology as accessible and available as possible not just for healthcare but also all the different walks of life that can benefit from the use cases of XR.

XR has been popularized by its extensive use in gaming, which still remains the primary sector where most adoption, deployment, and mass use of this technology remain. Healthcare falls second to gaming, with growing demand for XR applications in different aspects and facets of this sector. Although, the present scenario and XR adoption in healthcare cannot be quantified, the growing interest and drive within healthcare and allied organization to adopt this technology is surely a sign in the favor of XR taking over as one of the key transformative technologies for healthcare.

There is some evidence of XR applicability and efficacy for applications related to mental health management for varied conditions and scenarios. Besides these, experts see vast opportunity for XR application in areas such as social care and geriatric health management to improve quality of life for individuals who are struggling to

Extended Reality for Healthcare Systems, Volume 1
ISBN 978-0-323-98381-5, https://doi.org/10.1016/B978-0-323-98381-5.00009-X

communicate their health issues in any way or form. An interesting use case of VR is management of issues related to end-of-life. The support, choices, and assistance that this technology can provide top people during the end days of their life pave way for a person-centered approach to death management.

Furthermore, medical practice has also shown greater acceptance for XR adoption in areas such as helping patients to deal with pain and anxiety, and management of conditions related to phobias, cancer, and stroke. Having said that, we are still far from translating XR into applications that become a clinical reality, which will require focused research and development involving collaborations across interdisciplinary sciences.

Several studies around the impact of COVID pandemic on health services in the United Kingdom and around the world have predicted that organizations such NHS will have to face immense pressure for years to come and will require continued investment to sustain itself. The services that are expected to be most affected include hospital care and mental health management. Futuristic plans and measures will need to be put in place to ensure sustainable management of these services.

The most intuitive solution for improved management of increasing demand is to invest on workforce. In addition, another investment that can support such organization in unimaginable ways to invest on development of efficient clinical pathways, which can facilitate avoidance of unnecessary admissions and treatments.

A significant requirement that COVID pandemic has introduced to clinical pathways and delivery of services is that the patient may or may not be physically present. Therefore, telemedicine and teletherapy are areas being investigated and streamlined for maximum efficacy. Technologies such as XR can be instrumental in backing such clinical interventions, by improving patient experience and clinician's information gathering capabilities for assisted diagnostics and treatment.

To assess the economic viability of a technology for healthcare, an analysis of clinical pathway and settings for deployment, expected change and modeling of benefits with respect of costs need to be done. The financial aspects of the model are expressed as costs or financial savings. On the other hand, benefits may be expressed in financial terms or any measurable consequences such as benefits in quality of life. Some studies[1] have researched the cost-effectiveness of XR with respect to its benefits, and conclusions advocate that XR has a strong case for healthcare services and related applications.

This chapter will explore the benefits and costs associated with XR, specifically in the context of healthcare. It is noteworthy that one of the key drivers for technological adoption and innovation in XR is associated costs and established uses in healthcare. This chapter, thus, explores the economic impact of XR adoption for use cases in healthcare across three main focus areas. Fig. 14.1 illustrates that benefits rendered by the use of XR

1 https://www.xrhealthuk.org/the-growing-value-of-xr-in-healthcare Background.

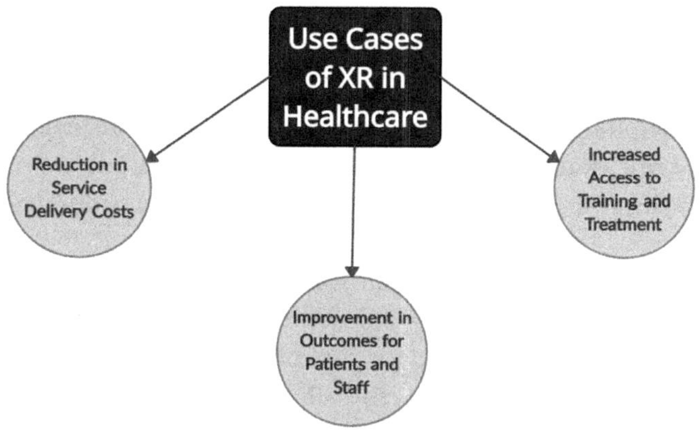

Figure 14.1 Benefits of XR in use cases for healthcare. *XR*, extended reality.

for varied use cases in healthcare can be broadly divided into three categories, namely, cost reduction of service delivery, improved patient outcomes and enhanced access to training and treatment. These benefits are covered in detail in the following sections.

14.2 Reduction in service delivery costs

Reducing the costs of delivery and providing value for money is one of the most notable benefits of XR. For the patients, XR provides a reliable and accessible option to therapies and treatment. This technology is not just capable of improving patient experience in the teleoperation mode, but also while the patient is inside the clinic. Therefore, XR can be seen as a pathbreaking technology when it comes to revolutionizing patient experience.

From a clinician's perspective, XR is capable of having a twofold benefit. Firstly, it can reduce the time required in administrative tasks that are involved in the therapeutic process by automating them. In addition, XR plays an instrumental role in improving patient engagement. Furthermore, one of the areas that has majorly benefitted with the introduction of XR is medical education. XR has provided more learning opportunities with a stark improvement in quality of education imparted to medical professionals.

Clinicians are under immense pressure owing to the growing demand from the system, particularly in pandemics such as COVID-19. Technologies such as XR can be extremely beneficial in improving the treatment capacity and patient handling capabilities of clinicians, allowing them to be in a much better position to deal with growing demands.

In entirety, technologies such as XR play a vital role in making treatment and public health programs accessible to one and all equally. Having said that, this technology still needs development of tools that can meet the growing demands of the use cases identified for XR in healthcare.

One of the case studies used for proving the cost-effectiveness of this technology with respect to its benefits is that of ProReal.[2] This visualization tool makes use of avatars and helps users to visualize situations, assess multiple perspectives, and assist problem-solving. Such a tool is known to be excessively useful for mental health management applications because patients with issues usually find it hard to communicate their issues and challenges. This visualization tool is typically used along with talking therapies to help individuals.

There are several identified benefits of using such a technology. Firstly, the use of digital imagery allows establishment of therapeutic relationship much faster than usual. Benefits are expected to be exemplified for patients who find it difficult to express. Thus, the use of such visualization tools will eventually reduce the number of sessions required for therapy.

Additionally, patient engagement and access are improved with proven reduction in stigma associated with such problems. One of the major cost saving aspects of this technology stem from the fact that the talking therapy can be performed over phone, and thus, this arrangement can successfully be delivered remotely. The reduced need to travel and reduced need to go face-to-face reduced costs by as much as 40%.

The use of this technology for real-life scenarios has demonstrated lower dropout rates and higher success rates. This technology has been tested for varied settings, which include schools, mental health centers, adult services that focus on treatment of issues such as eating and personality disorders, and prisons. Such technology-assisted therapy is also known to benefit people dealing with learning difficulties.

The use of XR at specific points in the core clinical pathway assists effective delivery of services and reduction of patient pressure on hospitals. This benefit is significant in view of the fact that every extra bed cost healthcare facility significantly (5 year forward view) and reducing any additional admissions reduce the operational costs of the healthcare infrastructure. This can be appreciated by the fact that statistics suggest a 0.5% cancellation in surgeries per day due to reasons such as patient nonattendance (Wong et al., 2018). However, with such low percentage, loss of operating theater time costs NHS, England, as high as £400 million per annum (Gillies et al., 2018). The use of technologies such as XR promises to reduce these costs significantly.

An example of this scenario is in-growing toenail surgeries, which are one of the most commonly performed minor surgeries performed by podiatrists. If this surgery is performed as an outpatient procedure, it is far less expensive in comparison with an inpatient procedure. However, due to anxiety and nervousness, most patients have to inevitably be referred to orthopedic department and may even involve the use of general anesthesia. As a result, the cost of the procedure rises. The use of VR therapy for anxious

[2] https://www.proreal.world/.

patients to pursue them has proven effective, and studies validate the efficacy and cost-effectiveness of using VR in cases such as these.

The other popular use of VR therapy lies in pain management where VR therapy is seen as an alternative to opioids. According to Delshad et al. (2018), the use of VR therapy instead of opioids in the United States resulted in cost savings of USD $5 per patient with reduced hospital stay, which escalates accordingly if we assume 15,000 admissions per annum and the cost savings will be more than enough to cover the costs of equipment and software licenses required for VR therapy. Furthermore, the benefits of VR therapy in this case are not just limited to costs considering the ill effects of opioids on human body.

Another study (Lloréns et al., 2015) evaluated the benefits of VR therapy in at-home, poststroke recovery. This study was performed in Spain and concluded that this form of intervention manifested 56% reduction in costs as a result of reduced face-to-face interaction with physiotherapists. Therefore, VR therapy can undoubtably be considered a significant contributor to driving remote or telehealth services for patients from the comfort of their homes.

14.3 Improvement in patient outcomes

The objective of any healthcare infrastructure is to deliver quality services to its patients and achieve desired patient outcomes. This does not just affect patients at the personal level, but it also goes a long way in improving the quality of life in the city and country. The use of XR has demonstrated significant improvements in patient outcomes while keeping the overall comparable costs low.

One of the known applications of VR lie in motivating people and enabling them to improve their physical activity.[3] In fact, Ng et al. (2020) states that the combined use of VR and AR can also manifest improvements in strength and physical performance. Improvements to physical activity is one of the clearest use cases of VR, and effective service delivery in this regard can convert into substantial savings.

A subarea that has been highlighted by recently published literature is falls prevention. The estimated cost of support provided by NHS to falls prevention is around £2.3 billion. Moreover, one of the NICE recommendations[4] clearly states that any individual assessed at risk or with a history of falls must be provided multifactorial interventions. Having said that, multifactorial interventions are way more expensive than exercise-based interventions (Winser et al., 2020). Corregidor-Sánchez et al. (2021) suggested that VR therapy can be viewed as a low-cost alternative to multifactorial interventions and are known to outperform the standard forms of therapy in term of efficacy.

[3] https://evidence.nihr.ac.uk/themedreview/moving-matters-interventions-to-increase-physical-activity/.
[4] https://www.nice.org.uk/guidance/cg161/chapter/recommendations multifactorial-intervention.

Staff care, retainment, and recruitment of new skilled staff are one of the primary objectives of any healthcare facility's development roadmap. In fact, the plan published by NHS in2020[5] included commitments toward measures for reducing discrimination, making optimal use of available staff, and finding innovative ways to deliver high-quality services while also taking good care of their staff. Having said that, most healthcare facilities suffer from major backlogs in terms of delivering elective care, which requires more staff and investments in staff retention and training.

A study[6] performed to analyze the effects of VR and its adoption on intensive care units (ICUs) used Rescape's DR:VR[7] for deployments in two ICUs during COVID. The results of the study were summarized as follows: 94% participants felt that VR therapy was an enjoyable and relieving experience, while 88% confirmed that their stress levels reduced after taking the VR experience. Solutions such as these are effective, accessible, and affordable and can potentially be used to manage burnout problems in healthcare staff.

Medical and surgical education has long been an untouched domain in terms of technological intervention. Thus, this field majorly relies on traditional forms of imparting education and skills training. However, if we look at the statistics provided by General Medical Council, UK,[8] then 1 out of 10 medical graduates feel that they are not prepared enough. Furthermore, this report also states that NHS, UK, spends as much as £2.5 billion a year on avoidable errors. Recent research in this field hints toward the fact that newer, innovative forms of imparting medical and surgical education can prove to be even better than traditional methods in terms of achieved outcomes. Portelli et al. (2020) performed a metaanalysis on differences in outcomes with traditional and technology-empowered training for laparoscopic surgery and reported lower error rates with the use of latter. Furthermore, Blumstein et al. (2020) highlighted that surgical performance can improve by as much as 230% with the use of technology-driven methods instead of traditional training.

14.4 Easier access to services and training

As a repercussion of the COVID pandemic and a consequence of a rapidly aging population, it is imperative for governments and healthcare organizations to develop and implement efficient, futuristic services. There are several use cases that demonstrate the effective use of XR for improving care delivery and education so that organizations such as NHS can meet the growing demands of the country. These use cases have been discussed in this section.

5 https://www.england.nhs.uk/ournhspeople/online-version/.

6 https://healthmanagement.org/c/icu/issuearticle/feasibility-and-potential-benefits-of-immersive-virtual-reality-in-the-intensive-care-unit.

7 https://www.rescape.health/virtual-reality-distraction-therapy-solution.

8 https://prescribingsafetyassessment.ac.uk/resources/be_prepared.pdf.

There are several device-led clinical services. One of these services is provided by Concept Health Technologies[9] in the form of PRinVR. This service provides support for pulmonary rehabilitation. This service allows real-time monitoring of the individual with the use of a smartphone, wearable sensor, and VR headset. It has been noticed that traditional exercises have more subscribers because of the shortage of staff who can provide physiotherapy sessions. Thus, this service allows delivery of exercise lessons to individuals within the convenient spaces of their homes. Moreover, they can be monitored all the time without the need for the clinician to spend excess time. In other words, the use of this service attests the improved service delivery with added benefits such as freeing up clinicians' time and elimination of the need for patients to travel to a facility for sessions.

There are several case studies that prove the economic and social benefits of using this solution in a clinical setting. A report[10] stated that the cost of PRinVR was £415 per patient as against £1200 per patient for conventional intervention. This dictates significant cost savings. In addition, it has also been found that people who opt for VR-based intervention benefit from the fact that this form of intervention is easily accessible. They no longer have to wait for months to get an appointment. In this respect, VR-based interventions can improve the operational efficacy of systems by making healthcare available and accessible to a greater number of people in lesser time. This service has gained increased attention in the COVID-19 pandemic where remote intervention became the preferred method of treatment for healthcare providers and patients all over the world (Houchen-Wolloff and Steiner 2020).

Another application of VR that benefits service access and delivery in the healthcare sector is in wayfinding. It is found that VR-based hospital wayfinding is much more effective than floor plans. Although the development of VR applications for this purpose requires investment, it provides good return on investment by saving staff time. In fact, according to a study (Halfer and Rosenheck, 2014), the total development and implementation costs for paper-based floor plans and VR-based wayfinding were comparable.

In terms of medical students training, standardized patients are typically used for helping students understand clinical decision-making and practice professional communication. It is critical to note that standardized patients are actually role players who are trained for 5 hours and employed for 21 h to portray specific scenarios and engage with medical students for their training purposes. Bosse et al. (2015) suggested that repeated training consumes more than 50% extra time. As a consequence, the use of virtual role players can trigger cost savings in addition to improving access and availability of diverse training scenarios. Other services such as HoloPatient are also being used to train nurses (Ditzel and Collins, 2021).

9 https://concepthealth.co.uk/home.html.
10 https://www.xrhealthuk.org/the-growing-value-of-xr-in-healthcare.

14.5 Affordability aspects of extended reality

Recent years have witnessed a growing interest and increased familiarity of XR in the general world population. This growing interest is attested by the increased purchases of haptic controllers and head-mounted displays (HMDs) by consumers for gaming. Moreover, there has also been increased interest in AR-based smartphone games.

Several aspects of these devices have improved over time. For instance, HMDs are far more comfortable to use now because they have inbuilt computing and there is no longer need to connect them to a computer for processing. In addition, there are many HMDs available in the market that can be connected to a smartphone as well. Having said that, the all-in-one untethered HMDs have replaced their counterparts because of their lower costs.

If we compare the costs of VR equipment as it has changed over the years, then around 25 years ago, a liquid crystal HMD and a PC to support a VR application would have costed more than £10,000. Now, adding the impact of inflation to this amount, the cost can be anything around £20,000 of today. On the contrary, VR equipment with similar and enhanced capabilities costed 10% of this amount in 2020. In fact, equipment such as Oculus Quest can be bought by consumers for around £300, and smartphone headsets are available at prices around £50. The cost of maintaining these devices or maintaining them has also lowered considerably.

From an organization's perspective, the per-person cost of these devices is further lower as the same device can be used by many individuals during a day. In comparison with their predecessors, the modern-day XR equipment is much more portable and robust, which makes it much easier than ever before to use this equipment for different people, in different rooms, healthcare settings, or facilities. Evidently, the costs involved in buying, deploying, or even renting out XR equipment is much lower than ever before, more so in comparison with staff costs. Therefore, this technology can adjudge as affordable for organizations intending to use them for diverse setting and many individuals.

To quantify the costs associated with VR or XR usage, it is vital to take into account the costs of HMD, Internet, and software, at the very least. If we look at the other end of the spectrum, a fully supported VR or XR system must be able to provide choice of multiple virtual environments, support for external equipment such as tablets and cameras, and backend storage options. These will, in all probability, raise the costs of the setup.

14.6 Challenges and limitations

Irrespective of the fact that XR has demonstrated its efficacy across use cases belonging to diverse subsectors of healthcare, it is yet to find its place in the healthcare

setting. Earlier, one of the biggest barriers in XR adoption was cost. With the evolution of XR hardware over the years, this limitation has largely been managed. However, they are still not low enough to drive large-scale adoption. In addition to this, other challenges have also been identified, which include technical limitations. Not all immersive solutions are available as mobile VR. The requirement of added infrastructure acts as a barrier as it makes the system all the more difficult to use and manage.

One of the most profound challenges in the adoption of XR is the lack of understanding. XR is not viewed and accepted as a healthcare technology yet. While lack of research in this domain is a contributing factor to this, technological unawareness is a more significant role player. Previous sections of this chapter have covered how XR has the technical capabilities and capacity to transform healthcare and the manner in which it is delivered. Moreover, there is supporting evidence to the claims. Despite this, the process of embedding this technology in the typical healthcare setting is yet to be seen.

One of the advantages of XR as a technology is that it includes several components, some of which are more functional yet optional and others are more affordable. This allows the users to customize their XR solution to suit their needs. Even with respect to hardware, the range of XR equipment is rather wide, from affordable add-on headsets that can work with mobile phones to advanced simulation systems. So, the users can access XR irrespective of their budget, space constraints, and custom requirements.

The XR spectrum of applications range across domains and budgets. Users can access a plethora of free apps. For example, many applications for targeting different aspects of well-being are available on the Oculus store. However, if the user wishes to seek more interactivity and richer content, the costs proportionately go higher. An important aspect of XR resource access is the fact that most of the resources such as software, support, and even devices can be taken on subscription basis.

It is noteworthy that most of the existing XR innovations and applications have relied on personal investment or donations. While this demonstrates growing interest in this technology, it does not let this service-based industry flourish in a well-defined commercial model. Services that are provided have to be supported post purchase, which is seemingly a daunting task for startups.

14.7 Conclusion

Immersive technologies require investment because of the complexity of the technology and the infrastructure required for it. This is one of the roadblocks in XR adoption since significant gaps lie between research, development, and procurement within health organizations such as NHS. The challenge with independent organizations is driven by the fact that healthcare services vary across the country, which makes an understanding of the target audience, quintessential. Although patients and clinicians are the most intuitive set of audience for this technology, it is providers and businesses that

actually hold the money strings. Therefore, comprehension of their requirements is also essential from a business point of view.

From a clinical perspective, one of most profound limitations is the lack of guidance or usage framework for XR in clinical settings. Although XR has demonstrated high efficacy for mental health management and is currently being used in NHS, organization such as NICE is yet to include a guidance on such usage. Having said that, organizations can rely on frameworks such as Evidence Standards Framework for Digital Health Technologies[11] by NICE and Digital Technology Assessment Criteria[12] by NHSX to identify products that are useful, useable, and within the realm if MHRA medical device regulations. However, to realize the potential of XR and streamline their route to market in the healthcare sector, it is critical to develop a framework that can ensure safety, operability, efficacy, and convenient procuring of XR products.

References

Blumstein, G., Zukotynski, B., Cevallos, N., Ishmael, C., Zoller, S., Burke, Z., SooHoo, N.F., 2020. Randomized trial of a virtual reality tool to teach surgical technique for tibial shaft fracture intramedullary nailing. Journal of Surgical Education 77 (4), 969−977.

Bosse, H.M., Nickel, M., Huwendiek, S., et al., 2015. Cost-effectiveness of peer role play and standardized patients in undergraduate communication training. BMC Medical Education 15 (183). https://doi.org/10.1186/s12909-015-0468-1.

Corregidor-Sánchez, A.I., Segura-Fragoso, A., Rodríguez-Hernández, M., et al., 2021. Effectiveness of virtual reality technology on functional mobility of older adults: systematic review and meta-analysis. Age and Ageing 50 (2), 370−379. https://doi.org/10.1093/ageing/afaa197. March 2021.

Delshad, S.D., Almario, C.V., Fuller, G., et al., 2018. Economic analysis of implementing virtual reality therapy for pain among hospitalized patients. Npj Digital Medicine 1 (22). https://doi.org/10.1038/s41746-018-0026-4.

Ditzel, L., Collins, E., 2021. Holograms in nursing education: results of an exploratory study. Journal of Nursing Education and Practice 11 (8). http://www.sciedu.ca/journal/index.php/jnep/article/view/19741/12542.

Gillies, M.A., Wijeysundera, D.N., Harrison, E.M., et al., 2018. Counting the cost of cancelled surgery: a system wide approach is needed. British Journal of Anaesthesia 122 (2), 691−694. https://doi.org/10.1016/j.bja.2018.08.002, 2.

Halfer, D., Rosenheck, M., 2014. Virtual education: is it effective for preparing nurses for a hospital move? The Journal of Nursing Administration 44 (10), 535−540. https://doi.org/10.1097/NNA.0000000000000112.

Houchen-Wolloff, L., Steiner, M.C., 2020. Pulmonary rehabilitation at a time of social distancing: prime time for tele-rehabilitation? Thorax 75 (6), 446−447. https://doi.org/10.1136/thoraxjnl-2020-214788.

Lloréns, R., Noé, E., Colomer, C., Alcañiz, M., 2015. Effectiveness, usability, and cost-benefit of a virtual reality-based telerehabilitation program for balance recovery after stroke: a randomized controlled trial. Archives of Physical Medicine and Rehabilitation 96 (3), 418−425. https://doi.org/10.1016/j.apmr.2014.10.019 e2.

11 https://www.nice.org.uk/about/what-we-do/our-programmes/evidence-standards-framework-for-digital-health-technologies.

12 https://www.nhsx.nhs.uk/key-tools-and-info/digital-technology-assessment-criteria-dtac/.

Ng, Y.-L., Song, Y., Kwon, K.H., et al., 2020. Toward an integrative model for online incivility research: a review and synthesis of empirical studies on the antecedents and consequences of uncivil discussions online. Telematics and Informatics 47, 101323. https://doi.org/10.1016/j.tele.2019.101323.

Portelli, M., Bianco, S.F., Bezzina, T., Abela, J.E., 2020. Virtual reality training compared with apprenticeship training in laparoscopic surgery: a meta-analysis. Annals of the Royal College of Surgeons of England 102 (9), 672–684. https://doi.org/10.1308/rcsann.2020.0178.

Wong, D.J.N., Harris, S.K., Moonesinghe, S.R., et al., 2018. Cancelled operations: a 7-day cohort study of planned adult inpatient surgery in 245 UK National Health Service hospitals. British Journal of Anaesthesia 121 (4), 730–738. https://doi.org/10.1016/j.bja.2018.07.002.

Winser, S.J., Chan, H.T.F., Ho, L., et al., 2020. Dosage for cost-effective exercise-based falls prevention programs for older people: a systematic review of economic evaluations. Annals of Physical and Rehabilitation Medicine 63 (1), 69–80. https://doi.org/10.1016/j.rehab.2019.06.012.

CHAPTER FIFTEEN

The future of XR-empowered healthcare: roadmap for 2050

Samiya Khan
Faculty of Science & Engineering, University of Wolverhampton, Wolverhampton, United Kingdom

15.1 Introduction

Extended reality (XR) is garnered immense attention particularly by technologist who work for the healthcare domain. Many widespread applications of XR have been identified with demonstrated results. Some of these applications include pain management, therapeutic interventions for anxiety management, and medical education, in addition to many others. Having said that, we are still far from a solution that can be adopted globally. The vision 2050 for XR healthcare is to develop an ecosystem that can facilitate development, clinical validation, and global adoption of XR solutions for healthcare.

Developing a roadmap in favor of the set objectives is a road full of challenges and limitations. One of the most critical challenges for XR adoption in healthcare stems from the criticality of use cases and associated data that need to be shared, stored, and processed by external systems. Therefore, security and clinical aspects of XR adoption must be clearly understood before a comprehensive roadmap and adoption strategy can be put in place. These aspects of XR-healthcare Vision 2050 have been investigated and discussed in later sections of the chapter.

In addition to the aforementioned, the challenges and prospects of XR for healthcare need to be evaluated. Some of the core challenges that plague XR adoption, like any other new technology, are lack of regulations and evaluation framework. Although the need and benefits of these systems are evident, it is difficult to classify them as medical devices, which delays regulatory approvals. The lack of a specific compliance and evaluation framework is also a system-level challenge that needs to be mitigated to accelerate research in XR healthcare. In the later sections, this chapter identifies the different challenges and the prospect of XR-healthcare amid these challenges to provide recommendations on these facets.

This chapter is organized in the following manner: Section 15.2 provides insights on the security and health safety considerations that need to be considered before adopting XR into healthcare. Section 15.3 provides an overview of the challenges associated with XR usage in healthcare and prospects of using this technology for varied clinical

Extended Reality for Healthcare Systems, Volume 1
ISBN 978-0-323-98381-5, https://doi.org/10.1016/B978-0-323-98381-5.00003-9

interventions. Section 15.4 provides recommendations on the different interventions required for global and widespread adoption of XR-healthcare solutions. Finally, Section 15.5 synopsizes the current challenges, potential of XR healthcare, and recommendations for effective use.

15.2 Health safety and security considerations

The healthcare sector has always been at the center of technological innovation. In the past few years, many transformative technologies have been adopted by healthcare. The latest trend in this respect is the convergence of multiple technologies and their cumulative use for benefits to the different stakeholders in the healthcare sector. Having said that, we are still discovering the benefits and risks associated with these systems as far as their use for healthcare sector is concerned. The risk aspect is magnified in the case of XR with an awaited clear definition and assessment of existing and expected risks. Research is underway in this sector, and one of the recent studies was performed by Department of Business Energy and Industrial Strategy, UK (2020)[1] to investigate the safety and risks associated with virtual reality (VR) systems when used in a domestic setting. In this section, we will look at the different categories of risks with respect to the UK healthcare sector and explore the novel risks that XR may introduce.

15.2.1 Clinical governance

One of the biggest challenges faced by XR adoption is the inability of regulators and clinicians to classify XR software and hardware. It is unclear whether XR devices can be considered medical devices or they are just devices that facilitate the use of a software, which is in actuality the medical device. The resultant delay in regulatory approval has been a deterrent in research, development, and deployment of XR, across the board. Regulatory compliances are all more difficult to achieve for businesses, making this a significant roadblock.

There have been discussions carried out by the US Department of Health and Human Services and FDA in 2020 on the use of XR for healthcare and the best evaluation practices that can be adopted for the same. Similar strategies must be formulated at the national level by countries to make a realistic assessment of how XR can be integrated with the existing system and used to its best capacity.

From the UK perspective, there are no NICE or MHRA guidance on XR usage yet, which is one of the key strategic gaps. Therefore, regulators need to collaborate with stakeholders to create an adoption plan for XR and other emerging technologies. In the

1 https://assets.publishing.service.gov.uk/government/uploads/system/uploads/attachment_data/file/923616/safety-domestic-vr-systems.pdf.

UK context, it is noteworthy that the Digital Technology Assessment Criteria (DTAC)[2] has been provided by National Institute for Health and Care Excellence (NICE).[3] It is a prerequisite for every organization willing to trial or deploy a digital technology within NHS England, in addition to other organizations such as MedCity and Public Health England. This standard can be used for XR adoption in the meantime.

15.2.2 Safety

Broadly, the interaction between the user and the system encompassing both primary and supplemental functions is generically referred to as usability. The equipment used for XR should offer high levels of social and physical comfort. This is specifically critical for clinical settings as there may be many hindrances such as background noise, other non-XR device interactions, and multiple light sources.

Besides this, multiple XR devices may be used within the same setting, and the operator may have to simultaneously interact with many people. To accommodate for these hindrances, the integration of the XR equipment with the surroundings should be seamless. For instance, a user may have to wear and remove the XR device multiple times during a treatment procedure. This should not cause any discomfort to the user. In other words, the process of moving in and out of a VR environment should be effortless without taking any excess time or causing clinical delays, from the user's end.

Within a clinical setting, the primary interaction is always the interaction that happens between the patient and the medical practitioner. The use of XR should preserve this interaction to maintain the flow of therapeutic intervention. An important aspect that needs to be considered when using XR in a clinical setting is that the prospective users will have diverse abilities and capacities. Therefore, the XR solution must be able to cater to these varying needs of comfort and skill. This aspect can be especially critical for route to mainstream usage for XR in healthcare. To have a preassessment of the accessibility benchmarks for XR solutions, usability studies on different groups of people with different abilities, in diverse age groups and dealing with different conditions, will need to be performed.

15.2.3 Cybersecurity

It is imperative that XR healthcare ecosystem can potentially include and use cloud systems, sensors, artificial intelligence (AI) algorithms, video cameras, different operating systems, and networking components. Therefore, such an ecosystem is at a heightened risk of cyberattack, majorly involving data associated with patients, their live locations, and behavioral information. The standard best practices with regard to confidentiality, privacy, and data security hold for XR application. However, considering the new data

2 https://www.nhsx.nhs.uk/key-tools-and-info/digital-technology-assessment-criteria-dtac/.
3 https://www.nice.org.uk/.

formats, connectivity requirements, and use cases of this technological intervention, advanced cybersecurity frameworks may need to be put in place to deal with new cyberthreats that may arise as a result of its adoption and usage.

The security concerns around use of XR have been validated by Casey et al. (2019) in their research that proved that adulterating camera controls, altering traffic, and uploading unintended files on XR headset can practically transform a human into a human joystick. The person can be tricked into taking an unintended path, falling and even causing injuries. Therefore, it is best to apply the cybersecurity protocols used for Internet of things (IoT) systems and medical devices to XR systems for now, appending them with any new aspects of security that are brought to light by research and clinical trials.

15.2.4 Managing assets and interoperability issues

The fact that the realm of XR deployment is relatively large gives rise to novel logistical concerns. Moreover, device maintenance and administrative procedures for each device may have to be varied on the basis of the skill set and geographical location of the user. Therefore, functional alignment of XR devices across user bases and geographical location must be included in strategic planning for product and service provisioning in this sector.

While using XR devices in the medical setup across different healthcare settings, there are several things that require attention. Firstly, XR devices are expensive and generally hard to procure for healthcare organizations because of the procurement policies and procedures. Besides this, a device may be rented out to multiple individuals, which makes maintenance of hygiene and regular disinfection essential.

Therefore, these aspects of XR usage must be incorporated in the maintenance procedures laid out for XR devices in the medical setting. An important thing to consider is that different organizations may have different hygiene standards and protocols. Therefore, the adopted guideline must abide and take into consideration the protocols set out by these organizations. Interoperability is a critical consideration with respect to asset acquisition and procurement. To ensure global reach and adoption of XR solutions, the solution must be interoperable and operational across systems, which may be standalone or distributed in nature.

15.2.5 Data governance

To look at the data governance requirements of the system, it is important to understand that XR devices have access to and process highly confidential patient data, which may not only include metadata. Therefore, data associated with health inferences and patient's movement or real-time location are also available to XR devices. Therefore, the definition of personal data with regard to XR is much broader than the conventional

definition, and the protection mechanisms also need to adjust to this alteration in definition.

Some categories of data such as cameras to capture the patient's eye movement or body sensors and triggers installed on the device collect data, which are targeted toward uniquely identifying the person with behavior, biometric information, room, physical attributes of the individual, mood, and gender presentation, to name a few. In other words, the standard definitions of personal data as provided by personal health information (PHI) or personally identifiable data (PII) (Iyengar et al., 2018) will need to adjust to accommodate the variable forms of data that XR uses, collects, and processes to uniquely identify individuals.

There has been some research on regulations required for addressing security concerns related to biometric information.[4] The regulations provided in this regard include providing information to individual on biometric data being processed, protection of biometric data with computational procedures such as encryption, tracking of data access pattern to databases that store biometric information and ensuring limited time period data retention at the database level.

Evidently, there is still a gap in data governance as far as security of complicated XR data is concerned. The problem becomes all the graver when we bring AI algorithms and their integrated use with XR in picture. Organizations such as NHS and XRSI are developing regulations for XR- healthcare to ensure privacy and safety to support research in this field (Research and Standards | XRSI—XR Safety Initiative, 2021).[5] More efforts in this domain are required to address the issue of security in XR at the global level.

15.2.6 Clinical validation

The ground rule for validating any technology is that its benefits should substantially be more than the risks associated with its usage. A critical risk with organizations involved in XR app development for healthcare is that their roots lie in the gaming industry. As a result, the developers are generally less acquainted to the risk-based approach that needs to be adopted for medical devices and evidence that need to be collected for efficacy—risk assessment. Therefore, clinical validation of XR solutions must include data such as evidence of efficacy on diverse use case models, risks, and side effects of the solution for specific patient groups such as patients with epilepsy or visual/hearing impairments, the psychological effects of overuse, and the generic impact of the solution on an individual's health and well-being. If the development team includes people who do not have a medical background, they must be adequately and appropriately trained on the requirements of clinical validations and regulatory standards for XR healthcare.

4 https://www.garanteprivacy.it/home/docweb/-/docweb-display/docweb/3590114.

5 https://xrsi.org/research-standards.

15.3 Prospects of XR healthcare

15.3.1 Challenges

There are several challenges that plague the adoption of this pathbreaking technology for real-life usage. The use of any technology in the medical sector requires to abide by several guidelines and standards because of the criticality of the application and its potential impact on human health. Although these guidelines are set up to ensure health safety of technology and devices, it also leads to several roadblocks and delays owing to technological complexity.

Firstly, the devices used for rendering and provisioning XR need to be classified by the healthcare providers, companies, and regulators. However, there is no clarity on whether XR devices qualify as medical devices or not. In scenarios such as these, approval from regulatory authorities can even take years, which deters research and development. Regulation compliance by businesses also remains one of the most crucial and challenging aspects of XR adoption in healthcare.

Another important aspect of research and development in this field is the interdisciplinary nature of this domain. Technological development in XR for healthcare requires collaborative efforts from technologists, clinicians, 3D content developers, and all the other stakeholders who might also have to be involved. For instance, to develop VR applications to alleviate anxiety in teenagers, special interest groups such as their parents, counsellors, and teachers may also have to be contacted. Although this aspect presents infinite opportunities, it also comes with unlimited challenges.

One of the key drivers of research and development is funding. Owing to the fact that XR for healthcare is in its early stages of development, the number of sources available to fund research of this domain is limited. There are also limitations with respect to carrying out pilot testing, which is crucial for chalking out a route to market for the solution. As a result, the opportunities for creating commercially viable solutions may be limited.

Other business-specific challenges with respect to XR-based healthcare solutions is the lack of clear pathways for sourcing and selling these solutions. Moreover, XR is a relatively new technology. Thus, there is no benchmarking or evaluation framework that can be used to standardize products for their quality. As a result, buyers cannot compare functionality and/or costs to assess products, making it challenging for businesses to pitch their products.

The COVID-19 pandemic has made it clearer that there is an evident shortage of clinical workforce. As a result, the clinical staff is under continuous and immense pressure. The use of digital technologies and upskilling the staff to create a digitally skilled workforce can help the medical sector in multiple ways and is the need of the hour. Firstly, this step will allow delegation of monotonous, noncritical tasks to tech-

supported devices, reducing pressure on clinical workforce. Besides this, it will also streamline healthcare processes and reduce the scope of human error.

However, in the existing setup, there is no strategic plan in place to upskill staff for implementation and adoption of XR solutions for therapeutic interventions. Other challenges surrounding the use of XR for healthcare applications include ethical approvals, compliance with quality standards and clinical governance protocols, and mitigating issues that might arise from data protection and cybersecurity facets of software development.

15.3.2 Opportunities

The opportunities and prospects of using XR for a wide variety of healthcare applications overpower the limitations and challenges around it. There is no limit to the innovation that can possibly be done in the field of XR for healthcare. However, there are some key sectors that can particularly benefit from the use of XR technology.

The COVID-19 pandemic has tested and bruised the healthcare sector of every country around the world. The impact of the pandemic has been most serious on the clinical workforce. As a result, workforce acquisition and retention has become challenging. In addition, workforce development and training to deal with new challenges that we face in the post-COVID-19 era has also become necessary. According to Healthcare UK Annual Review 2019–2020,[6] the world needs and is now looking at newer and better way to provision high-quality healthcare education and training. XR technology can play an instrumental role in providing highly accessible, cost-effective, and quality training to healthcare staff.

One of the other healthcare sectors that can greatly benefit from XR is rehabilitation and physiotherapy. Majority of the activities related to rehabilitation from musculoskeletal conditions are performed in clinics and hospitals. However, COVID-19 resulted in cancellation of appointments and the growing need to look at remote solutions that can facilitate high-quality treatment to patients within the comfortable surroundings of their homes. The existing remote solutions include teaching exercises to patients or their kin and leaving it up to them to carry these out at home at their convenience.

However, several patients have reported that there is a lack of motivation, and they typically find it difficult to maintain regularity and schedule in their planned treatment activities. XR-based solutions are capable of solving these issues by improving patient engagement. In addition, there will be added benefits such as reduced home visits, provisioning of supervised self-management, and cost savings. Therefore, the use of XR-based solutions for rehabilitation and physiotherapy can have a significant and beneficial impact on the long-term and sustainable management of these conditions. The same

[6] https://www.gov.uk/government/publications/healthcare-uk-annual-review-2019-to-2020.

principles of treatment apply to management of pain/emergency, mental health, and geriatric health issues.

XR has demonstrated great success in applications such as mental well-being and general health. Interestingly, the most promising use cases have emerged as a result of collaboration between academics, XR technologists, healthcare professionals, and creative game companies. It is critical to note that the games sector recorded £7 billion in the United Kingdom for the year2020.[7] The reason for this upsurge has its roots in the occurrence of COVID-19 pandemic, which led to lockdowns and restrictions on normal life.

Therefore, XR-based games have had an indirect contribution in maintaining mental well-being through this difficult time. Therefore, creative games companies and XR experts are seeking opportunities to demonstrate and validate the varied contribution of XR to mental well-being and health, in general. Strategic initiatives to drive research and development in this sector can reap multifaceted benefits to patients and stakeholders alike.

The technological world is gradually converging toward development of heterogenous systems involving synergistic integration of multiple technologies. Technologies such as 5G, IoT and AI have been extensively used in healthcare systems to support remote health monitoring and diagnosis, by allowing collaborative work among clinicians, healthcare practitioners, medical facilities, and patients. The integration of XR, particularly biofeedback technologies, can take existing remote healthcare systems to support a higher level of responsiveness and greater impact. While AI and IoT allow capturing of data and transforming them into actionable insights to improve diagnostic abilities, XR can complement them by improving patient experience and therapeutic efficiency.

15.4 Recommendations

The primary requirement to support research and development in the field of XR for healthcare is an end-to-end framework that can enable design, development, testing, and commercialization of XR solutions across different models of care. From the business perspective, this framework should be able to support development of XR products from conceptualization to funding and route to market. At the core of this framework is multilateral collaboration between different stakeholders to ensure innovation is supported with the right mix of talent, expertise, insight, and infrastructure.

To create the aforementioned XR healthcare ecosystem, it is important to understand the capabilities and capacities available within the country. Therefore, mapping of healthcare organizations to academic/research institutions and businesses to understand the scale, value, and potential of XR healthcare market is important. The academic and

7 https://www.nme.com/news/gaming-news/uk-gamers-spent-a-record-high-7-1billion-in-2021-3194084.

research organizations must ensure development of scalable, robust, and marketable solutions. Establishment of national-level platforms to facilitate distribution and quality assurance of these products should also be considered.

The different components of recommended XR-healthcare ecosystem are described in Fig. 15.1. The development of scalable and marketable XR solutions requires collaborative efforts from different stakeholders. To trigger such collaborations, coinnovation programs and R&D funds to support early-stage research projects must be made available. Existing resources being used to understand the impact of digital technologies or immersive technologies such as Playbooks must be made available to research centers so that minimum viable products can be developed in lesser time.

The creation of prototypes or minimum viable products aids in assessment of project impact. Therefore, creation of prototypes or pilot for testing impact should be encouraged for high-value projects. This will not only act as body of evidence, but it will also help in streamlining the commercialization strategy. From the UK perspective, there are many bodies that standardize clinical solutions, which include NICE, MHRA, and NHS Digital, to name a few.

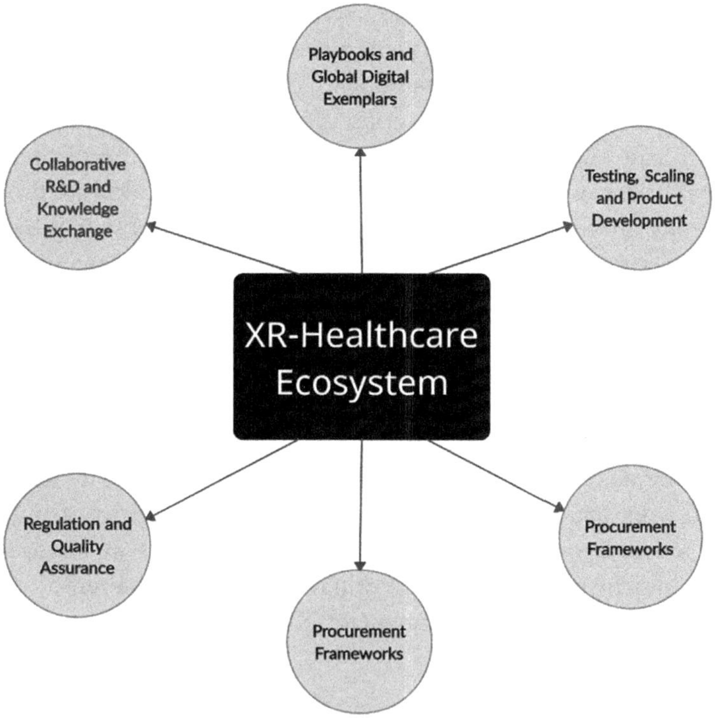

Figure 15.1 Recommendation framework.

Standard regulation and compliance framework should be put in place to enable businesses to benchmark their solutions. Similar benchmarking is also required with respect to procurement frameworks and qualification criteria for suppliers to register themselves with healthcare settings. A futuristic yet feasible component of the framework is setting up of new distribution models such as digital pharmacies, which can provision digitally available healthcare services to patients.

Lastly, there is a need for an XR healthcare community network that can connect academics, clinicians, healthcare providers, researchers, businesses, and all the other stakeholders involved in development of XR-healthcare solutions. This will allow them to share expertise and insights to create commercially viable products, which can solve real-world problems.

This network should also be linked to XR initiatives of the government to facilitate birth of innovative projects that can change the landscape of therapeutic interventions and training in the healthcare industry. International partnerships and collaborations can prove critical for early-stage innovations and independent evaluations, leading to development of world-leading XR healthcare solutions.

15.5 Conclusion

The COVID-19 pandemic has changed the needs, requirements, and expectations from the healthcare sector. The most profound change is the need to support patients remotely without compromising on the quality of service. Moreover, delivery of other medical care services such as therapy sessions, diagnostic services, and delivery of devices or medicines to ease their adherence to treatment are also some of the additional requirements that have paved their way to the priority lists of healthcare providers around the world. Finally, the healthcare sector is the heart of a country's infrastructure, and considering the criticality of their operations, there is no scope of failure. Digital transformation and adoption of technologies can support core healthcare staff to provide services in time, at a lower cost and without compromising on the quality of service.

XR has demonstrated clear benefits across different laterals of the healthcare sector. Moreover, there are several use cases that validate the accessibility, affordability, and flexibility of these solutions for varied healthcare application domains. More specifically, VR has found acceptance in the healthcare setting of several countries such as the United Kingdom[8] and is known to have provided paybacks in the form of improved patient outcomes and reduced costs of service delivery. Although there are obvious cost benefits of using XR, which include reduced cost of travel because of the remote nature of these interventions, a clear economic assessment of XR benefits is yet to be performed for recent XR innovations.

[8] ardengemcsu.nhs.uk/showcase/blogs/blogs/immersive-technologies-in-healthcare-the-rise-of-ar-vr-and-mr/.

The most significant problem with XR adoption in healthcare is the lack of an offering that can clearly fit into existing settings, governance structure, and work to serve multiple patients with varied intervention and support. Generically, this requires digital inclusion and codesigning of solutions. To facilitate large-scale adoption of XR solution in varied healthcare settings, it is important to scale out pilots and minimum viable products by large-scale testing and development of a generic evaluation and quality assurance framework.

In addition, it will also be useful to carry out extensive health economic assessments of XR solutions, which should be facilitated by regulatory organizations. XR solutions that pass the evaluation test must be supported with funding to scale up their solution to develop a full-fledged marketable product. This should also translate into seeking funding and investment opportunities to facilitate development of high-quality solutions.

References

Casey, P., Baggili, I., Yarramreddy, A., 2019. Immersive virtual reality attacks and the human joystick. IEEE Transactions on Dependable and Secure Computing 18 (2), 550–562.

Iyengar, A., Kundu, A., Pallis, G., 2018. Healthcare informatics and privacy. IEEE Internet Computing 22 (2), 29–31.

Index

Note: 'Page numbers followed by "f" indicate figures and "t" indicate tables'.

E

F

G

H

I

W

9780323983815